Soils and the Greenhouse Effect

Conference 'Soils and the Greenhouse Effect'

Organizing Committee

W.G. Sombroek (ISRIC, Wageningen; chairman)
A.F. Bouwman (ISRIC, Wageningen; secretary)
N. van Breemen (Wageningen Agricultural University)
G.P. Hekstra (Ministry of Environment - VROM)
A.C. Imeson (University of Amsterdam)
W.G. Mook (University of Groningen)
L.R. Oldeman (ISRIC, Wageningen)
J.W.M. LaRivière (I.H.E., Delft)
P. Westbroek (University of Leiden)
I.S. Zonneveld (ITC Enschede)

Advisory Group

F. Blasco (CNRS, Toulouse, France)
F. di Castri (CEPE, CNRS, Montpellier, France)
R.E. Dickinson (NCAR, Boulder, U.S.A.)
J.A. Eddy (UCAR, Boulder, U.S.A.)
M.D. Gwynne (UNEP, Nairobi, Kenya)
J. Kindler (Warsaw Technical University, Poland)
B.G. Rozanov (Moscow State University, U.S.S.R.)
W. Seiler (Fraunhofer Institute, G. Partenkirchen, F.R.G.)
B.H. Svensson (University of Agr. Sciences, Uppsala, Sweden)
K. Szesztay (Hungary)
Y. Takai (Tokyo University of Agriculture, Japan)

Editorial Committee

W.G. Sombroek (chairman)
G.P. Hekstra
J.W.M. LaRivière

Assistant Editor

M.-B. Clabaut

Soils and the Greenhouse Effect

The Present Status and Future Trends
Concerning the Effect of Soils and their Cover
on the Fluxes of Greenhouse Gases,
the Surface Energy Balance and the Water Balance

Edited by **A.F. Bouwman**

Proceedings of the International Conference
Soils and the Greenhouse Effect

organized by:
International Soil Reference and Information Centre
(ISRIC)

on behalf of:
the Netherlands' Ministry of Housing,
Physical Planning and Environment (VROM)

This publication is sponsored by:
The Commission of the European Communities (CEC)
The United Nations Environment Programme (UNEP)

Published by:
JOHN WILEY AND SONS
Chichester-New York-Brisbane-Toronto-Singapore

Copyright © 1990 by John Wiley & Sons Ltd.
                    Baffins Lane, Chichester
                    West Sussex PO19 1UD, England
Reprinted June 1992

Other Wiley Editorial Offices

John Wiley & Sons, Inc., 605 Third Avenue,
New York, NY 10158-0012, USA

Jacaranda Wiley Ltd, G.P.O. Box 859, Brisbane,
Queensland 4001, Australia

John Wiley & Sons (Canada) Ltd, 22 Worcester Road,
Rexdale, Ontario M9W 1L1, Canada

John Wiley & Sons (SEA) Pte Ltd, 37 Jalan Pemimpin   05-04,
Block B, Union Industrial Building, Singapore 2057

Library of Congress Cataloging-in-Publication Data:

International Conference Soils and the Greenhouse Effect (1989) :
   Wageningen, Netherlands)
   Soils and the greenhouse effect : the present status and future
trends concerning the effect of soils and their cover on the fluxes
of greenhouse gasses, the surface energy balance, and the water
balance : proceedings of the International Conference Soils and the
Greenhouse Effect / organized by International Soil Reference and
Information Centre (ISRIC) on behalf of the Netherlands' Ministry of
Housing, Physical Planning and Environment (VROM) : edited by A. F.
Bouwman.
        p. cm.
   "Sponsored by the Commission of the European Communities (CEC),
the United Nations Environment Programme (UNEP)."
   Conference held Aug. 14-18, 1989, Wageningen, Netherlands.
   Includes bibliographical references.
   ISBN-0-471-92395-8
   1. Soils and climate—Congresses. 2. Greenhouse effect,
Atmospheric—Congresses. 3. Soil ecology—Congresses. I. Bouwman,
A. F. II. International Soil Reference and Information Centre.
III. Netherlands. Ministerie van Volkshuisvesting, Ruimtelijke
Ordening en Milieubeheer. IV. Commission of the European
Communities. V. United Nations Environment Programme. VI. Title.
S596.3.I58 1989
574.5'26404—dc20   89-21519
                                                            CIP

British Library Cataloguing in Publication Data:

Soils and the Greenhouse Effect *(Conference; 1989;*
   Wageningen, Netherlands)
   Soils and the greenhouse effect.
   1. Climate. Effects of carbon dioxide
   I. Title   II. Bouwman, A. F.   III. International Soil
   Reference and information Centre   IV. Netherlands,
   Ministry of Housing, Physical Planning and Environment
   551.6

ISBN 0-471-92395-8

Printed and bound in Great Britain by
Antony Rowe Ltd, Chippenham, Wiltshire

Table of Contents

v

PART VII CONCLUDING ADDRESS

PART VIII

Foreword

The likelihood of significant changes in world climatic conditions taking place through increased atmospheric concentrations of so-called "greenhouse gases" has attracted considerable attention in recent years. Greenhouse gases can absorb thermal radiation and thus contribute to the warming of the atmosphere. Changing patterns of evapotranspiration and overall reflectance (albedo) exert an influence, too. The world soils and their land use are important sources of a number of greenhouse gases such as water vapour, carbon dioxide, methane, nitrous oxide and nitric oxide.

Many of the predicted global changes are latent and irreversible. They may not become apparent until long after the human actions that gave rise to them. The potential climate changes induced by the anthropogenic emission to the atmosphere of greenhouse gases, provides an example of the need for conscious management of the global environment. The solution to the problems the world is now facing are beyond the power of individual nations.

The sources and sinks of most of the greenhouse gases are poorly identified. As for their contribution to the global gas budgets many knowledge gaps exist. There is a need to study the fundamental processes underlying the emissions from soils and ecosystems. A great number of ecosystems are still undersampled. To improve estimates of sources and source regions, methods for stratification of ecosystems need to be developed. The existing soil, vegetation and climate geo-referenced data bases need to be improved to enable a proper extrapolation of measurement data to regional and global scales. The need to assess the sources of greenhouse gases is driven by the need to parameterize global and regional models of climate. The coupling of ecosystem emissions with atmospheric chemistry is an important controlling factor in climatic change.

The International Geosphere-Biosphere Programme (IGBP-Global Change) of the International Council of Scientific Unions (ICSU) is starting multidisciplinary studies on the interactions between changes in biogeochemical cycles and processes in Solid Earth, Terrestrial Ecosystems, Marine Ecosystems and the Atmosphere. Recently ICSU's Scientific Committee on Problems of the Environment (SCOPE) organized a Dahlem workshop on "Exchange of trace gases between terrestrial ecosystems and the atmosphere", which focused on ecosystem processes and methods of extrapolation mainly. Another important conference on this subject will be organized by SCOPE in Stockholm in 1990. The World Commission on Environment and Development (Bruntland Commission) has stipulated that governments should now strive for sustainable development. The Second World Climate Conference, to be held in 1990, will set the stage for further planning of scientific research and policy actions.

The present book is the result of an international Conference on "Soils and the Greenhouse Effect", 14-18 August 1989 in Wageningen, The Netherlands. It was organized by the International Soil Reference and Information Centre (ISRIC) on behalf of the Netherlands Ministry of Housing, Physical Planning and Environment (VROM) as a contribution to the IGBP-Global Change Programme.

The Conference was one of soil scientists and researchers from many other fields. The Conference aim was to identify research gaps in the field of the geographic distribution of the world soils and land cover types (natural vegetation, cropland or grazing land) and trends in land use on the one hand, and of greenhouse gas fluxes, evapotranspiration and albedo, on the other hand. It may be seen to form a significant contribution of soil science to the study of climatic change.

The Conference resulted in a number of conclusions towards further research on soil and land use related processes involved in the emission of greenhouse gases, and also land use policies which may contribute to reduced emissions. These conclusions are summarized in chapter 1.

The organization of the conference was made possible through funds of VROM. Financial support for the publication of this book and the conference background paper was received from the Commission of the European Community (CEC) and the United Nations Environment Programme (UNEP). Additional financial support was received from the Netherlands Royal Academy of Arts and Sciences (KNAW). Cooperating organizations were IGBP-Global Change, the International Society of Soil Science (ISSS) and the Unesco Man and the Biosphere Programme (MAB).

Dr. W.G. Sombroek
November 1989
Chairman of the Organizing and Editorial Committees

Acknowledgements

This volume is the result of the Conference "Soils and the Greenhouse Effect". It is a collective effort of a diverse group of scientists. The choice of topics and authors of the invited papers is the result of consultation with the members of the Advisory Group and the Organizing Committee of the conference. Thanks are due to the chairmen and rapporteurs of the Conference's four working groups: Marion Baumgardner, Norman Bliss, Robert Brinkman, Gerrit Hekstra, Anton Imeson, Wim Mook, Leendert Pleijsier, Pedro Sanchez, Bo Svensson, Mark Trexler and many others for writing the draft version of the Conference's conclusions and recommendations. The final version of the conclusions and recommendations is presented in chapter 1. Thanks are also due to thank Francis Bretherton for his excellent concluding speech, which is reproduced almost literally in chapter 18.

Many persons have contributed to Part II (Background). It is the text of the conference background paper, in which the viewpoints and suggestions as expressed during the conference itself are gratefully included.

Chapters 8-17 are the conference invited papers and chapter 19 contains all the extended abstracts of papers and posters which were presented during the conference. I am indebted to all the authors for timely submitting their texts, and for prompt response to questions and comments by reviewers. Special thanks are due to Ann Stewart of ITC Enschede for editorial assistance. I am also indebted to many colleagues who critically read parts of the manuscript and who made suggestions for major improvements. I very much appreciated the support of ISRIC during the organization of the conference and the preparation of the manuscript of this volume.

Lex Bouwman
November 1989
Editor

Executive Summary

1. The relative contribution to global warming of a one year's increase of atmospheric concentrations of carbon dioxide (CO_2), methane (CH_4) and nitrous oxide (N_2O) is 15, 5 and 1, respectively. Considering only biotic emissions the relative importance of the three gases would be 5, 4 and 1, respectively. In addition to impacts on the atmospheric radiative balance, CH_4 and N_2O play an important role in many atmospheric chemical reactions, while CO_2 is chemically inert.

2. Regarding the present status of knowledge, the sources and sinks of CO_2 are relatively well known, although the effect of future policies such as massive land use changes (deforestation, afforestation), on the partitioning of CO_2 between the atmosphere, biosphere and oceans needs investigation. Compared to the atmospheric increase of CO_2 by emissions from fossil fuel combustion, the sequestering capacity for CO_2 by massive plantations of new forests is only modest. Nevertheless, for soil and water conservation and for satisfying future demands for industrial and fuel wood, forest planting is indispensable.

The global budgets of CH_4 and N_2O are well known, but the individual sources and sinks of CH_4 and N_2O are less well known than for CO_2. The main cause for the increase of 1% in atmospheric CH_4 is its increasing global source strength, whereas diminishing sinks cause a minor increase. Major individual sources are wetland rice, natural wetlands, ruminants, termites, landfills for solid waste dumping, biomass burning, coal mining, oil and gas exploitation and gas distribution. All these sources are increasing at present.

Sources of CH_4 of which emissions may be reduced and which should receive priority attention from scientists and decision makers are:

- Wetland rice emissions (adaptation of agronomic practices mainly);
- Natural wetlands (drainage; this may, however, not be desirable in many places in view of nature conservation);
- Ruminants (reduction of stock-breeding);
- Landfills (harvesting of CH_4, recycling of solid wastes);
- Biomass burning (stimulation of agroforestry to substitute slash-and-burn agriculture; use of alternative sources of energy to replace fuel wood).

The causes for the increase of atmospheric N_2O are not sufficiently understood. Increasing use of nitrogenous fertilizers induces N_2O losses, and fossil fuel combustion is also a growing, though minor, source of N_2O. The individual sources of N_2O are not sufficiently known, but tropical savannas and rain forests are probably important source regions.

3. Ecosystems which are of major importance, either through supposed climate induced changes or through man induced changes are:

- Areas which are subject to deforestation or increasing intensity of shifting cultivation. Deforestation is a source of CO_2 and causes a reduction of the biospheric pool of carbon, while shifting cultivation constitutes a source of CH_4 and N_2O. Agroforestry provides an alternative to slash-and-burn agriculture.
- Paddy rice cultivation is likely to increase in Asia, Africa and South America. A number of options - mainly in the field of agronomic and water management practices - for controlling CH_4 fluxes are proposed, but these require further research before they can be implemented at farm level.
- Permafrost areas are likely to emit increasing amounts of CH_4 as global warming proceeds. Research is needed to improve the predictions of future reactions of these ecosystems.
- Landfills will probably show a marked increase as a consequence of growing urbanization in developing countries. Harvesting of CH_4 fluxes and prevention of emissions through recycling of wastes are feasible options.
- Areas which are in the process of desertification have an important effect on local and regional climate through changes in evapo(transpi)ration and albedo. Afforestation and improved range management are options which reverse desertification and at the same time provide sinks of CO_2, mitigating the annual increase of atmospheric CO_2.

4. Regarding methods for estimating gas emissions, integration of flux measurements with geographic mapping is needed. Geographic databases form the framework for extrapolating results of process studies and flux measurements to global scale ecosystems. In this respect steps should be taken to improve the availability, accessibility and exchange of data and information.

5. Sustainable development should be the goal of all land use policy options. Energy policy is a prerequisite for the successful implementation of land use policies. The proposed policy options are:

- Slash-and-burn agriculture should be reduced where socially and economically feasible, and agroforestry should be stimulated as an alternative.
- Landfills should be provided with facilities for harvesting CH_4 emissions; recycling of solid wastes to prevent emissions should be stimulated.
- Reduction of deforestation should be stimulated. Forest plantations for soil and water conservation and for covering the demands for industrial and fuel wood, should be given the highest priority.
- More efficient use of nitrogen fertilizers should be stimulated as a first step towards reduction of N_2O fluxes.

Part I

Conclusions
and
Recommendations

CHAPTER 1

Conclusions and Recommendations of the Conference Working Groups

1.1 INTRODUCTION

Four working groups met during the conference on "Soils and the Greenhouse Effect" to address the conference aims and to give concise and practical recommendations on future research. The aims of the conference were to:

- estimate the fluxes of greenhouse gases, evapo(transpi)ration and albedo for the major soils of the world and their land cover types;
- quantify global land use changes and their effect on fluxes of greenhouse gases, evapotranspiration and albedo;
- identify research gaps with regard to global geographic data coverage; measuring techniques; model development; collection, collation, storage, dissemination and processing of baseline data; techniques and networks for monitoring;
- recommend policy options for reducing greenhouse gas emissions.

The four working groups and their mandates are listed below.

Working group 1. Review present day knowledge of the emission of greenhouse gases, evapo(transpi)ration and the surface energy balance for world soils and their land covers. Indicate priority areas for research, with special attention to uncertainties in present knowledge of processes and geographic extent.

Working group 2. Evaluate changes in the patterns of greenhouse gas fluxes, evapo(transpi)ration and surface energy fluxes caused by changes in land use (such as changes in cropping patterns and intensities, urbanization, afforestation, deforestation) and other human influences on land and soil conditions (such as land drainage, waterlogging, desertification).

Working group 3. Inventory existing and new methods for estimating greenhouse gas emissions, evapo(transpi)ration and surface energy fluxes for world soils and their land covers. These methods include sampling and measurement, modelling, remote sensing, extrapolation of results using databases of soils, vegetation, land use and climate.

Working group 4. Assess policy options with respect to land use control, agriculture and forestry practices necessary for a reduction of greenhouse gas emissions by soils. Emphasis will be on recommendations for future research as well as remedies.

Some overlap in the working group discussions was inevitable. The conclusions are therefore grouped according to subject instead of as exact accounts of the separate groups' discussions.

Soils and the Greenhouse Effect. Edited by A.F. Bouwman
© 1990 John Wiley & Sons Ltd.

1.2 THE SELECTION OF PRIORITY
AREAS FOR RESEARCH

An attempt was made to rank sources of greenhouse gases according to their relative importance. The basis for such ranking includes a number of different considerations, such as the warming effect of the gases and also present-day knowledge of the strength of sources and sinks. Other important aspects in this respect are the future trends in gas fluxes, as influenced by either climate warming or land use changes. A last criterion is the feasibility of reducing emissions.

1.2.1 Criteria Related to Relative Importance and Present Knowledge

Relative contribution to greenhouse warming of biotic CO_2, CH_4 and N_2O. The relative contribution to greenhouse warming of one year's increase in atmospheric concentration of a gas can be estimated as concentration increase times relative thermal absorption potential. The contribution of biotic sources is found by multiplying the obtained value by the proportion of biotic emissions of the gas considered in the total budget. Values for these proportions (% biotic) are given in Table 1.1. They are intended only for order-of-magnitude estimates.

The effect of a reduction of net fluxes can be simulated by estimating the effect of eliminating a one year's increase in atmospheric concentration. The reduction of emissions has more effect for gases having long atmospheric residence times. Hence, it can be estimated by multiplying the relative contributions to increased thermal absorption (Table 1.1) by the respective residence times. If the far future has less weight (in view of supposed future possibilities and lower costs of correction with improvements in technology), the effect of a reduction of gas emissions on thermal absorption should be discounted over time.

Table 1.1 Atmospheric concentrations, increase, residence time and relative contribution to warming for the three major greenhouse gases

	CO_2	CH_4	N_2O
Concentration (ppm)	350	1.7	0.3
Increase (ppb y^{-1})	1750	19	0.75
Relative potential for thermal absorption (CO_2=1)	1	30	150
Relative contribution of 1 year's increase to increased thermal absorption	15	5	1
% Biotic	30	70	90
Residence time (y)	100	8-12	100-200

This discounting will most strongly effect gases with long residence times. Discounted residence times (at a discount rate of 3%) for CO_2, CH_4 and N_2O are 32, 8.5 and 33 years, respectively, leading to relative values for discounted importance for CO_2, CH_4 and N_2O of 9, 2 and 2. Clearly, considering biotic sources only, CO_2 is the most important greenhouse gas. CH_4 has a greater immediate impact than N_2O, but their cumulative effects discounted over their residence times are equal. If the discount rate is set at zero, N_2O would be more important than CH_4. While CO_2 is capable of only heat absorption, the other two gases also have impacts on atmospheric chemistry and this further increases their importance compared to that of CO_2.

Status of knowledge. Priority should be given to gases for which there are gaps in our knowledge concerning sources, sinks and processes.

The sources and sinks of CO_2 are relatively well-known, and good biosphere models are available. Our knowledge of CH_4 and N_2O, especially with respect to processes and identification of sources and sinks, is more limited. This clearly indicates that research concentrating on CH_4 and N_2O should have priority. The present increase of the atmospheric methane concentration is non-linear. Since there are biological and chemical sinks for CO_2, N_2O and CH_4, "natural" removal rates may be proportional to the concentration. The ongoing changes may cause a more severe imbalance, which could lead to a situation in which responses are no longer linear and increasing effects can be expected.

1.2.2 Criteria Related to Sources of Gases which are Increasing and Sources that can be Reduced

The following criteria were identified:

- Resources should be concentrated on the sources of greenhouse gases subject to the strongest anthropogenic perturbations.
- Research should aim at minimizing emissions of greenhouse gases. Resources should therefore be concentrated on those sources where emissions may be reduced through changed land use policy or introduction of adapted practices of soil, water and crop management.
- Areas liable to increased emission as a reaction to climate warming should be studied at an early stage.

Control of fluxes of greenhouse gases is very difficult in many instances, but it is possible to reduce gas emissions from some sources through land management or land use policies. On the basis of discussions in the working groups, ecosystems and areas for research priority were selected. The ecosystems that exert the greatest influence on global warming, indirectly via changing emissions of greenhouse gases or through direct climatic effects, are listed in Table 1.2.

Table 1.2 Fields or areas (of change) which merit research priority

Type/impact	Area/change	Possible measures for reduction of emissions
CO_2	Clearing of tropical forests	Reduction of deforestation, introduction of sustainable land use such as agro-forestry
CH_4	Increasing areas of paddy rice	Soil, water and crop management
CH_4	Waste production/landfill sites	Harvesting of methane, prevention of emissions
CH_4	Warming in permafrost areas; boreal forests and tundras (wetlands in northern latitudes)	
N_2O	Increasing N-fertilizer use in	Improved efficiency of cultivated lands N-fertilizer use
Direct impact on climate and C-cycle	Desertification	Forest plantations, rangeland management

1.3 WORKING GROUP 1 (STATUS)

1.3.1 General Conclusions

1. Increasing concentrations of greenhouse gases are caused by not only industrial activities, but also by changes in land use and fauna. The contribution of soils, land use (or vegetation) and fauna to the net fluxes of greenhouse gases are approximately 30% for CO_2, 70% for CH_4 and 90% for N_2O. To some extent land use also concerns fluxes of chlorofluorocarbons (CFCs), since landfills are sources of these gases. The exact strength of this source is not known. Although the gases CO_2, CH_4 and N_2O are discussed separately below, their sources should not be studied in isolation but in the context of regional or global biogeochemical cycles, considering exchange processes between terrestrial ecosystems and the atmosphere.

2. Virtually all changes in albedo and evapo(transpi)ration of land areas are attributed to land use (e.g., overgrazing) or land use changes (e.g., deforestation). Water vapour is an important greenhouse gas. It differs, however, in several basic aspects from CO_2, CH_4, N_2O. These latter gases are external to climate formation. Water vapour is involved in all hydrologic processes. It is internal to climate at a global scale, while CO_2, CH_4 and N_2O initiate and quantify the greenhouse effect as external forcings. Water vapour responds to changes in the atmosphere

by amplifying and redistributing the consequences in space and time. Present modelling techniques show no evidence that land use changes have a major global effect on circulation or temperature per se. However, important regional changes may occur, as in rainfall distribution patterns. Global climate models (GCMs) are increasingly efficient and reliable tools for analyzing the radiation and heat regime but GCMs at present are not suitable for analyzing the amplifying and redistributing role of water at regional and local scales.

3. The spatial distribution and extent of different ecosystems (including agricultural) are insufficiently known. Databases and maps of land use and vegetation are available, but generally in mutually incompatible forms. Specific data required for studies of greenhouse gas fluxes from terrestrial ecosystems can be derived only partly or imperfectly from available resource maps or databases (of soils, land use or vegetation, climate).

4. For extrapolation of data and for making predictions, the underlying processes of the emissions should be understood. These processes are largely microbiological. We thus need to investigate the microbiology and biochemistry of the environmental and soil factors influencing the processes. This will also throw light on how land use/soil management practices can influence the emission of greenhouse gases.

5. There should be a clear understanding of the contrast between short-term and long-term data on greenhouse gas fluxes. As these gases are part of biogeochemical cycles of elements, there are short-term cyclic variations (diurnal, seasonal) in their fluxes. The greenhouse effect, however, relates to long term data. Field studies and modelling should focus on integrating knowledge of detailed processes in long-term, large-area models.

1.3.2 Recommendations

Carbon dioxide. At present, better models are available for CO_2 than for N_2O and CH_4. The rates of emission from combustion of fossil fuel, deforestation and soil organic matter decomposition, as well as the absorption by oceans and photosynthesis, are very high compared with the amount sequestered by to the CO_2 fertilizing effect. Hence, even good models are largely uncertain. Model validation should receive attention with regard to both CO_2 sources and sinks. For large-area models, the databases (e.g., regarding biomass quantity) should be improved in order to produce better estimates of the effects of land use changes, such as deforestation or the conversion of savanna to arable land. The area estimates of land use changes also need improvement to provide a reliable basis for the biosphere models.

Longer-term effects of management practices, such as mulching or incorporation of crop residues in soils, on soil organic matter content also merit attention. There is some concern about anticipated climate change effects on soil carbon storage in tundra boreal peatlands and desert regions. Both are currently small sinks of atmospheric carbon. Northern peatlands currently sequester approximately 0.3 Gt C y^{-1} and formation of caliche in desert may amount to 0.05 Gt C y^{-1}. Each of these systems represent large stores of carbon. Climatic changes may reverse their role in the global carbon cycle. Their future response is very sensitive to precipitation and changes in the hydrological conditions. These changes cannot be adequately predicted at present, but this will improve in future. Soil scientists should be prepared to evaluate the importance of soils for run-off characteristics, water tables and moisture regimes, particularly in the important and sensitive peatlands and deserts.

Methane. The present view of the global methane budget is that 70 to 80% of the emissions stem from terrestrial ecosystems. Fossil methane from hydrocarbon exploitation complements the 100%. The major sources are believed to be known, but the accuracy of estimates of individual sources is poor.

Attention was drawn to the fact that to date the aggregate extent of small wetland areas may have been underestimated. Furthermore, temporarily water-saturated soils have been generally neglected, as well as wet forests in boreal regions. These systems contribute substantially to the methane budget. The construction of hydroelectric dams produces large, shallow water bodies. Similarly, losses of phosphate fertilizers from agricultural land contribute to the eutrophication of coastal zone and inland waters and thus to the formation of organic sediments. All these potential sources of methane require further investigation.

The extent of microbial methane oxidation as a regulating process is uncertain. It is important in several wetlands including rice fields and tundra areas. In natural wetlands, the level of the water table will be the main factor controlling methane oxidation. Methanogenesis and methane oxidation respond differently to temperature as a controlling factor, but accurate data are still lacking.

The mechanisms of methane emission from rice paddies and wetlands are not well understood. In soils used for wet rice cultivation, there is a complex regulation of methane formation. The various soil, crop and water management practices in wet rice cultivation affect mostly the soil redox potential. Methanogenesis, in turn, is also affected by redox conditions and in addition by soil chemical properties (such as free iron, organic matter) and soil fertility. The controlling soil parameters should be elucidated in order to supply information supporting mechanistic models that can be incorporated in global models, which can be used to optimize management for decreased methane emissions.

There is a shortage of CH_4 flux measurements. The number and accuracy of flux estimates for methane are as poor as for N_2O. Point measurements are available from only a few of the many types of wetlands. Consequently, extrapolation of the available measurement data to other areas to derive global emission rates is fraught with potential errors. In summary, the demand for more flux data should also include data on process-regulating factors fulfilling the requirements for a global extrapolation.

Our knowledge of the geographic distribution of the two major sources, wetland rice and natural wetlands, the is still poor. The total cultivated area of rice paddies is known, but little is known on soils used, soil and water management, types, amounts and mode of application of fertilizer and number of crops per year. There are contrasting estimates of wetland distribution by type, but there is agreement on the total extent.

Estimates of CH_4 emissions from landfill sites and biomass burning cover a wide range. These sources should receive attention, since they may show important increases in future, as discussed in more detail by working group 2.

Nitrous oxide. At present, the budget for the N_2O-exchange between terrestrial ecosystems and the atmosphere is largely unknown. Approximately 90% of the N_2O emitted is believed to be generated by soils (Table 1.3). Fossil fuel combustion probably produces only minor amounts of N_2O. An intensive search for sources to match the atmospheric sink is therefore needed. With regard to the possible soil sources, the working group recommended investigation of the ecosystems in Table 1.3. Study of the two microbiological processes (denitrification and nitrification) and measurement of emission rates should be related to edaphic and climatic factors. The actual scale of measurement should allow integration to a coarser level (e.g., extrapolation from chamber flux measurements to fields or regions) and ideally allow combination with global models. Special attention should be paid to areas with high rates of organic matter turnover, i.e., tropical regions as well as heavily N-fertilized agricultural areas.

The role of nitrogen-fixing leguminous plants, in particular in savannas and tropical forests, merits attention. These legumes may contribute to N_2O fluxes directly (through denitrification by free-living or symbiotic Rhizobia) or indirectly (by adding nitrogen to the N-cycle). This may explain the high N_2O fluxes reported for tropical forests and savannas.

Table 1.3 Ecosystems and their major soils relevant for N_2O emission to the atmosphere. The ecosystems are presented in order of estimated relative quantitative importance

Ecosystems	Major Soils
Tropical rain forests	Ferralsols, Acrisols
Highly fertile or heavily fertilized arable systems and grasslands	Cambisols, Luvisols, Fluvisols
Subtropical and tropical savanna	Nitosols, Acrisols, Planosols

Water vapour. Water cycles differ basically from the biogeochemical cycles of C, N and other major elements. For example, CO_2 and other greenhouse gases relate to soil and vegetation processes primarily in the dimension of soil chemistry. Soil chemical processes are relatively slow and can be tied rather closely to the soil and land cover classification schemes. Water and its renewal processes relate to soil and vegetation primarily in the dimension of soil moisture. Available soil water is extremely variable in space and time. It can be tied to ecosystems as a whole (climate-soil-vegetation/land use), rather than to soil and land cover alone. Much has already been done, both nationally and internationally, in the field of hydrologic research. There are still basic gaps in understanding the integrating and regulating role of the water cycles within the climate-soil-vegetation systems. The translation of such knowledge into model structures and data merits attention. This will allow future improvements in the quantification of precipitation, evapo(transpi)ration, run-off and other water-related terms of the regional and local climates.

1.4 WORKING GROUP 2 (TRENDS)

1.4.1 General Conclusions

1. Research should aim at reducing emissions of greenhouse gases from soils. Where associated with agriculture, crop and livestock productivity must be maintained or improved. Research on ecosystems should address their roles as sinks and sources of the different greenhouse gases. The response of soils to temperature rise should also be investigated, for example in soils occurring in permafrost areas.

2. Changes in land use can effect the fluxes of greenhouse gases, particularly CO_2, CH_4 and N_2O. The present state of knowledge does not permit making specific recommendations regarding land use practices for reducing gas emissions, except for decreasing deforestation and slash-and-burn agriculture. Intensive studies of the conditions leading to the formation and escape of CH_4 and N_2O are recommended, with the aim of creating the necessary scientific basis for improved land management practices.

3. While many of the processes involved in the carbon cycle are well understood, there are two important gaps in our knowledge of trends in the strength of sinks and sources of CO_2:

- The production of litter. Net primary production is likely to increase because of the CO_2 fertilizing effect. Moreover, plants will become more drought-resistant, and this is of particular interest in semiarid and arid zones. At higher CO_2 levels, however, soil nutrients may limit growth and diminish the CO_2-enhanced growth. This still remains to be quantified.

- Litter quality will be altered if the CO_2 fertilizing effect is not accompanied by increased input of nitrogen into the system. The C/N ratios will tend to increase and litter decomposition will proceed more slowly.

A consequence of these processes may be an increase of soil organic carbon. Worldwide, soils may become an important sink for carbon.

4. A number of the essential data for understanding and predicting trends in methane emissions are missing. First, methane emissions may change as a consequence of land use changes, particularly in areas of wet rice cultivation. The wetland rice fields are of special importance, since reduction of emissions is feasible through the introduction of certain management practices. More fundamental studies of the mechanisms of methane emission are required for estimating trends in CH_4 fluxes. Geographic data are lacking on harvested area of wetland rice, soil types used for rice, water management, levels, types and modes of fertilizer application, organic matter incorporation, cultivation practices, and also on the possibilities of emission control using soil, crop and water management.

5. CH_4 emissions from natural wetlands may increase in response to climate warming. Since climate warming will probably be strongest in northern latitudes, the boreal and tundra ecosystems are likely to show changing CH_4 and possibly N_2O emissions. Insufficient data are available for the tundra ecosystems to obtain the necessary information on present rates and trends of CH_4 fluxes in these vast areas.

The rates of reclamation of wetlands and the creation of new shallow wetlands in the realization of hydroelectric power schemes (as presently occurring in the Amazon Basin) are poorly known. Attention to permafrost areas is merited, since they are likely targets for an early reaction to a warming of the climate.

Data are lacking for proper estimates of methane emissions from landfills. Since the emissions are manageable by gas-pumping (methane harvesting) systems, such actions should be promoted. It may also be possible to prevent emissions from landfills by recycling of wastes. As a result of growing urban populations, particularly in the developing world, this source will probably become very important in the next century.

The burning of biomass, such as agricultural wastes, savanna fires and fuel wood, is an important and increasing source of CH_4. The estimates presented to date indicate considerable uncertainty and variability in amounts of biomass burned, types of burning and conditions under which burning takes place. Growing populations will increase the area under shifting cultivation and the demand for fuel wood. Hence, this source may continue to increase in the future.

Table 1.4 Projected direction of fluxes of greenhouse gases and physical properties relative to current and recommended trends of land use practices. A number of key trends in land use are considered and discussed in the text. The multiple response of changing land management practices needs to be addressed in making an assessment of the implications of changing management to offset greenhouse warming effects

	Climate and greenhouse gas component*					
	CH_4	N_2O	CO_2	H_2O	Albedo	Roughness
a. *Current land use trends*						
Increasing deforestation	#	+	+	-	+	+/-
Increasing acreage of paddy rice	+	+/-	o	o	o	o
Reduction of permafrost caused by climate warming	+	+	+	o	o	o
Increasing N-fertilizer consumption	o	+	o	o	o	o
Increasing desertification	o	-	+	-	+	-
b. *Modified trends*						
Reducing deforestation or reforestation	#	-	-	+	-	+
Reduction of slash-and-burn agriculture	-	-	-	+	-	+
Improved wetland rice management:						
- use of inorganic fertilizers	-	+	o	o	o	o
- straw addition	+	-	+	o	o	o
- compost addition	+	-	+	o	o	o
- water management practices	-	+	+	o	o	o
Improved efficiency of N-fertilizer use	o	-	o	o	o	o
Controlling desertification	o	o	+	+	-	+

* + indicates an increase in flux or property adding to greenhouse warming; - indicates a decrease in flux or property diminishing the greenhouse warming; o indicates no effect or negligible effect on the flux or property; # forest burning is a source of CH_4, but in clear cutting of forests no CH_4 is released.

6. The atmospheric sinks for N_2O are relatively well-known. The N_2O emission rates of the various temperate and tropical ecosystems, including heavily-fertilized agricultural areas, are not known accurately. It is therefore difficult to specify causes for the observed atmospheric increase. The growth in consumption of nitrogenous fertilizers is

probably a major source of increasing emissions. Fossil fuel combustion is also an increasing anthropogenic source. A third possible cause for increasing atmospheric N_2O is conversion of forest to grassland or arable land. Finally, climate warming may lead to increased N_2O fluxes. Elevated temperatures increase turnover rates of C and N, and consequently nitrification and denitrification processes will be accelerated and escape of N_2O from these processes will increase.
7. Trends in ecosystems which will alter climate and associated in greenhouse gas fluxes are listed in Table 1.4.a.

1.4.2 Recommendations

1. Worldwide, coordinated research and policy programs should be developed to monitor the changes in land use that trigger greenhouse gas emissions and to reduce gas emissions through improved management practices, particularly in agricultural systems.

2. In tropical regions, the area used for shifting cultivation is increasing. The increasing need for food production, industrial wood and fuel wood is a threat to the tropical forest areas. In many tropical regions, cleared forests are being replaced by unproductive grasslands, waste lands or other unsustainable systems. This trend is not only a cause of increasing CO_2 emissions (Table 1.4.a), but possibly results in increased fluxes of N_2O. Forests are important pools of carbon. Young, actively-growing forests may play a significant role in the global carbon cycle by sequestering part of the CO_2 injected into the atmosphere by fossil fuel combustion. Vast areas of forest plantations are needed to produce a significant reduction of CO_2, however, and this sequestering can be effective only during the active growth period of the forests, which is in the order of decades. Furthermore forest clearings may have an impact on local and regional climates. Loss of species diversity and accelerated erosion are additional negative effects of deforestation.

Sustainable land uses, such as agroforestry, provide alternatives to slash-and-burn agriculture and will provide sinks for CO_2, while emission of CH_4 and N_2O from biomass burning will be prevented. Measures are urgently needed to reduce the rate of tropical deforestation. Sustainable alternatives to shifting cultivation that are both productive and affordable must be developed.

3. Paddy rice cultivation is likely to increase in Africa and South America, and this process should be monitored. Agronomic techniques to reduce methane and nitrous oxide emissions without sacrificing rice yields need to be developed. Worldwide, the application of organic manures in rice is decreasing. Since the cultivated area is increasing, however, this source is probably growing globally. A number of options for controlling emissions are given in Table 1.4.b. This list is certainly incomplete, but it serves as an indication of management and land use options.

4. As global warming proceeds, many permafrost areas and wetlands in northern latitudes are likely to emit increasing amounts of greenhouse gases, particularly CH_4, and suffer serious degradation. Research is needed in these areas to understand processes and to predict the response of these systems to climate warming.

5. Landfills will probably show a marked increase in the coming decades. The increasing urban populations in the developing world will generate growing amounts of waste. This source deserves a high research priority, because there are ways to prevent emissions, for example by collecting methane from landfills as an alternative source of energy. Furthermore, the growing number of people in rural areas will intensify the use of natural resources. Consequently, biomass burning will remain an important source of methane in the future.

6. Areas threatened by desertification are critical in terms of climatic impact. Desertification influences the albedo because soil or degraded soil or rock horizons are exposed for very long periods of the year. Afforestation and improved range management are probably the most practical ways of reversing desertification processes in many Mediterranean and semi-arid areas. Afforestation would at the same time meet critical needs for fuel wood and industrial wood.

Improvement of the vegetation cover in semi-arid areas is recommended in order to modify the soil-atmosphere interaction characteristics and thus to alter micro- and mesoscale climatic conditions as an adaptive measure ahead of the greenhouse effect. Furthermore, it is important to obtain a better understanding of the processes which lead to desertification.

7. The only ecosystems in which N_2O emissions can be reduced by human intervention, are agricultural systems where N-fertilizers are applied. Depending on the soil type there are a number of options, including mineral fertilizers, timing and distribution of applications, mode of application (broadcasting, incorporation or injection of fertilizer), and liquid or solid forms. Measures should aim at prevention of N-losses in the form of N_2O and improving efficiency of N-use.

8. An efficient monitoring system should be set up, focusing on these critical areas (tropical rain forests, wetland rice, permafrost and desertification threatened areas) and soil management techniques should be developed and transferred to alleviate emissions of greenhouse gases, coupled with government policies that promote them.

1.5 WORKING GROUP 3 (METHODS)

1.5.1 General Conclusions and Recommendations

1. The production, consumption and emission of greenhouse gases are affected by physical, chemical and biological processes occurring in the soils covering the continents of the Earth. Natural phenomena and

changes in land use contribute to the magnitude of these processes. Their spatial distribution is commonly identified with separate maps of soils, vegetation, topography, land use and climate. These attributes, when properly synthesized, describe natural and identifiable delineations of the Earth's landscape known as ecosystems, which must be described in terms of the proportion of component landscape units. Field measurements are made to characterize concentrations and fluxes of the greenhouse trace gases for landscape unit components of ecosystems. International coordination is needed to integrate the flux measurements (and process studies) with the geographic mapping. The following elements of this coordination are recognized:

- Data requirements for process studies should underlie and motivate database development, particularly in terms of developing new data sources (e.g., new satellite sensors).
- Data collection and database development will continue using existing technologies, and efforts should be made to infer the maximum amount of information attainable from these databases.
- Existing databases will be used to develop a framework for extrapolating the results of process studies of local-scale landscape units to global-scale ecosystems.

2. Spatial data important for emissions (e.g., identification of wetlands which are a major source of methane) may be obtained directly by remote sensing, or inferred from soil mapping and land use databases. Many other important factors (e.g., soil nitrate as a substrate for N_2O, or straw incorporation as a promoter of methane production) cannot be estimated in this way, however. Information must instead be obtained by ground-based survey techniques. A combination of the more permanent mappable features and information on the magnitude of important but much more transient variables will always be required for any estimation of global or regional gas fluxes.

Research into processes has yielded much relevant information about the key variables involved in the formation of CH_4, N_2O, and other gases. Inevitably, the relationship between the magnitude of the gaseous fluxes and measurable soil characteristics must be simplified in order to make global or regional flux estimates. Provided that this is done, the existing extensive datasets (i.e., soil, land use, topography, vegetation) can be used for the scaling and integration needed for calculation on a regional or global scale.

3. Each category of data in the natural resource databases can be expanded into detailed variables. Soil databases commonly include many characteristics, such as structure, texture, bulk density, water retention and organic matter content. Climatic data include station location, precipitation amount and intensity, temperature, humidity and evaporation. It is necessary to develop procedures to infer the magnitude and associated variability of those parameters from these characteristics, by process-oriented studies and modelling. Delineation of land use categories should account for distinctions that are important for trace

gas emissions, such as separating fresh water wetlands from salt water wetlands.

4. Data integration methods can be used to define spatial strata for extrapolating trace gas measurements.

- The database layers are typically compiled separately caused by differing methods of data collection. Geographic information system technology can be used to integrate these layers using spatial overlays. The resulting detailed land units have all attributes of each input layer.
- The spatial resolution and attribute accuracy of each input database should be recorded as an additional set of layers in the database.
- Complex integration strategies that account for variations in data quality at different times and locations will be needed to interpret the data into the ecosystem strata to be used for extrapolating the trace gas fluxes. For example, it may be appropriate to integrate data from a life zones map at 1:25,000,000 scale with a soil map at 1:1,000,000 scale and a land use composition map at 1:1,000,000 scale that incorporates sampling from detailed sites mapped at 1:24,000 scale.

1.5.2 Specific Recommendations

Gas flux measurements. The methods used to date to study the surface fluxes (emission and deposition) of trace gases, each have advantages and disadvantages. These need to be considered when selecting a technique for a particular purpose (Table 1.5). A general description of these techniques is provided by Mosier (this volume) while more detailed descriptions and analyses of these techniques are available in Andreae and Schimel (1989).

Table 1.5 Gas flux measurement techniques

Method	Items to be measured							
	CO_2	CH_4	N_2O	CO	NO_x	HC	H_2O	Energy
Chamber	GC/IR	GC/IR	GC/IR	GC	Chemi	GC	IR	-
Eddy correlation	IR	TDL*	TDL**	TDL	Chemi	-	IR	Radio
Gradient	IR	TDL	?	-	Chemi	-	Psychr	Radio

* possible now if resources are available; ** should be possible in the future; # other species including NH_3, O_3, HCHO, and H_2 are active in controlling atmospheric chemistry; Key: GC - gas chromatography; IR - Infrared spectrophotometry; Chemi - Chemiluminescence; Psychr - Psychrometry; Radio - Radiometry

Chamber methods have been extensively used and allow a detailed study of soil processes related to the flux. This is especially useful when surface factors have a controlling influence on the emission or deposition. Thus, although the enclosure isolates the system from other components of atmospheric transport, it permits investigation of the most significant component.

In other cases, the chamber technique may be the only method applicable, for example where site conditions are restricting in terms of fetch (with the possible exception of the Denmead mass balance technique).

Micrometeorologic methods, i.e., methods where measurements are made in the *free atmosphere* within the constant flux layer (up to a few tens of meters above the soil surface) where the concentration gradients are steepest, allow an integrated measurement of fluxes representative of vegetation types that may extend over large areas. These methods must be applied within the requirements of fetch, etc. as discussed extensively in the micrometeorologic literature. Such measurements could complement chamber studies and allow larger scale exchange coefficients to be derived under conditions that allow natural reactions, if any, to occur in the atmosphere. Micrometeorologic methods can provide broad spatial coverage in terms of ecologic types and temporal coverage, but by nature have to be limited to "campaign" type studies to determine seasonal changes.

Within the micrometeorologic techniques, probably most success has been obtained with gradient techniques. These are still suitable and acceptable for application to grassland and cereal crops, if applied correctly and if required stability corrections are made.

Such corrections are not necessary for eddy correlation techniques which - in principle - provide a better alternative. For eddy correlation faster instrument response and sensitivity is needed, however. In the case of tall vegetation (forests), the eddy correlation technique is probably the only suitable micrometeorologic method, because the usual flux gradient techniques appear to break down over taller vegetation.

Application of eddy correlation techniques to airborne measurements broadens the scale of the flux measurements.

Generally, a combination of techniques is necessary to appropriately quantify the spatial and temporal variability of a system. This objective cannot be achieved by the exclusive use of any one method.

Isotope methods. The elements constituting the soil gases have more than one isotope. These isotopes can be used as natural tracers since the abundance ratios depend on factors such as processes of formation, parent material and age. The concentrations of the rare isotopes, whether radioactive or stable, thus may give information about the various sources of the gases. Table 1.6 contains a compilation of practical applications.

Table 1.6 Possible applications of isotopes for analysis of soil gaseous fluxes

Gas	$^{13}C/^{12}C$	$^{14}C/C$	$^{15}N/^{14}N$	$^{18}O/^{16}O$	$^{2}H/^{1}H$	$^{3}H/H$
CO_2	+ +	+ +		+		
CH_4	+ +	+ +			+ +	
N_2O			+	+ +		
CO		+				
C_xH_y		+				

+ possible, but not yet exploited; + + successfully used.

For regular isotope ratio mass spectrometry, approximately 1 ml of gas at NTP is needed. Applied to air samples, this amounts to collecting 5 liters for CO_2 analysis, 1 m^3 for CH_4 and 5 m^3 for N_2O analysis. Modern mass spectrometers, however, are equipped to analyze 10 microliters of gas at similar precision. The extremely useful small-scale analysis of CO_2, as well as the determination of atmospheric CH_4, has thus become possible.

Our knowledge of the isotopic abundance in gases of various sources is still limited. The ranges of values for each isotope and source have to be defined with more certainty in order to allow a more accurate determination of the relative contributions of the various sources to normal atmospheric concentrations. The information on source strengths available from isotope studies has been under-utilized, particularly in their use as an independent constraint for global tracer models.

General procedure for extrapolations. A general procedure for making global estimates of gas fluxes is outlined below:

- Calculate the global fluxes with the aid of the flux rates and the spatial extent appropriate for each species. Where possible, also estimate isotope fluxes to employ additional checks against observations.
- Compare the estimates derived in these steps with other estimates of global fluxes and pools, and evaluate global mass balances.

Recommendations with regard to impact assessment of the different trace gases:

- Contribute the results of the trace gas flux calculations as inputs to general circulation models of the atmosphere for estimation of the impacts of the gases on global temperature and precipitation patterns.
- Assess the feasibility of using indirect methods of evaluating the impact of greenhouse gases, where possible, to quantify the strength of their source (e.g., studying the primary productivity of ecosystems as an indirect indicator of CO_2 increase).

1.6 WORKING GROUP 4 (POLICIES)

Scientific targets need to be set for anthropogenic emission reductions required to stabilize greenhouse gas cycles. Policies intended to address the threat of climate change must focus on the terrestrial biosphere as well as the atmosphere and oceans. The implementation of specific measures to achieve such targets should take into account, to the extent present knowledge permits, interactions with measures to address similar issues, such as the stratospheric ozone layer, acid precipitation, eutrophication of coastal zone and inland waters and land degradation.

The need for policy strategies to manage the stresses on our global environment derives, however, from more deep-seated aspects of our human society. The major causes of the environmental problems are the continuing rapid increase in world population and the ever-increasing energy and resource use. A realistic strategy to address greenhouse warming and the underlying emissions of greenhouse gases must include consideration of options from a broad range of integrated economic and social perspectives, as well as technologic and agricultural ones.

Fossil fuel combustion is by far the most important source of CO_2. Hence, energy policy measures are a prerequisite for all other measures. Most natural ecosystems in many parts of the world have been transformed into managed systems and agricultural or urban areas. Any effort in the field of policies and scientific research must be accompanied by education of the citizens of the world on sustainable use of the environment, natural resources and food.

The recommendations cover two major areas: education and information exchange and sustainable land use.

1.6.1 Education and Information Exchange

Governments need to support the implementation of biological, chemical and physical monitoring and documentation capabilities related to soils and land use in order to improve the databases used in modelling global climate systems. Developing countries should be helped to set up such monitoring systems.

National governments should stimulate scientific research, for example by establishing scientific advisory groups that:

1. Estimate the order of magnitude of impact of national greenhouse gas emissions on the increasing atmospheric concentrations of CO_2, N_2O, CH_4, and CO.

2. Clarify and quantify the linkages between particular types of land use and water regimes and rising atmospheric concentrations of greenhouse gases. As part of such programs, governments should support efforts to improve the understanding of the role of terrestrial ecosystems in greenhouse gas trends, atmospheric chemistry and hydrologic cycles. This information should form the basis for future measures to react to

changing climate. It should further provide for the reconsideration of primary and secondary forest utilization practices.

3. Clarify the most important components of the nitrogen cycle in the build-up of N_2O concentrations, e.g., the use of nitrogen fertilizers in farming systems, the transport of nitrogen into the soils through air pollutants such as NO_x and NH_3, and the role of nitrogen-fixing leguminous plants and microorganisms in N_2O and NO fluxes.

Governments should support and stimulate the scientific community in better understanding the present status of the global climate, as well as realistic assessments of possible changes in the near future. National agencies that conduct observation programs and support basic research should collaborate with the scientific community by:

1. Providing their own observations, data and information in a timely fashion, including information on stations, instruments, methods of observation, calibration and validation, and any changes in these, along with any other information needed now and in the future to use such observations. In addition, governments need to initiate observations in areas not well covered in support of such research programs as the WCRP and IGBP, and the monitoring programs of, among others, WMO, UNESCO and UNEP.

2. Providing older historical data related to global change studies, especially long-time series of observations where these have not been published or otherwise made available. This will allow the present status of the Earth's climate system to be ascertained in the context of its variability in the recent past. Moreover, participation is also needed in studies of the data quality of past observations.

3. Supporting and strengthening those national data centers and information activities in their own countries that are part of the established data systems being operated under the auspices of ICSU, WMO, IOC, UNEP and other such agencies; moreover, national resources need to be increased to support timely data collection, collation and formatting, analysis and processing into data summaries and indices, and the design and establishment of datasets on environmental and related programs.

4. Take steps to maintain low-cost availability of this information to scientists and government decision makers. Copyright and commercial restrictions should be minimized, and data of national sensitivity could be made available at lower resolution but still in a useful format for global change studies.

1.6.2 Sustainable Land Use

The United Nations Commission on Environment and Development recommended, inter alia, that sustainable development be the goal of governmental policy. Sustainable development was defined as meeting the needs of the present without compromising the needs of future generations. Soil is a limited resource. Continuously increasing demands are being placed on it to feed, clothe, house and provide energy for growing populations. The nations of the world should agree to use their land on the basis of sound resource management. A land use policy should therefore allow and stimulate maximum sustainable utilization of land resources.

1. Governments should support actions to reduce slash-and-burn agriculture by improving farming practices, while recognizing economic and social realities in the regions concerned, as well as undertaking general forest management practices to reduce carbon emissions. They should also commit to rapid reforestation of waste or unproductive lands. Any tree removal should be compensated by tree planting schemes; these can greatly benefit soil protection efforts, as well as providing industrial wood supplies and alternative sources of energy. Efforts should be intensified for identifying products and services from forested land which do not entail the removal of large amounts of biomass, carbon and nutrients from the system (e.g. bark or fruit rather than lumber).

2. Planned expansion of agricultural and other land uses must be evaluated for the suitability of local soil, topography and climate conditions for the proposed land use.

3. Reduce where feasible the N-losses induced by nitrogen fertilization to reduce the potential for increased N_2O emissions. To allow for the need for the increased application of fertilizers in developing countries and to minimize the overall release of N_2O through such activities agronomic practices which aim at reducing N_2O losses should be promoted. In general terms, a more efficient use of nitrogen fertilizers is supposed to reduce these fluxes. A more efficient use of phosphate fertilizers should be stimulated to reduce losses from agricultural land. This would reduce eutrophication of coastal zone and inland waters and hence also the formation of organic sediments, which constitute important sources of CH_4.

4. The methane emission from landfills is expected to increase substantially in the coming decades, because of increasing urban populations in the developing countries. Policy efforts must aim at minimizing these losses of methane to the atmosphere by gas harvesting or recycling of solid wastes. Methane harvesting in turn may be an alternative source of energy.

5. Governments should provide for conservation of large intact areas of forest. Such conservation can be assured only by certain minimum critical conservation efforts, and by providing for economic diversification and employment opportunities in tropical forest regions, taking advantage of the biological wealth of these regions (e.g., through natural history oriented tourism, domestication of promising animals and

plants). Developed nations should consider compensating through international aid less developed countries that protect virgin forest areas.

6. Plantation of large areas of forests and introduction of agroforestry must be stimulated. This will be a way to sequester part of the continuous injection of CO_2 into the atmosphere. It will also satisfy the demand for hardwood and fuel.

7. Evaluate the impact of existing land use patterns and land use changes on soil moisture patterns. Where moisture regime issues are ignored or miscalculated, a frequent result is desertification, salinization, alkalinization, erosion or other types of ecologic degradation. Such land use degradation inevitably has significant consequences for biotic greenhouse gas emissions. National land and water use policies and integrated river basin development programs should therefore incorporate:

- Maps and databases qualifying and quantifying existing soil moisture regimes, and linking the results to factors of net primary production;
- assessment of foreseeable or reasonable ranges of future variation in agricultural, forestry and other land uses, as well as in natural conditions;
- balance surveys and simulation models identifying the impacts of the predictable anthropogenic and natural changes in effective soil porosity and other parameters of soil moisture regimes and on soil chemistry;
- analysis of cost factors of soil moisture regulation measures and cultivation technologies that could mitigate damage and enhance benefits associated with alternative future land and water use programs;
- in these programs, particular attention should be given to forest and grasslands, both rainfed and influenced by ground water. Under these conditions, the replacement of the natural vegetation by annual crops frequently leads to soil degradation. This can be mitigated only at costs exceeding by far, in a longer-term perspective, the initially envisaged gains.

The issues discussed above are generally interrelated. Policies selected to deal with them must be evaluated with care, otherwise side effects may well defeat the original purpose.

Funding mechanisms for policy and research measures will need to be found. While many of the named initiatives can be built into existing programs, significant funding for improving our understanding of global processes is called for immediately.

REFERENCES

Andreae, M.O. and D.S. Schimel (1989) Exchange of trace gases between terrestrial ecosystems and the atmosphere. Dahlem Konferenzen. Wiley and Sons, Chichester (in prep.).

Kaye, J.A. (1987) Mechanisms and observations for isotope fractionation of molecular species in planetary atmospheres. Review of Geophysics 25:1609-1658.

Mook, W.G. (1986) ^{13}C in atmospheric CO_2. Netherlands Journal of Sea Research 20:211-223.

Stevens, C.M. and A. Engelkemeir (1988) Stable carbon isotopic composition of methane from some natural and anthropogenic sources. Journal of Geophysical Research 93:725-733.

Whalen, M., N. Tanaka, R. Henry, B. Deck, J. Zeglen, J.S. Vogel, J. Southon, A. Shemesh, R. Fairbanks and W. Broecker (1989) Carbon-14 in methane sources and in atmospheric methane: the contribution from fossil carbon. Science 245:286-290.

Wahlen, M. and Y. Yoshinari (1985) Oxygen isotope ratios in N_2O from different environments. Nature 313:780-782.

Yoshida, N. and S. Matsuo (1983) Nitrogen isotope ratio of atmospheric N_2O as a key to the global cycle of N_2O. Geochemistry Journal 17:231-239.

ABSTRACT

Part II

Background

CHAPTER 2

Introduction

A.F. BOUWMAN

International Soil Reference and Information Centre (ISRIC)
P.O. Box 353, 6700 AJ Wageningen, The Netherlands

2.1 GENERAL

The steadily increasing atmospheric concentration of carbon dioxide has provided an unequivocal signal of the global impact of human activities. Moreover, a number of other atmospheric gases of biotic or industrial origin are increasing in concentration. The sources, sinks and dynamics of these gases are important for several reasons. Firstly, many gases affect the chemistry and physics of the atmosphere. They alter characteristics as diverse as the energy budget of the earth, concentrations of oxidants in the atmosphere and absorption of ultraviolet radiation in the stratosphere. Secondly, trace gases or their reaction products affect terrestrial biota directly in ways that can be more or less species specific and that range from enhancing productivity or competitive ability to causing substantial damage. Measurement and understanding of trace gas production and consumption will thus be useful for understanding terrestrial ecosystem dynamics as well as atmospheric processes.

There is concern about the effects of massive deforestation on the rainfall regime and the hydrology of large river systems. Water vapour (H_2O) is one of the major greenhouse gases. Besides its absorption of thermal radiation in the atmosphere, water vapour is of eminent importance in cloud formation and rainfall. Clouds affect the radiation balance. In addition, water vapour plays an important role in the energy transport between equatorial and temperate regions.

Human interference in the land cover brings about changes in the earth's albedo, which plays a major role in the surface energy balance and therefore has important (mostly local and regional) climatic impacts.

Present knowledge regarding the sources of disturbances of atmospheric chemistry and the impact of atmospheric chemicals on valued atmospheric constituents is presented in Table 2.1. The gases which will be considered are soil borne or related to land use and they affect the thermal radiative budget directly. These so-called "greenhouse gases" gases have asterixes in impact column B and in the source columns 2, 3, 5, 6 or 7 of Table 2.1. The gases selected in this manner are: carbon dioxide (CO_2), methane (CH_4), nitrous oxide (N_2O) and ammonia (NH_3/NH_4^+).

Soils and the Greenhouse Effect. Edited by A.F. Bouwman
© 1990 John Wiley & Sons Ltd.

Table 2.1 Sources of major disturbances to atmospheric chemistry and the impacts of atmospheric chemistry on valued atmospheric components. The o's and *'es indicate that the listed chemical is expected to have a significant *direct* effect on the indicated atmospheric property or that the expected source is expected to exert a significant *direct* effect on the listed chemical (data from Clark, 1986; Crutzen and Graedel, 1986; Mooney et al., 1987). *'es are items considered relevant (see text), while o's indicate sources and effects not relevant. Gases for which both the source of disturbance and the impact are relevant are printed in bold

Chemical species	Source of disturbances											Impact					
	1	2	3	4	5	6	7	8	9	10	11	A	B	C	D	E	F
C (soot)						*			o	o		*			o		
CO₂	o	*	*			*	*		o	o		*					
CO	o	*	*			*			o	o	o						
CH₄		*		o		*	*	*	o			*					
CₓHᵧ	o	*				*	*							o		o	
NOₓ	o	*				*	*		o	o	o		o	o	o		
N₂O	o	*				*	*		o	o		*					
NH₃		*	*	o		*	*	*	o			*		o			
SOₓ									o	o	o			o	o	o	o
H₂S	o		*		*	*											o
COS	o		*		*												
Organic S	o		*		*												o
Halocarbons										o		*					
Other halogens	o								o	o	o						o
Trace elements	o					*			o	o	o						
O₃													o	*	o		

Sources of disturbances:
1. Oceans and estuaries
2. Vegetation
3. Soils
4. Wild animals
5. Wetlands
6. Biomass burning
7. Crop production
8. Domestic animals
9. Petroleum combustion
10. Coal combustion
11. Industrial processes

Impacts:
A. Ultraviolet energy absorption
B. Thermal infrared budget alteration
C. Photochemical oxidant formation
D. Precipitation acidification
E. Visibility degradation
F. Material corrosion

Ammonia (NH₃) can absorb thermal radiation but because of its short atmospheric lifetime the role of this constituent is not significant in this respect.

To assemble a complete picture of the role of soils and land use in the greenhouse effect, it is necessary to attend to indirect effects as well. Indirect effects of a source change (i.e. soil change or land use change) are induced changes in chemical species 'A' which affect a given valued atmospheric component through an intermediate influence on chemical species 'B'.

Atmospheric constituents with biotic sources having indirect effects on concentrations of the greenhouse gases listed above are carbon monoxide (CO), nitric oxide (NO) and nitrogen dioxide (NO_2; NO and NO_2 as a group are denoted by NO_x), and ammonia (NH_3). Carbon monoxide is oxidized to CO_2 thereby affecting many other atmospheric constituents, such as ozone (O_3), hydroxyl radicals (OH) and CH_4. About 10% of atmospheric ammonia is oxidized to NO and NO_2. These gases plays a catalytic role in various photochemical reactions in which O_3, CH_4, CO and OH are involved. Thus, the complete list of relevant gases is:

$$H_2O, \ CO_2, \ CH_4, \ CO, \ N_2O, \ NO_x, \ NH_3$$

Possible causes of the increase in the concentrations of these gases are the increasing emissions by the various sources and in some cases a reduced sink strength. The fluxes of these atmospheric constituents will be discussed in more detail in chapter 4.

Regarding the most important greenhouse gases Table 2.2 provides data on the annual rise in concentration, heat absorbing capacity and the contribution to the supposed global temperature rise which occurred during the past 100 years. The most abundant greenhouse gas is CO_2, and because of its high annual emission and long atmospheric residence time its contribution to greenhouse warming is about 50%.

Table 2.2 Atmospheric concentrations of the major greenhouse gases, their rise, residence time and contribution to the global warming

Type	Residence time (y)	Annual rise (%)	1985 concen- tration	Radiative absorption potential [9]	Contribution to greenhouse warming [9]
CO_2	100 [1]	0.5^3	$345ppm^3$	1	50
CO	0.2^4	$0.6\text{-}1.0^7$	$90ppb^2$	n.a.	n.a.
CH_4	$8\text{-}12^5$	1^5	$1.65ppm^7$	32	19
N_2O	$100\text{-}200^6$	0.25^6	$300ppb^6$	150	4
O_3^{10}	$0.1\text{-}0.3^2$	2.0^2	n.a.[8]	2,000	8
CFCs[11]	$65\text{-}110^2$	3.0^2	0.18-0.28ppb	>10,000	15

[1] The atmospheric burden is 720 Gt, annual sources 5-7 Gt; if the atmospheric burden is equal to residence time times the source strength, lifetime must be about 100 years; [2] Ramanathan et al. (1985) (data for 1980); [3] Bolin (1986); [4] Cicerone (1988); [5] Cicerone and Oremland (1988); [6] Crutzen and Graedel (1986); [7] Bolle et al. (1986); [8] O_3 varies from 25ppb at surface to 70 ppb at 9 km (Ramanathan et al., 1985); [9] Enquete Kommission des 11. Deutschen Bundestages "Vorsorge zum Schutz der Erdatmosphäre" (1988); [10] O_3 = ozone; [11] Chlorofluorocarbons; data presented are for the two major CFCs.

Methane and nitrous oxide contribute less to the global warming, but the role of these gases is gaining importance caused by the observed high increase in concentration or their chemical reactivity. This latter aspect will be discussed in more detail below.

To address the global impact of changes in land use and soils the present knowledge (and lacunae) of greenhouse gas fluxes, evapotranspiration and the surface energy balance is summarized in chapters 3 to 7. In chapter 3 the major soil groupings, land cover types and trends in land use will be discussed. Sections 4.1-4.3 address the fluxes of CO, CO_2 and CH_4. In sections 4.5-4.7 the sources and sinks of nitrogen compounds (N_2O, NO_2, NO and NH_3) are reviewed. Chapter 5 addresses evaporation models and the impact of changing land cover on the water balance, and chapter 6 attends to the reflection characteristics of soils and vegetation and potential climatic impacts of land cover changes. Remote sensing techniques are discussed in chapter 7. In the remainder of this introduction a brief outline of the atmospheric chemistry of carbon and nitrogen compounds will be presented.

2.2 ATMOSPHERIC CHEMISTRY OF CARBON AND NITROGEN COMPOUNDS[*]

2.2.1 General

The troposphere is the part of the atmosphere nearest to the earth's surface and extends to 10 km in polar regions to 15-20 km in the tropics. The tropopause forms an abrupt change to the stratosphere. The stratosphere extends to about 55 km.

The principle oxidizing reagents in the lower atmosphere are ozone (O_3) and hydroxyl radicals (OH). Ozone is produced in the troposphere by the peroxyl radical oxidation of NO:

$$NO + RO_2 \rightarrow NO_2 + RO\cdot$$
$$NO_2 + h\nu \rightarrow NO + O\cdot$$
$$O\cdot + O_2 + M \rightarrow O_3$$

Ozone is also formed in the stratosphere through dissociation of molecular O_2 and it may be transported towards the troposphere. Tropospheric destruction of ozone constitutes the primary source of OH:

$$O_3 + h\nu \rightarrow O_2 + O^*$$
$$O^* + H_2O \rightarrow 2OH\cdot$$

---

[*] Logan et al. (1981), Levine et al. (1984), Crutzen and Graedel (1986), Crutzen (1987) and Cicerone and Oremland (1988) and the lecture presented by Crutzen at this conference were consulted in the compilation of sections 2.2.1-2.2.3

A further source of OH are oxygenated organic compounds, e.g.:

$$HCHO + h\nu \quad \rightarrow H\cdot + CHO\cdot$$
$$H\cdot + O_2 + M \quad \rightarrow HO_2\cdot$$
$$CHO\cdot + O_2 \quad \rightarrow HO_2\cdot + CO$$

followed by either: or:

$$HO_2\cdot + HO_2\cdot \rightarrow H_2O_2 + O_2$$ $$HO_2\cdot + NO \rightarrow NO_2 + OH\cdot$$
$$H_2O_2 + h\nu \rightarrow OH\cdot + OH\cdot$$

2.2.2 Methane (CH_4) and Carbon Monoxide (CO)

The carbon compounds, which are involved in the atmospheric carbon cycle are CO, CH_4, CO_2 and NMHC (non methane hydrocarbons). Carbon monoxide is not interacting in the atmospheric radiation balance and is oxidized to CO_2 relatively quickly, thereby influencing the concentrations of other constituents. CO_2 is chemically not reactive in the atmosphere.

Most of the methane present in the troposphere is oxidized to CO. All reaction paths proceed via the intermediate product formaldehyde HCHO, but the reaction sequences are different at high and low atmospheric NO_x concentrations:

$$CH_4 + OH\cdot \quad \rightarrow CH_3\cdot + H_2O$$
$$CH_3\cdot + O_2 + M \rightarrow CH_3O_2\cdot + M$$

High NO environment (>10ppt NO)   *Low NO environment (<10ppt* NO)
$$CH_3O_2\cdot + NO \rightarrow CH_3O\cdot + NO_2$$ $$CH_3O_2\cdot + HO_2\cdot \rightarrow CH_3O_2H + O_2$$
$$CH_3O\cdot + O_2 \quad \rightarrow HCHO + HO_2\cdot$$ $$CH_3O_2H + h\nu \rightarrow CH_3O\cdot + OH\cdot$$
$$HO_2\cdot + NO \quad \rightarrow OH\cdot + NO_2$$ $$CH_3O\cdot + O_2 \quad \rightarrow HCHO + HO_2\cdot$$
$$2[NO_2 + h\nu \quad \rightarrow NO + O\cdot]$$
$$2[O\cdot + O_2 + M \rightarrow O_3 + M]$$

net: net:
$$CH_4 + 4O_2 \quad \rightarrow HCHO + 2O_3 + H_2O$$ $$CH_4 + O_2 \quad \rightarrow HCHO + H_2O$$

Methylhydroperoxide (CH_3O_2H) is oxidized slowly, resulting in a residence time of about one week. It may thus be lost via rain or through reaction with the earth's surface or with aerosol particles, and in that case oxidation of CH_4 may lead to the loss of 1 OH and 1 HO_2. In NO *poor* environments the following subcycle may cause an additional loss of OH and OH_2:

$$CH_3O_2\cdot + HO_2\cdot \rightarrow CH_3O_2H + O_2$$
$$CH_3O_2H + OH\cdot \rightarrow CH_3O_2\cdot + H_2O$$

Further oxidation reaction of HCHO is equal for both low and high NO concentrations:

$$HCHO + h\nu \quad \rightarrow H\cdot + HCO\cdot$$
$$H\cdot + O_2 + M \quad \rightarrow HO_2\cdot$$
or: $$HCHO + OH\cdot \quad \rightarrow H_2O + HCO\cdot$$

Oxidation of HCO. is as follows:

$$HCO\cdot + O_2 \quad \rightarrow CO + HO_2\cdot$$

Hence, oxidation of methane constitutes an important source of atmospheric CO. The CO formed from CH_4 and the CO injected into the atmosphere by anthropogenic sources are oxidized as follows:

$$CO + OH\cdot \quad \rightarrow H\cdot + CO_2$$

Depending on the NO concentration the subsequent pathways are:

High NO environment (>10ppt NO)	*Low NO environment (<10ppt NO)*
$H\cdot + O_2 + M \rightarrow HO_2\cdot$	$2[H\cdot + O_2 \rightarrow HO_2\cdot]$
$3[HO_2\cdot + NO \rightarrow NO_2 + OH\cdot]$	$3[HO_2\cdot + O_3 \rightarrow OH\cdot + 2O_2]$
$3[NO_2 + h\nu \rightarrow NO + O\cdot]$	
$3[O\cdot + O_2 + M \rightarrow O_3]$	

net:

$HCHO + 6O_2 \rightarrow CO_2 + 3O_3 + 2OH\cdot$ $HCHO + 3O_3 \rightarrow CO_2 + 3O_2 + 2OH\cdot$

The important implications of the overall oxidation of CH_4 via CO to CO_2 are, that in the presence of sufficient NO there is a net production of 3.7 O_3 molecules and 0.5 OH radicals for each CH_4 molecule oxidized. In the absence of NO a net loss of 1.7 O_3 and 3.5 OH may occur (Crutzen and Graedel, 1986). Cicerone and Oremland (1988) estimated a loss of 1-2 HO_x molecules in NO poor environments. The global net result is a loss of OH in clean atmospheres and a gain of O_3 in polluted atmospheres. A potentially very important consequence of methane oxidation is that of CH_4, CO and OH perturbations. Since OH is the primary sink of atmospheric CH_4 and CO, and because in the oxidation of both CH_4 and CO, OH radicals are consumed, there is an instability in the system. Increases in CH_4 and CO can lead to depletion of OH, and this leads to further perturbations of CO and CH_4 (Cicerone and Oremland, 1988).

The reaction:

$$CH_4 + Cl \quad \rightarrow CH_3 + HCl$$

is very important in stratospheric chemistry, because it sequesters ozone destroying Cl atoms into HCl molecules which are inactive toward ozone (Cicerone and Oremland, 1988).

For the oxidation of other, more complex hydrocarbons (represented by RH), the reaction path is similar to that of methane in the first two reactions:

$$RH + OH\cdot \quad \rightarrow R\cdot + H_2O$$
$$R\cdot + O_2 \quad \rightarrow RO_2\cdot$$

Two possible consecutive paths (depending on the NO concentration) exist:

$RO_2\cdot + NO \rightarrow RO\cdot + NO_2$	$RO_2\cdot + R'OO\cdot \rightarrow ROOR' + O_2$
$RO\cdot + O_2 \rightarrow R_{-1}CHO + HO_2\cdot$	$ROOR' + h\nu \rightarrow RO\cdot + R'O\cdot$

2.2.3 Nitrogen Compounds

Tropospheric processes. From the above oxidation reactions of methane and carbon monoxide can be deduced that NO_x plays an essential role in the oxidation of CH_4 and CO. The reactions of NO and NO_2 in the atmosphere are diverse. They play an important catalytic role in many photochemical reactions. In the troposphere NO_x enhances the formation of O_3, while in the stratosphere the opposite is the case.

During daytime HNO_3 is formed according to the reaction:

$$NO_2 + OH\cdot + M \rightarrow HNO_3$$

while during nighttime a different path is followed:

$$NO_2 + O_3 \rightarrow NO_3 + O_2$$
$$NO_3 + NO_2 \rightleftharpoons N_2O_5$$
$$N_2O_5 + H_2O_{aq} \rightarrow 2HNO_3$$

During the photochemical breakdown of many non-methane hydrocarbons various organic nitrates may be formed. The most important of these, PAN (peroxyacetylnitrate: $CH_3C(O)O_2NO_2$) is an important reservoir of NO_x in urban areas, but may occur in the middle and upper troposphere as well (Levine et al., 1984). PAN is formed as an intermediate, and it decomposes according to the reaction:

$$CH_3C(O)O_2NO_2 \rightarrow CH_3C(O)O_2\cdot + NO_2$$

thereby liberating NO_x radicals.

Ammonia itself is not capable of absorbing thermal radiation. It is lost from the atmosphere through wet and dry deposition, but 10% (Crutzen, 1983) to 20% (Levine et al., 1984) of all atmospheric NH_3 will be oxidized by $OH\cdot$. The reaction shown below is rather slow compared to the estimated residence time of NH_3 in the atmosphere.

$$OH\cdot + NH_3 \rightarrow NH_2 + H_2O$$

The subsequent chemistry of NH_2 is uncertain:

or:
$$NH_2 + O_2 \rightarrow NH_2O_2$$
$$NH_2 + NO \rightarrow products\ (N_2,\ N_2O)$$
$$NH_2 + NO_2 \rightarrow products\ (N_2,\ N_2O)$$
$$NH_2 + O_3 \rightarrow products\ (NH\cdot,\ HNO,\ NO)$$

At NO_x concentrations below 60 ppt the oxidation of ammonia leads to enrichment of oxides, while with NO_x concentrations exceeding 60 ppt, the oxidation process could provide a sink of NO_x. Ammonia may also react with gaseous HNO_3 to form aerosol nitrate:

$$NH_{3(g)} + HNO_{3(g)} \rightarrow NH_4NO_{3(s\ or\ aq)}$$

Stratospheric processes. The primary source of stratospheric NO_x is probably the photolysis of N_2O. In the stratosphere N_2O is oxidized to NO via:

$$O_3 + h\nu \rightarrow O\cdot + O_2$$
$$O\cdot + N_2O \rightarrow 2NO$$

NO_x catalyses the destruction of ozone in the stratosphere according to the reaction:

$$O_3 + h\nu \rightarrow O\cdot + O_2$$
$$O\cdot + NO_2 \rightarrow NO + O_2$$
$$NO + O_3 \rightarrow NO_2 + O_2$$

net:
$$2O_3 \rightarrow 3\,O_2$$

Below 40 km ozone is formed via the reaction path:

$$O_2 + h\nu \rightarrow 2O\cdot$$
$$2[O\cdot + O_2 + M \rightarrow O_3]$$

net:
$$3O_2 \rightarrow 2O_3$$

The region between 10 and 40 km altitude contains almost all the ozone of the atmosphere. Below 25 km ozone is formed (under the influence of NO_x) as follows:

$$HO_2\cdot + NO \rightarrow OH\cdot + NO_2$$
$$NO_2 + h\nu \rightarrow NO + O\cdot$$
$$O\cdot + O_2 + M \rightarrow O_3$$

$$HO_2\cdot + O_2 \rightarrow OH\cdot + O_3$$

In this 10-40 km region of the stratosphere NO_x thus counteracts the set of reactions:

$$OH\cdot + O_3 \rightarrow HO_2\cdot + O_2$$
$$HO_2\cdot + O_3 \rightarrow OH\cdot + 2O_2$$

net:
$$2O_3 \rightarrow 3O_2$$

The following chain of reactions will then proceed:

$$OH\cdot + O_3 \rightarrow HO_2\cdot + O_2$$
$$HO_2\cdot + NO \rightarrow OH\cdot + NO_2$$
$$NO_2 + h\nu \rightarrow NO + O\cdot$$
$$O\cdot + O_2 + M \rightarrow O_3$$

no net effect

Above about 25 km the NO_x additions decrease the ozone concentration, whereas below 25 km NO_x protects ozone from destruction.

CHAPTER 3

Global Distribution of the Major Soils and Land Cover Types

A.F. BOUWMAN

International Soil Reference and Information Centre (ISRIC)
P.O. Box 353, NL 6700 AJ Wageningen, The Netherlands

ABSTRACT

A global data base of soils is an important tool in trace gas studies. Soil fertility and soil chemical and physical parameters play an important role in the production and emission of trace gases. Extrapolation of point measurement data for fluxes can only be done on the basis of a reliable soil data base. The best data base available is the FAO/Unesco Soil Map of the World which is available as a map or as a digital data base. The information on the map is however, not reliable in many parts of the world. Much new information has become available since its compilation. A major problem with the map is that it represents codes for soil associations, and information on texture of the topsoil for the major component of the association only.

Only experienced soil scientists are capable to translate information included in the descriptions of the map units into terms of controlling factors of trace gas fluxes. A new soil data base, the Soil and Terrain Digital Data Base of the World at 1:1,000,000 scale, is being developed. This data base will contain, where available, all the basic information needed for trace gas studies. In the final part of section 3.1 a brief description is given for 18 Major Soil Groupings, which are associations of the Major Soils occurring on the FAO/Unesco Soil Map of the World. These general descriptions include information on climate natural vegetation and broad physical and chemical soil properties.

There are many maps and digital data sets of vegetation available. Differences in the classification system used cause definitional problems which make intercomparisons difficult. Many maps and data sets present the theoretical climax vegetation and not actual land cover or land use. For most trace gases the type and intensity of use of land determines the flux rate and pattern. The lack of good data on land use explains the uncertainty of the estimates of sources of e.g., wetland rice cultivation in the case of methane, and cultivated land in the case of nitrous oxide. At present, there is no good system for classifying the different types of land use at a global scale. Some attempts to develop such a scheme have been made, but most classifications have local or regional value only.

There is great controversy concerning the global annual deforestation rate of tropical forests. Estimates of global forest destruction for permanent agriculture range from 10 to 20×10^{10} m^2 y^{-1}, much of it in the Amazon region. Global reforestation is about 14.5×10^{10} m^2 y^{-1}. There now remains only about 10×10^{12} m^2 of tropical forest. The figures reported for shifting cultivation show even greater variability than those for permanent clearing. Most of the data base is unreliable. Careful comparison learns that there is great controversy between the various estimates concerning the nature of changes (permanent clearing versus partial destruction or shifting cultivation; the latter process could account for an even greater extent of forest loss than the permanent clearing). There are also definitional difficulties with regard to the type of vegetation which is being cleared

Soils and the Greenhouse Effect. Edited by A.F. Bouwman
© 1990 John Wiley & Sons Ltd.

(primary forest, secondary forest; fallow forest, non-fallow forest; wet forest, moist forest, seasonal forest, dry forest; open forest, closed forest). Thirdly there are differences in the type of land use after clearing. Knowing the exact type is important for making an estimate of the carbon flows and changes in fluxes of other trace gases. These will have a different pattern when forest is cleared for grassland or for cropland. Even when focusing on a specific area, the deforestation figures given in the literature disagree. Possibly the interpretation of time series of NOAA-AVHRR data (in conjunction with vegetation classifications) can improve the estimates of changes.

3.1 INTRODUCTION

In the first part of this chapter global scale soil maps and data bases and the major soil grouping of the world according to the FAO/Unesco legend are discussed. In the second part the present knowledge of the distribution of land cover/use types is evaluated.

Reviews of various systems of soil classification are presented in most books on soil science (e.g. Buringh, 1979; Fitzpatrick, 1983). Two systems are of particular interest in the context of global ecological studies. These are the legend of the FAO/Unesco soil map of the world (1971-1981) and the USDA Soil Taxonomy (USDA, 1975). Both systems will be discussed. Furthermore, a number of global soil maps and global data bases will be compared.

The land cover forms an important factor in the radiation balance of the earth and in numerous biogeochemical cycles. The land cover is subject to modifications by natural cycles and through human activities. Direct effects of human activities are the clearing of the natural vegetation for agriculture, selective clearing during shifting cultivation in the tropics, afforestation, etc. One indirect effect of human activities is the forest die-back caused by air pollution particularly in industrialized regions of the world.

The published literature deals with vegetation and land use only and does not consider the soils distribution. In the first part of this chapter therefore, a matrix is compiled of the world soils, their land cover types and climates. The matrices are based on the FAO (1983) Resource Base of the World at scale 1:25,000,000, the Matthews (1983) 1° resolution land cover data base and the 0.5° resolution Holdridge life zone data base (Leemans, 1989).

3.2 THE MAJOR SOIL GROUPINGS OF THE WORLD

3.2.1 Soil classification systems

FAO Soil Map of the World. The FAO/Unesco (1971-1981) Soil Map of the World is a compilation of many national and regional soil maps. At the time of compilation, soil maps of many countries were available for limited areas only. All these maps had to be translated in order to achieve uniformity. Soils that were geographically related were combined in soil associations. Map units consist of soil units or associations of soil units occurring within the limits of a mappable physiographic entity. When a map unit is not homogeneous (when it does not consist of one soil unit as frequently occurs on a small scale map) it is composed of a dominant soil and of associated soils, the latter covering at least 20% of the area; important soils covering less than 20% of the area are added as inclusions.

The classification presented in FAO/Unesco (1971-1981) Volume I (legend) is not a systematic classification of soils. It is a classification of soil mapping units. The maps present more than 5000 map units or delineations. The original system distinguishes 28 groups of major soils, and 106 soil units. In the codes of map units the textural class of the major component of the association and the map unit slope class is indicated. On the map properties such as stoniness and lithic, petric, petrogypsic and saline phases are indicated with special map symbols. The FAO legend is especially designed for use on a very small scale. The map units are sufficiently broad to have general validity and contain sufficient elements to reflect as precisely as possible the soil pattern of a large region. A modification of the traditional genetic classification system combined with the concept of the diagnostic horizon was used. The only separations based on climate are the major soil groupings *Yermosols* and *Xerosols*, otherwise soil temperature and moisture are only used where they correlate with other properties such as in *Gleysols* and the separation of areas of permafrost. The system is easily transcribed to other classifications, and the legend is used in many countries at a national level too. Very recently a new version of the classification system was proposed (FAO, 1988). The new version distinguishes 28 major soils and 153 soil units. The properties of the major soils groupings is discussed in 3.2.3.

USDA Soil Taxonomy. The USDA (1975) soil taxonomy is a rather complicated systems which requires thorough training to use it. It is a morphometric system, which implies that all properties used to characterize soils can be measured in the field or laboratory. Soils are classified according to the absence or presence of such properties and their degree. In many situations this is a disadvantage, in particular when the laboratory analysis data are not available or unreliable. Soil Taxonomy distinguishes 10 soil orders, which are subdivided into suborders, great groups, subgroups, families and series. They are, in alphabetical order: *Alfisols, Aridisols, Entisols, Histosols, Inceptisols, Mollisols, Oxisols, Spodosols, Ultisols and Vertisols*. Some of the names used on the FAO/Unesco soil map of the world, such as *Histosols* and *Vertisols*, are similar to those proposed in Soil Taxonomy. Some of the concepts are similar, but different names are used: *Alfisols* are similar to *Luvisols, Aridisols* are almost identical to *Xerosols* and *Yermosols*; *Oxisols* to *Ferralsols*, and *Ultisols* to *Acrisols*. However, some major soils of the FAO/Unesco soil map of the world are not in the first category of Soil Taxonomy, e.g. *Solonchaks, Andosols and Gleysols*.

The second category is the suborder level, classified on the basis of diagnostic soil horizons and properties. Many of the FAO/Unesco diagnostic horizons are similar to the ones used in the USDA Soil Taxonomy. For example, a mollic epipedon (epipedon = surface layer) in the terminology of USDA (1975) is a mollic A-horizon in the FAO/Unesco (1971-1981) legend. There are also umbric, histic and ochric epipedons. For the subsurface horizons (B-horizons in FAO/Unesco) there is also similarity, e.g. the argillic, natric, cambic and oxic horizons. For global and continental studies only the USDA (1975) orders, suborders and possibly great groups are of relevance. Finer subdivisions are too detailed.

3.2.2 Global Soil Maps and Data Sets

A great number of global soil maps are available. Each of them offers specific advantages, which will be outlined shortly in this section. In recent years digital geographically referenced data sets are gaining popularity because of their flexibility of use, and their capability of combining and overlaying with other types of maps, such as vegetation maps.

Table 3.1 List of global soil maps and digital data sets

Reference	Scale	Classification	Nº of types
Maps			
FAO/Unesco (1971-1981)	1:5,000,000	FAO	126
USDA (1972)	1:50,000,000	USDA (1975)	117
FAO (1983)	1:25,000,000	FAO	18
Kovda et al. (1975)	1:10,000,000	Kovda et al. (1967)	280
CEC (1985)	1:1,000,000	FAO	126
Digital data sets			
Gildea & Moore (1985)	½x½ degree	FAO	106
Zobler (1986)	½x½ degree	FAO	106
UNEP/GRID	1:5,000,000	FAO	106
Wilson and Henderson-Sellers (1985)	1x1 degree	FAO*	27
Matthews & Rossow(1987)	1x1 degree	USDA (1975)	43
Bouwman (1989c)	1x1 degree	FAO (FAO,1983)	18

* the basis is the FAO 1:5,000,000 soil map of the world with additional data for USSR, eastern Europe, but data were translated into 3 colour, 3 texture and 3 drainage categories.

Most of the digital data bases listed in Table 3.1 are based on FAO/Unesco (1971-1981). The map shows codes for soil associations, with special symbols for topsoil texture of the major component of the association, and for the slope gradient class.

Most information is in the soil reports and not on the map, in a form difficult to translate into parameters such as soil water storage capacity and soil fertility. Only experienced pedologists can make the implied soil parameters accessible for use in e.g. climate models. A disadvantage of the FAO/Unesco map is that in many areas it is based on knowledge available in the 1960s. In the meantime much additional information has become available for regions such as the Amazon basin, the USSR, parts of Eastern Africa, Western Europe.

A very recent development is the compilation of a 1:1,000,000 Global Digital Data Base acronymed SOTER (Shields and Coote, 1988) which has started in some pilot areas in South America and North America. SOTER uses no soil classification in order to be universally applicable and to avoid problems of interpretation. Furthermore it includes attribute files containing all available soil data necessary to derive information on e.g. soil water storage capacity, rooting depth, drainage condition. Baumgardner (this volume) and Sombroek (this volume) provide detailed information on the development of SOTER.

Maps of the agro-ecological zones of the developing world were published by FAO (1978-1981) in the Agro-Ecological Zones project. Soil climatic data can be derived from the small maplets in the reports of the Soil map of the World (FAO/Unesco, 1971-1981). A map of soil moisture regimes was compiled by USDA (1971). A map of wetlands and soils with hydromorphic properties was produced by Van Dam and Van Diepen (1982). This map shows all the flat wetlands and in the accompanying report the suitability for rice cultivation is assessed. Aselmann and Crutzen (1989) compiled a data set of the world's wetlands using a 2.5x2.5 degree grid and Matthews and Fung (1987) also compiled a data set of the wetlands of the world. The latter data set is not geo-referenced, however.

3.2.3 The World's Major Soil Groupings

In this section short descriptions will be given of the major soil groupings occurring on the global map compiled by FAO (1983). Much of the information is from Fitzpatrick (1983), Buringh (1979) and FAO/Unesco (1971-1981). For each major soil grouping the USDA (1975) equivalent is given. In Table 3.2 a matrix is presented of the global areas of major soil groupings and land cover types. Table 3.3 shows a matrix of the global extent of the same soil groupings and the Holdridge Bioclimatic Zones (Holdridge, 1967).

For this compilation the following digital data bases were used: for soils the data base compiled by Bouwman (1989c), for land cover types Matthews (1983) and for bio-climates the data base compiled by Leemans (1989). The bioclimatic zones defined by Holdridge (1967) are listed in Table 3.4. Matrices of ecosystems for the 6 continents presenting bioclimatic zones, soil groupings, and for each ecosystem the % distribution of land cover, are available at ISRIC, Wageningen (not presented here).

Table 3.2 Matrix of Major Soil Groupings according to FAO (1983) and the major land cover types according to Matthews (1983). Areas in 1,000,000 ha. Descriptions of the Major Soil Groupings are given in section 3.3

Land cover types (Matthews, 1983)	1	2	3	4	5	6	7	8	9	10	11	12	13	14	15	16	17	18	total
1. Tropical evergreen rainforest, mangrove forest	576			301	121	42		79	51	11	30				5	6		1	1223
2. Tropical/Subtrop-evergreen seasonal broadleaved forest	164			42	13	22		54	19				1	1	5		6	1	331
3. Subtropical evergreen rainforest		4		1	3											6		1	19
4. Temperate/subpolar evergreen rainforest						14		11											39
5. Temp. evergreen seasonal broadleaved forest summer rain	42				7	8		18											81
6. Evergreen broadleaved sclerophyllous forest winter rain					1	6		9							2				47
7. Tropical/subtropical evergreen needleleaved forest	17					6	16	10									2		49
8. Temperate/subpolar evergreen needleleaved forest		20	232	70	57	22		57	26	276	12	13	3	14	29	7	1	5	909
9. Tropical/subtrop.drought deciduous forest	57	11	66		131	144	0	40				17		6	5	1	2	0	292
10. Cold-deciduous forest with evergreens	45	33	5		20	6		109	53		12	32	1	6	2	1	1		510
11. Cold-deciduous forest without evergreens	16				78	102		56	5		9	2	3	3	1	18			393
12. Xeromorphic forest/woodland	22	2		34	123	147	39	17	2		12	46	6		11	3	21	12	270
13. Evergreen broadleaved sclerophyllous woodland	18			2	3	9	1	20	1		9	18	5		19		15	17	170
14. Evergreen needleleaved woodland	8		8		24	38	16	68		50		6		9	2		4	2	247
15. Tropical/subtrop. drought-deciduous woodland	19			107	27	51	2	6	86		47	2	6		6		10	37	371
16. Cold-deciduous woodland			1	1	121	15			5										244
17. Evergreen broadleaved shrubland/thicket, dwarf-shrubland	3				3	117		2	5	6	2	10	6		0	7	9	60	130
18. Evergreen needleleaved/microphyllous shrubland/thicket					7	21						2		9	1		5	16	66
19. Drought-deciduous shrubland/thicket						20	6	12	6		9	2			1			41	84
20. (Dwarf) Shrubland, cold-deciduous subalpine/subpolar				1		8	3							9					45
21. Xeromorphic shrubland/dwarf shrubland	39	1			26	3		13	19		39	19	34	7	28	1	173	409	883
22. Arctic/alpine tundra, mossy bog	53	2	34	9	3	90	38	33		42			1	10					715
23. Grassland, 10-40% tree cover	8	40	1	212	379	214	23	9	64	1	73	15	19		10	15	6	17	642
24. Grassland, <10% tree cover	6	3	2	76	23	78	1	56			29	3	3		54	14	14	4	359
25. Grassland, shrub cover	1	14		16	16	31	128	52			205	25	54		34	1	101	122	929
26. Tall grassland, no woody cover	2	11		16	5	115	1	4			1		9		26				81
27. Medium grassland, no woody cover						2	16	9			18	1	2	1	3	1	4	11	79
28. Meadow, short grassland, no woody cover		41		5			84	30	9	2	11	19	17	42	6	2	28	67	604
29. Forb formations								3										1	27
30. Desert					47	240	1				159		44		2		49	845	1544
31. Ice					16	10													1640
32. Cultivated area	149	196	18	95	128	133	156	273	116	26	56	127	14	31	86	18	54	51	1726
Total	1245	378	368	988	1363	531	156	986	515	478	713	398	233	163	339	87	503	1721	14748

Table 3.3 Matrix of Major Soil Groupings of the World according to FAO (1983) and Bio-Climates according to Holdridge (1967). Areas in 1,000,000 ha. Descriptions of the major Soil Groupings are presented in section 3.3. The Holdridge Bioclimatic Zones are listed in Table 3.4

Bioclimates (Holdridge, 1967)	Major Soil Groupings (FAO, 1983)																		Total
	1	2	3	4	5	6	7	8	9	10	11	12	13	14	15	16	17	18	
1. Polar desert					216	47												2	265
2. Subpolar dry tundra					17	19													36
3. Subpolar moist tundra			3		154	57													221
4. Subpolar wet tundra					96	67													213
5. Subpolar rain tundra						10		5											15
6. Boreal desert					3	18		5		2									30
7. Boreal dry scrub		3	6		65	99	31	5		28								2	237
8. Boreal moist forest		98	251		231	341	29	10		112				25			13	8	1148
9. Boreal wet forest		3	27		56	23		62		191							1	5	412
10. Boreal rain forest					1			14		7									40
11. Cool temperate desert						38	4	3		1	7	1		1				93	164
12. Cool temperate desert scrub		89	3		1	109	36	38		1		3	16	5			8	119	345
13. Cool temperate steppe		4	4		5	152	269	5		4		34	18	5	1		47	74	786
14. Cool temperate moist forest	7	132	75		47	136	26	55		82		44	9	18		1	83	1	891
15. Cool temperate wet forest					3	32	1	31		29		5		13		8	3	2	122
16. Cool temperate rain forest	4					12				7		1							24
17. Warm temperate desert		22				32	4	2			13	1	14	2	1		1	40	93
18. Warm temperate desert scrub		12			1	85	17	2			2	7	12	2	7		15	69	211
19. Warm temperate thorn steppe					3	39	29	8	4		5	30	16	2	7	5	38	56	228
20. Warm temperate dry forest	7			9	39	61		40	11		1	73	28	12	7	19	28	10	384
21. Warm temperate moist forest	134			1	12	49		53				9	2	4	11			1	308
22. Warm temperate wet forest	2				1	2		8						1					15
23. Warm temperate rain forest																			0
24. Subtropical desert					1	148												437	648
25. Subtropical desert scrub	26	9			1	80	11	4			90	2	25	5	15	1	24	198	543
26. Subtropical thorn woodland	411	11		2	8	25	22	22	8		154	20	46	1	55	6	75	105	511
27. Subtropical dry forest	97		135	135	48	38	44	60	122		84	22	14	3	73	8	63	41	812
28. Subtropical moist forest	1		530	530	125	47	1	80	88	1	97	66	5	10	31	33	48	2	1493
29. Subtropical wet forest			26	26	11	9		35	9			19	3	2			5		190
30. Subtropical rain forest					2	3		1	1										8
31. Tropical desert	1				1	140		2			13			1	1		15	330	502
32. Tropical desert scrub	3				9	23					59	4	1		5		10	74	180
33. Tropical thorn woodland	9			8	13	9		13	11		78	24	6	1	22		4	27	178
34. Tropical very dry forest				4		27	4	53	106		59	26	3	4	67		3	14	356
35. Tropical dry forest	159			130	42	30	3	34	141		37	1	9	4	42		7	1	689
36. Tropical moist forest	310			129	75	6		8	13		11		2	3		9			588
37. Tropical wet forest	22					1													32
38. Tropical rain forest				1															0
Total	1193	379	370	975	1286	2106	531	953	515	478	712	398	237	155	339	90	491	1712	12920

Table 3.4 Holdridge life zone classification (Holdridge, 1967); table presenting the bioclimatic zones with climatic data (Holdridge, 1967). P = precipitation, PET = potential evapotranpiration, T = temperature

Life zone (Holdridge, 1967)	Latitudinal belt	Altitudinal belt	PET/P	P (mm)	T (°C)
1. Polar desert	polar	nival	<1.5	0- 750	<1.5
2. Subpolar dry tundra	subpolar	alpine	1-2	<125	1.5-3
3. Subpolar moist tundra	subpolar	alpine	0.5-1	125- 250	1.5-3
4. Subpolar wet tundra	subpolar	alpine	0.25-0.5	250- 500	1.5-3
5. Subpolar rain tundra	subpolar	alpine	0.125-0.25	500-1000	1.5-3
6. Boreal desert	boreal	subalpine	2-4	<125	3-6
7. Boreal dry scrub	boreal	subalpine	1-2	125- 250	3-6
8. Boreal moist forest	boreal	subalpine	0.5-1	250- 500	3-6
9. Boreal wet forest	boreal	subalpine	0.25-0.5	500-1000	3-6
10. Boreal rain forest	boreal	subalpine	0.125-0.25	1000-2000	3-6
11. Cool temperate desert	cool temperate	montane	4-8	<125	6-12
12. Cool temperate desert scrub	cool temperate	montane	2-4	125- 250	6-12
13. Cool temperate steppe	cool temperate	montane	1-2	250- 500	6-12
14. Cool temperate moist forest	cool temperate	montane	0.5-1	500-1000	6-12
15. Cool temperate wet forest	cool temperate	montane	0.25-0.5	1000-2000	6-12
16. Cool temperate rain forest	cool temperate	montane	0.125-0.25	2000-4000	6-12
17. Warm temperate desert	warm temperate	lower montane	8-16	<125	12-18
18. Warm temperate desert scrub	warm temperate	lower montane	4-8	125- 250	12-18
19. Warm temperate thorn steppe	warm temperate	lower montane	2-4	250- 500	12-18
20. Warm temperate dry forest	warm temperate	lower montane	1-2	500-1000	12-18
21. Warm temperate moist forest	warm temperate	lower montane	0.5-1	1000-2000	12-18
22. Warm temperate wet forest	warm temperate	lower montane	0.25-0.5	2000-4000	12-18
23. Warm temperate rain forest	warm temperate	lower montane	0.125-0.25	4000-8000	12-18
24. Subtropical desert	subtropical	premontane	8-16	<125	18-24
25. Subtropical desert scrub	subtropical	premontane	4-8	125- 250	18-24
26. Subtropical thorn woodland	subtropical	premontane	2-4	250- 500	18-24
27. Subtropical dry forest	subtropical	premontane	1-2	500-1000	18-24
28. Subtropical moist forest	subtropical	premontane	0.5-1	1000-2000	18-24
29. Subtropical wet forest	subtropical	premontane	0.25-0.5	2000-4000	18-24
30. Subtropical rain forest	subtropical	premontane	0.125-0.25	4000-8000	18-24
31. Tropical desert	tropical		16-32	<125	>24
32. Tropical desert scrub	tropical		8-16	125- 250	>24
33. Tropical thorn woodland	tropical		4-8	250- 500	>24
34. Tropical very dry forest	tropical		2-4	500-1000	>24
35. Tropical dry forest	tropical		1-2	1000-2000	>24
36. Tropical moist forest	tropical		0.5-1	2000-4000	>24
37. Tropical wet forest	tropical		0.25-0.5	4000-8000	>24
38. Tropical rain forest	tropical		0.125-0.25	>8000	>24

1. *Acrisols* (A) (USDA: *Ultisols*)
This Soil Grouping occurs mainly in Bioclimatic Zones 21 (warm temperate moist forest), 28 (subtropical moist forest) and 36 (tropical moist forest).
Characteristics of *Acrisols*:

- strong acidity;
- frequently high aluminium saturation
- low base saturation;
- very low availability of nutrients;
- topsoil organic matter easily lost;
- weak physical structure, high susceptibility to rainfall erosion;
- water may stagnate on the argillic B-horizon, causing impeded internal drainage;
- frequently low in trace elements.

Acrisols are formed in Wet Equatorial to Tropical wet-dry climates. The thick, acid and strongly leached upper horizons create many problems for utilization. In many places even after liming and heavy fertilizer dressings their productivity is low, although in for example Brazil good results have been reported. Deficiencies of micronutrients is common. Annual cultivation leads to rapid decomposition of the soil organic matter, and also to rapid structural deterioration, deficiencies of minor and major nutrients. Continued cultivation may lead to compaction and increased danger of water erosion. The required fallow periods are long under low management, with 1-2 years of cultivation followed by 20 or more years of rest. Under high management levels, shorter rest periods under grass or green manure crops are required.

2. *Chernozems* (C), *Phaeozems* (H) and *Greyzems* (M) (USDA: *Mollisols*)
Soil Grouping N°2 occurs mainly in Holdridge zones 8 (boreal moist forest), 13 (cool temperate steppe) and 14 (cool temperate moist forest).
Characteristics of *Chernozems*:

- inherent fertility good;
- excellent physical structure;
- high available water capacity;
- rich in organic matter;
- calcareous layer within 125 cm;
- moderate to high cation retention capacity.

Chernozems are developed almost exclusively in loess, but they also occur in other sediments. They have a dark grey to black A-horizon which is vermicular (i.e. 'full of worm casts'). *Chernozems* are confined largely to continental conditions from Humid Continental to Mid-Latitude steppe. Annual cultivation of these soils leads to rapid loss of soil organic matter. Required fallow periods are short.

Characteristics of *Phaeozems*:

- dark organic rich topsoil;
- good structure;
- high available water capacity;
- moderate to high cation retention capacity.

Phaeozems occur in more humid environments than *Chernozems* or *Kastanozems* (Soil Grouping N° 7) and on fine textured basic parent material. *Phaeozems* have a brown to grey mollic A-horizon rich in organic matter of 30-50 cm thickness over a brown subsoil or yellowish brown substratum. Generally evapotranspiration in summer exceeds precipitation and therefore *Phaeozems* are susceptible to droughtiness.

Greyzems are intergrades between *Chernozems* and *Luvisols* (see Soil Grouping No. 8). They have the dark humus rich mollic top (A) horizon of the *Chernozems* and the argic B-horizon (horizon with clay illuviation) of *Luvisols*. They occupy a zonal position at the transition from tall grass steppes to lands with deciduous forest. *Greyzems* are formed in warm continental areas under grassland cover. Topography in general is gently sloping.

3. *Podzoluvisols* (D) and *albic Luvisols* (La) (USDA: *Alfisols and Eutroboralfs* respectively). This Soil Grouping occurs mainly in Holdridge zones 8 (boreal moist forest) and 14 (cool temperate moist forest).
Characteristics of *Podzoluvisols*:

-    poor drainage;
-    moderately acid soil reaction.

Podzoluvisols occur in cool humid continental regions. In the natural state, *Podzoluvisols* occur under a forest vegetation. A raw litter layer tops a dark but thin topsoil over a bleached subsurface layer which extends into a brown clay illuvial horizon. They are confined to flat or gently sloping areas where moisture can accumulate in the upper part of the soil.

Characteristics of *albic Luvisols*
The subgroup of *albic Luvisols* (see also Groupings 8 and 9) consists of soils with a bleached horizon in between the humus rich topsoil and a heavier subsoil. This bleached horizon may exceed 20 cm in thickness. These soils occur in humid climates with a marked dry season. *Albic Luvisols* are moderately to highly suitable for cropping, their suitability decreasing along with the thickness of the bleached horizon. With high levels of management these soils are always good.

4. *Ferralsols (F)* (USDA: *Oxisols*)
Ferralsols occur in the Holdridge Zones 27 and 28 (subtropical dry and moist forest), 35 and 35 (tropical dry and moist forest).
Characteristics of *Ferralsols*:

-    strong acidity;
-    very low availability of nutrients;
-    high aluminium saturation and low base saturation;
-    low cation retention;
-    no reserves of weatherable minerals;
-    organic matter (predominantly in topsoil), is easily lost during cultivation;
-    good physical structure and low inherent susceptibility to rainfall
     erosion.

Ferralsols are formed in Wet Tropical, Trade wind Littoral and Tropical Wet-Dry climates, with mean annual temperatures of generally over 25°C and mean annual precipitation of over 1000 to 1200 mm. They have a deep solum, usually several meters thick, over weathering rock, with diffuse or gradual horizon boundaries. High iron content in the ferralic B-horizon, and the good internal drainage are responsible for distinct red (hematite) or yellow (goethite) matrix colours, usually without mottles. Annual cultivation of *Ferralsols* leads to rapid loss of organic matter. Since organic matter is responsible for a great part of the cation retention, the latter will also decrease. Under a high level of management limited rest periods are required to control pests and diseases. Although inherent erodibility of *Ferralsols* is low, their usually sloping to rolling topography may induce high run off and soil loss rates.

5. *Histosols* (O) and *Gleysols* (G) (USDA: *Histosols* and *Aquepts/Aquents*, respectively)
Soil Grouping Nº 5 is not typically occurring in any Holdridge bioclimatic zone.

Histosols are typical wetland soils, generally with high water tables and with an organic
layer of at least 40 cm thickness. A more common name is organic soils or peat soils.
Histosols frequently occur in association with *Gleysols* and usually have a flat topography.
Natural drainage is a problem common to these soils, but can easily be corrected.

Gleysols (G) occur on level land, in many cases with a high water table. Their properties,
such as texture, physical and chemical properties, vary widely. The *Gleysols* (and also
Fluvisols) formed in deposits of the tributaries of the Amazon for example are formed
in very poor material deposited by the rivers carrying material derived from very poor
eroded *Ferralsols*, *Acrisols* and acid rocks. In other regions however, *Gleysols* and
Fluvisols may be among the best soils available if their drainage is well managed. With
intermediate and high input levels there is no need for rest periods.

6. *Lithosols* (I) (in new FAO legend: *Leptosols*) (USDA: *lithic* subgroups)
The formation of *Lithosols* has no relation to climate, but major occurrences are in
Holdridge zones 8, 13, 14,24 and 31.
The group of *Lithosols* comprises all the soils which are shallow (less than 30 cm deep)
independent of the parent rock, climate or physiographic position.

7. *Kastanozems* (K) (USDA: *Mollisols*)
Main bioclimatic zone Nº 13 (cool temperate steppe).
Characteristics of *Kastanozems*:

- high organic matter content in the topsoil;
- good structure;
- high available water capacity;
- high inherent fertility, high cation retention capacity.

Kastanozems are found predominantly in semi arid climates in the middle latitude steppe
conditions. Grassland of medium height is their natural vegetation, their overall
topography is gently sloping. There is a fairly wide range of variability within this soil
group as determined by differences in parent material, age and topographic position. The
northern Eurasian *Kastanozems* may have dark brown topsoils of 50 cm thickness, grading
into a cinnamon or pale yellow massive to coarse prismatic subsoil. In the drier south the
topsoil in much shallower and colours are lighter throughout the profile. Usually these
soils are found on elevated situations, but they may form a continuum to *Solonetz* or
Solonchaks (Soil Grouping 13) in depressions and also to *Chernozems* (Soil Grouping No.
2).

8. orthic *Luvisols* (L) and *Cambisols* (B) (USDA: *Alfisols* and *Inceptisols*, respectively)
About half of this Soil Grouping occurs in Holdridge zone Nº 14 (cool temperate moist
forest), with minor occurrences in zones 8, 13, 20, 21, 27, 28, 35 and 36.
Characteristics of *Luvisols*:

- inherent fertility moderate;
- moderate to high cation retention capacity;
- organic matter content low to moderate;
- physical structure weak in the topsoils, moderate but unstable in the argillic B-
  horizon;
- moderate to high available water capacity.

These soils are formed under humid conditions with a marked dry season. The precise
temperature regime and total precipitation are variable. *Luvisols* can be found in Tropical

Wet-Dry, Humid Subtropical, Mediterranean and Humid Continental climates. In addition they are found in semi arid climates where they were formed under more humid conditions and are now fossiled to drier climatic conditions. *Luvisols* have a brown to dark brown topsoil over a (greyish) brown to strong brown or red subsoil (which shows clay illuviation). In subtropical *Luvisols* a calcic layer or pockets of soft powdery lime may be present below the reddish brown subsoil. Soil colours are less reddish in cool climates than in warmer regions. *Luvisols* constitute one of the major soils for food production, particularly maize, sorghum, groundnuts, and other food crops. At a low input level, a fallow period of about 2 years in 3 is required to keep the soils in a good condition. At high levels of management, the required fallow period is shorter (Young and Wright, 1980).

Characteristics of *Cambisols*:
-   medium textured;
-   good structural stability;
-   high porosity, good water holding capacity;
-   good internal and external drainage.

This group of soils comprises a variety of soils in the tropics and temperate regions. The typical soil profile has a dark coloured surface layer over a yellowish-brown subsoil. The latter subsoil may in places also have intense red colours. Textures are loamy to clayey. *Cambisols* range from shallow to moderately shallow soils occurring in cool upland regions (South America) to deep soils in old alluvium in the Ganges flood plain. Generally the organic matter status is moderate to good. With high inputs almost continuous cultivation is possible. Note that the above characteristics are generalizations, and that numerous exceptions exist.

9. *Nitosols* (N) and *ferric Luvisols* (L) (USDA: *Alfisols/Ultisols*)
This Grouping occurs primarily in Holdridge Zones 27 and 28 (subtropical dry and moist forest) and 34 and 35 (tropical very dry and dry forest)
Characteristics of *Nitosols*:

-   high clay content;
-   good physical structure;
-   high available water capacity;
-   solum mostly deep;
-   moderate to high cation retention;
-   nutrient availability high in *eutric Nitosols*, low in *dystric Nitosols*;
-   *dystric Nitosols* have a problem of acidity;
-   moderately high reserves of weatherable minerals;
-   generally higher organic matter levels than in other freely drained soils;
-   moderately high erosion hazard.

Nitosols are strongly weathered halloysitic or kaolinitic soils (1:1 lattice clays) formed in mostly basic parent material, the principal feature being the steady increase in clay with depth to a maximum in the middle layer and below that remaining uniform for some depth. The solum extends normally down to a depth of 150 cm or more. Most *Nitosols* are red soils. *Nitosols* are among the best tropical soils. At low input levels *eutric Nitosols* may well be cultivated for at least 1 out of 2 years, while *dystric Nitosols* need a longer rest period. *Nitosols* respond well to inputs and under high levels of management they can be continuously cropped.

Characteristics of *ferric Luvisols*: these soils are the tropical lateritic podzolic soils with high base saturation. They are widespread in the tropical savanna zone. They have an horizon with clay illuviation with a moderate but unstable structure. The physical structure is commonly weak in the topsoils. Organic matter content and inherent fertility are moderate.

10. *Podzols (P)* (USDA: *Spodosols*)

Major occurrences in Holdridge Zones 8 and 9 (boreal moist and wet forest) and 14 (cool temperate moist forest)

Characteristics of *Podzols*:

- chemically poor;
- good drainage;
- acid soil reaction;
- low cation retention capacity;
- generally low available water capacity.

Podzols are usually formed in coarse to medium textured unconsolidated deposits, often containing a high proportion of boulders and stones. A typical zonal *Podzol* has a pale grey, strongly leached subsurface layer between a dark topsoil with spongy partly humified organic matter, and a brown to very brown subsoil. At the drier end of the climatic range for zonal *Podzols* the subsoil has a high chroma due to illuviation of oxides of alumina and iron, while in more humid regions the subsoil is darker and has a higher content of translocated organic matter. Where the parent material consists of pure quartz a subsoil in which the illuviated material is entirely organic. They generally occur in any topographic situation where aerobic conditions prevail and water is allowed to percolate freely through at least the upper and middle parts of the soil profile. The range of climatic conditions under which *Podzols* are formed is quite wide. They are most widespread under a tundra or marine climate with rainfall variation from 450 to 1250 mm per annum. They also occur under Humid Continental cool summer climate and Humid Tropical climate. In both of these environments the parent material is highly siliceous producing conditions which favour *Podzol* formation. Under high levels of management these soils may be improved a lot. Tropical *Podzols* occur either in the cold tropics or in the lowland rainforest areas. The latter *Podzols* are formed in sands and their suitability for cultivation is probably low. These *Podzols* have a bleached light grey to white eluvial horizon varying in thickness between 20 cm to more than 2 metres (Giant *Podzols*). The illuvial subsoil is commonly dark brown and irregular in depth.

11. *Arenosols* (Q) and sandy *Regosols* (R) (USDA: *Psamments*)

This Soil Grouping occurs in many subtropical and tropical zones, the major being 25-28 and 32-34

Characteristics of both *Arenosols* and sandy *Regosols*:

- high sand content;
- low organic matter levels;
- low cation retention and nutrient availability;
- few or no reserves of weatherable minerals;
- rapid permeability, high susceptibility to leaching;
- low available water capacity.

Arenosols and sandy *Regosols* are formed in coarse textured materials, exclusive of recent alluvial deposits of aeolian, colluvial or alluvial origin. In many cases they seem to have formed from a previous soil or by the leaching of clay formed in material with high contents of quartz such as granite. The greatest extent of these soils is in the Tropical Wet-Dry and tropical Desert and Steppe climates. Their topography varies from flat to moderately sloping. Annual cultivation of these poor soils leads to rapid depletion of their organic matter. This causes a rapid decline of the cation retention capacity. In humid climates leaching is severe, in semi-arid and arid zones drought hazard is serious. Due to the rapid deterioration the period of sustainable cultivation is short and required fallow periods are long.

12. *chromic Luvisols* (Lc) and *Cambisols* (B) (USDA: *Alfisols* and *Inceptisols*, respectively)
The main bioclimatic zones for this Soil Grouping are 20 (warm temperate dry forest) and 27 (subtropical dry forest)
Characteristics of *chromic Luvisols*:

- good physical structure;
- moderate organic matter content;
- moderate to high cation retention and inherent fertility;
- moderate to high available water capacity.

Chromic Luvisols are soils also known under the name Terra Rossa. They are usually soils found in areas with mediterranean conditions with a gently sloping to rolling topography. Their potential for agriculture is moderate to high potential. Generally their susceptibility to erosion is high. More characteristics of *Luvisols* are given under soil grouping No. 8.

For a description of *Cambisols* see Soil Grouping No.8.

13. *Solonchaks* (Z) and *Solonetz* (S) (USDA: *Salorthids* and *Natr*-Great Groups)
This Grouping occurs in many bioclimatic zones, but major occurrences are in the dry zones 11 and 12 (cool temperate desert and desert scrub), 17-20 (warm temperate forest, desert scrub, thorn steppe and dry forest) and 24 and 25 (subtropical desert and desert scrub)
   Solonchaks are grouped because of their high salinity. They are soils with very little profile development. Due to their salinity they are not suitable for cultivation.
   Solonetz have a natric horizon in common, which is a subsoil with a clay illuviation from the topsoil, having a sodium (Na) saturation of over 15% at the cation exchange complex. The structure in the subsoil is columnar. These soils are considered virtually not suitable for cultivation.

14. *Andosols* (T) (USDA: *Andepts*)
This Soil Grouping occur in many bioclimatic zones
Characteristics of *Andosols*:

- high to moderate content of organic matter in the topsoil but often throughout the solum;
- good physical structure;
- good drainage;
- high available water capacity;
- high cation retention;
- problems of phosphorous fixation.

Andosols are developed in volcanic ash. They typically have a dark topsoil of 20-50 cm thickness (thinner and thicker topsoils also occur) over a brown subsoil or substratum. *Andosols* are generally very good soils for cropping when they are well managed. In places *Andosols* may have fertility problems, usually due to phosphorous fixation. They frequently occur on steep slopes and this feature makes them highly susceptible to erosion. In the field the soil material may be smeary and feel greasy or unctuous; it may become almost liquid when rubbed ('thixotropy'). *Andosols* occur in a wide variety of climates and under various vegetation types.

15. *Vertisols* (V) (USDA: *Vertisols*)
The group of *Vertisols* occurs mainly in the Holdridge zones 26-28 (subtropical thorn woodland, dry forest and moist forest) and 34-35 (tropical dry and very dry forest)
Characteristics of *Vertisols*:

- very high clay contents, mainly montmorillonitic;
- high nutrient availability;
- relatively high organic matter content;
- high cation retention;
- low water available capacity;
- slow permeability when wet, leading to low infiltration and high run off.
- soil is very hard when dry, causing problems of cultivation and seedbed preparation.

Vertisols have a topsoil consisting of both the surface mulch (or crust) and the underlying structure profile that changes only gradually with depth. *Vertisols* in general have favourable chemical properties, but are problem soils in their physical qualities. Under high levels of management however, some of the physical problems can be overcome and high intensities of cultivation may become possible. *Vertisols* may be highly suitable for paddy rice cultivation, but lack of drainage may induce salinization.

16. *Planosols* (W) (USDA: *Alfisols, Ultisols, Aridisols*)
This Soil Grouping occurs mainly in Holdridge zone 28 (subtropical moist forest)
A typical soil profile of *Planosols* has a dark topsoil over a leached subsurface layer on top of a subsoil with clay increase. *Planosols* have slowly permeable subsoils, which may cause problems of drainage. Hydromorphic properties are general, and their topsoils may be of poor physical and chemical properties. *Planosols* with a leached topsoil are moderately suitable for cultivation, but when their topsoil is richer, these soils are good.

17/18. *Xerosols* (X) and *Yermosols* (Y) and shifting sands (USDA: *Aridisols, Psamments*)
Major occurrences are in the Holdridge zones 11-13 (cool temperate desert, desert scrub and steppe), 25-27 (subtropical desert scrub, thorn woodland and dry forest) and 32 (tropical desert scrub)
    In the new FAO legend (FAO, 1988) *Xerosols* and *Yermosols* have been combined in the *Xerosols*). This soil group comprises soils of semi arid (*Xerosols*) to arid (*Yermosols*) climatic conditions. *Xerosols* and *Yermosols* may be very fertile soils but due to the lack of rainfall they are of little or no value for agriculture. *Xerosols* usually occur in areas with a growing period of less than 75 days. With irrigation these soils may be classified among the best soils.

3.3 THE GLOBAL LAND COVER DISTRIBUTION

3.3.1 Vegetation Classification Systems

The vegetation cover of the earth basically has two important aspects: variation in time and variation in space. One of the important tools for determining vegetation's change in space is mapping. Determining its variation in time is done by repeated mapping. The latter approach has not been used as widely as it could have been, but it is gaining importance through the development of remote sensing techniques.
    Vegetation cover is classified according to a number of criteria based on the vegetation itself, its environment or a combination of both. A large number of classification methods have been used. A selected

number of classifications and maps which can be used on global and regional scales are listed in Table 3.5.

Table 3.5 A number of vegetation classification systems for small scale mapping

Main criteria used	Units distinguished	Reference
Properties of the vegetation		
Physiognomic properties	dominant life form	Unesco (1973)
Floristic properties	dominant species/associations of species	
Properties outside the vegetation		
Environment	climate mainly; topography, soil, landform, combinations	Köppen (1936) Thorntwaite (1948) Gaussen (1954) Holdridge (1967) Walter (1973) Schroeder (1983)
Geographical location		Hueck & Seibert (1972)
Successional stage		Clements (1916, 1928)
Combination of vegetation and environmental properties		
Overlay of vegetation and environmental properties		Küchler (1965)
Ecosystem classification		Ellenberg (1973)

Most systems are based on the potential vegetation or climax vegetation. Potential vegetation is commonly mapped on the basis of bioclimatic parameters and this procedure rests on the idea that typical communities develop in response to climate, particularly temperature and precipitation. Seasonal effects which may influence ecosystem structure, landscape and edaphic factors are not used as criteria. Classifications based on bioclimatic factors can be used to predict primary production. Examples of such classification systems are those proposed by Holdridge (1967) and Schroeder (1983). Mueller-Dombois (1984) discussed of the value of the Holdridge classification system for vegetation mapping. It is based on precipitation, potential evapotranspiration and temperature, and can therefore also be used as a mere bioclimatic system as shown in Table 3.3. and Table 3.4.

Actual vegetation is mapped usually on the basis of physiognomy or floristics. Physiognomy provides a better estimate of phytomass since it considers variations due to succession and habitat (Mueller-Dombois, 1984). The most widely known physiognomic classification is the system developed by Unesco (1973) which was designed for mapping of vegetation on a 1:1,000,000 scale. It gives guidelines for graphical representing cultivated areas. Dent (1978) supplemented this with a classification (coding) system for cultivated areas (for use in soil mapping and land evaluation). The Faculty of Geography of Moscow State University (Rjabchakov, 1986) produced a map of actual land use of the world at a 1:15,000,000 scale. Apart from the Unesco (1973) method, the Institut de la Carte Internationale de la Vegetation (1988) uses the physiognomic or 'landscape' classification method proposed by Gaussen (1954) for mapping at a 1:1,000,000 scale.

A selected number of global maps of vegetation are listed in Table 3.6. Apart from these, a number of continental and regional vegetation maps were prepared, such as Hueck and Seibert (1972) of South America at a 1:8,000,000 scale; Institut de la Carte Internationale du Tapis Vegetal (1982, South America 1:5,000,000, Africa 1:5,000,000, based on Unesco); White (1981; Africa 1:5,000,000); Küchler (1965, North America 1:7,500,000); the Mediterranean Region (Unesco, 1:5,000,000) and maps produced by the Institut de la Carte Internationale de la Vegetation (1988) on a 1:1,000,000 scale for India, Madagascar, Sri Lanka, Cambodia, Mexico. The maps made by the Institut de la Carte Internationale de la Vegetation are based on the system proposed by Gaussen (1954); this institute also produced a number of bioclimatic maps.

Table 3.6 List of global vegetation maps and digital data sets

Reference	Map scale	Classification used	Number of types	
Maps				
Brockmann-Jerosch(1918)	1:100,000,000	Physiognomic-environmental	10	
Schmithüsen (1968)	1: 25,000,000	Physiognomic-environmental	144	
Odum (1959)	?	Physiognomic-environmental	12	
Whittaker (1970)	1: 50,000,000	Physiognomic-environmental	25	
Duvigneaud (1972)	?	Physiognomic-environmental	17	
Whittaker and Likens (1975)	?	ecosystems	13	
Schroeder (1983)	?	Bioclimatic		42
Rjabchakov (1986)	1:15,000,000	Actual land use		
Digital data sets				
Olson et al. (1983)	0.5x0.5° grid (1:30,000,000)	Physiognomic-environmental	43	
Matthews (1983)	1°x1° grid	Physiognomic (Unesco, 1973)	32	
Emanuel (1985)	0.5°x0.5° grid	Bioclimatic(Holdridge, 1967)	37	
Leemans (1989)	0.5°x0.5° grid	Bioclimatic(Holdridge, 1967)	38	
Wilson and Henderson-Sellers (1985)	1°x1° grid	Land cover	53	

Ecosystems are self contained units in terms of primary production and nutrient cycling (Burton, 1987). Most descriptions of ecosystems are based on the dominant species, but deciding which characteristics should be used to determine ecosystem boundaries, presents problems. To some extent this problem is one of scale (see Table 3.6) as discussed by Mueller-Dombois (1984).

The physiognomic vegetation types shown on maps are also called biomes. A biome is a group of terrestrial ecosystems on the same continent, which are similar in physiognomy, in the main features of their environment and in some characteristics of their animal communities. Biomes in similar environments common to more than one continent are biome types or, in the terminology of Walter (1985), zonobiomes.

It is virtually impossible to draw a sharp distinction between any of these ecosystem types or their subdivisions. Each division is arbitrary. Most boundaries are gradual. Only tundra and boreal forest have some continuity throughout the northern hemisphere. Other ecosystems such as tropical rain forests and temperate grasslands occur isolated within different biogeographical regions and may be expected to have ecologically equivalent, but often taxonomically unrelated species (Ajtay et al., 1979). The terms 'forest', 'woodland', 'grassland' and 'savanna' are difficult to specify, as each author has his own concept of what they constitute.

The Unesco (1973) classification has been used in a modified form by Whittaker and Likens (1975), Ajtay et al. (1979), Matthews (1983), Goudriaan and Ketner (1984), and many other authors. Furthermore it has been applied in the map unit descriptions of the FAO/Unesco Soil Map of the World (1971-1981) and this makes linking to soil geography an easy task. Comparisons between the various studies are therefore possible without much problems of data interpretation and conversion.

3.3.2 Global Land Cover Data Sets

Historically the study of the spatial distribution of plant communities has been approached in terms of vegetation mapping. Small scale vegetation maps were compiled using field observations and other small and large scale thematic maps. The final product is generally a subjective composite of information reflecting the unrecorded choices of the compiler.

A feature of most vegetation classification systems is that they represent a rather theoretical potential vegetation or climax vegetation. A general aspect often neglected is the actual situation resulting from modifications by natural cycles and man's activities.

A number of authors have attempted to compile vegetation and land use data sets for the world. A selection of these studies will be discussed here. Hummel and Reck (1979) produced digital data files with resolution ranging from $0.4° \times 0.9°$ latitude $\times$ longitude for albedo studies. The results of Olson et al. (1983) were based on this study. Their map has a 1:30,000,000 scale and a $0.5° \times 0.5°$ resolution on digital format. Examples of older studies are Whittaker and Likens (1975) and Ajtay et al. (1979). The most recent and possibly best documented data bases were developed by Matthews (1983) and Wilson and Henderson-Sellers (1985).

Table 3.7 Area coverage for the major land cover types according to: A.Whittaker and Likens (1975); B. Ajtay et al. (1979); C. Olson et al. (1983); D. Matthews (1983); E. Houghton et al. (1983)

Land cover type	Area (10^{12} m^2)				
	A	B	C	D	E
1. Tropical rain forest	17.0	10.3	[12.1	12.3	11.6
2. Tropical seasonal forest	7.5	4.5	[12.1	6.7	7.7
3. Temperate forest	12.0	7.0	12.0	[20.3	10.0
4. Boreal forest	12.0	9.5	11.5	[20.3	11.4
5. Woodland, shrubland and interrupted woods	8.5	4.5	4.7	25.2	14.8
6. Savanna(incl. tropical grassland)	15.0	22.5	24.6[3]	[27.4[4]	12.5
7. Temperate grassland	9.0	12.5	6.6	[27.4[4]	17.7
8. Tundra, alpine	8.0	9.5	13.6	7.3	7.0
9. Desert, semidesert	18.0	21.0	8.5	[15.6	[21.3
10. Extreme desert	24.0	24.5	16.4	[15.6	[21.3
11. Cultivated land	14.0	16.0	15.9	17.6	15.0
12. Swamps, marshes, coastal land	2.0	2.0	2.5		
13. Bogs and peatland		1.5			
14. Other types	2.0	4.0		2.4	5.2[5]
Total	149.0[2]	149.3[2]	129.7	134.8	133.6

[1] data presented here are calculated from the data presented by Houghton et al. (1983);
[2] includes areas of ice, lakes and rivers; [3] includes "warm or hot shrub and grassland" and "tropical woods" (remnants with fields and grazing)[4] denoted as "grassland";[5] pasture.

Table 3.8 Comparison of areal estimates (in 10^{12} m^2) of pre-agricultural and present land cover distribution

Cover type	Pre-agricultural	Present	% Reduction
Total forest	46.28	39.27	15.2
Tropical rain forest	12.77	12.29	3.8
Other forest	33.51	26.98	19.5
Woodland	15.23	13.10	13.8
Shrubland	12.99	12.12	6.7
Grassland	33.90	27.43	19.1
Tundra	7.34	7.34	0.0
Desert	15.82	15.57	1.6
Cultivation	0.93*	17.56	—

*limited areas with long use histories and for which reliable vegetation data could not be acquired have been designated as cultivated land. From: Matthews (1983)

In Table 3.7 a number of global land cover maps and data sets are compared. Wilson and Henderson-Sellers (1985) used a system with 53 classes based on the expected requirements of land surface parameterization schemes for use in global circulation models. It was especially designed to provide information about the gross surface properties. Matthews (1983) used the Unesco (1973) vegetation classification scheme, the database by Olson et al. (1983) is based on a land systems grouping while the other references used the scheme proposed by Whittaker and Likens (1975). The latter system is also based on Unesco (1973). Comparison is possible but difficult.

There is great disagreement about the areas of the land area occupied by each ecosystem. Estimates of the coverages in the tropics may differ by a factor of 3.7 (highest and lowest literature estimate as presented in Brown and Lugo, 1982, Table 8 page 173). In many cases the map does not include documentation of the sources used in compilation or of the relative weight given to various sources which may conflict. Where source information is provided, it is not uncommon to find, that the map has borrowed heavily from a single source (Matthews, 1983).

The estimated areas of tropical rain forests show a considerable range. The forest area calculated by Whittaker and Likens (1975) is about 25% higher than for all other studies. As far as woodlands and deserts are concerned, it is clear that the various authors have used different concepts.

The area of cultivated land is similar in all the maps. Matthews estimated cultivation intensities within ecosystems and her figure for the global cultivated area corresponds well with the FAO (1985) reports from local and national sources (although regional discrepancies may exist).

This simple comparison highlights the need for critical evaluation of vegetation definitions and classifications. Apart from the areal extent of land cover types, the description in terms of carbon stocks are a subject of deep disagreement (Brown and Lugo, 1984).

Houghton (1983) and World Resources Institute (1987) (Table 18.3, on page 272, prepared by Houghton) calculated land use changes between 1700 and 1980 using the FAO Production Yearbooks (FAO, 1949-1985) supplemented with a model for the increase in agricultural land based on population growth.

Estimates made by Matthews (1983) concerning the pre-agricultural and present land cover distribution are shown in Table 3.8. The figures presented are not consistent with those of other authors. For example, Postel (1984) reported that up till now 13.3% of all the tropical forests have been logged, while Matthews reported a decline of 3.75% for tropical rain forests since pre-agricultural times. According to the data presented by Houghton (1983) the decline of the extent of tropical rain forest and tropical seasonal forests since 1700 is 11.4%.

3.3.3 Tropical Forest Resources and Distribution

A selection of literature data on the areal extent of tropical forests is presented in Table 3.9.

There are few reliable estimates of the actual area covered by tropical rain forest. Moreover, the available estimates vary widely because of the differences in criteria used to define this ecosystem. The statistics used are often unreliable. The great mass of existing information has not been "organized" at a national level in many tropical countries. Even when national syntheses exist it is not possible to regroup them together because the classifications and concepts used differ from one country to another (Lanly, 1982).

It may be useful to compare the above data with those of Matthews (1983) (see Table 3.2). Conversion of the Matthews (1983) types to tropical open and closed forest classes, yields areas of 1640×10^{10}ha for closed tropical forest (including types 1, 2, 3, 6 and 7) and 290×10^{10} ha for open tropical forest (type 9) + 800×10^{10} ha (areas of savanna, including types 12, 13 and 15, i.e., woodlands, shrublands and thickets).

Table 3.9 The extent of tropical forests by various authors

Reference	Extent (10^{10} m^2)			
Type	Asia	Latin America	Africa	Total
Persson (1974)				
Closed forest	290.4	576.8	195.9	1067.8
Total forest	406.0	734.1	754.9	1895.0
Myers (1980)				
Moist forest	271.4	641.6	151.4	1064.4
Postel (1984)				
Moist forest	305.0	629.0	217.0	1201.0
Total forest	445.0	1212.0	1312.0	2969.0
Lanly (1982)[*]				
Closed forest	305.5	678.7	216.6	1200.8
Open forest	30.9	217.0	486.4	734.4
Total forest	336.5	895.7	703.1	1935.2
World Resources Institute (1988)[*]				
Closed forest	269.4	679.4	217.6	1166.4
Open forest	26.8	207.2	481.1	715.1
Total	296.2	886.6	698.7	1888.5

[*] data for the same 16 countries (Asia), 23 countries (Latin America) and 37 countries (Africa); see Lanly (1982) for the list used of temperate and tropical countries.

3.3.4 Temperate Forest Resources and Distribution

Data for the temperate forest resources were taken from World Resources Institute (1988). The distribution is presented in Table 3.10.

The extent of temperate forests is greater than that of tropical forests and a great proportion of the temperate forests is formed by the boreal forest in the USSR and North America.

Table 3.10 The extent of temperate forests (10^{10} m^2)

	Temp. Africa	Temp. North America	Temp. Latin America	Temp. Asia	USSR	Europe	Temp. Oceania
Closed forest	3.7	459.4	121.9	161.7	791.6	137.0	52.1
Open forest	2.8	275.1	nd	38.3	137.0	21.9	67.4
Total	6.5	734.5	nd	200.0	928.6	158.9	119.5

Source: World Resources Institute (1988); nd = no data available.

3.4 LAND COVER CHANGES

3.4.1 Causes of Deforestation

Most of the present agricultural and urban land was once under forest cover. Traditionally there has been a need for development of land resources for economic benefit. Deforestation has been a necessary evil. An important issue to be considered is whether additional deforestation is necessary.

A list of principal causes of deforestation is given in Table 3.11. A cause of forest die-back which is playing a role of growing importance is the forest degradation due to air pollution. The observed consequences of this phenomenon will be discussed briefly in section 3.4.3.

Table 3.11 Causes of deforestation and forest degradation

"Natural"	Fires
	Acid precipitation
Traditional	Shifting cultivation
	Fuel wood
	Harvesting forest products
Economic	Agriculture
	Ranching
	Plantation crops
	Timber and commercial wood
	Infrastructure and urbanization
Socio-political	Strategic
	Population migration
	Speculation

Historically forest dwellers have sought their livelihood and basic necessities of life by harvesting forest products. In addition to this harvesting of minor products in the humid tropical forest zones the system of shifting cultivation is widespread. This system involves incomplete clearing and cultivation allowing the forest to regenerate during fallow periods. This system is sustainable if the shifting cultivators maintain a fallow period long enough. This is often not the case. Jackson (1983) estimated that there are 200 million people who rely solely on forest lands for their living. Some reports indicate that perhaps 50% of all present tropical deforestation is caused by permanent clearing of land that was previously used for shifting cultivation (e.g. Myers, 1980; Houghton et al., 1985, 1987). In addition to shifting cultivation, deforestation in developing countries for providing fuelwood to meet household energy needs and industrial wood is an important cause of forest loss.

National resources planning often involves deforestation for alternative land utilization such as agriculture, ranching, plantation crops, urbanization. Population migration from densely populated areas to forest land is a major socio-political reason for the present large scale deforestation in some regions, e.g. Sumatra and Amazonia. Often the forest is cleared to establish propietary claims (Lal, 1986). The result is a rapid depletion of the existing forest reserves in favour of new land development and colonization schemes.

Fearnside (1987) lists causes of deforestation in the Amazon. He distinguishes underlying and proximal causes. A number of proximal causes of deforestation in the Amazon basin mentioned by Fearnside are: tax incentives, tax penalties, interest incentives, subsidies and loans.

3.4.2 Deforestation and Reforestation in Tropical Regions

Present deforestation rates for the tropics are difficult to estimate. Lack of infrastructure, accessibility and scarcity of trained manpower in many countries in the tropics contribute to the scarcity of reliable surveys of the existing forest resources and conversion rates. Inadequate communication is also a major hindrance in updating the field records particularly in those countries where fast deforestation occurs. Therefore, the presently available estimates for forest conversion rates in the tropics are unreliable and obsolete.

A recent study of tropical deforestation was carried out by FAO/UNEP (1981a, 1981b and 1981c) for tropical Africa, tropical Asia and tropical America. The results of these analyses are summarized in Lanly (1982). The data used are local and country land use statistics for 63 countries complemented with 1972-1978 LANDSAT imagery for 13 additional countries. The deforestation rates as presented by Lanly are summarized in Table 3.12. The deforestation rates estimated by Lanly are 0.623% for tropical America, 0.615% for tropical Africa and 0.596% for tropical Asia. The total rate of deforestation for the tropics is 0.58% or 11.3×10^{10} m^2 y^{-1}, whereas reforestation amounts to 0.9×10^{10} m^2 y^{-1}. In his estimate of deforestation Lanly (1982) has not included partial or selective felling of trees or managed productive forests and including this would give even higher deforestation figures.

Since there is disagreement about the definitions regarding classes of forests, the conversion rates will disagree as well, even when focusing on a specific area (Henderson-Sellers, 1987). The question of how much tropical forests have been cleared is very difficult to resolve and the interpretation of LANDSAT and other satellite is unlikely to provide a complete answer.

Table 3.12 Average annual deforestation and reforestation (10^7 m^2) in tropical zones in the periods 1976-1980 and 1981-1985. nd = no data

	Area (10^7 m^{2}y^{-1})						
	Tropical America		Tropical Africa		Tropical Asia		Total
	76-80	81-85	76-80	81-85	76-80	81-85	81-85
Closed broadleaved forests	3807	4006	1319	1318	1767	1782	7106
Closed coniferous forests	312	309	8	7	35	30	346
Closed bamboo forests	nd	nd	6	6	13	14	20
Total closed forests	4119	4339	1333	1331	1815	1826	7496
Open forests	nd	1272	nd	2345	nd	190	3805
Total open + closed forests	nd	5611	nd	3676	nd	2016	11303
Reforestation**		409		85		418	912

* refer to Lanly (1982) for a list of countries.
** from World Resource Institute (1988) for the 1980s.

Table 3.13   Rates of tropical deforestation reported by selected authors

Reference	Area of permanent clearing (10^{10}m^{2}y^{-1})
FAO (1983)	6.0
Seiler and Crutzen (1980)	3.0
Myers (in Houghton, 1985)	10.5
Lanly (1982)	11.3

A selection of estimated deforestation rates is given in Table 3.13. The major differences between the analyses by Myers (in Houghton, 1985) and other estimates is found in the fate of fallow lands. Shifting cultivation is the traditional form of cultivation. In this system forests are cleared and cultivated during a limited number of years and then left to regrow during sufficient time to regain most of the original soil fertility. Shifting cultivation is nowadays being replaced by more permanent forms of land use, referred to as "sedentary shifting cultivation" or "forest farming". The land may remain permanently cleared by the arrival of new colonists or as a result of the inability of forests to reestablish as a consequence of soil nutrient depletion.

3.4.3 Deforestation and Reforestation in Temperate Regions

Data on deforestation in temperate regions are not available for many countries (See World Resources Institute, 1988:286-287). However, the reforestation rates for all temperate countries for which data are available is presented in Table 3.14. These figures represent both afforestation of previously not forested land and re-afforestation of land which was under forest cover within the previous 50 years. It is therefore difficult to compare these estimates of reforestation with deforestation data. The total extent of reforestation in temperate regions is $13.5 \times 10^{10} m^2 y^{-1}$.

Table 3.14 Average annual deforestation and reforestation ($10^7 m^2$) in temperate zones.[1] (n.d. = no data available)

	Temp. Africa	Temp. North America	Temp. South America	Temp. Asia	USSR	Europe	Oceania
Reforestation[2]	111[3]	2495	40	5233[4]	4540	1031[5]	112[6]

[1] Countries for each zone are listed in FAO (1949-1985) Production Yearbooks; [2] Reforestation includes areas of previously unforested lands and which was under a forest cover within the previous 50 years; [3] Only data available for Algeria, Egypt, Libya, Morocco and Tunisia; [4] only data available for China, Israel, Japan, Jordan, Korea Dem. People's Rep., Korea Rep. and Turkey; [5] No data available for Albania, Denmark, German Dem. Rep., Greece, Iceland, Luxembourg, Maltha and Romania; [6] Excl. Papua New Guinea, which is included in tropical Asia, after Lanly (1982).

3.4.4 Forest Die-Back Caused by Air Pollution

Almost all the work on so called acid precipitation related to atmospheric pollutants has so far been carried out in Europe and North America. The results of a SCOPE conference on acidification of tropical ecosystems (Rodhe and Herrera, 1988) point out that with the exception of China, little evidence exists of serious regional acidification effects of anthropogenic emissions of sulphur and nitrogen.

There are four main hypotheses concerning the cause of the forest decline caused by so called acid precipitation:

- acid rain, causing increased leaching of cations from soils and consequently soil acidification, increased aluminium concentration which may at high levels become toxic to plant roots;
- ozone, direct leaf injury due to high ozone concentrations in the troposphere;
- nitrogen excess; an excess of nitrogen may cause a disturbed nutrient balance in the soil. The best known symptom is potassium and magnesium deficiency, decreasing photosynthesis capacity, weakening rooting systems, greater frost susceptibility, greater susceptibility to pests and diseases;
- stress; during long periods forests have been influenced by air pollutants; the combined effect of all pollutants is greater than the sum of the individual effects.

Most data published on forest decline and on the effects of air pollution on forest growth have mainly been descriptive or hypothesizing with a significant lack of experimental support (Andersen and Moseholm, 1987).

Novel forest decline in Europe was first observed in silver fir (*Abies albas*) in rural areas in the southern part of the Federal Republic of Germany in the beginning of the 1970's. Subsequently the decline was described in the early 1980's for Norway spruce (*Picea abies*) and recently for deciduous trees, i.e. European beech (*Fagus sylvatica*).

In the Federal Republic of Germany the forest inventory for 1986 showed no further decline between 1985 and 1986. In 1986 19% of the forested area was moderately to severely damaged. In Austria and Norway about 5% of the forested area is reported to show moderate to severe loss of vitality. In Denmark 3% of the forested area shows a decreased vitality. In The Netherlands an investigation showed that 21% of all forests has a severe loss of vitality (Staatsbosbeheer, 1986). For the United States, Eastern Europe and the USSR similar data on forest decline are known although no exact decline figures can be given.

Damage to cultivated areas is less pronounced. Cultivated soils usually have a higher buffering capacity. The relative contribution of atmospheric deposition to the total soil acidification in cultivated soils is 5 to 35 % in calcareous soils, while its contribution is 7 to 55 % in noncalcareous mineral soils. The atmospheric contribution in forest soils on calcareous soils is 5 to 45 % and 56 to 99 % in noncalcareous soils (Staatsbosbeheer, 1987).

There is recent evidence of a direct effect of soil pollution on plant leaves. At the present atmospheric concentrations of O_3, SO_2 and HF yield reduction of agricultural crops amounting to 5% can be expected (Staatsbosbeheer, 1987).

3.5 CONCLUSIONS

The best soil map with global coverage available at present is the FAO/Unesco (1971-1981) Soil Map of the World. This map is also available in a digital form. However, in a number of areas the information on this map is not accurate. New information is now available for many regions of the world. A global Soil and Terrain Digital Data Base at 1:1,000,000 scale is in development, which will include all information available on the soil, in separate attribute files, so as to enable its application in e.g. climate studies and trace gas studies. This work will be completed in 15-20 years. In the mean time the existing FAO/Unesco Soil Map of the World can still be used. Updates for specific important regions such as the Amazon basin, where many new data on soils have become available since the compilation of the FAO/Unesco map, will prove very useful.

Tropical deforestation is the present major global land cover change. The data pertaining to land use change are difficult to obtain and exceedingly difficult to verify, particularly data from developing countries. Estimates of forest destruction range from 10 to 20×10^{10} m^2 y^{-1}, much of it in the Amazonian region. Most of the data base is unreliable. Great disagreement exists concerning the nature of changes (permanent clearing versus partial destruction or shifting cultivation; the latter process could account for an even greater extent of forest loss than the permanent clearing), while definitional differences add to the difficulty of comparing estimates. At present about 14.5×10^{10} ha of reforestation takes place globally, although part of this reforestation is on land previously under forest vegetation and does not add to the total forest area.

The quantification of the geographic distribution of vegetation and land use involves the use of some sort of classification system. A great number of classifications have been proposed, each having its own specific criteria for separating vegetation types. Of the most important systems, i.e. the physiognomic-environmental and the bioclimatic classifications, the first is most suited for prediction of phytomass and is most widely used for description of units covering a large area of land. The latter is most suitable for prediction of potential production figures. If maps on a global or continental scale are compared, one aspect becomes clear immediately: it is very hard to compare maps produced on the basis of different criteria.

The Unesco (1973) system is one of the most widely accepted systems. The global vegetation data set made by Matthews (1983) is probably the best documented and has the advantage that it is based on Unesco. Concerning the bioclimatic systems there are a number of useful classifications, i.e. Holdridge, Gaussen and Schroeder. The Holdridge system has a number of disadvantages (Mueller-Dombois, 1984), but is as practical as the other classifications, in particular to describe bioclimatic conditions. The maps produced by the Instut Internationale de la Carte Vegetal (on the basis of Gaussen) are very useful, but their global coverage is still incomplete.

For classifying actual vegetation and land use, a system has to be developed in close cooperation with remote sensing specialists. Apart from classification problems there is an enormous lack of data, especially in the third world where most changes occur. Intensive satellite monitoring, e.g. with the NOAA-AVHRR sensor, will improve the capacity of measuring changes regionally and globally and the digital sensor data will have to be linked to one of the vegetation/land use classification systems.

ACKNOWLEDGEMENTS

I thank Wim Sombroek, Dick Creutzberg and Sjef Kauffman of ISRIC and Hans Groenendijk for their valuable comments.

CHAPTER 4

Exchange of Greenhouse Gases Between Terrestrial Ecosystems and the Atmosphere

A.F. BOUWMAN

International Soil Reference and Information Centre (ISRIC)
P.O. Box 353, NL 6700 AJ Wageningen, The Netherlands

ABSTRACT

The release of CO_2 from the terrestrial biota, including soil emissions and clearing and burning of forests, have contributed significantly to the present atmospheric CO_2 concentration. At present, however, the main anthropogenic source of CO_2 is the combustion of fossil fuels. Future emissions of CO_2 resulting from fossil fuel combustion are highly uncertain. The estimated CO_2 emission for 1987 is 5.7 Gt C y^{-1}. (Gt = gigaton; 1 Gt = 10^{15} g). In addition, biotic sources of CO_2 (mainly deforestation) contribute about 1 to 3 Gt C y^{-1}, including 0.2 to 0.9 Gt C y^{-1} from soil carbon loss. Major problems in assessing the role of the terrestrial biota in the CO_2 cycle are the lack of knowledge of the present (and future) magnitude of the CO_2 fertilizing effect, uncertainty in estimates of the extent of forest clearing and carbon stocks of forests cleared. The various analyses of deforestation are difficult to compare because of definitional differences. Sinks of CO_2 are the atmosphere (55%), oceans (30%) and terrestrial biota (15%). The annual increase of the atmospheric concentration is about 0.5% or 3.6 Gt C y^{-1}.

The increase of the human world population correlates well to the atmospheric increase of 1% y^{-1} of methane (CH_4). This indicates that the rise in atmospheric CH_4 is most likely related to anthropogenic activities. The present annual release rate is 300 to 700 Tg CH_4 y^{-1} (Tg = terragram; 1 Tg = 10^{12} g) which is a sum of all source estimates. The major individual sources are rice paddies (60-140 Tg), wetlands (40-160 Tg), landfill sites (30-70 Tg), oceans/freshwater lakes/other biogenic (15-35 Tg), ruminants (66-90 Tg), termites (2-42 Tg), exploitation of natural gas and coal mining (65-75 Tg), biomass burning (55-100 Tg). The total sink of CH_4 includes: reaction with OH radicals in the troposphere (260 Tg y^{-1}), transport to the stratosphere (60 Tg y^{-1}) and oxidation in arid soils (16-48 Tg y^{-1}). Clearly sources and sinks are not balanced in this budget, and the allocation among the individual sources and sinks is still not well known. The strong increase of methane over the past 200 years is primarily caused by increasing emissions and a minor cause is the depletion of OH radicals, which are responsible for oxidation of atmospheric CH_4 and carbon monoxide (CO). This OH depletion is caused by increasing CH_4 and CO emissions.

The major sources of CO are known, but their magnitudes are still uncertain. The background concentration of CO is increasing at a rate of 0.6-1% per year or higher, but estimates are uncertain due to fluctuations of sources and sinks and the relatively short residence time of CO in the atmosphere. Estimates of global emission range between 1270 and 5700 Tg y^{-1}, the major sources being: biomass burning (800 Tg), fossil fuel burning (450 Tg), oxidation of hydrocarbons including methane (960-1370 Tg). The primary sinks are oxidation to CO_2 (3000 Tg) and soil uptake (450 Tg). The budget is not completely balanced, indicating the uncertainty in these estimates.

Soils and the Greenhouse Effect. Edited by A.F. Bouwman

The global N_2O emission from cultivated soils ($3 \, T \, g \, N_2O$-N y^{-1}) and from natural soils ($6 \, T \, g \, N_2O$-N y^{-1}) form the major sources in the N_2O budget. These estimates are still highly uncertain. Particularly natural ecosystems are undersampled. One of the major sources are the tropical forests of the earth, where in places extremely high fluxes have been measured. Savannas show high fluxes and because of their vast extent they also form an important global source. All N_2O is eventually transferred to the stratosphere where it is oxidized to NO. The size of this sink is $10.5 \, T \, g \, N_2O$-N y^{-1}, while the atmospheric increase is about $2.8 \, T \, g \, N_2O$-N y^{-1}.

The importance of NO and its relation to N_2O production was recognized in recent years. The production ration of NO:N_2O is a highly uncertain factor in all such calculations as it is very sensitive to abiotic controls such as oxygen pressure. Recent research indicates that NO production during nitrification of NH_3 in aerobic soils can exceed that of N_2O. NO is catalytic in various atmospheric photochemical reactions and also affects the oxidation reactions of CH_4 and CO.

4.1 CARBON DIOXIDE

4.1.1 Introduction

Carbon dioxide (CO_2) is the most abundant so-called greenhouse gas. Its current atmospheric concentration is approximately 345 ppm and its annual increase is 0.5%. Estimates of the total size of the global biomass carbon pool are: 835 Gt C (Gt = gigaton; 1 gigaton = 10^{15} g) (Whittaker and Likens, 1975), 560 Gt C (Ajtay et al., 1979), 594 Gt C (Goudriaan and Ketner, 1984). The other carbon pools are: atmosphere (720 Gt C), oceans (38000 Gt C), fossil reserves (6000 Gt C) (Goudriaan and Ketner, 1984) and caliche (petrocalcic horizons in arid and semi-arid regions; 780–930 Gt C; Schlesinger, 1982; 1985).

Table 4.1 Estimates of the pool of carbon in world soils

Approach	Reference	Carbon pool (Gt C)
Vegetation types	Bolin (1977)	700
	Schlesinger (1977)	1456
	Bolin et al. (1979	1672
	Ajtay et al. (1979)	1635
Soil types	Bohn (1976)	3000
	Ajtay et al. (1979)	2070
	Post et al. (1982)	1395
	Schlesinger (1984)	1515
	Buringh (1984)	1477
	This study	1700
Modelling	Meentemeyer et al. (1981)	1457
	Goudriaan and Ketner (1984)	1400

Net primary production by terrestrial biota is 60 Gt C y^{-1}, litterfall 40-50 Gt y^{-1}, the total stock of litter 60 Gt C (Ajtay et al., 1979).

The estimated total CO_2 emission from fossil fuel combustion was 5.3 Gt C y^{-1} in 1984 (Rotty, 1987) and 5.7 Gt y^{-1} for 1987 (CDIAC, 1989), while the annual release caused by deforestation is estimated to be 0.3-1.7 Gt C y^{-1} (Detwiler and Hall, 1988) to 1 to 2.6 Gt C y^{-1} (Houghton et al., 1985). Projections for the year 2050 for the emission from fossil fuel combustion range from 2 to 20 Gt y^{-1} (Keepin et al., 1986). According to the different scenarios the atmospheric CO_2 concentration will reach a level of 440 to 660 ppm in the year 2050. In equilibrium the sinks for atmospheric carbon dioxide are the atmosphere (71%), oceans (18%) and the terrestrial biota (11%) (Goudriaan, 1988; pers. comm.). Due to lack of time for redistribution at present, oceans absorb only 40% of the annual carbon injection into the atmosphere. The resulting increase of the atmospheric concentration is 0.5% or 3.6 Gt C y^{-1}.

On a world-wide basis the amount of carbon in decaying plant litter and soil organic matter may exceed the amount of carbon in living vegetation by a factor 2 to 3. There have been several attempts to estimate the storage of soil organic matter in world ecosystems with basically four different approaches: the vegetation groups, the soils, the life zone groups and modelling. Some of these estimates are listed in Table 4.1. Bohn (1976) presented a figure for the carbon pool which does not fit within the range of estimates by other authors. His estimate is partly based on the world soil map by Gaussen and Hädrich (1965) since the FAO Soil Map of the World had not yet been completed. The estimation made by Post et al. (1982) is based on the world life zone groups according to Olson (1983) and analyses of about 2700 soil profiles representing virtually every major ecosystem. Samples from each soil horizon or at standard depths were analyzed for carbon content (not incl. litter) and bulk density. The global estimates of soil carbon are much more accurate than estimates of the area assigned to the different ecosystems as was discussed in chapter 3.

4.1.2 Soil Organic Matter Fractions

Organic matter in soils is represented by plant debris or litter in various stages of decomposition through to humus and also includes microbial biomass. Soil organic matter exists in many forms. Kononova (1975) separated the soil organic matter into fresh and incompletely decomposed plant residues and a stable component (humus). Humus in turn can be subdivided into strictly humus substances (humic and fulvic acids) and other products of advanced decomposition of organic residues and products synthesized by microbes. Since a large portion of the soil carbon is found above a depth of 50 cm, Schlesinger (1984) suggests that nearly half of the carbon in a typical profile is relatively labile. The remainder is found in the lower sections of the soil profile and is more stable. Spycher et al. (1983) found that the light fraction (density <1.6 g cm^{-3}) of the soil organic material in a temperate forest was 34% and they suggest that this light fraction is equivalent to the labile components of the organic matter. Kortleven (1963) also divided soil carbon into humus and a fraction of relatively stable carbon, that is composed of recalcitrant humus, charcoal and other forms of elementary carbon. Janssen (1984) divides soil organic matter into 'young' and 'old' soil organic matter. 'Young' soil organic matter is material younger than 1 year. The lifespan of humus and charcoal according to Goudriaan and Ketner (1984) are 10 to 50 years and 500 years respectively. Paul and Van Veen (1978) use the terms 'physically protected' and 'non-protected' soil organic matter, the former having the lowest decomposition rate.

The stability and related age of soil organic matter is closely related to the pedogenic factors under which the soil is formed. Martel and Paul (1974) reported that the mean residence time of organic carbon in a chernozemic soil was 350 years. The mean residence time of the $ZnBr_2$ residue (= organic material excluding the light material) was 500 years. The oldest fraction found was the NaOH extract with 1900 years. Jenkinson and Rayner (1977) used a soil sampled in 1881 with an equivalent age of 1450 years between 0 and 23 cm, 2000 years between 46 and 46 cm and 3700 years between 46 and 67 cm. Scharpenseel and Schiffmann (1977) found a linear relationship of mean residence time with depth, the average increase per cm being 46 years for Chernozems. Soils formed under different conditions showed different patterns of residence time with depth.

4.1.3 Models of Litter Decomposition and Soil Organic Matter Accumulation

A variety of different models have been developed for calculating plant residue decomposition rates and soil organic matter levels:

1. Models with decomposition rates which are constant in time. Basic formulae are:

$$dC/dt = h \times A - k \times C \tag{4.1}$$
$$C_E = h \times A/k \times (1 - e^{-kt}) \tag{4.2}$$

   where: C = quantity of organic carbon;
            C_E = steady state level of organic C;
            t = time;
            k = decomposition constant;
            h = humification constant;
            A = addition of organic material.

The decomposition of soil organic matter often shows an initial period of rapid loss of labile constituents such as soluble carbohydrates, followed by a longer period of ever-decreasing decomposition rates during which more recalcitrant components such as lignin are oxidized. Examples of models based on this first order decomposition process are Henin and Dupuis (1945), Jenny et al.

(1949), Greenland and Nye (1959), Kortleven (1963) and Olson (1963).

2. Models with decomposition constants which are variable in time. An example of this approach is Janssen (1984), who proposed:

$$k = 2.82 \times (a + t)^{-1.6} \qquad (4.3)$$

where: a = apparent age of the added material.

3. Models with different fractions of soil organic matter, each having a different decomposition rate. As was reported by Minderman (1968) it is necessary to know the decomposition rate of the separate chemical constituents of litter and soil organic matter to calculate the decomposition of the total mass of soil organic matter. Models of this type gained popularity after the introduction of computer simulation models, which enable repetitive execution of complex calculations. Representative models of this type are Jenkinson and Rayner (1977), Van Veen and Paul (1981) and Parton et al. (1987).

4.1.4 Factors Affecting the Stability of Organic Matter in Soils

Mineralogy. Base status and clay content are often related. Soils derived from basic rocks are usually more fertile than their acidic counterparts, which gives rise to higher annual inputs of organic matter to the soil. A basic environment accelerates the decomposition of litter (short term effect), but the mixing with soil components, which retards organic matter turnover, increases the retention and leads to higher organic matter contents (Oades, 1988). In acidic soils, the initial decomposition of debris is retarded, but subsequent oxidation of organic matter proceeds relatively quickly because of lack of stabilizing mechanisms. Jones (1972) indicated, that in savanna soils parent material has an effect on soil carbon contents, but this effect may be attributed to the clay content, which in turn is related to the parent material.

Soil texture. Heavy textured soils generally show lower decomposition rates than sandy soils. This can be explained by the adsorption of organic matter on clay surfaces in combination with the spatial arrangements of substrates and organisms with the soil pore system. In heavy textured soils the movement of meso and micro life may be restricted. The protection from predation by larger organisms (amoeba, eelworms) may also be caused by restricted movement of these organisms (Van Veen et al., 1985). The movement of enzymes may also be restricted. Many authors (e.g. Kortleven, 1963; Jenkinson, 1977; Schimel et al., 1985a, 1985b) reported that finer, clayey soils tend to be more preservative than the coarser, sandy soils. Jenkinson (1977) indicated that this texture effect is smaller under both near neutral and acidic conditions than under more basic conditions. Jones (1972) found a good positive correlation between the clay content and soil carbon for well drained soils. For poorly drained soils the correlation was less good. Vertisols (see USDA, 1975; FAO/Unesco, 1972) showed a linear relationship between soil carbon and clay content between 35 and 80% clay.

Clays and organic molecules are both negatively charged. Cations may build bridges between clay particles and organic molecules. A substantial proportion of this protected organic matter is found in stable aggregates. Not only clay bridging, but also interactions between charged and uncharged polymers with clay surfaces are involved in determining the stability of micro-aggregates.

Soil structure. Organic matter is undoubtedly stabilized by physical processes. The protective action of clay is an illustration (Jenkinson and Rayner, 1977; Jenkinson, 1977). Probably soil organic matter is a continuum, with physically protected but decomposable materials at one end and free organic matter, which is highly resistant to biological attack at the other (Jenkinson and Rayner, 1977). Tiessen and Stewart (1983) showed, that during decades of cultivation particle size fractions of 0.5 to 10 μm have lost the smallest amount of organic matter, while organic matter associated with fine clay (<0.5 μm) was rapidly depleted. Apparently new biomass is separated mainly in the clay and fine clay fractions. Cultivation of virgin grasslands enhances the accessibility of organic matter to microbial attack by disruption of soil structure. This explains considerable losses of soil organic matter during cultivation of such soils (Voroney et al., 1981).

Soil fertility. Although the effect of calcium on the formation of stable humus compounds was known, in the numerous organic matter studies where the influence of soil texture was assessed (Jenny et al., 1968; Jones, 1972; many others) the base status was usually not taken into account.

The soil's base saturation may be an important factor in stabilizing organic matter in clay soils through clay bridging (Oades, 1988). The higher organic matter content of calcareous soils compared with non calcareous soils is in part due to Ca-humates. Removal of Ca from a soil stimulates the decomposition of organic matter. Addition of Ca inhibits the release of CO_2 and stabilizes soil structure. These phenomena may not occur in acid soils receiving Ca-fertilization, where organic matter decomposition is stimulated (Oades, 1988).

The nutrient status of a soil, which is influenced by fertilizer applications and nutrient uptake by plant roots, may also have an effect on the quality of the microbial population. At higher mineral nutrient levels bacteria are probably preferentially stimulated. At lower levels a shift towards more fungi-dominated populations might occur (Van Veen et al., 1987).

Vegetation-soils effect. Although forests produce more organic matter and more organic debris reaches the soil surface, the decomposition is faster in forest soils than in grassland soils and resulting organic matter contents are lower. Most organic matter in forests is added as litter and is decomposed before it is incorporated in the soil. In grasslands additions are in the form of a root system which is well distributed throughout the complete soil profile. It is thoroughly mixed before decomposition.

For temperate conditions a rule of thumb is, that soil carbon decreases in the following order: natural grassland (prairie)- forest- cultivated land. The total biomass of a forest is greater than the biomass of prairies, but recycling of organic matter in prairies proceeds faster. The quantity of organic material added to the soil in prairies is 2-4 times greater than in forest areas. In addition, less leaching occurs in prairies and leguminous species present in the plant population in prairies can make nitrogen available for the formation of stable humus compounds. The organic carbon content is higher in cultivated soils than in soils under forest or grass, because:

- Most of the biomass produced on arable land is harvested;
- The organic substances of crops are less resistant against decomposition. Substrates are less well decomposable and possibly more re-immobilization of N occurs in rangeland soils than in cultivated soils (Schimel et al., 1985a). Decomposition rates are higher in cultivated soils. Possibly the microbial species composition was changed with the importance of fungi declining (see also Voroney et al., 1981);
- The average temperature on the soil surface in a forest is lower than in cultivated fields;
- Losses due to leaching are smaller in forests.

Climate. Temperate and tropical soils should be treated separately in the assessment of the soil carbon content under natural conditions. In tropical regions with a udic (moist) soil moisture regime (see USDA, 1975) the biological activity continues during most of the year. Hence, equilibrium after clear cutting of forests or cultivation of grasslands, will be reached after a shorter period than in temperate regions. In the 22% of the tropics with udic soil moisture regimes neither temperature nor moisture limits the biomass production and decomposition of organic matter. Forests in these areas produce about 5 times more biomass and soil organic matter per year as comparable temperate forests, but the rate of decomposition of organic matter is 4 times faster than in temperate regions (Jenkinson and Ayanaba, 1977). If the decomposition process in the tropics is 4 times faster than in temperate regions, the annual input of organic matter must be 4 times higher to maintain a soil at the same C-content once steady state conditions have been achieved. Oades (1988) reported that in mediterranean Australian soils decomposition proceeds 2 times faster than in England. In dry tropical areas, the biological activity is reduced during parts of the year, thereby extending the period needed to reach equilibrium. In temperate regions the biological activity is greatly reduced during the winter . In the 78% of the tropics, which have an ustic (alternating moist and dry season) or aridic soil moisture regime (USDA, 1975), the lack of moisture during parts of the year has a similar effect.

Temperatures during tropical rainy seasons are similar to -but seldom as high as- the corresponding summer temperatures in temperate regions (Sanchez, 1976). The carbon content in arid and semi-arid soils may even increase after agricultural use with irrigation. The effect of temperature and moisture stress on metabolism and composition of bacteria, fungi and yeasts isolated from soil have provided evidence, that these two factors may act independently (Van Veen, 1987).

The soil's preservation capacity. Though the microbial biomass is a relatively small and labile fraction of the soil organic matter complex, its turnover is a rate determining process in the cycling of organic matter and nutrients. Soil texture and structure affect microbial biomass turnover and the related cycling of carbon and nitrogen. Van Veen (1987) related the soil conditions to the soil's preservation capacity. Each set of soil physical, chemical, hydrological and climatic conditions gives the soil characteristic capacities to preserve both organic matter and microorganisms. Preservation of organic matter could be the result of protection of microbial biomass and organic matter against predation or amelioration of harsh environmental conditions. Organic matter additions in excess of the soil's preservation capacity will be decomposed at an accelerated rate (Jenkinson and Rayner, 1977; Van Veen and Paul, 1981). Biomass formed in excess of a soil's preservation capacity for microorganisms is assumed to die at a relatively high rate (Van Veen et al., 1984; Van Veen, 1987). Biomass and its immediate products of decay are considered to form a fairly tightly closed system from which only small proportions of the products leak out as stabilized materials.

4.1.5 Soil Carbon Loss due to Mineralization following Interference in Natural Conditions

Modern agricultural practices, such as monocultures with limited return of crop residues to the soil and the use of chemical fertilizers, have been reported to cause a serious decline of soil organic matter levels (Allison, 1973; Diez and Bachtaler 1978; many others). Agricultural conversion disrupts the steady state conditions that exist in many natural communities. Annual organic carbon additions are drastically reduced when forests are brought under cultivation. The depletion of organic

Table 4.2   Mean loss of soil carbon after agricultural conversion for different ecosystems

Type	N⁰ of studies	Mean loss (%)	Range(%)
Temperate forest	5	34.0	3.0-56.5
Temperate grassland	24	28.6	-2.5-47.5
Tropical forest	19	21.0	1.7-69.2
Tropical savanna	1	46.0	n.a.

Source: Schlesinger (1986)

carbon and nitrogen can be attributed to changes in the magnitude of biological and physical processes in the soil. The rate of decomposition increases during cultivation due to increased microbial activity under a more favourable soil moisture and temperature regime. Incorporation of plant residues and pulverization of soil structure caused by tillage operations would increase the rate of mobilization of microorganisms (Voroney et al., 1981) and thus increase the accessibility of soil organic matter and its mineralization.

The data listed in Table 4.2 are from a literature review by Schlesinger (1986) of long term paired plot experiments. The soils were sampled to various depths. Steady state soil carbon levels have been reached presumably in most experiments. The losses of soil humus listed in Table 4.2 are likely to be equivalent to the 'light' or labile fraction of the soil organic matter. Mann (1986) calculated the carbon loss from 650 studies from temperate and tropical zones. The carbon loss from the top 30 cm varied from 0 to 20%. The greatest average loss was from the taxonomic class Borolls (USDA, 1975). The loss of carbon from soils with a high initial carbon content tended to be higher than from soils poor in carbon.

From long term experiments on the behaviour of soil carbon in different soils under cultivation in Denmark can be concluded that soil carbon contents decrease by about 25% on fertilized (NPK) and on unmanured and unfertilized soils (Dam Kofoed, 1982). The decrease was slightly lower (15 to 25%) on heavily manured soils. After prolonged periods of cultivation a new steady state level was reached. Jenkinson and Rayner (1977) indicated, that soil carbon remained equal in the Rothamsted classical experiments if no manure is applied and annual additions of organic matter are 1.1 t C y^{-1}, exclusively from crop residues. In their experiments the addition of farm yard manure enhanced the build up of soil organic matter substantially.

Organic matter levels have dropped to less than 50% after cultivating virgin grasslands on Chernozems (Van Veen and Paul, 1981; Voroney et al., 1981). The loss of soil carbon from the plough layer (Ap horizon) may be 90% of the total loss (Voroney et al., 1981). The degree of physical protection of soil organic matter decreased from 50% in virgin grasslands to 20% (in the top 15 cm) and 40% (below 15 cm). Schimel et al. (1985a) found a decrease in soil carbon after 44 years of cultivation of grasslands on Haploborolls and Argiborolls (USDA, 1975) of 24 to 56% depending on the soil's texture and the rate of erosion. The lowest loss rates were found in places were a deposition of soil material was recorded. The amounts of organic C increased and proportional losses decreased downslope.

Brams (1971) showed the general effects of cultivation on organic carbon contents in two West African soils. Cultivated soils had 40-60% of the soil carbon present in ferralitic soils under secondary forest. After 2 years of rice cultivation and 3 years of intertilled crops soils contained 30% of the original carbon present in alluvial soils with secondary forest cover. Jones (1972) calculated from data of the Guinea savanna, that soils under cultivation or under fallow following cultivation, have a mean carbon content slightly more than 50% of the content in virgin soils. Lugo et al. (1986) reported losses after clearing of Central American subtropical wet forest soils of 46% after 10 years of cultivation, 70% after 100 years of cultivation. In dry areas losses were lower. In all cases in relatively short periods of time (decades) most of the losses could be regained by abandonment of cultivation and secondary forest succession or use as pasture.

Shifting cultivation seldom results in substantial soil organic matter depletion; usually the carbon contents are maintained at 75% of natural equilibrium levels (Greenland and Nye, 1960:105), but in conditions of decreasing rehabilitation periods soil carbon contents may drop to 50% of the original levels (Sanchez, 1976).

Soils under cultivation generally have higher bulk densities than soils under natural vegetation. This makes comparison of virgin and cultivated soils a difficult task (Voroney et al., 1981, Mann, 1986).

The soil carbon loss from a forest directly after clearing was assessed by Edwards and Ross-Todd (1983). They found an increase of the annual C evolution in the first year after cutting of only 79 kg C ha^{-1} for clear cut mixed deciduous forest with only sawlogs removed and only 169 kg C ha^{-1} for clear cut forest with removal of all woody material. This indicates, that a forest soil where forest is allowed to regrow after clearing and is not cultivated, will not produce substantial quantities of CO_2. Alternating drying and wetting of a bare soil with fresh decomposable organic material, such as after wood clearing, can lead to considerable 'priming effects' and organic matter destruction. The degree of pulverization or destruction of soil aggregates is the primary factor in determining the C-loss and CO_2 evolution.

Only a portion of the decomposition is accounted for by measuring CO_2 output (Van Veen and Paul, 1981). If X is the amount of CO_2-C evolved during decomposition and Y is the efficiency of the use of C for biosynthesis expressed as a percentage of the total C-uptake under aerobic conditions and non-cell metabolite production is negligible, the actual amount decomposed is: $X(1+Y/(100-Y))$ (Paul and Van Veen, 1978). Microorganisms use carbon compounds for biosynthesis forming new cellular or extracellular material and as energy supply. In the latter process carbon compounds are converted into CO_2 and to a lesser extent low molecular weight compounds.

In Table 4.3 CO_2 evolution given by various authors for cropped and forested soils are presented. The data shown are quite consistent, although in all the experiments with native vegetation and crops it is difficult to account for the root respiration and calculate the production of C purely from the soil organic matter decomposition.

Table 4.3 Total CO_2 evolution (from root + soil respiration) from different temperate soils calculated from various reports

	C-loss	
	g C $m^{-2}d^{-1}$	10^2 g C $m^{-2}y^{-1}$
Cropped soils/bare soil		
Oats[1]	2.7	
Cabbage[1]	1.9	
Bare soil[5]	0.8	
Bare soil (winter)[2]	0.4	
Bare soil (summer)[2]	1.6-1.9	
Barley[2]	2.7-5.4	
Wheat after summer fallow[3]	3.4-3.8	23.5-25.5
Wheat[4]	1.8	6.4
Crop rotation (6 y)[8]	0.19-0.38	0.7-1.4
Crop rotation, heavily manured (8 y)[8]	0.56	2.0
Forest soils		
Spruce forest[6]	1.3	4.9
Mixed forest[7]	2.3	8.5
Mixed deciduous forest[9]	2.3	8.4
Mixed oak forest, summer[10]	0.55-0.95	
Mixed oak forest, winter[10]	0.14-0.22	
Oak forest[11]	2.2	7.9
Cedar swamp forest[11]	2.1	7.4
Fen forest[11]	1.9	7.1

[1] Lundegard(1927) quoted in: Buyanovski et al. (1986);[2] Monteith (1964) quoted in: Buyanovski et al. (1986);[3] De Jong and Schappert (1972);[4] Buyanovski et al. (1986);[5] Koepf(1953) quoted in: Minderman and Vulto (1973);[6] Hager (1975) quoted in Baumgartner and Kirchner (1980);[7] Bouten et al. (1984);[8] Janssen (1984);[9] Edwards (1975);[10] Froment (1972);[11] Reiners (1968).

Assuming a total decline of soil carbon of 75 tons C ha^{-1} during a 15-year period after forest clearing (see Table 4.4) this figure can be compared with experimental data on CO_2 production in cultivated soils. The period necessary to reach equilibrium and the soil humus content at equilibrium conditions depend on additions of organic material (direct effect) and fertilizers (indirect through increased dry matter production). In case of clear cutting of a forest and cultivating the soil, the initial decomposition rate will be high since first the easily oxidizable substances are decomposed. The break-down will proceed ever slower until the equilibrium level is reached and the annual additions equal the decomposition. For simplicity an average carbon loss of 5 tons ha^{-1}y^{-1} (= 5×10^2 g m^{-2}y^{-1}) or 1.4 g C day^{-1} m^{-2} is assumed in this example. Table 4.3 suggests, that this loss rate fits well within the ranges for bare soils.

The variation in the above presented data is enormous, probably because the lack of systematic research on the carbon loss from pedogenetically uniform units or soils with similar textural and structural properties.

4.1.6 Soil Carbon Loss caused by Erosion

A total soil loss caused by rainfall erosion of 5 to 10 tons ha^{-1}y^{-1} with an average carbon content of 3% means 0.15 to 0.30 tons C-loss ha^{-1}y^{-1}. This is only a low carbon loss rate but in areas of extreme rainfall erosion these figures may be considerably higher.

Compared with the heavy losses through oxidation the above figures are low (see Table 4.4). However, after prolonged periods of cultivation the cumulative effect of erosion may be dramatic (Voroney et al., 1981; Schimel et al., 1985a; Schimel et al., 1985b). Moreover, organic matter is among the first constituents removed due to its weight. Eroded material is about 5 times richer in organic matter than the residual soil material (Allison, 1973). Erosional removal of organic matter following forest cutting is probably unimportant as long as vegetation is allowed to regenerate immediately (Bormann et al., 1974). Bouwman (1989b) reported that rainfall erosion of tropical soils used for cultivation may increase the carbon loss from soils considerably.

4.1.7 Estimation of the Global Soil C Pool

From literature data generalized soil carbon profiles have been compiled for the major soils distinguished on the FAO/Unesco Soil map of the world (FAO/Unesco 1971-1981). Data from Fitzpatrick (1983), Kononova (1975), Sanchez (1976) and Buringh (1983) have been used. The bulk density used in Table 4.4 is 1.5 g cm^{-3} for all soil units. The maximum thickness of the profiles is 1.0 m (after Houghton et al., 1983; Schlesinger, 1977). In Table 4.4 living roots, partly decomposed material, dead leaves and litter are not included in soil carbon.

The soil carbon profiles are based on the assumption that soil carbon decreases in the top 1.0 m by 40-60 % when forest or grassland is converted to cropland and by 25-35% when forest is converted to grassland. The depletion rate used is 25-40% for conversion of grassland to cropland. In this respect no difference in speed of decomposition between temperate and tropical regions is considered. The above assumptions are similar to those proposed by Buringh (1984). Depletion rates used by other researchers are discussed in 4.1.8.

The resulting soil carbon profiles for the different land cover types are presented in Table 4.4. The carbon profiles are not exact averages applicable to any particular country. They are based mainly on assumptions and generalizations.

Table 4.4 can be used to make a rough estimate of the total C-pool in soils by combining it with the areas for each soil and land cover type as given in Table 3.2. This resulted in an estimate of the global soil C pool of 1600 to 1800 Gt C, excluding carbon in organic soils. This figure fits in the range of estimates given in Table 4.1. The range is caused by difficulties in transcribing ecosystems from Table 4.4 to the system of land cover types defined by Matthews (1983) and used in Table 3.2.

Table 4.4 Generalized organic carbon profiles for the major soil units of the FAO/Unesco (1971-1981) 1:5,000,000 Soil Map of the World

FAO Major Soil	Natural Ecosystem	Generalized soil carbon profile in natural (typical) ecosystem C (%) at depth (cm)					Total soil C (10^6 g C ha^{-1})			
		0-20	20-40	40-60	60-80	80-100	Forest Prim.	Sec.	Grass land	Crop land
Acrisols	rain forest	3	2.5	0.6	0.5	0.4	220	165	160	110
Cambisols	forest	5	3	1	0.5	0.1	290	215	210	140
Chernozems	grassland	4	3	2	1	0.5			315	235
Podzoluvisols	forest	4	0.5	0.5	0.5	0.1	160	120	120	100
Rendzinas	forest	5.8	3.0	0	0	0	130	100	90	65
Ferralsols	rain forest	3	1	0.5	0.2	0.1	160	120	100	70
Gleysols	grassland	10	5.8	0.5	0.25	0.1	350	260	350	200
Phaeozems	grassland	3	2.5	1.5	1	0.2	250	200	225	150
Lithosols	forest	2	0.5	0	0	0	75	50	50	40
Fluvisols	grassland	1	1	0.5	0.5	0.5	100	75	100	60
Kastanozems	grassland	2.9	2.3	1.5	1	0.5	300	225	250	190
Luvisols	forest	6	4	1	0.5	0.5	200	150	150	115
Greyzems	grassland	3	2.5	1.5	0.8	0.2	300	250	240	180
Nitosols	forest	3	1	1	0.8	0.5	200	150	150	115
Histosols*										
Podzols	forest	2	0.5	0.2	0.7	0.7	100	75	100	60
Arenosols	various	1.8	0.5	0.1	0	0			70	50
Regosols	forest	3	0.5	0.1	0.1	0	110	85	85	60
Solonetz	grassland	1.8	0.6	0.2	0.1	0.1			85	65
Andosols	forest	8	5	3	2	1	330	250	250	185
Rankers	forest	2	0.5	0	0	0	75	55	50	40
Vertisols	grassland	1.2	1.0	0.5	0.1	0.1	190	145	135	115
Planosols	forest	3	0.1	0.5	0.5	0.1	150	115	115	85
Xerosols	s.des./desert	0.8	0.4	0.1	0.1	0.1			50	30
Yermosols	desert	0.8	0.4	0.1	0.1	0.1			50	30
Solonchaks	grassland	1.8	1.2	0.5	0.1	0.1			110	80

* Histosols as C pools are considered separately in 4.1.8.

4.1.8 Release of CO_2 from Soils

Release of CO_2 from mineral soils. Since pre-historic times the global conversion of forests to agriculture has caused considerable losses of soil carbon as CO_2 to the atmosphere. Estimates range from about 537 Gt C for the total flux since pre-historic times (Buringh, 1984) to 36 Gt C for the loss since the mid-1800s (Schlesinger, 1984). The wide range in estimates of the carbon release from soils due to land use changes is shown in Table 4.5.

Houghton et al. (1983) assumed losses of 35%, 50% and 15% in the top 1.0 m for agricultural conversion of tropical forests, temperate forests and boreal forests respectively. Houghton et al. (1987) applied loss rates of 25% after cultivation of forests in temperate regions and 20% after forest clearing and subsequent regrowth. In the tropics Houghton et al. (1987) assumed loss rates of 30% for conversion to cropland and 25% for

Table 4.5 Estimates of the carbon release from soils circa 1980.

Reference	Soil C release (Gt C y^{-1})
Bolin (1977)	0.3
Schlesinger (1977)	0.85
Buringh (1984)	1.5 - 5.4
Schlesinger (1984)	0.8
Houghton (1987, pers. comm.)	0.2 - 0.5
Bouwman (1989b)	0.1 - 0.4
Detwiler and Hall (1988)	0.11- 0.25

conversion to grassland. Palm et al. (1980) used losses of 30% and 20% for conversion of primary respectively secondary forest to permanent agriculture and 25% and 17% loss for conversion to primary respectively secondary forest in shifting cultivation in South East Asia. Soil carbon levels recovered to 90% of primary forest levels after forest regrowth. In Buringh (1984) a major source of CO_2 is the transition of forest to urban land and this causes the extraordinary high estimate of global soil CO_2 release.

Release of CO_2 caused by drainage of organic soils. Organic soils increase in mass by vertical accretion or by paludification (the lateral spreading of peat). Armentano (1980) used a global area of natural Histosols of 450×10^{10} m^2. With an average carbon accumulation of 300 kg ha^{-1} y^{-1} total C sequestering is 0.135 Gt C y^{-1}. Duxbury (1979, quoted in Armentano, 1980) assumed that at present 7 to up to 35×10^{10} m^2 of wetlands have been drained globally. An assumed C release from drained wetlands by oxidation of the organic material of 10 t C ha^{-1}y^{-1}, yields a global annual C release of 0.05 to 0.35 Gt C. A different approach is based on a subsidence rate of 1-3 cm, a bulk density of 0.5 g cm^{-3}, a C content of 30%, yielding a release of 0.02-0.07 Gt C y^{-1}. Drainage of 10^{10} m^2 of Gleysols causes an extra release of 0.01 Gt C y^{-1}. The resulting global release from Gleysols and Histosols ranges between 0.03 and 0.37Gt C y^{-1}.

Armentano and Menges (1986) analyzed organic soil wetlands of the temperate zones, occupying about 350×10^{10} m^2. Based on an average storage rate of 200 kg C ha^{-1}, the annual accumulation before disturbance was 0.06-0.08 Gt Cy^{-1}. They estimated that the total area drained in the period 1795-1980 was 8.2×10^{10} m^2 for crops, 5.5×10^{10} m^2 for pasture and 9.4×10^{10} m^2 for forests. In the tropics about 4% of the wetland area has been reclaimed in the period 1795-1980. The annual shift (loss of sink strength and gain of source strength) in the global C balance is 0.063-0.085 Gt C due to reclamation of Histosols in temperate regions. Including tropical Histosols, the global shift would be 0.15 to 0.184 Gt C y^{-1}.

4.1.9 Land Use Changes and the Global Carbon Cycle

There are two approaches to the assessment of the role of terrestrial biota in the CO_2 budget. Firstly, a number of researchers have developed dynamic models of the global carbon cycle including the CO_2 injection into the atmosphere by fossil fuel combustion and forest clearing, the oceans as a CO_2 sink and the fertilization effect of CO_2. Examples of this approach are Goudriaan and Ketner (1984) and Esser (1987 and this volume). Secondly, there are bookkeeping models which account for rates of deforestation and forest volumes and yield the CO_2 release. The best known representatives of this approach are Bolin (1977), Woodwell et al. (1978), Houghton et al. (1983, 1985, 1987) and Detwiler and Hall (1985, 1988). This bookkeeping approach yields the release of CO_2 from deforestation.

In the final part of this section bookkeeping analyses of the effect of massive forest plantations will be discussed. Finally, other potential important biotic sources and sinks of CO_2 will be reviewed. Van Breemen and Feytel (this volume) discuss soil CO_2 fluxes in more detail.

There are a number of uncertainties in global analyses of CO_2 fluxes:

-   The biomass of the forests which are being cleared is an uncertain factor. Estimates of the biomass of tropical forests are extremely variable. Whittaker and Likens (1975) use 202 t C ha^{-1} and 156 t C ha^{-1} for tropical rain forest and tropical seasonal forest, respectively. The C densities based on destructive sampling estimated by Brown and Lugo (1982) are 172-185 t C ha^{-1} for tropical lowland rain forests and moist forests, 146 for tropical lowland dry forest, 146-161 for subtropical lowland rain forests and moist forests plus tropical premontane forests, and 40 t C ha^{-1} for subtropical dry forests. Their later estimate (Brown and Lugo, 1984), based on forest volumes of tropical forests, amounts to 55-90 for broadleaf closed forests, 31-70 for conifer closed forests and 11-30 for open forests. The weighted averages for tropical forests are 188 (Whittaker and Likens, 1975), 124 (Brown and Lugo, 1982) and 53 (Brown and Lugo, 1984).
-   The area of forest clearing. This aspect is discussed in chapter 3.
-   The stimulation of growth by increased CO_2 partial pressure in the atmosphere. Some authors (e.g. Woodwell et al., 1978) suggested that this fertilization effect is not great enough to compensate for the vast amounts of CO_2 injected into the atmosphere by deforestation. This aspect will be discussed in more detail below.

Dynamic models. A number of dynamic carbon models have been developed. Examples are Goudriaan and Ketner (1984), who used a simple matrix of conversions of land use and a 12 layer ocean model, and Esser (1987; this volume) who used soil, climate and land use data sets and an ocean consisting of 2 layers, all regionalized on a 2.5° latitude × 2.5° longitude grid. The results of both dynamic models suggest that at present the biosphere is a net sink of CO_2 (see Table 4.6).

Table 4.6 Comparison of the annual flux of carbon from terrestrial biota by various authors. All figures in GT C y^{-1}

Reference	Present release caused by forest clearings	Present net release including CO_2 fertilizing effect
Bookkeeping models		
Houghton et al. (1983)	1.8-4.7	
Houghton et al. (1987)	1.0-2.6	
Detwiler & Hall (1988)	0.4-1.6*	
Dynamic models		
Goudriaan & Ketner (1984)	7	0 to -0.5
Esser (1987)	2.7	- 0.1

* the estimate reported is for the tropics only; the authors estimate on the basis of literature data that the temperate regions release -0.1 to +0.1 Gt C y^{-1}.

The CO_2 induced increase of net primary production is greater than the loss of CO_2 caused by deforestation. Goudriaan and Ketner (1984) suggest that charcoal formation in the process of biomass burning is another important sink of carbon. The point in time where the biosphere turned into a sink of CO_2 is around 1970 according to both dynamic models. Further deforestation however, will suppress the CO_2 fertilizing effect (Esser, 1987). The decreasing vitality of forests caused by acid precipitation in industrialized regions may also become increasingly important.

Bookkeeping models. Houghton et al. (1983) used the biomass figures presented by Whittaker and Likens (1975) and for land use changes both data in the FAO Production Yearbooks (1949-1978) and data presented by Myers (1981). The estimated total annual release rates from the terrestrial biota are 1.82 and 4.70 Gt C y^{-1} respectively. In a third data set Houghton et al. (1983) assumed that the rate of agricultural expansion since 1950 has been proportional to the growth of populations in tropical regions. By assuming that 40% of the required increase in food production was achieved through increasing agricultural output on existing cropland, this third data set yielded a global release rate of 2.6 Gt C y^{-1}.

Houghton et al. (1983) also assessed the effect of a reduced soil response to agricultural clearing. Assuming only 20% soil carbon loss after forest clearing the total annual flux was reduced by only 11%, but the net flux from harvested and regrowing forest decreased by almost 50%. In another test zero soil carbon loss was assumed following the clearing of forests. The effect of this assumption is a reduction of about 10% of the total CO_2 flux.

The analysis was also tested for the low estimate of biomass per unit area of tropical forests (Brown and Lugo, 1982). With lower carbon stocks of forests less carbon will be released, but net biomass accumulation will also be lower during forest regrowth. Hence, the total net release does not proportionally reflect the differences in forest biomass.

Houghton et al. (1985, 1987) repeated their analysis with reduced soil carbon loss rates (see 4.1.8) and including data on shifting cultivation and clearing of fallow forests (forests that had been previously in shifting cultivation) for permanent cultivation. Using the FAO/UNEP data the spatial distribution of the CO_2 emission could be mapped. Carbon stock figures were taken from both Brown and Lugo (1982) and Brown and Lugo (1984) in separate analyses. The range in carbon release rates calculated by Houghton et al. (1985,1987) is 1.0 to 2.6 Gt C y^{-1}. In this figure the contribution of the tropics is 0.9 to 2.5 Gt C y^{-1} of which 0.4 to 0.8 Gt C y^{-1} is due to permanent deforestation of fallow forests. In relation to previous calculations (Houghton et al., 1983) the range is narrower, whereby the maximum flux decreased from 4.7 to 2.6 Gt C y^{-1}. The soils contribution to this emission is about 20% or 0.2 to 0.5 Gt C y^{-1} (Houghton, 1987, pers.comm.).

Detwiler and Hall (1988) analyzed the annual release due to forest clearing using all available statistical land use data and carbon stock data. Land use data used are from FAO Production Yearbooks (FAO, 1949-1985), Seiler and Crutzen (1980), and Myers (in Houghton, 1985). The carbon density data used are from Brown and Lugo (1982) and Brown and Lugo (1984) respectively. The resulting annual release is 0.3 to 1.7 Gt C, most of which is from tropical regions. Included in this range is soil organic matter loss of 0.11 to 0.25 Gt C.

Forest Plantations. Massive forest planting is also mentioned as a potential sink of CO_2. Mature forests are in equilibrium with respect to CO_2. Young actively growing forests accumulate carbon. Forest plantations can thus be implemented to reduce the atmospheric CO_2. Based on high accumulation figures of 6240 kg C $ha^{-1}y^{-1}$ in new forests about 465×10^6 ha of plantations would sequester 3 Gt CO_2-C surplus of sources over sinks (Sedjo, 1989).

On the basis of projected needs for fuel wood, industrial wood and need for environmental plantations, Wiersum and Ketner (1989) estimate that in the year 2000 an area of 200×10^6 ha of newly planted forest would sequester 0.6 Gt C y^{-1}. These forests accumulate 2000-3000 kg C $ha^{-1}y^{-1}$ during their productive life of 30 years. These authors further estimate that an area of 30×10^6 ha of fertilized energy wood plantations would accumulate 4500 kg C ha^{-1} y^{-1} during 8 year rotations. These plantations would replace present agricultural land which may be taken out of farming, mainly in the USA and Europe. The energy wood produced can substitute 0.12 Gt C y^{-1} of fossil fuel. The sum of all these plantations, 230×10^6 ha or about half of the total land area of Europe, would have to be increased by a factor of 4 to 5 to sequester the current atmospheric CO_2 increase. Moreover, these plantations only sequester CO_2 during their period of effective growth. For comparison: at present about 14.5×10^6 ha of forest is being planted world-wide (World

Resources Institute, 1988). The effect of such massive forest plantations needs further analysis using dynamic global carbon models.

Other sinks and sources of CO_2. Aspects of the carbon cycle, that would influence the CO_2 release rate are suppression of forest fires (reduction), expansion of paddy rice (reduction), expansion of no tillage agriculture (reduction), drainage of wetlands (increase), increase of the intensity of shifting cultivation (increase) and increase of the area of arid lands (increase). These changes will be of large local importance, but at a global scale insignificant.

Another possible source of atmospheric CO_2 is the digestion by termites. Termites process material equivalent to about 28% of the earth's net primary production and about 37% of the net primary production in areas where they occur. Thereby amounts of CO_2 are released of 13 Gt C y^{-1} (Zimmermann et al., 1982). With all uncertainties the above emission of CO_2 by termites could range from 6.5 to 26 Gt C y^{-1}. Later studies by other authors revealed however (see section 4.2), that the biomass consumption rate assumed by Zimmermann et al. (1982) is much too high. Termite induced CO_2 release is only important where the population increases. Although CO_2 would probably be released as the result of any decomposition process, termites serve to accelerate carbon cycling. The ecological areas that should show the largest increase in emissions from termites are tropical wet savanna and areas that have been cleared or burned and cultivated land in developing countries (see also 4.2).

4.1.10 Conclusions

There is a number of uncertainties in the estimates of CO_2 release from terrestrial biota. This is expressed by the variation in the estimates produced by the various dynamic carbon models, such as Goudriaan and Ketner (1984), Esser (1987 and this volume), and bookkeeping models, such as Houghton et al. (1983, 1987).

- The largest biotic source of carbon dioxide is forest conversion to agriculture, most of which occurs in the tropics. Yet the greatest uncertainties are the rate of deforestation and the type of land use after clearing. Definitional controversy on land use and vegetation types is also responsible for discrepancies between estimates.
- A second important cause of unreliability of the estimates is the volume and carbon stock of tropical forests. However, the effect of carbon stocks on the net flux of CO_2 is small compared to the effect of the rate of forest clearing.
- A third uncertain factor is the soil carbon loss caused by land cover changes and the amount of charcoal formed during biomass burning and forest burns.

The use of one single vegetation and land use classification method would improve the comparison between estimates. This aspect is referred to in chapter 3, where differences and points of similarity of the various methods are discussed. With respect to the forest biomass, intensified field work linked with the use of remote sensing techniques, especially radar, would improve characterization of the distinguished ecosystem types.

To improve estimates of soil carbon pools and the annual release of carbon from soils, modelling of decomposition combined with ^{14}C techniques will be necessary to single out pool size and release rate.

Finally, for monitoring of tropical regions on regional and global scales a number of remote sensing techniques are available. The NOAA-AVHRR sensor data as discussed in chapter 7, combined with radar techniques and field observations, will prove a very helpful tool in the future.

4.2 METHANE

4.2.1 Introduction

The presence of methane (CH_4) in the atmosphere has been known since the 1940's, when strong absorption bands in the infra-red region of the electromagnetic spectrum were discovered which were caused by the presence of atmospheric CH_4. Concentrations of CH_4 in the troposphere vary from about 1.7 ppmv in the Northern hemisphere to about 1.6 ppmv in the Southern hemisphere (Rasmussen and Khalil, 1986; Steele et al., 1987).

The first evidence for an increase of methane in the atmosphere was given by Rasmussen and Khalil (1981). This was later confirmed by Blake et al. (1982). The methane concentration has doubled during the past 200 years (Ehhalt, 1988). Its average temporal increase during the 1978-1983 was about 18 ppbv per year (Blake and Rowland, 1986) or 1.1% per year (Bolle et al., 1986). Steele et al. (1987) observed a slowing down of the growth rate of the methane concentration to 15.6 ppb y^{-1} in 1983/1984, or globally somewhat less than 1%.

The atmospheric burden of CH_4 is about 4700 Tg (Tg = Terragram; 1 Tg = 10^{12} g) (Wahlen et al., 1989) to 4800 Tg (Cicerone and Oremland, 1988). With a residence time of 8.1-11.8 years (Cicerone and Oremland, 1988) to 8.7-8.8 years (Wahlen et al., 1989) the total sources in quasi steady state must be 500 ± 95 Tg CH_4 (Cicerone and Oremland, 1988) to 580-590 Tg y^{-1} (Wahlen et al., 1989).

The observed increase over 1970-1980 requires a yearly excess of 70 Tg CH_4 of sources over sinks (Blake et al., 1982) to 50 Tg CH_4 (Bolle et al., 1986) or 40-46 Tg (Cicerone and Oremland, 1988). Transport to the stratosphere may amount to 60 Tg (Crutzen and Gidel, 1983).

Khalil and Rasmussen (1983) showed that during summer in the Northern hemisphere CH_4 concentrations are lower and that a sharp rise occurs in fall. This phenomenon suggests a fall source at Northern latitudes. These seasonal variations are generally consistent with seasonal and latitudinal variations of concentrations of hydroxyl radicals (OH).

The major removal process for methane is the atmospheric process presented by reaction (1), with the production of OH driven by photochemistry as in (2) and (3) (after Blake et al., 1982):

$$OH. \quad + \quad CH_4 \quad \rightarrow \quad H_2O \quad + \quad CH3 \qquad (1)$$
$$O3 \quad + \quad h\nu \quad \rightarrow \quad O \quad + \quad O2 \qquad (2)$$
$$O \quad + \quad H_2O \quad \rightarrow \quad OH^. \quad + \quad OH. \qquad (3)$$

The increase of methane over the past 200 years is probably due to the increase of emissions (70%) while about 30% may have been be caused by depletion of OH (Khalil and Rasmussen, 1985). This OH depletion is caused primarily by emissions of carbon monoxide (CO) from various anthropogenic sources.

The role of methane in the atmosphere is a complex one:

1. CH_4 has absorption bands in the infrared region of the spectrum.
2. It is oxidized in the troposphere by the free radical OH.
3. CH_4 is a sizeable source of CO through its oxidation by OH. The further oxidation of CO is discussed in chapter 2.
4. methane is a source of water vapour in the stratosphere as a result of its oxidation to CO_2 and H_2O.
5. stratospheric methane can react with Cl radicals, forming HCl which slows the rate at which Cl and ClO destroy stratospheric ozone.

Methane is produced during microbial decomposition of organic materials under strictly anaerobic conditions. Natural wetlands and wet rice cultivation are places where anaerobic conditions prevail. In landfill sites oxygen is consumed in the course of time and fermentation of the organic wastes will become the dominant process. Methane is also formed in the digestive tract of ruminating animals and in the guts of various insects, of which the termites are the most important. CH_4 is also formed in the abiogenic process of burning of biomass.

Recent investigations showed that 21 ± 3 % (Wahlen et al. 1989) to about 30% (Manning et al., 1989) of the total source is fossil or 'dead' methane. This fraction of dead methane includes about 30 Tg of CH_4 from wetlands or old peat layers (Cicerone and Oremland, 1988), the remainder is from coal mining operations and exploration of natural gas.

Estimates of the sources recognized so far are listed in Table 4.7. Man may control about 50% of today's total source. Since the world population has quadrupled over the last two centuries, the anthropogenic sources should also have increased by a factor 4. This should have resulted in an increase of the total source of more than a factor 2, which is consistent with the atmospheric increase (Ehhalt, 1988).

4.2.2 Controlling Factors of Methane Production in Soils

A review of CH_4 formation, controlling factors and biochemical pathways can be found in Oremland (1988). In this section a number of soil processes and conditions important in methanogenesis will be discussed briefly.

Table 4.7 Sources of atmospheric CH_4 from various reports

Source	CH_4 emission $(Tg\ CH_4\ y^{-1})$
Rice fields	60 -140[a]
Natural wetlands	40 -160[a]
Landfill sites	30 - 70[b]
Oceans/freshwater lakes/other biogenic	15 - 35[c]
Intestines of ruminants	66 - 90[d]
Termites	6 - 42[e]/2 - 5[f]
Exploration of natural gas	30 - 40[c]
Coal mining operations	35[c]
Biomass burning	55 -100[c]
Other nonbiogenic	1 - 2[c]
Sum	334 -714
Total source [g]	400 - 640
Total sinks [g]	300 - 650

[a] Aselmann and Crutzen (1989); [b] Bingemer and Crutzen (1987); [c] Seiler in Bolle et al. (1986); [d] Crutzen et al. (1986); [e] Fraser et al. (1986); [f] Seiler (1984); [g] Cicerone and Oremland (1988) report a total source derived from residence time, atmospheric burden and annual increase; the sum of the individual sources therefore differs from this separate estimate.

Reduction processes in soils. Organic matter in soil is decomposed by microbes in consecutive steps in which the hydrogen acceptors (= oxidizing agents) are used up one after the other in a thermodynamically determined sequence of the following processes: aerobic respiration, nitrate reduction, general fermentations, sulphate reduction, methane fermentation. When the inorganic hydrogen acceptors have been consumed, the remaining organic matter will continue to be degraded by microbial oxido-reduction processes that ultimately lead to formation of a mixture of CO_2 and CH_4, the ratio of the two gases depending on the degree of oxidation of the initial organic material. One of the main pathways in CH_4 formation in nature proceeds via acetic acid, which yields equal amounts of CO_2 and CH_4.

The redox potential of a soil decreases after flooding as a consequence of sequential reduction of oxidized compounds by organic matter. A positive correlation exists between the soil reduction potential E_h and methane emission. For the equilibrium:

$$Ox + ne + m\ H^+ \rightleftharpoons Red$$

the relation which describes the redox potential is:

$$E_h = E_o + 2.303\ RT/nF\ \log\ (Ox)/(Red) - 2.303\ RTm/nF\ pH \quad (4.4)$$

where: R = gas constant;
       T = temperature (K);

F = 96500 Coulomb eq^{-1};

E_o = equilibrium potential.

For Fe^{3+}/Fe^{2+} the reaction is:

$$Fe(OH)_3 + 3 H^+ + e \rightleftharpoons Fe^{2+} + 3 H_2O$$

for T = 25°C:

$$E_h = 1.06 - 0.059 \log (Fe^{2+}) - 0.177 \text{ pH} \qquad (4.5)$$

Thus, at increasing Fe^{2+} activity or higher pH the E_h will be low. However, not all soils have the same steady state E_h under flooded conditions. High levels of available Fe^{3+}, high content of organic matter, low NO_3^- MnO_2 and O_2 and high temperatures favor E_h decrease (Ponnamperuma, 1981). The type of organic manure may also have an effect on the fall of E_h (Swarup, 1988).

Fermentation is the major biochemical process in organic matter degradation in a flooded soil. The main products of this fermentation are ethanol, acetate, lactate, propionate, butyrate, molecular hydrogen, methane and carbon dioxide. More hydrogen gas and less methane are found in wetland soils planted to rice at later growth stages of rice than in an unplanted rice field. H_2 usually does not accumulate in significant amounts in flooded soils (Yoshida, 1978).

Soil pH. In acid soils flooding will cause an increase of the soil's pH, while flooding will decrease the pH in alkaline soils. In acid soils the pH increase is caused by reduction of Fe^{3+} to Fe^{2+} mainly. In alkaline soils accumulation of CO_2 is the main cause for a pH decrease (Ponnamperuma, 1985). Williams and Crawford (1984) showed that the optimum pH for methanogenesis is 6.0 for peat soils with actual soil pH values of 3.8 to 4.3.

Substrate and nutrient availability. CH_4 fluxes may correlate with availability of oxidizible substrate (DeLaune et al., 1986), or peat depth and nutrient enrichment from e.g. groundwater (Harriss and Sebacher, 1981) or nitrogen fertilizer (Cicerone and Shetter, 1983). Nutrient availability and hydrogen-mediated interactions between nitrogen fixing cyanobacteria and methanogenic bacteria may be important for methane production (Svensson, 1986).

Nitrate and sulphate or their reduction products repressed methane formation (Jacobsen et al., 1981). The effect of nitrate is twofold: first it delays methane formation until the reduction of nitrate is complete and the redox potential is lowered sufficiently for further anaerobic reactions to proceed. Secondly nitrate exerts a toxic effect on methanogenesis.

Holzapfel-Pschorn and Seiler (1986) reported that in early stages of rice growth methane emission is similar to emissions from unplanted fields. In both situations the flux shows a peak short after inundation due to mineralization of soil organic matter. A second maximum occurs in the planted fields in the physiologically most active period of the rice plants (Holzapfel-Pschorn and Seiler, 1986; Schütz et al., 1989). This second peak was attributed to a supply of organic matter in the form of root exudates. These exudates are easily decomposable substrates which are preferentially released during the vegetative stage of growth. Results presented by Swarup (1988) indicate that the second peak may also be related to the E_h of the soil, which was at its minimum after 30 days of crop growth. Schütz et al. (1989) reported that CH_4 emission rates are increased considerably in the early stage after application of organic matter. Very high applications did not increase CH_4 emission, probably due to the formation of toxic products of fermentation. Schütz et al. (1989) measured reduced methane fluxes after incorporation of $(NH_4)_2SO_4$ or urea in the soil. In the case of $(NH_4)_2SO_4$ this was attributed to the presence of sulfate. Due to the addition of sulfate methanogens may be outcompeted by sulfate reducing bacteria. For the reduction of the methane emission by urea no simple explanation could be given.

Temperature. Methane flux was poorly correlated with temperature in the study by Sebacher et al. (1986). Fluxes from Northern areas were higher than could be expected from other experiments in temperate regions. A proposed explanation was the presence of low temperature adapted methanogens, a phenomenon earlier reported by Svensson (1984).

Holzapfel-Pschorn and Seiler (1986) reported a marked influence of soil temperature on the CH_4 flux. They even found doubling of emission rates at a temperature increase of 20 to 25°C, which was later confirmed by Schütz et al. (1989). Diurnal variation of the CH_4 emission is correlated with temperature. Amplitudes of temperature and CH_4 fluxes are high in early stages of growth and lower during the second half of the growing season when the soil is shaded by rice plants (Schütz et al., 1989).

Sulphate concentration and presence of sulphate reducing bacteria. Sea water sulphate and sulphate reducing bacteria may interact with methanogenesis by competition with or inhibition of methanogenesis in sediments (or both). Methanotrophic bacteria may even oxidize CH_4 causing low concentrations in the soil or water column and subsequently low fluxes to the atmosphere. Bartlett et al. (1985, 1987) found CH_4 and SO_4^{2-} concentrations to be negatively correlated. A possible conclusion is that methane emission from salt wetlands, which usually contain considerable amounts of SO_4^{2-}, are lower than from fresh water wetlands. The competition between methanogens and sulphate reducing bacteria may not be important in salt marsh sediments due to the utilization of non-competitive substrates such as methanol and methylated amines or dihydrogen as electron donors (Oremland et al., 1982; Wiebe et al., 1981). Sulphate (Jacobsen et al., 1981) and sulphides may be toxic to methane formation. Williams and Crawford (1984) reported inhibition of methanogens by acetate. They suggested that methane (and other products) in peats at some depth may have been produced many years before and that its presence is inhibitory to further methanogenesis.

Organic matter in paddy soils. The organic matter content of rice soils ranges from over 30% in peaty soils to 0.8% and less in certain mineral soils. In Africa the organic matter content is used as an indicator of natural fertility for rice cultivation. The presence of too much organic matter may become a limiting factor. Organic matter application to wetland soils may lead to organic acid production which has an injurious effect. Application of ammonium sulphate may reduce this latter effect (Yoshida, 1978).

Changes in soil organic matter content have been reported under wet rice cultivation. These changes are most marked in soils which in their natural status are more or less freely drained, but are only minor or absent in the majority of poorly drained lowland soils under rice (Moormann, 1981).

The differences of paddy soils from other aquatic environments are:

1. the paddy system is not continuous due to incorporation of organic matter, green manure, crop residues, etc.;
2. the paddy system is heterogeneous: the reduced layer of flooded rice is situated between the oxidized surface soil and the oxidized subsoil. Within the reduced layer there are patches of aerobic sites due to excretion of oxygen by rice roots.

4.2.3 Factors Determining the Methane Flux

Water depth. The depth of the water layer over the soil may control methane fluxes. Sebacher et al. (1986) observed that water depths greater than 10 cm do not promote methane emission. Microbial CH_4 oxidation in aerobic water columns deeper than 10 cm may occur (de Bont et al., 1978; Delaune et al., 1983). Sebacher et al. (1986) also found that emission rates were linearly related with water depths up to about 10 cm.

Profile of the anaerobic environment. In steady state undisturbed methanogenic ecosystems show a characteristic stratification. The top layer is aerobic, the second layer is more reduced with Fe^{3+}, Mn^{4+} and NO_3^- still present. Below that a zone of sulphate reduction is found and finally a zone of methane generation is found. This layering coincides with the value of the redox potential, which has been discussed above. In the oxidized layer consumption may occur of the methane produced in the zone of CH_4 formation. The CH_4 concentration in interstitial water increases with depth in peat soils (Williams and Crawford, 1984; Dinel et al., 1988). Williams and Crawford (1984) suggested that the CH_4 measured at some depth was formed many years ago and now inhibits further methanogenesis.

Depth of permafrost. Svensson and Rosswall (1984) reported that in acid peats in subarctic mires methane formation is restricted to the upper horizons in the peat, and that there is no correlation between the depth of the permafrost layer and methane fluxes. Maximum concentrations of CH_4 occurred in the layers below this maximum production level. This may be caused by freezing, which starts at the surface and proceeds down to the permafrost layer. Gases in the peat will be entrapped. As the ice melts in spring, a release of CH_4 will occur leading to the profile discussed above.

Methane oxidation. Although methane flux rates appear to be a function of the total amount of methane in the soil, the vertical distribution of the gas also plays a role. Both the magnitude and the depth in the sediment of maximum methane concentration appear to increase as sediment temperatures increase. This suggests that the balance between the microbial processes of methanogenesis and methane consumption controls methane concentrations near the surface and thus in large part the flux of methane to the atmosphere (Bartlett et al., 1985). Holzapfel-Pschorn et al. (1986) reported that during a rice crop 67% of the methane produced is being oxidized and only 23% is actually emitted. In the absence of rice plants 35% of methane production is actually emitted, but methane production was much lower.

Process of methane release into the atmosphere. Possible ways of release are:

1. *ebullition*: methane loss in the form of bubbles from sediments should be a common and significant mechanism accounting for between 49 and 64% (Bartlett et al., 1988) to 70% (Crill et al., 1988) of total flux.
2. *diffusion*: diffusional loss of methane across a water surface is a function of surface water concentration of methane, wind speed and methane supply to the surface water (Sebacher et al., 1983).
3. *transport through plants*: typically in aquatic plants parts of the parenchyma break down leaving large lacunae for gas storage and transport. The phenomenon of methane transport through aerenchyma has been reported for rice (de Bont et al., 1978; Seiler, 1984; Cicerone and Shetter, 1983) and for other aquatic plants by Sebacher et al. (1985). More than 95% of the total methane release from paddy soils is through diffusive transport through the aerenchyma system of the rice plants (Seiler, 1984; Holzapfel-Pschorn et al., 1986). Transport of CH_4 from paddy soils into the atmosphere by rising bubbles is only important in unplanted fields. Studies by de Bont et al. (1978) show, that the presence of rice plants enhances the escape of methane from soil. Older rice plants at the ripening stage released about 20 times more CH_4 than 2 week old seedlings. Holzapfel-Pschorn et al. (1986) reported that rice plants, and not weeds, stimulate methane emission. Rice paddies emitted about twice as much methane as unplanted fields.

High concentrations of dissolved methane in root mats of floating meadows suggest high in situ methanogenesis with restricted flow where water can become oxygen deficient; other possible causes are trapping and subsequent dissolution of bubbles (Bartlett et al., 1988). Air samples from gas spaces in stems and leaves indicate elevated methane concentrations over concentrations in ambient air. Emission through plants would also be expected to show great diurnal variations tied to environmental changes and variations in respiration and photosynthesis rates.

Textural stratification in rice paddies. Puddling, the wet tillage of rice soils, causes a stratification of the soil material. Coarse materials settle first and are covered by finer silt and clay. Medium textured soils show a clear stratification upon puddling. In sandy soils the clayey cover is thin or absent. In very fine soils stratification may be difficult to observe. Trapping of gases may cause a build up of a vesicular structure. The presence of fine clay layers and algae at the soil surface restricts the escape of gases and contributes to the formation of vesicles.

4.2.4 Geographic Distribution of Paddy Soils

The harvested area of paddy rice is about 144×10^6 ha of which 95% is located in the Far East (FAO, 1985; see Table 4.8). This area corresponds to approximately 9.5% of the total global cultivated area. The harvested area of paddy rice has increased from 86×10^6 to 144×10^6 ha between 1935 and 1985 (see Table 4.8) which is an annual average increase of 1.05%. Between 1950 and 1985 the average annual increase has been 1.23%. The last few years however, the expansion of the total acreage of paddy rice is decreasing. In Table 4.8b the geographic distribution of paddy rice cultivation in Asia is presented. The area of paddy rice includes so called wetland rice and dryland rice. Wetland rice is grown in puddled soil and may be irrigated and continuously inundated or rainfed and almost permanently inundated.

Rice is grown on a wide variety of soils, predominantly belonging to the Gleysols, Fluvisols, Luvisols, Acrisols, Nitosols (for a definition of these soil concepts the reader is referred to FAO/Unesco (1971-1981). Vertisols and Histosols are of minor importance. Special mention should

Table 4.8a Global area harvested of wet and dryland rice 1935-1985 (areas in $10^7 m^2$)

	1935	1950	1960	1970	1980	1985
Africa	1850	2900	2880	3960	4894	5467
N/C America	540	1040	1280	1428	2076	1914
S America	1190	2300	3880	5741	7258	6122
Asia	82000	87600	110940	122302	128393	129977
Europe	220	300	350	395	366	388
Oceania	10	30	40	50	123	140
USSR	148	nd	100	356	637	667
World	85958	94170	119470	134232	143747	144675

Source: FAO (1952-1986) Production yearbooks 1951-1985 (nd = no data)

be made of the Gleysols, which occur to a great extent in wetland areas and are wet during part of the year. This circumstance makes them attractive for rice cultivation. More detailed information on the distribution and classification of paddy soils can be found in Moormann and van Breemen (1978) and Moormann (1981).

Table 4.8b Areas of irrigated rice, deepwater rice, dryland rice in 10^7 m^2. Shallow refers to water depths of up to 30 cm; intermediate to water depths of 30 cm to 1 m, both in bounded fields. Double cropped areas are counted twice. Source: Huke (1982)

	IRRIGATED		RAINFED			DRYLAND	TOTAL
	Wet season	Dry season	Shallow	Inter- mediate	Deep		
Southeast Asia							
Burma	780	115	2291	1165	173	793	5317
Indonesia	3274	1920	1084	534	258	1134	8204
Kampuchea	214		713	170	435	499	2031
Laos	67	9	277			342	695
W. Malaysia	252	212	92			10	566
Philippines	892	622	1207	379		415	3515
Sabah	8	4	9			21	42
Sarawak	6	4	46	11		60	127
Thailand	866	320	5128	1002	400	965	8681
Vietnam	1326	894	1549	977	420	407	5573
Total SE. Asia	7685	4100	12396	4238	1686	4646	34751
South Asia							
Bangladesh	170	987	4293	2587	1117	858	10012
Bhutan			121	40		28	189
India	11134	2344	12677	4470	2434	5937	38996
Nepal	261		678	230	53	40	1262
Pakistan	1710						1710
Sri Lanka	294	182	210	22		52	760
Total S. Asia	13569	3513	17979	7349	3604	6915	52929
S. Korea	1120		99			12	1231
N. Korea	500					150	650
China	33676*			1880		606	36162
Total Asia	56550	7613	32354	11587	5290	12329	125723

* Huke (1982) indicated 23968 and 9690 for 1st and 2nd crop, respectively.

4.2.5 Fluxes of CH$_4$ from Rice Paddies

For a Californian rice paddy Cicerone et al. (1983) reported average daily emissions of 0.25 g CH$_4$ m^{-2} day^{-1}. The total accumulated emission over the 100-day growing season was 22 to 28 g CH$_4$ m^{-2}. During a 2 to 3 week period before the harvest the emissions reached 5.0 g CH$_4$ m^{-2}

day^{-1}. Dramatic variation in methane flux through the growing season was observed. Seiler et al. (1984) measured emissions of 12 g CH_4 m^{-2} over the growing season in a Spanish rice paddy. Their low emission rate was attributed to inflow of sulphate containing mediterranean water, which would have inhibited methanogenesis. Measurements in an Italian rice paddy yielded a figure of 27-81 g CH_4 m^{-2} during a vegetation period (Holzapfel Pschorn and Seiler, 1986) while Holzapfel-Pschorn et al. (1986) report a rate of emission of 36.3 g m^{-2}.

Methane emission rates as presented by a number of researchers are shown in Table 4.9. Extrapolation of these emission rates to a global scale is difficult, since the effect of variations in agricultural practices, number of crops per year and other factors discussed above are highly uncertain.

Table 4.9 CH_4 emission over the growing season (about 100-150 days) from wet rice soils as determined by a number of researchers

Flux rate (gCH$_4$m^{-2}d^{-1})	Annual emission (g m^{-2}y^{-1})	Fertilizer treatment (kg ha^{-1})	Reference
	210		1
0.15-0.18	42	140 kg ha^{-1} (NH$_4$)$_2$SO$_4$	2
0.22-0.28	25	220 kg ha^{-1} (NH$_4$)$_3$PO$_4$-(NH$_4$)$_2$SO$_4$ (16-20-0) + topdressing of 113 kg urea-N at planting time	3
0.1	12	160 kg N as urea + 40 kg N as NH$_4$NO$_3$ after tillering	4
0.2-0.58	54	various treatments/unfertilized	5
0.15-0.42	36.3	not documented	6
8×10^{-3}-8×10^{-4}	7		7
0.16-0.38	17-42	unfertilized, over 4 years	8
0.47-0.60	53-68	6-12 T straw + 200 kgN urea/(NH$_4$)$_2$SO$_4$	8
0.1-0.3	35	CaCN$_2$	8
0.12-0.15	14-16	200 kg N as (NH$_4$)$_2$SO$_4$	8
0.18-0.21	19-22	100/50 kg N as (NH$_4$)$_2$SO$_4$	8
0.19-0.38	21-42	100-200 kg N as urea	8
0.23-0.68	24-77	3-12 T rice straw ha^{-1}	8

[1] Koyama (1963); [2] Cicerone and Shetter (1981); [3] Cicerone et al. (1983); [4] Seiler et al. (1984); [5] Holzapfel-Pschorn and Seiler (1986); [6] Holzapfel-Pschorn et al. (1986); [7] Minami and Yagi (1987) the low fluxes measured were caused by the fact that E_h values in the soils studied did not reach low values which favour methane formation; the soils were only recently, inundated (Minami, pers. comm.); [8] Schütz et al. (1989).

In Table 4.10 the geographic information from Table 4.8a and 4.8b is combined with flux estimates provided by Schütz et al. (1989). The latter authors related methane release rates to temperature. The methane emission for intermediate and deep water rice is taken from Bartlett et al. (1988) who reported CH_4 emissions from a tropical open water lake.

About 90% of the world's harvested area of paddy rice is located in Asia. Of the total harvested area in Asia about 50% is irrigated (permanently wet) and another 39% is wetland rainfed paddy rice (almost continuously wet). In the estimate of Table 4.10 the period of inundation of rainfed paddies is assumed to be 80% of the growing period. The resulting 1985 global emission from rice paddies based on the data in Table 4.10 is 53 to 114 Tg CH_4 y^{-1}. In this estimate the areas

Table 4.10 CH_4 emissions from wetland rice per continent

Continent	Area	CH_4 release g $m^{-2}d-1$**	Growing period (days)	Total emission (Tg y^{-1})	Type*
Asia	64163	0.5-0.8	90-120	29-62	irrigated
	32354	0.5-0.8	90-120	12-25	rainfed, 0-30 cm
	11587	0.5-0.8	90-120	4-11	rainfed, 30-100 cm
	5290	0.03	150	<1	deepwater, > 100 cm deep
Africa	3650	0.5-0.8	90-120	2-4	irrigated
	1820	0.5-0.8	90-120	1-2	rainfed
N/C America	1914	0.3-0.6	120-150	1-2	mostly irrigated
S America	6122	0.5-0.8	90-120	3-6	irrigated
Rest	1195	0.3-0.6	120-150	0-1	irrigated
Total	128095			53-114	

* for rainfed paddy rice a period of inundation of 0.8 of the growing season was assumed;
** from Schütz et al. (1989)

of dryland rice for continents outside Asia are included. The 1989 release of CH_4 from paddy rice land is thus 60 to 120 Tg CH_4 y^{-1} (on the basis of an increase in the harvested area of somewhat more than 1%).

This global estimate has a narrower range and as a whole is somewhat lower than estimates by e.g. Seiler in Bolle et al. (1986), whose estimate was 70-170 Tg y^{-1}, Aselmann and Crutzen (1989), who reported a range of 60-140 Tg (see Table 4.7) and Schütz et al. (1989). The latter authors report a global CH_4 flux from rice paddies of 47-145 Tg y^{-1}. The reason for the lower estimate in Table 4.10 is probably the separate calculation of rainfed and irrigated areas. In rainfed rice considerable extents have deep and intermediate water depths. Methane release rates are much lower from these fields than for the well managed shallow rice paddies where measurements have been made (see Table 4.9).

Scharpenseel (1988, pers. comm.) estimated from organic matter additions and decomposition that the global CH_4 release from rice paddies is 110 Tg y^{-1}.

4.2.6 Geographic Distribution of Natural Wetlands

Prior to estimating areal extents of wetlands, it is interesting to note that the scale of the data set used is of influence on the result. Van Diepen (1985) analysed the extents of soils with hydromorphic properties in a region in Ivory Coast derived from the 1:5,000,000 Soil Map of the World (FAO,1978-1981) and from 1:2,000,000, 1:500,000, 1:200,000 and 1:50,000 maps respectively. The proportion of hydromorphic soils was 0, 3, 9, 11 and 17-29 % respectively. This indicates the degree by which small scale maps may deviate from more detailed maps. Using global low resolution data sets probably leads to underestimation of the wetland CH_4 source. On the other hand, wetlands or hydromorphic soils mapped as such on small scale maps may also contain aerobic or well drained soils. In that way unmapped wet soils in dry areas would be compensated.

Matthews and Fung (1987) distinguished 5 types of wetlands. The areas are presented in Table 4.11. Aselmann and Crutzen (1989) distinguished 6 types of natural wetlands including lakes. Although the total area corresponds well between the estimates, regional disagreement between the data suggests that the present knowledge of the geography of wetlands is still incomplete.

Table 4.11 Areas of wetlands by different authors (areas in $10^{10} m^2$)

a. Matthews and Fung (1987)	
Forested bog	208
Nonforested bog	90
Forested swamp	109
Nonforested swamp	101
Alluvial formations	19
Total	526
b. Aselmann and Crutzen (1989)	
Bogs	187
Fens	148
Swamps	113
Marshes	27
Floodplains	82
Shallow lakes	12
Total	569

Part of the wetlands are only temporarily flooded and therefore are only active in methanogenesis during part(s) of the year. Matthews and Fung (1987) estimate the total area of wetlands at 2750×10^6 ha. Of the 2750×10^6 ha about 1383×10^6 ha is distinguished exclusively on Operational Navigation Charts (1:1,000,000 scale). The remainder, or 1367×10^6 ha, is the area where wetlands coincide on Matthews (1983), FAO (Zobler, 1986) and the ONC charts on the same location (corroboration cases).

About 19% or 530×10^6 ha of the total global wetland area estimated by Matthews and Fung (1987) is considered actually inundated. The duration of the flooding period and the extension of the flooded area depend on the prevailing climatic and hydrologic conditions. Large parts of the marshlands in the Amazon and Congo Basins are only flooded during half of the year and even shorter flooding periods may occur in other wetland areas. Furthermore, considerable portions of the marshes may in reality consist of unvegetated open water, with lower CH_4 emission rates than vegetated areas (Cicerone et al., 1983).

Yet another, independent analysis made by Van Dam and Van Diepen, shows that the global area of soils with hydromorphic properties occurring on level land is 1261×10^6 ha. This figure was derived from dominant soils from the FAO Soil Map of the World (1971-1981). A higher figure would result if associated soils and inclusions would also be taken into account. And use of more detailed maps would probably also result in more extensive wetlands (Van Diepen, 1985).

In Table 4.12 wetland areas from the FAO/Unesco Soil Map of the World are combined with data from other sources. The figure of 1367×10^6 ha for the corroboration cases presented by Matthews and Fung (1987) corresponds with the 1283×10^6 ha of not cultivated hydromorphic soils and wetlands (Table 4.12) and with Van Dam and Van Diepen (1982). About 30% of the area of hydromorphic soils in Table 4.12 is in the tropics, and this correlates well with the 35% calculated by Matthews and Fung (1987). However, a further analysis is required to check regional agreement between all data sources. The total extent of marshes and swamps is 210×10^6 ha (Matthews and Fung, 1987; and chapter 3 Table 3.2), a figure higher than that used by Bolle et al. (1986) who used 160×10^6 ha, but lower than the 260×10^6 and 250×10^6 ha given by Clark (1982) and Olson et al. (1983) respectively. Tundras occupy 695×10^6 ha (Matthews, 1983; and Table 3.2), the total extent of wetlands being 905×10^6 ha. However, Histosols (peat soils) cover a larger area than that covered by wetlands since those soils are not necessarily in swamps or may be drained. Other hydromorphic soils and potential sources of methane are the Gleysols and Fluvisols. For these soils the period during which anaerobic conditions prevail determines whether they are potential sources or not. Part of the Gleysols and Fluvisols are cultivated and either drained and not potential CH_4 sources, or are used for paddy rice.

In this repect it is interesting to note that in the period 1795-1980 an area of 23×10^6 ha or 6-7% of the wetland area in temperate zones has been drained. For the tropics this was only 4% (Armentano and Menges, 1986). This relatively small decrease in the global wetland area has probably had more effects on the CO_2 balance than on CH_4 fluxes.

Table 4.12   Areas (in 10^{10} m²) of soils with hydromorphic properties estimated from FAO/Unesco (1971-1981), Matthews (1983) and FAO (1983)

Type	Tropics	Temperate zones	Total area
Marshes + swamps	120	90	210
Other histosols (excl. those in marshes and swamps)	25	175	200
Gleysols	104	51	155
Tundras		545	545
Other wet soils: Fluvisols (uncultivated)	102	71	173
Total	351 (27%)	932 (73%)	1283
Fluvisols (cultivated)	78	47	125
Gleysols (cultivated)	52	93	145

4.2.7 Fluxes of CH_4 From Wetlands

Harriss et al. (1982) found that fresh water peat soils in waterlogged conditions are a net source of methane to the atmosphere with seasonal variations in the emission rates of less than 0.001 to 0.02 g CH_4 m^{-2} day^{-1}. During drought conditions they measured that swamp soils consume atmospheric methane at rates of less than 0.001 to 0.005 g CH_4 m^{-2} day^{-1}. This illustrates the complexity of processes which regulate the net flux of methane between wetland soils and the atmosphere. Their results raise questions concerning the generally accepted estimates of global methane emissions from wetlands.

Methane is usually found at very low concentrations in reduced soils if sulphate concentrations are high. Possible reasons for this phenomenon are:

1. competition for substrates between sulphate reducing bacteria and methanogens;
2. inhibitory effect of sulphate or sulfide on methanogens;
3. a possible dependence of methanogens on the products of sulfate reducing bacteria;
4. methane can be oxidized by aerobic and anaerobic methanotrophic bacteria (see review of aspects of both types of oxidation by Cicerone and Oremland, 1988).

This suggests that methane release is higher from fresh water environments than from saline water (Smith et al., 1982). A possible reason for this is the lower sulphate concentration in freshwater environments so that sulphate reduction is less important in these environments (see 4.2.2).

In Table 4.13 methane emission rates for natural wetlands from various reports are listed. As in rice paddies the temporal and spatial variation is extremely high (Harriss et al., 1982; Harvey et al., 1989). Soil water content, temperature and other sesasonal climatological factors are all potentially critical factors in determining whether a wetland soil acts as a source or a sink of atmospheric methane.

Recent global flux estimates for natural wetlands are 110 Tg y^{-1} (Matthews and Fung, 1987) and 40-160 (Aselmann and Crutzen, 1989). This range of estimates may narrow if estimates of geographic extents of different types of natural wetlands improve and the number of flux measurements increases, particularly in undersampled regions.

Harriss (1988, pers. comm.) noted that future global warming in northern areas may increase flux rates from natural wetlands considerably due to increased phytomass production and accelerated fermentation. This aspect will be discussed in other research papers in this volume (e.g. Burke et al.).

4.2.8 CH$_4$ Production by Herbivorous Animals

Ruminants. Crutzen et al. (1986) estimated that the CH$_4$ production by domestic animals is 74 Tg CH$_4$, with an uncertainty of 15%. Cattle contribute 74% (54 Tg), buffaloes (6 Tg) and sheep (7 Tg), the remainder stems from camels, mules, asses, pigs and horses. Humans produce less than 1 Tg CH$_4$. The world's wild ruminants may produce between 2 and 6 Tg CH$_4$ y^{-1}. The total global emission from the domestic and wild animals is thus 66 to 90 Tg. Since the data in the case of livestock are comparatively well documented, the CH$_4$ contribution from livestock is a fairly safe estimate.

Termites. Agricultural activities following deforestation, such as clearing, burning and cultivation, influence the activity and abundance of termites. Termites occur on about 68% of the earth's land surface (Zimmermann,. 1982). Human activities such as conversion of tropical forests to grazing land and arable land tend to increase the density of termites. The ecological areas that should have the largest methane emissions from termites are tropical wet savannas, areas that have been cleared or burnt, and cultivated land in the (sub-) tropics.

Methane has been found in the guts of various xylophagous insects including scarab beetles, wood-eating cockroaches and various lower termites (*Reticulitermes, Cryptotermes, Coptotermes*). The digestion of these insects is primarily dependent on anaerobic decomposition by symbiotic bacteria in the higher termites (family *Termiditidae*) and by

Table 4.13 CH$_4$ emission rates from natural wetlands

		Flux rate (10^{-3} g m^{-2} d^{-1}) Range		Mean	Annual emission (g m^{-2} y^{-1})	Reference
1. Freshwater environments						
SWAMPS						
Waterlogged	Michigan, USA	4.6	13	110	40*	Baker-Blocker (1977)
Cypress, waterlogged	S.Carolina,USA			10	3.7	Harriss and Sebacher (1981)
Cypress, waterlogged	Georgia, USA	11	256	90	32.9	Harriss and Sebacher (1981)
Cypress, fertilized, waterlogged	Florida, USA			970	354	Harriss and Sebacher (1981)
Cypress, unfertilized	Florida, USA	8.2	265	67	24.5	Harriss and Sebacher (1981)
Dismal swamp, waterlogged	Virginia, USA	-20	1			Harriss et al. (1982)
Dismal swamp, drought	Virginia, USA	-1	-5			Harriss et al. (1982)
Tropical flooded forest	Manaus, Brazil	164	219	192	70*	Bartlett et al. (1988)
Tropical flooded forest	Brazilian Amazon	2	760	110	40*	Devol et al. (1988)
MARSHES						
Panicum spp.	Louisiana, USA			440	160	Delaune et al. (1983)
Spartina cynosuroides	Virginia, USA			49*	18.2	Bartlett et al. (1987)
Boreal marsh	Alaska, USA	101	111	106	39	Sebacher et al. (1986)
Tropical floating grass mats	Manaus, Brazil	158	302	230	84*	Bartlett et al. (1988)
Tropical floating grass mats	Brazilian Amazon	0	5200	590	215*	Devol et al. (1988)
BOGS						
Bog	Minnesota, USA	126	206	156	57*	Harriss et al. (1985)
Bog	Minnesota, USA	19	174	47	17*	Harriss et al. (1985)
Bog	Minnesota, USA	33	468	194	71*	Harriss et al. (1985)
FENS						
Fen	Minnesota, USA	3	5	3	1*	Harriss et al. (1985)
Fen + Bog	Minnesota, USA	60	1943	419	151*	Harriss et al. (1985)
Shoreline fen	Minnesota, USA	158	171	165	60*	Harriss et al. (1985)
Sedge meadow	Minnesota, USA			664	242	Harriss et al. (1985)
Alpine fen	Alaska, USA	274	301	289	105	Sebacher et al. (1986)
Subarctic fen	Quebec, Can.	19*	46*		0.1-0.6	Moore and Knowles (1987)

Table 4.13 continued

TUNDRAS						
Ombrotrophic,non waterlogged	Sweden				0.1	Svensson (1976)
Ombrotrophic,waterlogged	Sweden				1.4	Svensson (1976)
Minerotrophic, waterlogged	Sweden				30.5	Svensson (1976)
Coastal tundra, waterlogged	Alaska, USA	34	266	119	43*	Sebacher et al. (1986)
Moist tundra,not waterlogged	Alaska, USA	0.3	12.5	4.9	2.0*	Sebacher et al. (1986)
Meadow tundra, waterlogged	Alaska, USA	9.4	77.6	40	15*	Sebacher et al. (1986)
Eriophorum (1987)	Alaska, USA				8.1	Whalen and Reeburgh (1988)
Carex (1987)	Alaska, USA				4.9	Whalen and Reeburgh (1988)
Intertussock (1987)	Alaska, USA				0.6	Whalen and Reeburgh (1988)
Moss (1987)	Alaska, USA				0.5	Whalen and Reeburgh (1988)
Eriophorum (1988)	Alaska, USA				11.4	Reeburgh (1989, pers. comm.)
Carex (1988) (1988)	Alaska, USA				0.8	Reeburgh (1989, pers. comm.)
Intertussock (1988)	Alaska, USA				3.9	Reeburgh (1989, pers. comm.)
Moss (1988)	Alaska, USA				4.4	Reeburgh (1989, pers. comm.)
OPEN WATER						
Tropical lake	Manaus, Brazil	-3	57	27	9.9*	Bartlett et al. (1988)
Open water	Brazilian Amazon	0	830	120	43.8*	Devol et al. (1988)
Tidal creek waters	Virginia, USA			2.2*	0.82	Bartlett et al. (1985)
2. Salt water environments						
Salt marsh, tall spartina	Georgia, USA			1.2	0.4	King and Wiebe (1978)
Salt marsh,interm.spartina	Georgia, USA			15.8	5.8	King and Wiebe (1978)
Salt marsh, short spartina	Georgia, USA			145.2	53.1	King and Wiebe (1978)
Salt marsh	Louisiana, USA			15	5	Smith et al. (1982)
Brackish spartina patens	Louisiana, USA			200	73	Delaune et al. (1983)
Salt spartina alterniflora	Louisiana, USA			12	4.3	Delaune et al. (1983)
Salt meadow	Virginia, USA			1*	0.43	Bartlett et al. (1985)
Short spartina alterniflora	Virginia, USA			4*	1.3	Bartlett et al. (1985)
Tall spartina alterniflora	Virginia, USA			3*	1.2	Bartlett et al. (1985)
Brackish spart. cynosuroides	Virginia, USA			79*	29.0	Bartlett et al. (1987)
Salt spartina mixed	Virginia, USA			16*	5.6	Bartlett et al. (1987)

* = calculated from data in the cited article.

Protozoa in the lower termites (all other families). Their digestion efficiency is usually 60 % (Zimmermann et al., 1982).
A first estimate of the potential production of CH_4 by termites was made by Zimmermann et al. (1982). Their annual production of 150 Tg CH_4 y^{-1} was based on the laboratory measured ratio of total gas evolved to food consumed by the termites. Accompanying releases of other greenhouse gases estimated using the same procedure were 5.4×10^{15} g C y^{-1} as $CO_{2,\ 0.4\times1013}$g C y^{-1} as CO and 7×10^{11} g y^{-1} of dimethylsulfide. The species used were *Reticulitermes tibialis*, fam. *Rhinotermitidae*; *Gnathamitermes perplexus*, fam. *Termitideae*; *Nasutitermitinae* (unidentified), fam. *Termitidae*. Zimmermann et al. also calculated that with an assumed uncertainty of 50% and additional uncertainty in the significance of termites in the various ecosystems of the world, the methane emissions could range from 75 to 310 Tg CH_4 y^{-1}. According to Zimmermann et al. (1982) the global area occupied by termites accounts for 68% of the earth's land area with 77% of the terrestrial net primary production of biomass. The world's termite population according to Zimmermann et al. is 2.4×10^{17}. This population processes 33×10^{15} g dry weight of organic matter which is the equivalent of 28% of the earth's annual net primary biomass production and an average of 37% of the net primary production in areas where termites occur.

Rasmussen and Khalil (1983) found a methane production by termites of 50 Tg y^{-1} (ranging between 10 and 90 Tg y^{-1}). They based their estimate on laboratory measured emission rates for the species *Zootermopsis angusticollis* only which is found in the American Rockies. They state that the uncertainty in the estimates of CH_4 emission by termites is in the production per termite and in the global number of termites and suggest that the disagreement with the data of Zimmermann et al. (1982) may be due to the different species observed. Their conclusion was that the CH_4 emission by termites is probably not more than 15% of the total global yearly emissions.

Seiler (1984) reported a much lower methane production by termites (2 to 5 Tg CH_4 y^{-1}). This estimate was based on in vivo measurements with several species, including soil feeders, grass feeders, wood and dung feeders, grass harvesters and fungus grown termites. Seiler calculated the ratio of methane emitted from termite mounds to carbon ingested bu termites of 6×10^{-5} to 2.6×10^{-3} (depending on the species) and a used total consumed biomass of 7×10^{15} g dry matter (about 1/4 of the figure used by Zimmermann et al.,1982). Collins and Wood (1984) give a global figure for the dry matter consumption by termites of 3.4×10^{15} g y^{-1}.

Seiler also reported characteristic values for the CH_4 to CO_2 emission ratio which appear to be typical for each termite species. Seiler concludes, that since the bulk of termites lives in ecosystems not affected by humans (contrary to Zimmermann et al., 1982), it is unlikely that the total methane emission has changed significantly during the last decades.

Fraser et al. (1986) arrived at a global production of 14 Tg CH_4 y^{-1} by termites (with a range of 6 to 42 Tg). Fraser et al. measured in the laboratory (using methods described by Khalil and Rasmussen, 1983) the methane production per termite. The species used by Fraser et al. are *Mastotermes darwiniensis*, *Nasutitermes exitiosus*, *Coptotermes acinaciformis*, *Coptotermes lacteus*, *Zootermopsis angusticollis* and *Coptotermes formosanus*. All the above species are wood feeders, none are fungus builders.

The highest estimates of termite methane production may be exaggerated. The subfamily of *Macrotermitinae* (fungus growing termites) is dominant in many ecosystems of the Ethiopian and Indo-Malayan regions (Collins and Wood, 1984). Since most of their digestion is performed aerobically by fungi, this subfamily is unlikely to produce much methane. Soil feeders, which occur in most tropical regions, use degraded soil organic matter. Therefore, if methane is produced by soil feeders, this is likely to occur in very small amounts. Air turbulence occurring during the measurements is known to cause increased activity and higher CO_2 and probably CH_4 production (pers.comm Dr. O. Bruinsma, 1987). The lowest figures (Seiler, 1984; Collins and Wood, 1984; Fraser et al.,

1986) are probably more realistic. Crutzen et al. (1986) estimated tentatively that the world's CH_4 emission by insects has an upper limit of 30 Tg CH_4. One aspect not accounted for in all the experiments with termites is the possible microbial breakdown of CH_4 occurring in the soil of termite mounds.

4.2.9 CH_4 Emission from Biomass Burning and Landfill Sites

Biomass burning and organic wastes as CH_4 sources are discussed in Schütz et al. (this volume), but a short discussion of present knowledge will be presented here.

Biomass burning. One of the major abiogenic sources of CH_4 is the burning of biomass, such as agricultural wastes, savanna fires, burning due to shifting cultivation and burning of fuel wood. Crutzen et al. (1979) measured CH_4 to CO_2 ratios in several fire plumes and estimated a total CH_4 emission due to burning of 25 to 110 Tg y^{-1}. If data for the burning of agricultural wastes are included the resulting CH_4 to CO_2 ratio in fire plumes is 1:53. The resulting current global methane production is 53 to 97 Tg CH_4 y^{-1} on the basis of a total amount (for 1975) of $48-88\times10^{14}$ g y^{-1} dry matter of biomass burnt is applied (Seiler, 1984). Biomass burnt in 1950 and 1960 was $37-76\times10^{14}$ and $42-67\times10^{14}$ g dry matter y^{-1}, respectively, giving CH_4 emissions of 41-74 and 47-84 Tg CH_4 y^{-1}, repectively.

Landfills. Methane is formed during anaerobic decay of municipal and industrial organic matter collected and dumped in landfills. Worldwide this source may add 30-70 Tg CH_4 to the atmosphere (Bingemer and Crutzen, 1987). The estimate is based on an amount of 85×10^6 T C y^{-1} of biodegradable carbon dumped worldwide, of which less than 20% in developing nations, and of $19-37\times10^6$ T C y^{-1} of carbon in industrial wastes, almost entirely from developed countries. These figures are uncertain by about 30%. The global emission rate is based on a production of 0.5 kg CH_4 per kg C and it is assumed that materials such as lignins and plastics do not produce CH_4.

To reduce the uncertainty in this estimate more measurements on CH_4 release from landfills and better statistics in different parts of the world are required. By far the largest contribution currently comes from the industrialized countries. This source has been increasing in recent years, but it is stagnating at present. In the coming decades very large increases are expected due to strong population growths and strongly increasing urbanization in the developing countries (Bingemer and Crutzen, 1987).

4.2.10 Methane hydrate destabilization

The possibility of destabilization of methane in hydrates by warming has been raised by various reports (see review by Cicerone and Oremland, 1988). These hydrates are solid structures, composed of strong cages of water molecules surrounding CH_4 molecules. These hydrates are most prevalent at depth in permafrost and in sea sediments. Arctic region hydrates may be vulnerable to warming while hydrates in coastal permafrosts may be releasing CH_4. Cicerone and Oremland (1988) report a range of 0-100 Tg CH_4, with a figure for the present release of 5 Tg. The authors call their own estimate questionable, and further investigation is required to understand the effect of warming on emissions of methane from hydrates.

4.2.11 Oxidation of Methane in Dry Soils

Methane oxidation is also discussed in Schütz et al. (this volume), and therefore a short discussion will be presented here.

Only a few measurements of the methane uptake in soils have beensummary of present knowledge will be presented here. carried out. Soil methanotrophic bacteria can grow with methane as their sole energy source. Other soil bacteria which consume methane are e.g. Nitrosomonas species (Seiler and Conrad, 1987, quoting various authors). The uptake of methane occurs in well aerated soils. Harriss et al. (1982) observed that a methane emitting a swamp changed into a sink for methane after it dried up. Seiler (1984) quoting research data by Seiler et al. observed a destruction of methane at the soil surface in semi-arid climates. The destruction rates varied between 3×10^{-4} and 24×10^{-4} g m^{-2} h^{-1} during the dry season with soil temperatures of 20 to 45°C. Keller et al. (1983) observed CH_4 uptake in temperate and tropical rain forests. For higher latitudes they reported loss rates of 1.2×10^{10} to 1.6×10^{10} molecules cm^{-2} s^{-1} with an average daily uptake of 2.5×10^{-4} g CH_4 m^{-2}. Methane decomposition was also observed at the surface of several types of soil in Germany. Hao et al. (1988) measured methane fluxes in tropical savannas in the dry season. They found no methane consumption there. Steudler et al. (1989) measured CH_4 comsumption rates of 0.13 mg CH_4-C m^{-2}h^{-1} in hardwood plots and 0.11 mg CH_4-C m^{-2}h^{-1} in pine plots. The estimated global consumption by temperate forests based on these consumption rates is 0.6-9.31 Tg CH_4-C y^{-1}. From literature data Steudler et al. (1989) predict that tropical forests contribute 1.26-2.53 Tg CH_4-C y^{-1}. Seiler (1984) estimated, that the global methane consumption must be at least 20 Tg y^{-1}. Seiler and Conrad (1987) reported a global methane consumption in soils of 32 ± 16 Tg y^{-1}.

4.3 CARBON MONOXIDE (CO)

4.3.1 Introduction

As was mentioned before, carbon monoxide does not interact in the atmospheric radiative balance, but it influences the concentrations of other atmospheric greenhouse gases such as CH_4, CH_3Cl, CH_3CCl_3 and $CHClF_2$ (F22). Moreover, the oxidation of CO is an important source of CO_2.

A rise of tropospheric CO concentrations would lead to a decrease of OH (Khalil and Rasmussen 1984a, 1984b, 1985) and this would influence ozone concentrations (see section 2.2.2). As the major tropospheric sink for many gases (particularly CH_4 and chlorinated hydrocarbons) is oxidation by OH, a rise in the CO mixing ratio would cause increasing concentrations of these gases in the troposphere.

The major sources of CO are known, but their magnitudes are still uncertain as indicated in Table 4.14. The background concentration of CO is increasing. Reported rates of increase range from 0.6-1% y^{-1} (Bolle et al., 1986) to 2-6% y^{-1} (Khalil and Rasmussen, 1984b), but fluctuations of sources and sinks and the relatively short residence time of CO in the atmosphere make these estimates uncertain. No positive trend is observed in the Southern Hemisphere (Cicerone, 1988).

4.3.2 Sources and Sinks of CO

Estimates of the sources and sinks of carbon monoxide made by various authors are listed in Table 4.14. The estimates for CO production due to biomass burning are, although quite similar, based on completely different statistics. Logan et al. (1981) applied clearing rates due to

Table 4.14 Estimates of sizes of possible sources and sinks of carbon monoxide by various authors

	Tg CO y^{-1}		Reference
	Range	Average	
Sources			
Vegetation	20- 200	110	Crutzen (1983)
	50- 200	130	Logan et al. (1981)
Soils	3- 30	17	Conrad and Seiler (1985)
Biomass burning	145-2015	660	Logan et al.(1981)
	240-1660	840	Crutzen et al. (1979)
	400-1600	800	Crutzen (1983)
Oceans	20 -80	40	Logan et al. (1981)
Fossil fuel burning	400-1000	450	Logan et al. (1981)
Oxidation natural NHMC[1]	280-1200	560	Logan et al. (1981)
Oxidadion anthropg. NHMC[1]	0- 180	90	Logan et al. (1981)
Oxidation of CH_4	400-1000	810	Logan et al. (1981)
		400	Khalil & Rasmussen, 1984a; 1984b)
Sinks			
Atmospheric oxidation of CO to CO_2	3000		Crutzen (1983)
-do-	1600-4000	3170	Logan et al. (1981)
Transport to stratosphere	190- 580	170	Crutzen (1983)
Microbial oxidation in soils	190- 580	450	Crutzen (1983)
-do-		250	Logan et al. (1981)

[1] NHMC = non methane hydrocarbons; natural NMHC's are isoprenes, C_5H_8 and terpenes, $C_{10}H_{16}$, produced in forest environments.

shifting cultivation of forest and woodland of 8×10^{10} to 36×10^{10} m^2 y^{-1} and 5×10^{10} to 32×10^{10} m^2 y^{-1} respectively. Crutzen (1983) used a range of 21×10^{10} to 62×10^{10} m^2 y^{-1} for burning due to shifting cultivation and 8.8×10^{10} to 15.1×10^{10} m^2 for deforestation due to colonization. Crutzen based his calculations on a ratio of CO/CO_2 production of 0.14. A recent study (Sachse et al., 1988) revealed increasing atmospheric CO concentrations over the Amazon, apparently from biomass burning.

Bartholomew and Alexander (1981) observed, that absorption of carbon monoxide occurs in most soils. Dry soils, that are producing CO turn into a net sink of CO after irrigation (Conrad and Seiler, 1982). CO absorption was stopped after heat-sterilization of the soil material, while the CO production was enhanced by heat-sterilization. Apparently the CO production is a chemical process, contrary to the CO oxidation in soils which results from microbial activity. Therefore, the CO producing soils can be found in arid and semi-arid zones, i.e. zones with predominantly Yermosols and Xerosols.

Since both CO production and oxidation occur simultaneously, it is very difficult to estimate the fluxes separately. Conrad and Seiler (1985) developed a technique to estimate the CO production and consumption rates at different soil temperatures and moisture contents. Measurements in arid subtropical soils demonstrated a strong dependence of CO production on the soil surface temperature, while the CO consumption was independent of surface temperature. This indicates that the CO production occurs at the surface while consumption occurs predominantly in subsurface layers at lower temperatures. In temperate climates where relatively humid soil conditions prevail, the CO production is insignificant and CO consumption very active. No data are available for humid tropical soils. Seiler and Conrad (1987) expected that soils in these regions are net sinks of CO. The global production of CO by soils is 17 Tg y^{-1} (ranging from 3 to 30 Tg of which 1 to 19 Tg y^{-1} is produced in dry tropical areas). CO consumption ranges from 300 to 530 Tg y^{-1} of which 70 to 140 Tg y^{-1} is oxidized in the humid tropics (Seiler and Conrad, 1987).

The global sources range between 1270 and 5700 Tg CO y^{-1} with an average of 2920 Tg. The global sink strength ranges between 1960 and 475 Tg CO y^{-1} averaging 3600 Tg. In this budget the role of the oceans as sources and sinks are neglected. The model is not completely balanced, indicating the uncertainty in the estimates given in Table 4.14.

4.4 CONCLUSIONS SECTIONS 4.2 AND 4.3

There is strong evidence, that the total CH_4 source has increased during the last decades. There is a strong correlation between the growth of the human world population and the atmospheric CH_4 concentration, indicating that this increase is most likely related to anthropogenic activities (Bolle et al., 1986; Ehhalt, 1988). The role of termites has apparently been over-emphasized in the past.

The production of methane in soils and wetlands is extremely sensitive to environmental conditions. Therefore, as for the nitrogenous trace gases, the variation in time and space of methane fluxes is extremely high. The currently used methods of measuring methane fluxes are point measurements. Extrapolation of results from such measurements to smaller scales is fraught with potential errors. The development of methods to estimate methane flux for largeer areas would greatly improve the quantification of the methane sources.

Since a number of ecosystems are undersampled, more flux data are needed. There are a number of fields where the present knowledge is inadequate:

1. The geographic distribution of soils used for wet rice cultivation;
2. geographic quantification of types of wetland rice cultivation, the number of crops per year, soil and water management, application rates of organic and anorganic fertilizers;
3. the role of soil parameters and soil and water management practices in the production and emission of methane;
4. the geographic distribution of the different types of salt and fresh water wetlands;
5. the relation between the type of wetland and methane fluxes;
6. other aspects such as water depth, temperature, influence of plants and their stage of development both for rice and natural wetlands;
7. fluxes from landfill sites, especially quantities of organic waste which is decomposed anaerobically in landfill sites are virtually unknown. Furthermore quantities of methane lost from biogas installations in both developed and developing countries.

Such information, linked with intensified measurement of fluxes, should yield more reliable estimates of regional and global methane production than currently possible. A secondary result of the above investigations is the capability to assess consequences of climate change for methane fluxes, especially those from natural wetlands.

Soils and vegetation are no major sources of carbon monoxide. Soils, however appear to degrade CO in considerable quantities. So far estimates are no more than rough guesses. More research is needed in this respect.

4.5 NITROUS OXIDE

4.5.1 Introduction

Nitrous oxide (N_2O) is capable of absorbing infrared radiation, but it is inert in the troposphere. In the stratosphere it is destroyed by reaction with atomic oxygen and in this process nitric oxide (NO) is formed. This gas reacts with ozone (O_3), leading to an overall reduction of the latter important atmospheric constituent. NO is also involved in the oxidation of CH_4 and CO. The contribution of N_2O to the supposed greenhouse warming over the past 100 years is about 5% (see Table 2.2). The atmospheric burden is about 1500 Tg N_2O-N (Tg = Terragram; 1 Tg = 10^{12} g). McElroy and Wofsy (1986) reported an accumulation of 2.8 Tg y^{-1} and a removal rate by stratospheric protolysis of 10.5 Tg y^{-1}.

As the lifetime of N_2O in the atmosphere is 100 to 200 years, changes in its concentration will have a long term effect. The budget presented in table 4.15 shows a slightly different figure for the stratospheric loss, but it provides a good overview of the global sources and sinks.

Release of oxides of nitrogen from soils (N_2O, NO and NO_2) is known to occur during biological denitrification (see 4.5.2), chemical denitrification (4.5.3) and nitrification (4.5.4).

Table 4.15 Global budget of tropospheric N_2O (all figures in Tg N y^{-1})

Sources	
Fossil fuel burning	2 ± 1
Biomass burning	1.5 ± 0.5
Oceans, estuaries	2 ± 1
Fertilized soils	1.5 ± 1
Natural soils	6 ± 3
Plants	<0.1
Gain of cultivated land	0.4 ± 0.2
Total production	14 ± 7
Sinks	
Stratospheric loss	9 ± 2

Source: Seiler and Conrad (1987).

4.5.2 Biological Denitrification

Biological denitrification is the dissimilatory reduction of nitrate (NO_3^-) or nitrite (NO_2^-) to gaseous forms of nitrogen by essentially anaerobic bacteria producing molecular N_2 or oxides of N when oxygen is limiting (adapted from SSSA, 1984). The species involved in denitrification are largely limited to the genera *Pseudomonas*, *Bacillus*, and *Paracoccus*, although *Thiobacillus denitrificans* and *Chromobacterium*, *Corynebacterium*, *Hyphomicrobium* or *Serratia* species will catalyze the reduction. Denitrifying bacteria are aerobic, but nitrate is used as electron acceptor for growth in the absence of oxygen (Alexander, 1977). Recent investigations have shown that *Rhizobium* bacteria and bacteroids can also denitrify (O'Hara and Daniel, 1985).

Denitrification occurs only at low oxygen pressures. It has been widely accepted as the cause of poor efficiency of nitrogen use in flooded soils, but it may also play a significant role in well aerated soils. Between 10 and 30% of the applied nitrogen is commonly lost by gaseous loss mechanisms of which denitrification is believed to be the major. However, recent research on NH_3 volatilization from flooded soils and research on gaseous nitrogen losses suggests that denitrification and nitrification proceed concurrently with NH_3 loss; the relative importance of these two loss mechanisms may change substantially over very short periods (Simpson and Freney, 1986).

The general pathway of the reduction of nitrate during denitrification is (see e.g. McKenney et al., 1982):

$$NO_3^- \rightarrow NO_2^- \rightarrow NO \rightarrow N_2O \rightarrow N_2 + H_2O$$

The energy for these reactions is supplied by the decomposition of carbohydrates.

Many microorganisms are capable of reducing nitrate (NO_3^-) to nitrite (NO_2^-), but not all are able to denitrify. Most but not all denitrifying bacteria can reduce N_2O to N_2. Ammonia may inhibit the further reduction of NO_2^- by denitrifiers (see 4.5.3).

There is still uncertainty about NO being an intermediate product of denitrification (see e.g. Poth and Focht, 1985). The highly reactive nature of NO makes its detection difficult. Both NO and N_2O gases may escape from the soil before being reduced. The ratio of N_2 to N_2O in the gases evolved from soil depends on such factors as soil pH, moisture content, redox potential (E_h), temperature, nitrate concentration and content of available organic C.

Rhizobia are bacteria which in a symbiotic relationship with leguminous plants make an important contribution to nitrogen input in soil. Rhizobial denitrification is potentially significant for losses of nitrogen and particularly for N_2O emissions. Free-living *Rhizobia* and symbiotic *Rhizobia* are widespread in many agricultural and natural ecosystems.

Symbiotic *Rhizobia* in root nodules have been demonstrated to possess the two opposed physiological processes of N_2 fixation and denitrification. Although root nodule denitrification is energetically unfavourable for the host plant, losses of N_2O have been reported for a number of leguminous field crops (see literature review by O'Hara and Daniel, 1985). The advantage of denitrification may be in the removal of NO_3^-, NO_2^- and N_2O, which are inhibitory to N_2 fixation. Under anaerobic conditions bacterial respiratory NO_3^- reduction can supply ATP for nitrogenase activity of root nodules (Zablotowicz and Focht, 1979).

For denitrification to occur, nitrate has to be available, and this will at the same time suppress the N_2 fixation capacity. In natural ecosystems with leguminous plants and in agricultural systems with leguminous crops nitrate will probably not be present in significant amounts.

Free-living *Rhizobia* have been found in soils where leguminous host plants have never grown. When nitrate is offered to soils, growth rates of bacteria are essential in competitive conditions. *Rhizobia* have relatively slow growth rates compared to e.g. Pseudomonas. Because of this slow growth *Rhizobia* may be better colonizers of infertile soils than are fast growers, and important properties of *Rhizobia* in this respect are: ability to denitrify, slow growth, slow metabolism, ability to assume dormancy (O'Hara and Daniel, 1985). Most slowly growing *Rhizobia* denitrify (Daniel et al., 1982) and this capability may play an important role in the persistence of *Rhizobia* in poor soils.

Losses through Rhizobial denitrification will depend in many situations on the nitrification activity, which produces NO_3^-. In soils with pH below 5.5 nitrification activity is negligible and Rhizobial denitrification may also be expected to be low. However, in soils above pH 5.5 nitrification will proceed and under such conditions soils there is potential for substantial losses through Rhizobial denitrification. Significant denitrification may occur in aerobic as well as anaerobic soils. More details on rhizobial denitrification can be found in e.g. O'Hara and Daniel (1985).

4.5.3 Chemo-denitrification

Chemical denitrification is the reduction of nitrite or nitrate by chemical reductants producing N_2 or oxides of N (adapted from SSSA, 1984).

High NO_2^- concentrations have been attributed to inhibition of the nitrite oxidation which is presumed to result from ammonia toxicity to *Nitrobacter*. Several investigators have noted that gaseous loss of nitrogen (via NO, N_2O or N_2) may accompany temporary NO_2^- accumulation. High concentrations of NO_2^- are sometimes found in anaerobic soils where NH_3 or NH_4^+ type fertilizers are applied at higher doses (Stevenson et al., 1970). Nitrite accumulation is enhanced by the application of phosphate (Minami and Fukushi, 1983). Nitrite ions react chemically with organic molecules forming nitroso-groups (-N=O) which are unstable. Gaseous products (N_2, N_2O) can be formed from these groups. Another reaction path that may be followed is the dismutation of HNO_2 whereby NO or NO and NO_2 may be formed (Slemr et al., 1984; Stevenson et al., 1970). For the latter reactions acidic conditions are required.

Keeney et al. (1979) found a nitrite accumulation at high temperatures in anaerobic systems and attributed this to thermophyllic species of *Bacillus* and *Clostridium*, which are nitrate respirers but not denitrifiers. The gas production consisted predominantly of NO in the closed anaerobic system, leading to the conclusion that chemo-denitrification reactions do not contribute significantly to N_2O production under anaerobic conditions.

4.5.4 Nitrification

Nitrification is the biological oxidation of ammonium (NH_4^+) to nitrite or nitrate (NO_2^-, NO_3^- respectively), or a biologically induced increase in the oxidation state of nitrogen (adapted from SSSA, 1984). The following genera are known to occur in soil: *Nitrosomonas*, *Nitrosococcus*, *Nitrospira* and *Nitrosolobus* (oxidize ammonium to nitrite) and nitrobacter (oxidize nitrite to nitrate). Of these, only *Nitrosomonas* and *Nitrobacter* are encountered frequently and these two are undoubtedly the major nitrifying chemoautotrophic bacteria.

Except for poorly drained or submerged soils, NH_3 formed through ammonification is readily converted to NO_3^-:

$$NH_3 \; \overset{+H^+}{\underset{-H^+}{\rightleftharpoons}} \; NH_4^+ \; \overset{Nitrosomonas}{\rightleftharpoons} \; NO_2^- \; \overset{Nitrobacter}{\rightleftharpoons} \; NO_3^-$$

where subreactions are:

$$NH_4^+ \; \overset{-\frac{1}{2}O_2}{\underset{-H^+}{\rightleftharpoons}} \; NH_2OH \; \overset{+\frac{1}{2}O_2}{\underset{-2H^+}{\rightleftharpoons}} \; NOH \; \underset{-H^+}{\rightleftharpoons} \; NO_2^- \rightleftharpoons NO_3^-$$

chemical biological
denitrification   denitrification

$$N_2O + NO$$

Evidence for the formation of NH_2OH as an intermediate was provided by Yoshida and Alexander (1970). Minami and Fukushi (1986) suggest that in well aerated soils NH_2OH may react with NO_2^- forming N_2O. The reaction may occur chemically as well as biochemically, but this is not the main mechanism for N_2O production.

The energy released during the formation of nitrite (272 kJ) and nitrate (79 kJ) is used by the *Nitrosomonas* and *Nitrobacter* organisms for carrying out their life functions. The relative populations, their past histories as well as changes in soil conditions (e.g. temperature) may introduce temporary accumulation of nitrite in the soil (C.C. Delwiche, 1987, pers. comm).

There is a differential effect of temperature on the nitrifiers, with *Nitrobacter* being more sensitive to low temperatures than is *Nitrosomonas*. In cold conditions this may lead to a NO_2^- accumulation in the soil, which may have a toxic effect on plants. Nitrifying organisms may contribute significantly to NO and N_2O emissions from soils. In experiments with nitrifying and denitrifying bacteria (*Nitrosomonas europeae* and *Alcaligenes faecalis* respectively) Levine et al. (1984) showed, that nitrification rather than denitrification is the primary biological process leading to the formation of N_2O and NO. The ratio of NO : N_2O for aerobic conditions (with low oxygen pressures as found in soil) found by Levine et al. was 2; in air the ratio was 0.13 to 0.29 and at an oxygen mixing ratio of 0.5% the ratio NO : N_2O ranged between 2 and 4. These results are similar to those reported for nitrifiers by Lipschultz et al., 1981. This ratio was found to be 0.01 in denitrification.

Bremner and Blackmer (1981) reported that soils evolve N_2O even when they are well aerated and when the moisture content is low (i.e. under conditions known to inhibit denitrification) and that N_2O emissions from aerated soils are correlated not with the NO_3^- concentration but with nitrifiable N-contents (e.g. Minami and Fukushi, 1983). In well aerated soils emissions of N_2O are greatly increased by the addition of nitrifiable forms of N such as ammonia, urea, alanine, etc. (Bremner and Blackmer, 1978), but are not significantly affected by the addition of nitrate, glucose, or both. This finding agrees with Breitenbeck et al. (1980) and Seiler and Conrad (1981) and others, but conflicts with findings of amongst others, Mulvaney et al. (1984).

Several studies have indicated, that N_2O is a by-product of NH_4^+ oxidation as well as of nitrate reduction by heterotrophic microorganisms. Yoshida and Alexander (1970) showed that N_2O could be produced by *Nitrosomonas europaea*. Focht (1974) doubted this could actually occur in soils where the present *Nitrobacter* bacteria, if present, would immediately oxidize NO_2^- to NO_3^- thereby limiting the possibility of formation of N_2O.

Poth and Focht (1985) observed that *Nitrosomonas* bacteria produce N_2O under oxygen limiting conditions and that N from nitrite but not nitrate is incorporated into nitrous oxide. In their study nitrification was not a direct source of N_2O. *Nitrosomonas* is a nitrifier which, under conditions of oxygen stress, uses nitrite as a terminal electron acceptor to produce nitrous oxide. They refer to this process as nitrifier denitrification. Denmead et al. (1979b) reported a simultaneous increase of soil NO_3^- and N_2O production after moistening the soil. They concluded that nitrification and N_2O production occur simultaneously. Parton and Mosier (1988) similarly report a simultaneous increase of NO_3^- and N_2O production and suggest that nitrification or nitrifier denitrification (see Poth and Focht, 1985) is dominant over denitrification until soils become very wet or saturated.

4.5.5 Factors Controlling N_2O Fluxes

Partial oxygen pressure. The partial oxygen pressure in the soil atmosphere depends on the oxygen exchange with the atmosphere. Thus the soil's oxygen status is closely related with the soil water status on the one hand and oxygen consumption by plant roots and microorganisms on the other. Denitrification is negligible at soil moisture contents below about 2/3 of the water holding capacity, but is appreciable in flooded soils. The process may occur in anaerobic microsites within the otherwise aerobic medium in well drained soils, such as pores filled with water or sites within structure aggregates (Dowdell and Smith, 1974). Anaerobic microsites can also exist in locations of high microbial activity where oxygen is being consumed and CO_2 produced, thus making conditions more favourable for denitrification (Parkin, 1987).

A tendency for increased N_2O release with improved aeration status suggests, that the reduction of N_2O may be slowed down (but not stopped) in soil at low oxygen pressures (Fillery, 1983). Possible explanations for this phenomenon are competition for electrons, preferential inhibition of N_2O reduction (by N_2O reductase), or a general slow-down of the denitrification process (allowing N_2O to move away unreduced). Alternatively, it might be argued that where the oxygen status of a soil is low, this is because the microbial demand for electron acceptors (of which oxygen is energetically the most favourable) outstrips the diffusive supply. Nitrous oxide may then act as an alternative electron-acceptor with diffusive properties little different from oxygen. The reduction of N_2O (and hence a reduction in the N_2O fraction in the gaseous products) will be favoured by just those factors which lead to a low oxygen status. Another possibility is the stimulation by enhanced aeration of nitrification of NH_4 whereby N_2O may be one of the products.

Soil water status. Letey et al. (1980) found that the release of N_2O from the soil environment to the atmosphere is stimulated under fluctuating oxygen pressures (alternating drying and wetting cycles). When the soil is wetted N_2O will be produced more rapidly than it is reduced; if the soil dries fast enough, N_2O reduction to N_2 is prevented and rapid diffusion is possible. Mean N_2O fluxes were generally higher during periods of irrigation in summer (alternating dry and wet) than in periods of frequent rains in winter (continuously wet). Parton and Mosier (1988) saw the production of N_2O and NO_3^- increasing simultaneously along with increasing soil water at low soil water contents, indicating that nitrification was dominant; at very high soil water contents only N_2O production increased on adding water, which would suggest that denitrification is dominant.

Mosier et al. (1981, 1986) showed brief peaks in the N_2O flux after precipitation events and Denmead et al. (1979b) found a marked response of N_2O emission to small additions of water. This would indicate that N_2O production takes place close to the soil surface, where the essentially aerobic conditions are unlikely to favour denitrification. In contrast, Mosier and Parton (1985) report only low N_2O emissions after rain events and Cates and Keeney (1987a) found no N_2O flux peaks following precipitation events in three prairie soils. They attributed this to nitrification being the dominant process involved.

Probably nitrifying and denitrifying microorganisms both contribute to N_2O production in aerobic soils (Minami and Fukushi, 1987) whereby nitrification is dominant in the topsoil (Seiler and Conrad, 1981; Denmead et al., 1979) and denitrification in the subsoil during periods of high soil water content (Goodroad and Keeney, 1985).

Flooding. Terry et al. (1981) found, that the major gaseous product prior to flooding was N_2O, while the major product after flooding was N_2. These results support the conclusion of Denmead et al. (1979a) and Sahrawat and Keeney (1986) that flooded soils contribute less N_2O to the atmosphere than drained soils (see also Goodroad and Keeney, 1984 in Table 4.16). By contrast, Minami (1987) measured appreciable N_2O emissions from flooded rice fields. Colbourn and Harper (1987) showed that drainage decreases total production of nitrogenous gases, but the balance was shifted towards N_2O in autumn and winter in a clay soil. The net result was a higher N_2O evolution from drained soils (see Table 4.16 and 4.17).

Terry et al. (1980) observed that N_2O emission by an organic soil is reduced to virtually zero after flooding. Drained organic soils however, may contribute considerably to atmospheric nitrous oxide (see also data from Goodroad and Keeney, 1984 in Table 4.16). The data in Table 4.16 suggest that open water and undrained wetlands are only small sources of N_2O.

Diurnal and seasonal variation; influence of temperature. There is a strong diurnal variation of N_2O emission rates (Ryden et al., 1978; Denmead et al., 1979b; Keeney et al., 1979; Blackmer et al., 1980; Conrad and Seiler, 1983; Minami, 1987) and seasonal variability (e.g. Keeney et al., 1979; Bremner et al., 1980). Generally the amplitude of variation is greater at higher temperatures and with larger fertilizer doses. This indicates that the timing of measurements may influence the result to a great extent. Blackmer et al. (1982) reported that there is no single time during a 24 hour period that is always satisfactory for assessing the amount of N_2O evolved during that period.

The optimum temperature for the denitrification process is 25°C and above, while the process is slow at 2°C. Denitrification is still rapid at elevated temperatures and will proceed to about 60 to 65°C but not at 70°C (Alexander, 1977:281). In nitrification the temperature optimum lies between 30 and 35°C, while below 5°C and above 40°C the activity is very low (Alexander, 1977:253). Important losses may also occur at lower temperatures, especially in grassland above 8°C and when cattle slurry is applied, i.e. with an abundant and mobile C-source even at lower temperatures (S.C. Jarvis, 1987, pers. comm). Keeney et al. (1979) reported, that while the rate of denitrification was low at temperatures below 15°C, the amount of N_2O (44-50% of the total gas production) was equivalent to that evolved at 25°C. Denitrification during late autumn and early spring in temperate climatic zones could account for a significant portion of the N_2O released per year, particularly in water saturated soils or soils with high water contents (Schmidt et al., 1988).

Both the rate of emission and the form of the products of denitrification / nitrification depend on temperature. Several authors (Bremner et al., 1980; Duxbury et al., 1982; Goodroad and Keeney, 1984; Goodroad and Keeney, 1985; Schmidt et al., 1988) reported appreciable emissions during the spring thaw. Nitrous oxide trapped under the frozen layer may be released (Bremner et al., 1980; Goodroad and Keeney, 1985) but N_2O production during denitrification at low temperatures may also be considerable.

Distribution of soil organic matter. A good supply of readily decomposable organic matter for energy supply is a prerequisite for the occurrence and good progress of nitrification and denitrification processes.

Chemical status of the soil. Many of the denitrifying bacteria are sensitive to high hydrogen ion concentrations. However, under certain conditions denitrification may still be rapid at a pH of 4.7.

In acid environments, nitrification proceeds slowly even in the presence of adequate supply of substrate, and the responsible species are rare or totally absent at great acidities. Typically the rate falls off markedly below pH 6.0 and becomes negligible below 5.0, but some soils nitrify at 4.5 due to presence of acid-adapted strains or chemical differences between the habitats (Alexander, 1977:252).

The composition of the gaseous products is governed by the pH in denitrification. Frequently N_2O makes up more than half of the nitrogenous gases evolved from acid habitats. Differences in gas composition associated with pH may largely be the result of acid sensitivity of the enzyme system concerned in N_2O reduction (Alexander, 1977:281). The effect of NO_3^- on denitrification is strongly related with the soil pH. Inhibition of the N_2O reduction to N_2 occurs at all NO_3^- levels at low pH; at a higher soil pH the inhibition of N_2O reduction is temporary, although N_2O remains a significant product for a longer period at higher nitrate levels (Fillery, 1983). Possibly a delay of N_2O reductase occurs until a threshold N_2O level is reached (Fillery, 1983). Sahrawat et al. (1985) concluded that an elevation of pH in acid forest soils enhances nitrification, but the ratio N_2O produced to NO_3^- produced is not influenced or decreases after liming.

Yoshida and Alexander (1970) showed that both phosphate and high soil pH enhance N_2O production in cell suspension of *Nitrosomonas europaea*. Bremner and Blackmer (1981) and Minami and Fukushi (1983) demonstrated that N_2O production increases in soils treated with ammonium-yielding fertilizers. The latter authors also showed that in soils treated with ammonium, phosphate has an additional effect on the N_2O emission. They concluded that calcium carbonate and phosphate create favourable conditions for both enzyme activity responsible for N_2O production and the nitrifying population.

Keller et al. (1988) observed increased N_2O production immediately after fertilizing Oxisols (Latosolo amarelo; Manaus, Brazil) under *terra firme* forest with NO_3 and NH_4. Losses of applied fertilizer were much lower for the NH_4 fertilizer than for the NO_3 fertilizer (see Table 4.17). The reaction to P fertilization was only small due to fixation of phosphate in these leached tropical soils.

Land use. Plants may reduce denitrification (and the N_2O fraction) by depleting the anorganic N-pool (Haider et al., 1985). Where there is little nitrate available microorganisms may be compelled to reduce more N_2O. On the other hand root exudates may stimulate denitrification and roots may also create anaerobic conditions by depleting oxygen in the rhizosphere. An important negative interaction exists between leaching of nitrate and denitrification, whereby the land use plays an important role. Stimulation of one process will reduce the other. Mosier et al. (1986) reported a marked influence of the growth and development pattern of the crop on the loss of fertilizer N as N_2O.

4.5.6 Spatial Variability of N_2O Fluxes

Spatial variability of N_2O emissions from soils has been recognized by several investigators (Ryden et al., 1978; Rolston et al., 1978; Breitenbeck et al., 1980; Bremner et al., 1980; Mosier et al., 1981; Duxbury et al., 1982; Folorunso and Rolston, 1984; Colbourn et al., 1984; Goodroad and Keeney, 1985; Parkin, 1987; Colbourn and Harper, 1987). Duxbury et al. (1982) reported that the spatial variability is reduced when fluxes were summed over important flux periods. The accuracy of the field measurements of N_2O emission is more limited by sampling methods than by analytical problems. Folorunso and Rolston (1984) calculated that in a Typic Xerorthent (USDA, 1975) 350 measurements are required to estimate the N_2O emission within 10% of the true mean in a 3 by 30 m plot. To achieve an accuracy of 50% the number of measurements should be 14 for N_2O. Contrary to the above variability, Conrad et al. (1983) found the spatial variability of N_2O flux to be insignificant compared to the effects of fertilization. They observed that the N_2O emissions never vary more than a factor 4. Cates and Keeney (1987a) reported that spatial variability of the N_2O flux from prairie soils is generally much lower than the values reported for agricultural soils.

High specific rates of denitrification may be associated with anaerobic microsites resulting from high O_2 consumption rates. This suggests that the patchy distribution of particulate organic matter or of the humus layer is a significant factor influencing the high spatial variability of N_2O fluxes from soils (Parkin, 1987; Schmidt et al., 1988).

4.5.7 Emission Rates of N_2O

General. The presented studies have generated a considerable amount of data. The greatest difficulty in the assessment of N_2O emissions is the extrapolation of the measurements to field conditions elsewhere, due to the complex and time dependent interactions of temperature, microbial populations, supply of organic carbon, oxygen diffusion, water content, nitrate concentration and the root systems. Important parameters, which are very difficult to measure in the field are the proportion of soil in a field profile that is anoxic, and the effect of soil organic carbon on microbial processes.

Several studies have not provided for frequent sampling of individual study sites or for intensive sampling after rainfall; most studies are only concerned with maximum N_2O fluxes after fertilization or after irrigation rather than average emission rates. In many cases even the source or process involved is not determined. Nevertheless, in the following sections an attempt will be made to assess the flux rates for soils and various land use types.

Natural undisturbed soils. Table 4.16 lists published data for N_2O flux from undisturbed natural soils.

Undisturbed tropical moist forests may be a significant source of N_2O (Kaplan, 1984; Keller et al., 1986). Nitrous oxide released from soils near Manaus (Brazil) is much higher than the global average, reflecting the rapidity of the N-cycle in tropical areas. Keller et al. (1988) observed that between 2.2 and 2.4% of all N in litterfall is emitted as N_2O in *terra firme* forest and Tabanuco forest respectively. The emission data in Table 4.16 suggest a variation of 10 to 30 μg N_2O-N $m^{-2}h^{-1}$, or 0.09 to 0.26 g N_2O-N $m^{-2}y^{-1}$. These figures are much higher

Table 4.16 N_2O emission rates from uncultivated lands and natural ecosystems

Temperate ecosystems

Soil / texture	Ecosystem	Flux (range) (μg N m^{-2}h^{-1})	Annual emission (kg N ha^{-1})	Method*	Reference
Fine sandy loam over clay	uncropped land	2.0	0.2		CAST (1976)
Fine sandy loam over clay	grassland (dry)	2.1	0.2	o -	Denmead et al. (1979b)
	grassland (moist soil)	25.0–104.0	2.2–9.1	o -	Denmead et al. (1979b)
Fine loamy mixed Haplargid	native prairie (summer)	10.0	0.9	c -	Mosier et al. (1981)
Loess loam	grass	0.5–2.5	0.0–0.2	c -	Seiler & Conrad (1981)
Silt loam	native prairie	0.8–2.9	0.1–0.3	c -	Cates & Keeney (1987a)
Loess over glacial till	burnt tall grass prairie	2.0–2.0	0.2–0.2	c -	Goodroad & Keeney (1984)
Loess over glacial till	unburnt tall grass prairie	3.0–2.0	0.3–0.2	c -	Goodroad & Keeney (1984)
Sand	meadow	2.0–13.0	0.2–1.1	c -	Seiler & Conrad (1981)
Organic soil	wet meadow	31.0–31.0	2.7–2.7	c -	Goodroad & Keeney (1984)
Organic soil	drained marshes	65.0–149.0	5.7–13.1	c -	Goodroad & Keeney (1984)
Organic soil	undrained marsh	1.0–1.0	0.1–0.1	c -	Goodroad & Keeney (1984)
	undrained marsh	4.0–6.0	0.4–0.5	c -	Smith et al. (1983)
Open water	open water	1.0–4.0	0.1–0.4	c -	Smith et al. (1983)
Loess	mixed forest	1.0–3.0	0.1–0.3	c -	Seiler & Conrad (1981)
Sand, 1-3 cm humus layer	temperate deciduous forest	4.5–10.5	0.4–0.9	c -	Schmidt et al. (1988)
Grey-brown podzol, 1-5 cm humus layer	temperate deciduous forest	3.5–9.5	0.3–0.8	c -	Schmidt et al. (1988)

Table 4.16 continued

Pseudogley soil, 1-3 cm humus layer	temperate deciduous forest	5.5	75.0	0.5	6.6	c -	Schmidt et al. (1988)
Grey brown podzol, 1-2 cm humus layer	temperate deciduous forest	5.5	7.5	0.5	0.7	c -	Schmidt et al. (1988)
Typic Dystrochrepts, (old tropical Acrisols)	temperate deciduous forest	1.5	3.5	0.1	0.3	c -	Schmidt et al. (1988)
Typic/Dystric Eutrochrepts, (Vertic Cambisols)	temperate deciduous forest	2.5	4.5	0.2	0.4	c -	Schmidt et al. (1988)
Loess over glacial till	deciduous forest	5.0	15.0	0.4	1.3	c -	Goodroad & Keeney (1984)
Loess over glacial till	coniferous forest	28.0	36.0	2.5	3.2	c -	Goodroad & Keeney (1984)
Tropical ecosystems							
Sandy loam (60-66% sand)	tropical savanna, dry season	2.0	4.0	0.2	0.4	c -	Johansson et al. (1988)
Sandy loam/sand	trop. savanna, dry season		6.0		0.5	c -	Hao et al. (1988)
Sandy loam/sand	trop. savanna, watered soil		16.0		1.4	c -	Hao et al. (1988)
Oxisol (Latosolo amarelo)	tropical forest (undisturbed)		30.0		2.6	c -	Keller et al. (1986)
Oxisol (Latosolo amarelo)	trop. secundary forest (20-30 yrs)		25.0		2.2	c -	Keller et al. (1988)
Oxisols/Ultisols, ridges/slopes	tropical Terra firme forests		11.0		1.0	c -	Livingston et al. (1988)
Spodosols/Psamments, Valley bottoms	tropical Campinarana forest		10.0		0.9	c -	Livingston et al. (1988)

* c = closed chamber; o = open chamber; 1 = N_2 and N_2O measured (C_2H_2 inhibition); 2 = ^{15}N labelling; - = N_2 measured exclusively.

than those reported for tropical savannas and temperate forests and suggest that tropical forests are the major source of atmospheric N_2O.

The global area of tropical rain forests from Table 3.2 and Table 3.7 is 1123 million ha, and tropical seasonal forests occupy 331 million ha. The total area of tropical forests is thus about 10% of the global land area, and this indicates the potential importance of tropical forests in the N_2O budget. The limited number of available flux measurement data does not allow for a global exptrapolation.

Dry season emissions for savanna regions reported by various authors correlate well and range between 0.01 to 0.03 g N in a dry period of about 7 months. Assuming that the data provided by Hao et al. (1988) for a watered savanna soil are representative for wet season conditions, the wet season flux would be 0.06 g N_2O-N m^{-2} y^{-1}, giving a total of 0.07 to 0.1 g N_2O-N m^{-2} y^{-1}. Unfortunately no figures for different soils and climatic conditions are available. In Table 3.2 the land cover types No. 12, 13 and 15 are savannas (Ketner, pers. comm.) with a global extent of about 796 million ha. Adding to this the 1983 million ha of grasslands with partial tree or shrub cover (No. 19, 23, 24 and 25 in Table 3.2) which occur predominantly in the tropics, yields a global area of 2779 million ha of savanna and savanna-like ecosystems. Although their annual N_2O emission may be lower than the flux from tropical forests, these ecosystems are important sources of N_2O because of their widespread occurrence.

A reason for the reported high fluxes in tropical wet forests and savannas may be the presence of leguminous plants in these ecosystems. These plants live in a symbiotic relationship with *Rhizobium* bacteria, which are capable to fix atmospheric molecular nitrogen (N_2). This fixed nitrogen is added to the vegetation and may enter the soil system through litterfall. This nitrogen, after being nitrified to NO_3^-, can be denitrified by soil denitrifiers. Although free living *Rhizobia* are slow growers, they may play an important part as denitrifiers in these infertile environments. Soils in most tropical rain forests are acid, a property known to be unfavourable to nitrification. In addition, there is a strong competition for the small quantities of available nitrogen between soil bacteria and plant roots. The potential for N_2O losses from denitrification by symbiotic *Rhizobia* must therefore be considered in these tropical forest and savanna ecosystems.

Temperate forests produce more N_2O than prairies. Goodroad and Keeney (1984) report lower emissions for temperate deciduous forest than for coniferous forests. The data from Schmidt et al. (1988) are in line with Goodroad and Keeney (1984) and suggest that there is a marked influence of soil type and thickness of the humus layer. However, the spread is too large to generalize the presented figures.

In general undrained marshes show the lowest N_2O fluxes followed by native prairie soils, although Mosier et al. (1981) measured flux rates from prairies which are considerably higher than those reported by other authors. Drained marshes however may produce considerable quantities of N_2O. Goodroad and Keeney (1984) report fluxes of 65 to

149 μg N_2O-N $m^{-2}h^{-1}$, Duxbury et al. (1982) presented extremely high fluxes for drained cultivated organic soils, with fluxes of between 180 and 1900 μg N_2O-N $m^{-2}h^{-1}$ (see Table 4.17).

Cultivated soils and fertilizer induced N_2O losses. Table 4.17 and Figures 4.1-4.4 show a selection of measured N_2O emission rates from cultivated soils reported in literature. Cultivated fields show a wide flux range of about 2 to 1880 μg N $m^{-2}h^{-1}$. Extremely high figures were presented by Ryden et al. (1978), Ryden et al. (1979) and Ryden and Lund (1980). A common aspect of the latter reports is that the soils were kept at field capacity during the measurements. Other extreme high emission rates are for drained organic soils and for fields receiving fertilizer applications higher than 500 kg N ha^{-1}. Such high fertilizer application levels for grasslands are rare, but may occur in Western Europe.

In Figure 4.1 the N_2O fluxes from both cropped fields and grasslands are presented. Apparently there is no correlation between the level of fertilizer application and the N_2O flux. In Figure 4.2 the data for grasslands are presented separately, with different symbols for the type of fertilizer and soil drainage. Most of the published flux data apply to poorly drained soils, conditions which favour denitrification. Apparently there is no good correlation between type or level of fertilizer application and the N_2O flux. However, if extreme fertilizer applications of over 500 kg N ha^{-1} and the high fluxes for 0 N fertilizer levels are excluded, the correlation appears to be better, but the small number of data do not permit regression analysis.

Figures 4.3a and 4.3b show the data for crops. Most of these measurements were carried out in well drained plots, conditions which favour nitrification. In cropped fields usually less than 200 kg N ha^{-1} is applied. In figure 3b the data for fertilizer applications of over 250 kg N ha^{-1} are excluded, as well as N_2O fluxes from poorly drained cropped fields. This reduces the variatiability considerably. The mathematical relation which can be calculated with the method described by Snedecor and Cochran (1980:165) from this reduced data set is:

$$N_2O = 1.878536 + 0.00417 \times N \qquad (4.6)$$

where: N_2O = N_2O emission (kg N_2O-N ha^{-1} y^{-1});
      N = N fertilizer level (kg N ha^{-1}).

Figures 4.2 and 4.3a seem to indicate that the type of fertilizer is of only little influence on N_2O fluxes.

Figure 4.4 shows the data for grass and crops combined, excluding the fertilizer applications of over 500 kg N ha^{-1} for grass, of over 250 kg N ha^{-1} for crops and data for poorly drained soils used for crops. The regression line for this data set is:

$$N_2O = 1.454114 + 0.007496 \times N \qquad (4.7)$$

Fertilizer induced N_2O losses. The data listed in Table 4.19 are graphically presented in Figure 4.5. Loss of fertilizer N as N_2O usually occurs within a few weeks after fertilization. Some authors suggest that the emission rate is higher for ammonia yielding mineral fertilizers than for nitrate (Conrad et al., 1983; Breitenbeck et al., 1980; Bremner and Blackmer, 1981). This phenomenon would be independent of the type of counter ion. Contrary to this, an effect of fertilizer type is not apparent from Table 4.18 and Figure 4.5. Other authors report lower fractions of fertilizer lost as N_2O for elevated applications than for low levels of N-fertilizer application (e.g. Breitenbeck et al. (1980).

Bolle et al. (1986) assumed average N_2O losses induced by fertilizer use of 0.04% for nitrate, 0.15-0.19% for ammonia and urea, and 5% for anhydrous ammonia. These values would be independent of climate and should thus be globally representative (Conrad et al., 1983).

Table 4.17 N_2O emission rates from grasslands and croplands in relation to the fertilizer type and application rate. The data are grouped according to the fertilizer type for grasslands and crops separately

Soil / texture	Remarks	Crop/ type	N-fertilizer applied (kg N ha⁻¹)	Flux range (μg N m⁻² h⁻¹)	Annual emission (kg N ha⁻¹ y⁻¹)	Method1	Reference
1. Grassland / grazing land							
NO FERTILIZER							
Fine loamy, mixed Typic Hapludalf ?		grazing land	0	3.8	0.3	c -	Cates & Keeney (1987b)
Clay loam		rye grass	0	14	1.3	c -	Breitenbeck et al. (1980)
Sandy loam over clay loam	stagnogley, 4.1% C	grassland	0	9 / 31	0.8 / 2.7	g -	Eggington & Smith (1986b)
Clay loam/silt loam		grass sward	0	9 / 10	0.8 / 1.0	c -	Webster & Dowdell (1982)
Sandy loam over clay loam	stagnogley, 4.1% C	rye grass	0	15 / 66	1.3 / 5.8	g -	Eggington & Smith (1986a)
Sandy loam over clay loam	stagnogley, 4.1% C	rye grass	0	2.2 / 8	0.2 / 0.7	g -	Eggington & Smith (1986a)
Organic (peat) soil		grass	0	183 / 1107	16.0 / 97.0	c -	Duxbury et al. (1982)
NO₃							
Sandy loam over clay loam	stagnogley, 4.1% C	grassland	250	30 / 128	2.6 / 11.2	g	Eggington & Smith (1986a)
Sandy loam over clay loam	stagnogley, 4.1% C	grassland	250	6 / 23	0.5 / 2.0	g	Eggington & Smith (1986a)
Loam; Typic Xerorthent	constantly wet,25oC	rye grass	300	21 / 49	1.8 / 4.3	c 2	Rolston et al. (1978)
Silt loam	2.3 % C; well drained	rye grass	400	50 / 70	4.0 / 6.0	c -	Webster & Dowdell (1982)
Clay loam	4% C;drainage restricted	rye grass	400	70 / 90	6.0 / 8.0	c -	Webster & Dowdell (1982)
Sandy loam over clay loam	stagnogley, 4.1% C	grassland	700	153	13.4	g	Eggington & Smith (1986b)
NH₄NO₃							
Heavy clay, Denchworth s. (stagnogley)	Nov79-June'80	grassland	210	7	0.6	c 1	Colbourn et al. (1984a)
Loam, ochraqualf		rye grass	250	40	3.5	o 1	Ryden (1983)
Loam, ochraqualf		rye grass	500	91	8.0	o 1	Ryden (1983)
UREA							
Loam, fine loamy mixed Haplargid		grassland	450	10	0.8	c -	Mosier et al. (1981)
ORGANIC							
Sandy loam over clay loam	stagnogley, 4.1% C	grassland	298	6 / 23	0.5 / 2.0	g	Eggington & Smith (1986a)
Sandy loam over clay loam	stagnogley, 4.1% C	grassland	700	38	3.3	g	Eggington & Smith (1986b)
Sandy loam over clay loam	stagnogley, 4.1% C	grassland	1230	30 / 128	2.6 / 11.2	g	Eggington & Smith (1986a)
2. CROPS							
NO FERTILIZER							
Loam, Ochraqualf	soil acted as a sink	weeds	0	-7	-0.6	c -	Ryden (1983)
Ustic Torriorthents		barley	0	14	1.2	c -	Mosier et al. (1982)
Silt loam		alfalfa	0	26 / 48	2.3 / 4.2	c -	Duxbury et al. (1982)
Silt loam		weeds	0	10 / 19	0.9 / 1.7	c -	Duxbury et al. (1982)
Organic (peat) soil		sugar cane	0	180 / 550	7.0 / 48.0	c -	Duxbury et al. (1982)
Organic (peat) soil		fallow	0	670 / 1880	59.0 / 165.0	c -	Duxbury et al. (1982)

Table 4.17 continued

NO_3									
Clay loam	high organic C (5%)	uncropped	125		16		1.4	c -	Breitenbeck et al. (1980)
Clay loam	high organic C (5%)	uncropped	250		16		1.4	c -	Breitenbeck et al. (1980)
Loam; Typic Xerorthent	constantly wet, 23°C	uncropped	300	7	24	0.6	2.1	c 2	Rolston et al. (1978)
Loam; Typic Xerorthent	constantly wet, 23°C	uncropped	300	62	110	5.4	9.6	c 2	Rolston et al. (1978)
Fine loamy pachic Haploxeroll	irrigated	artichokes	430	220	310	19.6	26.9	o 1	Ryden & Lund (1980)
Fine loamy pachic Haploxeroll	irrigated	lettuce-celery	620	230	480	20.2	41.8	o 1	Ryden & Lund (1980)
Fine loamy pachic Haploxeroll	irrigated	cauliflower	680	310	330	26.8	29.2	o 1	Ryden & Lund (1980)
NH_3/NH_4									
Fine montmorillonitic Aridic Argiustolls		corn	200		29		2.5	c -	Mosier & Hutchinson (1981)
Clay loam	high organic C (5%)	bare soil	125		24		2.1	c -	Breitenbeck et al. (1980)
Clay loam	high organic C (5%)	bare soil	250		27		2.4	c -	Breitenbeck et al. (1980)
Fine loamy pachic Haploxeroll	irrigated	celery	335	70	105	6.1	9.2	o 1	Ryden et al (1979)
UREA									
Clay loam	high organic C (5%)	bare soil	125		22		1.9	c -	Breitenbeck et al. (1980)
Clay loam	high organic C (5%)	bare soil	250		27		2.4	c -	Breitenbeck et al. (1980)
NH_4NO_3									
Ustic Torriorthents	sandy loam, 1% C	barley	58		25		2.2	c -	Mosier et al. (1982)
Heavy clay soil, Lawford series	ploughed, 2 yrs.		70/140	6	11	0.5	1.0	c -	Burford et al. (1981)
Heavy clay, Denchworth s. (stagnogley)	ploughed, 2 yrs.		70/140	10	64	0.9	5.6	c -	Burford et al. (1981)
Heavy clay soil, Lawford series	dir. drilled, 2 yrs.		70/140	17	24	1.5	2.1	c -	Burford et al. (1981)
Heavy clay, Denchworth s. (stagnogley)	dir. drilled, 2 yrs.		70/140	61	98	5.4	8.6	c -	Burford et al. (1981)
Ustic Torriorthents	sandy loam, 1% C	barley	112		28		2.5	c -	Mosier et al. (1982)
Silt loam		barley	132	25	33	2.2	2.9	c -	Duxbury et al. (1982)
Organic (peat) soil		corn	170	860	1740	76	152.0	c -	Duxbury et al. (1982)
Organic (peat) soil		sweet corn	170	970	820	85.0	72.0	c -	Duxbury et al. (1982)
Heavy clay, Denchworth s. (stagnogley)	Nov79-June'80	winterwheat	210	19	23	1.7	2.0	c 1	Colbourn et al. (1984a)
Ustic Torriorthents	sandy loam, 1% C	barley	224		38		3.3	c -	Mosier et al. (1982)
ORGANIC									
Ustic Torriorthents	sandy loam, 1% C	barley	71		29		2.5	c -	Mosier et al. (1982)
		corn	130	27	43		3.8	c -	Duxbury et al. (1982)
Fine loamy, mixed mesic Hapludalf	168/13 in manure/NH_4NO_3	maize	181		41	2.4	3.6	c -	Cates & Keeney (1987b)
Fine loamy, mixed mesic Hapludalf	168/13/56 in manure/NH_4NO_3/urea; maize		237		59		5.2	c -	Cates & Keeney (1987b)
Ustic Torriorthents	sandy loam, 1% C	barley	356		113		9.9	c -	Mosier et al. (1982)
Ustic Torriorthents	routinely fertilized	cropland		2	16	0.2	1.4	c -	Seiler & Conrad (1981)

1 c = closed chamber; o = open chamber; 1 = N_2 and N_2O measured (C_2H_2 inhibition); 2 = ^{15}N labelling; - = N_2 measured exclusively; g = gradient (diffusion) method.

2 the first number refers to the period November 1977-June 1978; the second figure refers to November 1978- June 1979, respectively. Fertilizer applications 70 kg N & 140 kg N as NH_4NO_3 in spring 1977 and sapring 1978 respectively. Winterwheat sown in October 1977, oilseed rape in October 1978.

Table 4.18 N$_2$O loss induced by N-fertilizer use as a % of the fertilizer gift for various types of fertilizer. The data are grouped according to the fertilizer type

Soil / texture	Remarks	Crop	N-fertilizer applied (kg N ha^{-1})	Induced loss (% of fertilizer)	Method[1]	Reference
UREA						
Silt loam	flooded	rice	90-180	0.01-0.05	c -	Smith et al (1982)
Clay loam	high organic C (5%)	bare soil	125	0.14	c -	Breitenbeck et al. (1980)
Clay loam	high organic C (5%)	bare soil	250	0.12	c -	Breitenbeck et al. (1980)
Fine loamy mixed Ustollic Haplargid		shortgrass prairie	450	0.6	c -	Mosier et al (1981)
NH$_3$						
Silt loam; fine silty mixed Pachic Ultic Haploxerolls		fallow	55	0.05	o -	Cochran et al (1981)
Silt loam; fine silty mixed Pachic Ultic Haploxerolls		fallow	110	0.07	o -	Cochran et al (1981)
Fine montmorillonitic Argiustolls	Low org.C (0.76%)	corn	200	1.25	c -	Mosier & Hutchinson (1981)
Silt loam; fine silty mixed Pachic Ultic Haploxerolls		fallow	220	0.09	o -	Cochran et al (1981)
NH$_4$						
Andosol		wheat	80	0.24	c -	Minami (1987)
Alluvial soil		wheat	80	0.18	c -	Minami (1987)
Andosol		wet rice	90	0.33	c -	Minami (1987)
Sandy clay loam	loess	grass	100	0.025	c -	Conrad et al. (1983)
Sandy loam	loess, 2-2.6 % C	meadow	100	0.03	c -	Seiler & Conrad (1981)
Pararendzina (sandy clay loam)	loess	meadow, grass	100	0.053	c -	Conrad et al. (1983)
Sandy clay loam	loess	clover	100	0.065	c -	Conrad et al. (1983)
Sandy loam/sandy clay loam	loess loam, 0.8 % C	grass	100	0.07	c -	Seiler & Conrad (1981)
Sandy loam	loess	grass	100	0.07	c -	Conrad et al. (1983)
Sand		mixed forest	100	0.09	c -	Seiler & Conrad (1981)
Brown soil, sandy loam	loess	beet, plants removed	100	0.153	c -	Conrad et al. (1983)
Brown soil, sandy loam	loess	beet, plants removed	100	0.216	c -	Conrad et al. (1983)
Alluvial soil		wet rice	100	0.33	c -	Minami (1987)
Sandy clay loam	loess	grass	100	0.376	c -	Conrad et al. (1983)
Andosol		wet rice	100	0.55	c -	Minami (1987)
Andosol		rape	150	0.09	c -	Minami (1987)
Alluvial soil		rape	150	0.06	c -	Minami (1987)
Clay loam	high organic C (5%)	bare soil	125	0.18	c -	Breitenbeck et al. (1980)
Clay loam	high organic C (5%)	bare soil	250	0.12	c -	Breitenbeck et al. (1980)
Clay; Oxisols (Latosolo amarelo)		tropical forest	200	0.1	c -	Keller et al (1988)
Andosol		carrot	200	0.26	c -	Minami (1987)
Alluvial soil		carrot	200	0.31	c -	Minami (1987)
Fine montmorillonitic aridic Argiustolls; clay loam		barley	200	0.4	c 2	Mosier et al. (1986)
Fine montmorillonitic Aridic Argiustolls	irrigated	corn	200	1.3	c -	Hutchinson & Mosier (1979)
Fine montmorillonitic aridic Argiustolls; clay loam		corn	200	1.5	c 2	Mosier et al. (1986)

Table 4.18 continued

NH₄NO₃

Soil	Amendment	Vegetation	Rate	Flux	Method[1]	Reference
Ustic Torriorthents	sandy loam, 1% C	barley	56	0.7	c -	Mosier et al. (1982)
Ustic Torriorthents	sandy loam, 1% C	barley	112	0.4	c -	Mosier et al. (1982)
Ustic Torriorthents	sandy loam, 1% C	barley	224	0.4	c -	Mosier et al. (1982)
Loam; Ochraqualf		grassland	250	1.3	o 1	Ryden (1981)
Loam; Ochraqualf		rye grass	250	1.4[2]	o 1	Ryden (1983)
Sandy loam			336	0.25[2]		McKenney et al (1980)
Loam; Ochraqualf		rye grass	500	1.6[2]	o 1	Ryden (1983)

NO₃

Soil	Amendment	Vegetation	Rate	Flux	Method[1]	Reference
Sandy clay loam	loess	clover	100	0.001	c -	Conrad et al. (1983)
Sandy clay loam	loess	grass	100	0.007	c -	Conrad et al. (1983)
Sandy loam/sandy clay loam	loess loam, 0.8 % C	grass	100	0.01	c -	Seiler & Conrad (1981)
Sand		mixed forest	100	0.01	c -	Seiler & Conrad (1981)
Sandy loam	loess	grass	100	0.017	c -	Conrad et al. (1983)
Brown soil, sandy loam	loess	beet, plants removed	100	0.018	c -	Conrad et al. (1983)
Sandy loam	loess, 2-2.6 % C	meadow	100	0.05	c -	Seiler & Conrad (1981)
Sandy clay loam	loess	grass	100	0.071	c -	Conrad et al. (1983)
Pararendzina (sandy clay loam)	loess	meadow, grass	100	0.073	c -	Conrad et al. (1983)
Clay loam	high organic C (5%)	bare soil	125	0.04	c -	Breitenbeck et al (1980)
Loamy sand		grassland	200	0.11	c -	Armstrong (1983)
Clay loam		grassland	200	0.15	c -	Armstrong (1983)
Clay; Oxisols (Latosolo amarelo)		tropical forest	200	0.5	c -	Keller et al (1988)
Clay loam	high organic C (5%)	bare soil	250	0.01	c -	Breitenbeck et al (1980)
Sandy loam over sandy clay loam	stagnogley,4.1% C	grassland	500[3]	0.32-1.34[2]	g	Eggington & Smith (1986a)

ORGANIC MANURES

Soil	Amendment	Vegetation	Rate	Flux	Method[1]	Reference
Fine loamy mixed Typic Hapludalf	168/13 kg N from manure/NH₄NO₃	maize	181	1.80[2]	c -	Cates & Keeney (1987b)
Fine loamy mixed Typic Hapludalf	168/13/56 kg N from manure/NH₄NO₃/urea	maize	237	2.05[2]	c -	Cates & Keeney (1987b)
Ustic Torriorthents	sandy loam, 1% C	barley	71	0.8	c -	Mosier et al. (1982)
Ustic Torriorthents	sandy loam, 1% C	barley	356[3]	1	c -	Mosier et al. (1982)
Sandy loam over sandy clay loam	stagnogley,4.1% C	grassland	1528[3]	0.01-0.07[2]	g	Eggington & Smith (1986a)

[1] c = closed chamber; o = open chamber; 1 = N_2 and N_2O measured (C_2H_2 inhibition); 2 = ^{15}N labelling; - = N_2 measured exclusively; g = gradient (diffusion) method.
[2] calculated with the data provided by the author(s).
[3] fertilizer gifts spread over 2 years (Eggington & Smith, 1986a; Eggington & Smith, 1986b).

Global estimate of the N_2O emission from cultivated fields. Bolle et al. (1986) assumed average loss rates of 0.5-2% based on fertilizer consumption figures.

The solubility of N_2O in water is relatively high. Considerable N_2O fluxes may occur from surface water draining from fertilized agricultural fields (Dowdell et al., 1979) and even cleared forests (Bowden and Bormann, 1979). Minami and Fukushi (1984) measured losses as N_2O of 200 to 1190 μg N m^{-2}h^{-1} in drainage water when calculated for the total draining surface. Minami and Fukushi also demonstrated that at low N_2O concentration standing water in rice paddies may act as a sink for N_2O. Amounts comparable to the N_2O gas emission may be lost through denitrification / nitrification of mineral fertilizers leaching from fields into groundwater or surface freshwater ecosystems (Bolle et al., 1986).

The assumption that the amount of N_2O emitted from ground water or surface waters enriched with N-fertilizer leached from cultivated lands, is equal to direct fluxes from cultivated lands, yields a total N_2O loss if 1-4% of the global N-fertilizer consumption. In 1986/78 this was 72 Tg N (FAO, 1988) and the total N_2O emission due to application of nitrogen fertilizers is thus 0.7 to 3.0 Tg N y^{-1} (based on Bolle et al., 1986).

However, the conditions under which N_2O is lost are as variable as the edaphic conditions in cultivated fields (Tables 4.17, 4.18; Figures 4.1-4.5). The doses in which fertilizer N are applied are usually less than 200 kg N ha^{-1}. At lower levels of application the fraction lost shows a lower range of 0.1 ± 0.08% independent of the type of fertilizer. In that

Table 4.19 Regional nitrogenous fertilizer consumption, land use and N_2O emissions (in Tg y-1)

	Total N use (10³ kg)	Arable land[1] 1000 ha	Pasture 1000 ha	kg N ha-1 Arable land[1]	kg N ha-1 Agric. land[2]	N_2O flux[4] (eq. 1 for crops) arable land area[1]
Africa	2,010,032	183,214	779,134	10.9	2.1	0.24 - 0.46
N. & C. America	12,533,670	274,122	359,590	45.8	19.5	0.45 - 0.68
S. America	1,835,648	138,878	456,431	13.0	3.0	0.19 - 0.35
Asia[3]	23,398,400	455,220	644,349	62.9	25.1	0.83 - 1.12
W Europe	11,161,531	87,079	64,054	119.0	68.6	0.20 - 0.21
E Europe[3]	4,549,493	53,476	21,504	85.1	60.7	0.11 - 0.13
USSR	11,475,000	232,290	373,133	49.4	18.9	0.39 - 0.58
Oceania	406,300	48,182	455,157	8.1	0.8	0.06 - 0.12
World	72,370,080	1,472,401	3,153,352	49.1	15.4	2.29 - 3.65

Sources: N-fertilizer consumption: FAO fertilizer yearbook vol. 34, 1984; Land use data from World Resources Institute (1988).
[1] arable land includes permanent crops; [2] agricultural area includes arable land, permanent crops and permanent pastures; [3] excluding USSR; [4] the N_2O flux is calculated with equation 1 on the basis of areas and fertilizer consumption for arable land.

case the 1987 loss induced by global N-fertilizer use would be 0.02 to
0.15 Tg y^{-1}. In addition to this N_2O loss there is a background or natural
N_2O emission from the cultivated lands, which amounts to 1 to 2 kg
N_2O-N ha^{-1} y^{-1} (see Figure 4.1). With a global cultivated area of 1500x10^6
ha the total background emission would be 1.5 to 3 Tg and addition to
the fertilizer induced loss yields 1.52 to 3.15 Tg N_2O-N y^{-1}. The
background fluxes may deviate considerably from the above rates,
making this estimate rather uncertain.

Another approach is estimation of the N_2O fluxes from the N-
fertilizer consumption per unit area. For this purpose the regional
fertilizer use is listed in Table 4.19. Equation 2 (not used in Table 4.19),
used with data for the total agricultural area (arable land, permanent
crops and pastures), yields a global annual N_2O emission of 5.6 (4.5-
6.8) Tg N_2O-N. It must be realized that much of the area included in the
estimate for permanent pastures is natural, uncultivated land. Fluxes for
these lands may be much lower than for cultivated lands and applying
equation 2 would result in an overestimate.

The last column in Table 4.19 shows regional estimates of the N_2O
emission from arable land plus permanent crops estimated with equation
1 and on the basis of fertilizer consumption for the arable land area.
This method probably gives a better estimate of global N_2O emission
from cultivated lands, since crops globally receive higher doses of
fertilizer except for Western Europe, where grasslands may also receive
very high N-applications. The resulting mean global loss estimated in
Table 4.19 from cultivated land is 3.0 (2.3-3.7) Tg N_2O-N. It is probably
more meaningful than the above figure and it correlates well with the
estimate derived from the fertilizer induced N_2O losses.

The analysis presented here does not consider use of organic manures, which in many
agricultral systems may be the sole source of nutrients. Moreover, the use of regional
averages of N-fertilizer usage is in actual fact an infringement of all statistical rules.
Geographically oriented data bases of soils, land use, fertilizer application and climate
(also suggested by Matthews, this volume) would imply significant improvements
compared to the presented simplified approach.

4.5.8 Other Sources of N_2O

Increasing emissions of N_2O due to the expansion of the global
agricultural area may be significant although natural N_2O fluxes from
tropical rain forests (where most clearing and conversion to cropland is
taking place) is comparable to emissions from arable land. However,
recent experiments indicate that tropical forest land cleared for pasture
produces much more N_2O than the original forest (Vitousek and Matson,
1989, pers. comm.). It is not certain however, whether these elevated
fluxes will be maintained over prolonged periods of time.

4.5.9 Global Sources and Sinks of Nitrous Oxide

Most studies presented in the previous sections are very useful to gain
insight in the processes of biological and chemical denitrification and

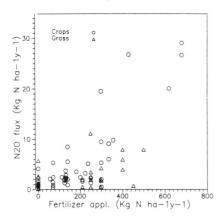

Figure 4.1. The effect of the level of N-fertilizer application on N₂O fluxes from cultivated fields. In this figure all the available data for crops and grass are presented.

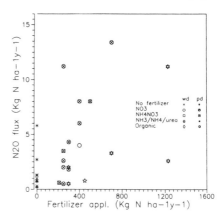

Figure 4.2. The effect of the level and type of N-fertilizer application on N₂O fluxes for grasslands only. For an explanation of the symbols see Figure 4.3a and 4.5.

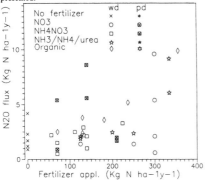

Figure 4.3a. The effect of the level and type of N-fertilizer application on N₂O fluxes for crops only. wd = well drained soils; pd = poorly drained soils.

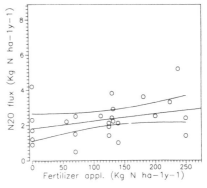

Figure 4.3b. The effect of the level of N-fertilizer application on N₂O fluxes for crops only. The data for fertilizer application of over 250 kg N/ha and data for poorly drained soils are excluded. The curves show regression lines with confidence interval (according to Snedecor and Cochran, 1980: 165) on the basis of this reduced data set.

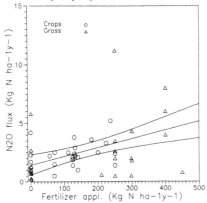

Figure 4.4. The effect of the level of N-fertilizer application on N₂O fluxes for crops and grass. Data for fertilizer applications of over 500 kg N/ha for grass, of over 250 kg N/ha for crops and data for poorly drained cropped soils are excluded. A regression line with confidence interval (according to Snedecor and Cochran, 1980: 165) is drawn on the basis of this reduced data set.

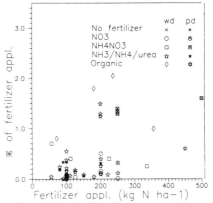

Figure 4.5. The effect of fertilizer level on the N₂O loss as % of the N fertilizer application. wd = well drained soils; pd = poorly drained soils.

nitrification in soils. However, field conditions are so variable and influenced by so many factors, that the presented data on N_2O fluxes from soils are very difficult to interpret. For extrapolations to smaller scales and to a global scale the available data must be accompanied by models which describe the controlling factors of N_2O fluxes. At present however, the limited knowledge of individual sources does not permit to construct a reliable budget of sources.

The total global emission from all possible sources including oceans and freshwater, fossil fuel burning and lightning is 12 - 15 Tg N y^{-1} (Bolle et al., 1986), to 14 ± 7 Tg N y^{-1} (Seiler and Crutzen, 1987). An example of one budget is given in Table 4.15. There is evidence that tropical forests are the major N_2O source. Kaplan (1984) and McElroy and Wofsy (1986) suggest that tropical forests, in particular tropical wet and moist forests, are emit 7 to 8 Tg N_2O-N annually. The use of fossil fuel as a source of N_2O is uncertain since it was reported that measurements made so far are incorrect. The fossil fuel combustion produces considerably less than the 1.5 ± 1 Tg N_2O-N y^{-1} presented in Table 4.14 (Muzio and Kramlich, 1988). The analysis in this chapter suggests that cultivated fields emit 2.4 to 3.7 Tg N2O-N y-1. The above figures would make the budget presented in Table 4.15 look quite different.

Another important source of N_2O is the burning of biomass, such as the burning of agricultural wastes, savanna fires and burning in shifting cultivation. This source may add 1-2 Tg N y^{-1} to the atmosphere (Crutzen, 1983). This is a tentative estimate, since N_2O release rates, quantities of biomass burned annually and conditions under which biomass is burned are not well known. The increasing populations in developing countries will probably cause an overall increase of this source in the future (see also 4.2.9).

4.6 NITRIC OXIDE AND NITROGEN DIOXIDE

4.6.1 General

Nitric oxide (NO) and nitrogen dioxide (NO_2) have no absorption bands in the infrared part of the spectrum. They may however, be involved in a number of atmospheric reactions, which affect the concentrations of other gases contributing to greenhouse warming. Both gases catalyze the ozone destruction in the atmosphere and they are involved in the oxidation of CH_4 and CO (see 2.2.2).

The main sources of NO_x are combustion of fossil fuels (40%), biomass burning (25%), the balance coming from lightning and microbial activity in soils. The global sources are listed in Table 4.20.

4.6.2 Production of NO in Soils

Three mechanisms may lead to production of NO in soils: microbial denitrification and nitrification, and chemodenitrification. Photolysis of nitrite has been suggested as a mechanism for NO production in flooded rice soils (Galbally et al., 1987). It is uncertain which mechanism is responsible for NO or N_2O production under field conditions. There is however evidence that nitrification is the main process in many soils (e.g. Slemr and Seiler, 1984; Johansson and Granat, 1984; Anderson and Levine, 1987; Kaplan et al., 1988; Johansson et al., 1988). The NO/N_2O production ratio is 1-5 by nitrifying bacteria and 0.01 in denitrification (Lipschultz et al., 1981; Anderson and Levine, 1986).

Table 4.20 Global tropospheric sources of NO_x (all figures in Tg N y^{-1})

Surface sources	Mean	Range
Fossil fuel combustion	21	14 - 28
Biomass burning	5.1	3.6- 6.7[*]
Nitrification/denitrification in soils	8	4 - 16[•]
Atmospheric sources		
Lightning	8	2 - 20[§]
NH_3 oxidation		1 - 10[‡]
From stratosphere	0.5	
Jet aircraft	0.25[§]	
Total	50	25 - 99

Unless indicated otherwise the data are from Logan (1983); [*] Galbally (1985); [•] Levine et al. (1984) estimated biogenic production at 10 Tg N y^{-1}; [§] Levine et al. (1984) gave a source of 1.8 T to 18 Tg N y^{-1}; [‡] Crutzen (1983) estimated that 10% of all atmospheric NH_3 is oxidized, i.e. 12-15 Tg N y^{-1}.

4.6.3 Emission Rates of NO

The extent of biogenic NO_x production is highly uncertain. Lipschultz et al. (1981) estimated the biogenic NO source at 15 Tg N y^{-1} on the basis of laboratory experiments with nitrifying bacteria found in soils. Levine et al. (1984) used both nitrifying and denitrifying bacteria. They calculated, that soil nitrifiers rather than denitrifiers are the major source of NO and NO_2. Assuming a production ratio of $NO:NO_2 = 2:1$ at low oxygen levels as usual in soils, and a global N_2O production of 5 Tg N y^{-1}, the estimated NO emission is 10 Tg N y^{-1}. The authors call their estimate conservative. Using the global N_2O production from Table 4.15, the NO_x production would be much higher. The production ratio of $NO:N_2O$ is a highly uncertain factor in all such calculations.

So far the number of NO measurements is low compared with the data set available for N_2O. In Table 4.21 a list of emission rates is presented. Most measurements show that NO from soils may contribute a large portion to the global source of NO_x in the atmosphere. Table 4.21 shows that site to site measurements differ by more than an order of magnitude. In some areas a net uptake of NO_x was measured. Most of this variation can be explained by extreme spatial, temporal and diurnal variability and possibly also by difficulties of measurement techniques

Table 4.21 NO emission rates from cultivated and natural land in relation to climate and fertilizer treatment

Soil / texture	Crop/ecosystem	Treatment	Flux (μg N m^{-2}h^{-1}) Range	Mean	Reference
Cultivated land					
1. Temperate climates					
	cropland, Sweden	unfertilized	1.1	61.2	2.2	Johansson and Granat (1984)
	cropland, barley	120 kg N ha^{-1}	36.0	68.0		Johansson and Granat (1984)
	cropland, USA	–	-32.4	100.8		Delany et al. (1986)
	cropland, USA	unfertilized	0.01	241.2	6.1	Anderson and Levine (1986)
	cropland, USA	fertilized			23.8	Anderson and Levine (1986)
	cropland, wheat	70 kg N ha^{-1}	14.0	163.0	54.0	Duffy et al. (1988)[*]
	cropland, sunflowers	80 kg N ha^{-1}	0	234.0	20.0	Duffy et al. (1988)[*]
	cropland, bare soil	70 kg N ha^{-1}	-37	497.0	94.0	Duffy et al. (1988)[*]
Flooded soil	flooded rice, Australia	80 kg N ha^{-1} (urea)	0.7	3.4		Galbally et al. (1987)
	grassland, Sweden	200 kg N ha^{-1} grazed	40.0	61.0		Johansson and Granat (1984)
Various soils	grassland,Australia		5.4	26.3	12.6	Galbally and Roy (1978)
	grassland, Germany	100 kg N ha^{-1}	-50	500.0	140.0	Slemr and Seiler (1984)
	pasture, U.K.	fertilized	0	129.6	28.8	Colbourn et al. (1987)
	sward, U.K.	unfertilized	-43.2	93.6	1.8	Colbourn et al. (1987)
2. Subtropical climates					
	bare soil, Spain	100 kg N ha^{-1}	0	12600	796	Slemr and Seiler (1984)

Table 4.21 continued

Natural / uncultivated land

1. Temperate climates
Various soils

ungrazed grassland, Australia	2.2	9.4	5.8	Galbally and Roy (1978)
coniferous forest, Sweden	0.4	2.9	1.4	Johansson (1984)
grassland, Germany	-21.6	50.4	7.9	Slemr and Seiler (1984)
grassland, USA	0.1	234	10.8	Williams et al. (1987)

2. Tropical climates

Oxisol(Latosolo amarelo) rainforest, rainy season	33.1	57.6	39.6	Kaplan et al. (1988)
savannna, dry season	10.8	54.0	28.8	Johansson et al. (1988)
savanna, wet season	7.2	900		Johansson and Sanhueza (1988)
savanna, 1 year	34.0	148	36-234	Johansson and Sanhueza (1988)
tropical cloud forest	0.4	7.2	1.8	Johansson et al. (1988)

* unpublished results of Galbally in: Duffy et al.,1988).

and timing of measurements. From the figures in Table 4.21 it becomes clear that fluxes in the tropics are 3 to 30 times higher than in temperate ecosystems. As for N_2O fluxes, rates are highest during wet periods (Johansson and Sanhueza, 1988).

4.6.4 NO_x Deposition Rates

The process of NO_x deposition and uptake of NO_x by plants is not well understood so far. Probably both processes occur at a local scale only. The deposition rates show a large scatter. Since NO_x fluxes are important at a local to regional scale, a global budget cannot be made. Galbally and Gillett (1988) estimated that NO_x is deposited within 1000 to 2000 km from the source area, in the wet and dry season, respectively. Tentative regional budgets for a number of continents for NO_x were compiled by the latter authors.

4.7 AMMONIA

Ammonia (NH_3) can absorb infrared radiation, but its role in the radiative balance is not significant due to its short atmospheric residence time. It is known however, as a major air pollutant with a major role in the process of soil acidification.

4.7.1 Sources of NH_3

The total global annual production of NH_3 is in the order of 117 to 150 Tg N. Individual sources of NH_3 are animal excretion, natural soils, mineral fertilizer use, biomass burning, coal burning and emissions during industrial N-fertilizer production. These will be discussed in short below.

Industrial emissions. These emissions occur mainly during the production of ammonia and fertilizers. Emission rates for NH_3 fertilizer production are 0.7 kg N/ton NH_3 produced. For NPK fertilizer production the emission is 10 kg N/ton N produced. The global annual N-fertilizer production is 74 Tg N; with an assumed average emission of 4 kg N/ton fertilizer N, the global emission amounts 29×10^{10} g N y^{-1} as NH_3.
Coal burning. On the basis of an emission factor of 2×10^3 g N per ton of coal and an annual use of coal of 3000 Tg, the annual emission is 4 -12 Tg N (Soderlund and Svensson, 1976).

Biomass burning. Nitrogen in biomass is present in a reduced form largely as proteins. Crutzen (1983) assuming an N-content of 1% estimated the maximum NH_3 emission at 60 T to 70 Tg y^{-1}. This figure is highly uncertain and represents only the upper limit.
N-fertilizer use. About 30% of all N-fertilizer is produced as urea and 20% as salts of ammonia. The rest consists of various compound and mixed fertilizers. NH_3 emission depends on such factors as urease activity, soil temperature, soil moisture content, soil pH, rate of absorption at the cation exchange complex, wind speed, type of fertilizer and way of application. Emission generally decreases in the order $(NH_4)_2SO_4$, NH_4NO_3, $(NH_4)_3PO_4$. Since measurement of the NH_3 emission from fertilized fields is difficult,

Crutzen (1983) assumed an average emission factor of 5% giving a global annual emission of 3.7 Tg N.

Natural soils. Denmead et al. (1976) showed, that in a situation of natural vegetation the leaves absorb almost all NH_3 emitted from the soil surface. There are no data for NH_3 emission from natural soils.

Animals. In areas with intensive animal husbandry the emissions of NH_3 have a local effect predominantly. Important NH_3 volatilization occurs both directly from deposited urine in the fields by grazing animals (Hutchinson et al., 1982) and also from the returns of stored slurry from housed animals. Only a minor part is transported through the troposphere and most of the NH_3 is redeposited (dry and wet deposition). Soderlund and Svensson (1976) estimate the total human and animal NH_3 emission at 20 T to 35 Tg N y^{-1}.

Plants. Evolution of ammonia and amines from senescing plants has been reported by several authors. O'Deen and Porter (1986) found a crop loss of 1.6 kg N ha^{-1} accounting for 3.1% of the total N in the plant material. Too few data are available to make a safe global estimate, however.

4.7.2 Sinks of NH_3

Crutzen (1983) estimated, that about 10% of all atmospheric NH_3 (12 T to 15 Tg N) reacts with OH. to forming NO or NO_2. The annual dry deposition is 72 to 151 Tg and wet deposition amounts to 30 to 60 Tg N (Søderlund and Svensson, 1976).

4.8 CONCLUDING REMARKS SECTIONS 4.4, 4.5, 4.6 and 4.7

The production of nitrogenous oxides in soils is highly dependent on environmental conditions. Diurnal, seasonal and spatial variability have been reported to be extremely high. From all the presented data much can be said about the degree of variability. However, estimates of source strength of ecosystems cannot be given with much reliability.

Extrapolation of data obtained with current measuring methods to smaller scales is fraught with potential errors. Current measurements are point measurements resulting from process studies mainly. To give statistically reliable estimates of fluxes from larger areas, other measurement techniques are available, such as remote sensing and micrometeorological techniques.

All flux measurements should be accompanied by fundamental studies of the controlling factors of N_2O emissions. Much information can be deduced from the spatial variability of soil conditions such as climate and soil drainage. The relation between soil morphometric properties, soil taxonomic classes and fluxes should be given more attention to improve estimates of these fluxes.

Ammonia is only of local importance. Its effect on greenhouse warming is globally unimportant due to its short atmospheric residence time.

4.9 TECHNIQUES FOR MEASUREMENT OF TRACE GAS FLUXES

Mosier (this volume) and Andreae and Schimel (1989) present complete descriptions of all measurement techniques currently available or in development. A very short summary is given here.

With regard to instrumentation in the measurements of trace gas fluxes the following division can be made:

- chamber techniques;
- tower techniques;
- aircraft and satellite techniques;
- isotope methods;
- mass balance and soil gas gradient methods.

In the chamber technique a sealed box is placed on or inserted into soil. The confined atmosphere above the soil is sampled and its composition determined. For N_2O this method can be combined with acethylene (C_2H_2) inhibition of N_2O reduction in soil to measure total N-loss ($N_2 + N_2O$) or with isotope techniques. Chambers may be closed, vented or with a continuous air flow. Often the box is placed over the soil surface for short periods only to prevent errors occurring in open or closed systems. During the past few years measurements of trace gas fluxes have been made almost exclusively with chamber techniques.

Chamber methods are used for point measurements of fluxes. They yield much information on the underlying processes of trace gas fluxes. Since the spatial variability of trace gas fluxes is extremely high, chamber methods apply to very small areas only. Depending on the degree of variability and the size of the chamber the coverage is up to 10 m. They can be used repetitively to increase temporal coverage and in array to extend spatial coverage.

For both tower and aircraft techniques the following micrometeorological methods are available at present:

- For vertical fluxes: eddy correlation, flux gradient methods, Bowen ratio (energy balance) and eddy accumulation. The backgrounds of these methods are discussed in detail in Andreae and Schimel (1989).
- For measuring horizontal fluxes the flux from an emission area is measured across a vertical plane. It can be considered as a plume measurement (Denmead, 1983).

Tower measurements apply to spatial scales of up to 100 m. Towers enable measurements over prolonged time periods, but their disadvantage is their immobility. Fundamentally towers are suitable for process studies only.

Aircrafts address spatial scales of 10-100 km, but over short time periods (up to 1 h). Aircrafts are costly in use. Examples of aircraft measurements are MacPherson et al. (1987); Austin et al. (1987).

Other methods for measuring gaseous losses in general from soils are isotope methods, soil gas concentration gradient methods and mass balance methods:

- Direct measurement of ^{15}N-gaseous emission: in this method the radio-isotope ^{15}N is used to label nitrogen additions to the soil and measure the ^{15}N labelled emitted gas. Direct methods using ^{15}N can be used only where substrate for denitrification is added at a high level of ^{15}N enrichment and require accumulation of evolved gases into a confined atmosphere. The costs of this method are high, but its sensitivity of estimating N-loss is greater as compared to methods not using N-tracer (Hauck, 1986).

- Nitrogen balance: deduction of denitrification losses from the balance of a nitrogen budget, accounting for crop uptake, soil residue and leaching. This method again needs ^{15}N and has the same characteristics as the first.
- Soil gas concentration gradients: measurement of vertical concentration gradients in the soil profile and diffusion coefficients of N_2O. The data presented by Seiler and Conrad (1981) and Denmead et al. (1979) suggest however, that N_2O production takes place at or close to the soil surface and that vertical N_2O profiles cannot be used for reliable determination of N_2O emission rates.

ACKNOWLEDGEMENTS

I am indebted to R.A. Houghton (Woods Hole Research Center, Woods Hole, Massachusetts, USA), W.H. Schlesinger (Department of Botany, Duke University, Durham, USA) and J.A. van Veen (Ital, Wageningen, The Netherlands) for commenting on earlier versions of section 4.1, R.J. Cicerone (Atmospheric Chemistry Division, NCAR, Boulder, USA), R.D. Delaune (Laboratory for Wetland Soils and Sediments, Baton Rouge, Louisiana, USA), R.C. Harriss (NASA Langley Research Center, Hampton, Virginia, USA), M.A.K. Khalil (Institute of Atmospheric Sciences, Oregon Graduate Center, Beaverton, USA) and K. Minami (National Institute of Agro-Environmental Sciences, Ibaraki, Japan) for their suggestions for improvement of earlier versions of section 4.2 and J. Arah (Edinburgh School of Agriculture, UK), C.C. Delwiche (Land Air and Water Resources, University of California, Davis, USA), J. Freney (CSIRO, Canberra, Australia), S.C. Jarvis (Institute for Grassland and Animal Production, Hurley, UK), J.C. Katyal (International Fertilizer Development Center, Muscle Shoals, Alabama, USA), M. Kimura (Faculty of Agriculture, Nagoya University, Japan), K. Minami (National Institute of Agro-Environmental Sciences, Ibaraki, Japan), R.D. Keeney (University of Wisconsin, Madison, USA), P.A. Leffelaar (Department of Theoretical Production Ecology, Wageningen Agricultural University, The Netherlands), A.R. Mosier (USDA Agricultural Research Service, Fort Collins, Colorado, USA) and K.L. Sahrawat (ICRISAT, Patancheru, India) for their valuable comments on earlier versions of section 4.5. I also wish to thank Hans van Baren and Piet van Reeuwijk of ISRIC, and Maurits LaRivière (IHE, Delft, The Netherlands) for critically reading the manuscripts of sections 4.1, 4.5-4.9 and 4.2-4.4, respectively.

CHAPTER 5

Estimating the Effect of Changing Land Use on Transpiration and Evaporation

A.F. BOUWMAN

International Soil Reference and Information Centre (ISRIC)
P.O. Box 353, 6700 AJ Wageningen, The Netherlands

ABSTRACT

A number of methods for studying or predicting evapotranspiration are discussed. These include regional water balance studies, the Penman-Monteith equation, the Priestley-Taylor and the Makkink equations, wet canopy evaporation models and models of the Planetary Boundary Layer (PBL). PBL models can be used to predict the saturation vapour deficit after land cover changes. Coupling of PBL models with e.g. the Penman-Monteith equation to estimate the latent heat flux, allows for studying the influence of land cover changes on vapour and heat exchange processes and on the stability of the lower atmosphere.

Denudation of once forested land causes an increase of the ratio of sensible heat flux to latent heat flux. The ultimate result of this change in the energy balance is that water originally lost through evapotranspiration, now has to drain superficially. This will cause an increase of the erosion hazard. In addition, less vapour will remain to evaporate into the atmosphere and less vapour will be available for cloud formation. Hence, massive land denudation will provide a positive feedback to any tendency to aridity, it will imbalance local rainfall and the global energy balance.

5.1 INTRODUCTION

Water vapour (H_2O) has strong absorption bands in the infrared region of the spectrum and is therefore one of the major greenhouse gases. Apart from its absorption of infrared radiation in the atmosphere, water vapour is of eminent importance in cloud formation. Clouds affect the radiation balance and, moreover, water vapour plays an important role in the energy transport between equatorial and temperate regions.

There is concern about the effects of land use changes taking place on the rainfall regime and the hydrology of large river systems. The role of vegetation, in particular forests, in the hydrologic cycle is multiple. Forests protect soils against rainfall (and wind) erosion, and their positive effect on soil structure and the sponge effect of the forest litter layer cause high rates of infiltration. Runoff is usually low in forests.

Transpiration and evaporation close, on a global scale, the loop in the hydrologic cycle over land by returning an amount equal to about 20%

Soils and the Greenhouse Effect. Edited by A.F. Bouwman
© 1990 John Wiley & Sons Ltd.

of the solar energy absorbed in the atmosphere plus surface over land (Dickinson, 1986).

The water balance equation for a catchment can be written in general form as (according to e.g. Pereira, 1973):

$$E = R - Q - dS - dG - L \qquad (5.1)$$

where: E = evapotranspiration
       R = precipitation
       Q = runoff
       dS = change in water stored in the root zone
       dG = change in water stored below the root zone
       L = net loss of ground water other than the runoff

Assuming that in the long run there are no changes in water storage in the root zone and ground water, three variables remain to be quantified: the evapotranspiration, runoff and rainfall. This simplification shows that there are two approaches of studying the effects of land cover changes on the hydrologic cycle, in the first place by considering the water balance of a catchment, whereas the second approach involves analysis of the surface energy balance. The aim of this paper is to provide background information on the various methods of predicting transpiration and evaporation and to review the effect of changing land use on climate through changing evapotranspiration.

5.2 WATER BALANCE STUDIES

One way of examining changes in the hydrologic cycle caused by changes in vegetation or land use is the measurement of precipitation and runoff in water balance studies. The analysis of the incidence of floods and extreme discharges also provides insight in changes in the hydrologic cycle resulting from shifts in land use.

Deforestation has generally been held responsible for increases in both the water yield caused by lower evaporation losses and in the flood peaks due to increased storm flows. Many studies indicate that the water outflow is increased after conversion of forest to annual cropping, particularly in areas of high precipitation (Russel, 1981; Edwards and Blackie, 1981). In a review of 94 catchment experiments Bosch and Hewlett (1982) concluded that: (i) coniferous and eucalypt forests give an increase of about 40 mm in water yield per 10% reduction in tree coverage; (ii) deciduous hardwoods give a 25 mm increase in water yield per 10% reduction in coverage; (iii) the effect of clear-cutting of forests is greatest in high rainfall areas, but it is shorter due to more rapid regrowth. Calder and Newson (1979) found that, using a simple model and assuming 50% canopy coverage, the water yield is reduced after afforestation. This reduction in water yield would be greatest in areas of high rainfall, but as surface runoff is also high in such areas, the difference in water outflow resulting from afforestation would be comparatively small.

Deforestation does not increase the water yield under all circumstances. Replacing rain forest by tea in Kenya resulted in reduced water use, but the water yield not influenced (Edwards and Blackie, 1981). Rakhmanov (1970b) used data for 53 large river basins in the USSR and found that as the proportion of remaining forested area increased, there was a strong positive trend in the water yield. Similar results were obtained from studies in other parts of the USSR (Rakhmanov, 1970a) and the USA (Lull and Sopper, 1966).

For large catchments a reduced forest coverage does not always lead to reduced evapotranspiration. For example, Morton (1984) compared evapotranspiration for a number of large basins in agricultural areas with those for forested areas in two regions in Canada and concluded that there was no significant difference between water budget estimates of evapotranspiration (estimated as rainfall minus runoff) for agricultural and forested basins. A problem with large catchments, however, is that it is difficult to control treatments, and to estimate rainfall and streamflow accurately (Bosch and Hewlett, 1982).

Forest clear-cutting appears to result in some cases in an increase in water table (e.g. Peck and Williamson, 1987; Burch et al., 1987) until the regrowing forest has resumed pre-cutting levels of interception and evapotranspiration. The effect of forest cutting also depends on the process of clearing - through its effect on soil infiltration rates (Lal, 1981) - and by the type of vegetation replacing the forest: for example infiltration rates of rainwater in pasture may be considerably lower than in primary forests (Salati et al., 1983).

An interesting case for discussion is the Amazon basin, where deforestation is continuing at a high rate for some decades. The removal of the Amazon jungle can be expected to produce extreme local and regional climatic effects. The area of the Amazon Basin is 600 to 700 million ha and its water outflow is 5.5×10^{12} m^3 y^{-1}, which constitutes about 20% of the Earth's total freshwater supply (Salati and Vose, 1984).

Gentry and Parody (1980) reported an increasing height of the flood crest of the river Amazon at Iquitos during the period 1970-1978. During that period there had been no significant changes in precipitation, and they concluded that the change in the Amazonian water balance was the result of increased runoff caused by deforestation in the Peruan and Ecuadorian part of the Amazon basin. Their conclusions were debated by Nordin and Meade (1982), who reported that for the Rio Negro at Manaus (Brazil) during 1942-1956 river flood stages were high, between 1957 and 1969 average flood stages were low, while between 1970 and 1979 the measured flood stages were high again. Their conclusion was, that statistically no increased runoff and consequent flood stages have occurred at Manaus in that period.

5.3 TRANSPIRATION AND WET CANOPY EVAPORATION MODELS

5.3.1 Transpiration

Transpiration or dry canopy evaporation is regulated by the opening and closing of stomata. The stomata respond not only to internal stress caused by constraints in the water supply from the soil, but also to external environmental factors such as solar radiation, temperature, vapour pressure deficit and atmospheric carbon dioxide concentration. In the Appendix a description is presented of the *Penman-Monteith* equation, the *Priestley-Taylor* equation and the *Makkink* equation for estimating transpiration. A more general discussion is given below.

Penman-Monteith equation. In the Penman-Monteith equation the vegetation layer is described in a very simple way, i.e. the canopy is considered as one 'big leaf' to which a canopy resistance r_c is assigned. This big leaf has the same albedo and surface roughness as the actual vegetation. The canopy resistance r_c is a complex function of net radiation, saturation deficit and other parameters such as soil moisture (see Appendix). The value of r_c cannot be measured directly and it is often obtained through the Penman-Monteith equation by measuring all other quantities. The value of the aerodynamic resistance r_a must also be estimated, so that values of r_c may be biased due to errors in r_a.

An important consequence of the interrelation between the latent heat flux density M_{LH} and the vapour pressure deficit $\delta(e)$ (see Appendix) is that the Penman-Monteith equation does not express the latent heat flux M_{LH} in terms of independent quantities. It is therefore a descriptive equation, and cannot be used to predict M_{LH}. The interrelation between M_{LH} and $\delta(e)$ can be accounted for by coupling e.g. the Penman-Monteith equation to a Planetary Boundary layer Model.

The use of the Penman-Monteith equation is somewhat restricted by the amount of information about wheather and vegetation required for its solution. A number of applications of the Penman-Monteith equation allow for the dependence of stomatal resistance on environmental conditions (e.g. Stewart and De Bruin, 1985; Dolman et al., 1988; Stewart, 1988). Recently Dickinson (1984) and Wetzel and Jy-Tai Chang (1987) developed methods to estimate evapotranspiration during the drying phase (soil moisture limiting conditions). Most of these models are, however, not applicable on a routine basis.

Priestley-Taylor and Makkink equations. At first sight the Priestley-Taylor and the Makkink formulae (see Appendix) are purely empirical. However, recently a similar relation (see 5.3.3) was found on the basis of the concept that evapotranspiration and saturation deficit are dependent variables. If at the surface water vapour and heat are brought into the lower atmosphere, the saturation deficit is changed and in turn this will affect evapotranspiration. On a regional scale the evapotranspiration of a well watered terrain with a short vegetation is primarily determined by net radiation and also by the temperature through the term $\Delta/(\Delta + \gamma)$. Therefore, the Priestley-Taylor equation and the Makkink equation describe fairly well the evapotranspiration of e.g. grass on a regional scale if there is no short of water. For a discussion the reader is referred to De Bruin and Holtslag (1987).

5.3.2 Wet Canopy Evaporation

Wet canopy evaporation rates may amount to many times the transpiration rates. Recent investigations show that wet canopy evaporation in coniferous forests may under certain conditions account for up to 70% of total rainfall (Dolman and Oosterbaan, 1986) while deciduous forests may show wet canopy evaporation in the order of 20-30% of total rainfall (Nonhebel, 1987). Interception expressed as a fraction of the annual precipitation decreases with increasing rainfall (Calder and Newson, 1979). Interception and wet canopy evaporation in forests is inversely related to intensity and duration of the rainfall (Rutter et al., 1971). The quantity of water lost through wet canopy evaporation also depends on the canopy storage which is in the order of one to several mm (see section 5.4)

The duration of rainfall is a much more important constraint on wet canopy evaporation than the net radiation (Morton, 1984). The energy required for wet canopy evaporation is attributed to advection (Shuttleworth and Calder, 1979; Calder, 1982; McNaughton and Jarvis, 1983; Morton, 1985). Even during the night very high rates of wet canopy evaporation were observed (Pearce et al., 1980).

A positive temperature gradient with height, observed when canopies are wet or in the early morning on dry days, shows that downward flux of sensible heat occurs, indicating that radiant input is insufficient to support the rate of evaporation. (Shuttleworth et al., 1985). Wet canopy evaporation from forests is found to exceed the evaporation equivalent of the net radiation by a large amount during the winter months in England and South Wales. When averaged over several years the total evapotranspiration loss from the forested area in two forest stands in the U.K. is about 12% higher than the total radiant energy input (Shuttleworth and Calder, 1979). Calder et al. (1986) observed that the energy required to evaporate the estimated annual evaporation of 1481 mm (composed of 886 mm transpiration and 595 mm wet canopy evaporation) from a tropical rain forest in West-Java is almost identical with the annual input of net radiation. This is in agreement with Shuttleworth et al. (1984b), who concluded that in wet months (precipitation 270 mm per month) evapotranspiration from Amazonian rain forest will approach and possibly exceed the accepted estimates of potential evapotranspiration, while the evapotranspiration in dry months (precipitation 30 mm per month) will fall to about 70% of the radiation equivalent.

5.3.3 Models of the Planetary Boundary Layer (PBL)

To estimate the effect of land cover changes on transpiration and evaporation, the changes in saturation deficit must be predicted, that will occur as a result of the land cover change. The interrelation between evapotranspiration and vapour pressure deficit can be described with a model for the planetary boundary layer (PBL, see Appendix), which was developed by Perrier (1980) and McNaughton and Jarvis (1983). The PBL height depends on surface heating, and it shows a diurnal variation from 100 to 200 m in the early morning to heights up to 1 to 2 km in the late afternoon in summertime. Under cloudy conditions and also in wintertime the variation in height is much less. Within the PBL, turbulent motions facilitate vertical transport of heat, vapour and momentum. The atmosphere above is stably stratified and vertical fluxes are small.

Advection within the PBL may enhance or depress the evaporation rate. Dry air advection occurs when the air introduced over an area is drier than the local saturation deficit, thereby enhancing evaporation, and moist air advection occurs when the air is wetter than the local saturation deficit, thereby suppressing evaporation.

If the PBL is heated from below by sensible heat, relatively warm air is near the surface, while at greater height the air is cooler and more dense. The PBL is then unstably stratified, and this state occurs during daytime. The PBL is stable if $M_{SH} < 0$, i.e., when the surface is cooling, which occurs during nighttime. Finally, when M_{SH} is small and the wind speed large, the PBL is neutrally stratified.

The energy balance equation

$$M_R{}^* + M_G + M_{SH} + M_{LH} = 0 \qquad (5.2)$$

shows that for a given net radiation ($M_R{}^*$) and soil heat flux (M_G), the latent heat flux (M_{LH}) determines directly the sensible heat flux (M_{SH}) and thus the turbulent state of the PBL. Water vapour also affects air density and in this way evapotranspiration directly influences the stability of the lower atmosphere.

For the PBL a relation is found which is identical to the Priestley-Taylor equation (see Appendix). Within the surface layer which is typically 1/10th of the PBL height, gradients of temperature and specific humidity are allowed and here the Penman-Monteith equation applies (De Bruin, 1987).

5.4 EVAPOTRANSPIRATION FOR DIFFERENT LAND COVER TYPES

From the discussion of the different models of transpiration and evaporation a number of general aspects become apparent: the latent heat flux is determined by the available energy and by the drying effect of over-passing air, which depends on vapour pressure deficit and wind speed. The available energy is $M_R{}^* - M_G$, the drying term is $\rho\, C_p\, \delta(e)/r_a$.

Forests have a much higher roughness length than low vegetation cover types. Because of this great roughness, the aerodynamic resistance r_a of forests is much lower than for grasslands (see Bolle, this volume, equation 16.11), the difference may amount to a factor 5-10 (Table 5.1). Therefore energy exchange processes proceed much faster in forests. Hence, the impact of the drying term $\rho\, C_p\, \delta(e)/r_a$ is much higher for forests than in low vegetation types such as grasslands. This difference is not compensated by higher net radiation (as a consequence of lower albedo). This shows that forest evapotranspiration depends primarily on the drying effect of the atmosphere, while the latent heat flux from low vegetation, such as crops or grass, is determined mainly by net radiation.

In this respect it is important to distinguish between transpiration and wet canopy evaporation. Transpiration from grassland may not be dissimilar from forest transpiration (Roberts, 1983), while wet forest canopies can evaporate significant quantities of water during rainfall, even during the night. Forests have a significant canopy storage capacity, even during the winter. Reported storage capacities are 1-3 mm for needle-leaved species, 0.2 mm in coniferous forests in winter and 0.8 mm in summer (Nonhebel, 1987); 2.2 to 8.3 mm in tropical rain forests (Herwitz, 1985). Storage in crops and grasslands is negligible. Forest transpiration is very low in winter, but evaporation from wet forest canopies may still be significant. Wet canopy evaporation does not play a significant role in grasslands and crops. Factors determining the rate of wet canopy evaporation are the rainfall intensity and duration of each storm (Rutter et al., 1971). The relative amounts of water transpired and evaporated over a year depend on the proportion of the time that the canopy is wet. When the canopy is kept wet by frequent small storms, the evaporation of intercepted water is by far the larger evaporative component of the water balance.

The turbulence around the canopy of forests accelerates the exchange of sensible heat and water vapour with the atmosphere. Over forests a vapour pressure deficit can be maintained by advection discharging water vapour. On the contrary, the vapour pressure over grasslands increases as a consequence of evapotranspiration.

Forests, in particular evergreen forests, generally evaporate during much longer parts of the year than crops. Plants are able to extract water at various depths to sustain a flow of water towards the stomata, whereby soil hydraulic properties, soil moisture status and atmospheric demand are major determinants. The pattern of water utilization is different for the various land cover types. Forests can extract water from greater depths than annuals. However, in forests soil moisture exerts a negligible control over transpiration, and the $\delta(e)/r_c$ feedback is a much more important constraint on transpiration (Roberts, 1983). Crops deplete soil moisture seasonally, and for crops and grassland soil moisture is a major constraint to transpiration.

All hydrometeorological studies in tropical rain forests indicate, that recycling of water vapour is an important component in the hydrologic cycle. Values of evapotranspiration in literature vary from 75% to 15%

of rainfall (Salati et al., 1983; Salati and Vose, 1984; Salati, 1987). Friedman (1977) in viewing part of the Amazon from the air, observed that great pillars of clouds sometimes appear to arise from the very top of the forest and where the trees thin out there is less cloud. Presumably cloud formation is in part caused by the high surface area presented by the forest canopy and high leaf induced instability. The moist atmosphere beneath it is possibly of influence as well (Salati et al., 1983). It is difficult to quantify regional evapotranspiration, because over the 600–700 million ha conditions are not homogeneous. An interesting conclusion was reported by Calder (1982), who suggested that very high rates of wet canopy evaporation may not be supportable from large scale forests, such as those in the Amazon Basin. However, as the storm size limits the extent of wet forest at any particular time, local advection of energy from areas outside the wetted region may still be possible.

In dry areas 'actual' and 'potential' evaporation rates are poorly related. Just after rainfall events one may expect actual rates to increase and the potential rates to be low caused by a decrease of the saturation deficit. In dry periods, potential evaporation will be high, while actual rates fall to low levels. Transpiration may be less than evaporation from bare soils or short vegetation caused by increased stomatal resistance and by turbulence around the canopy leading to an increase of the sensible heat flux. De Bruin (1988) mentions a number of common features important in dry regions:

1. Vegetation is sparse in arid regions. As a consequence soil evaporation is important, particularly just after rainfall events. Presently there are no methods to estimate evaporation and heat flux from bare soils requiring routine data only. Examples of work in this field are Menenti (1984), who studied evaporation in desert areas and based his estimates on vapour transport in the soil layer between the surface and the water table; and Ten Berge (1986) who presented a coupled PBL-surface model and discussed application of remote sensing for e.g. the estimation of evaporation.
2. Soil heat flux is of the same order as M_{LH} and cannot be neglected as is often done for temperate regions.
3. Total evaporation is small compared to the potential rate. This is caused by the low leaf area index.
4. The use of the concept of potential evaporation determined with the 'old' Penman equation is questionable in (semi-) arid regions (De Bruin, 1988).

McNaughton and Jarvis (1983) compiled generalized values for the various parameters illustrative of processes involved in forest and crop evapotranspiration (Table 5.1). Table 5.1 shows that forest water use in summer may under certain conditions be less than water use by agricultural crops. Forests have a much higher roughness length and can evaporate much more water under wet conditions.

Baumgartner (1965) and Baumgartner and Kirchner (1980) presented generalized energy balance terms for a number of land cover types. These generalizations illustrate the role of the different cover types in the partitioning of energy fluxes. The potential evapotranspiration is determined by assuming that net radiation is used up only for evaporation.

Table 5.1 Generalized summertime values of dry and wet canopy evaporation, $M_{LW}/(M_R^*-M_G-M_C)$ [#], Bowen ratio (β; $\beta = M_{SH}/M_{LH}$), canopy resistance (r_c) and aerodynamic resistance (r_a) for wet and dry canopies of temperate forest and crops (including grassland and field crops). Maximum values in parenthesis

Measure	Forest		Crops	
	Dry	Wet	Dry	Wet
Evaporation (mm h^{-1})	0.3(0.7)	0.2(0.9)	0.6(1.4)	0.1(0.6)
$M_{LW}/(M_R^* - M_G - M_C)$ [#]	0.2-0.6	0.6-4	0.7-1.2	0.7-1.2
β	0.5-4	± 0.5	± 0.5	± 0.5
Minimum r_c (s m^{-1})	40-100	<5	20-60	<5
r_a (s m^{-1})	5-10	5-10	20-200	20-200

[#] M_C = heat stored in and below the canopy.
Adapted from McNaughton and Jarvis (1983) with permission of Academic Press Inc.

Table 5.2 Generalized energy balance terms (annual averages) for earth surfaces with different land cover types. E_{pot} = potential evapotranspiration in mm y^{-1}; M_R^*, M_{LH} and M_{LH} in W m^{-2}; ρ = albedo (in %); z_o = roughness length (m)

Land Cover Type	M_R^*	M_{SH}	M_{LH}	E_{pot}	ρ	β	z_o
Coniferous Forest	80	27	53	1000	10	0.5	1.5-2
Deciduous Forest	67	20	47	900	15	0.4	1.6-2.2[#]
Savanna	67	27	40	800	25	0.6	0.05
Grassland	60	20	40	750	20	0.5	0.05
Cropland	67	27	40	800	25	0.7	0.3
Bare sand	47	27	20	600	30	1.3	0.001-0.01
Desert	47	40	7	600	30	6	0.001-0.01

[#] in deciduous forests there is a difference in roughness between winter and summer
Data from Baumgartner (1965, 1970) and Nonhebel (1987)

Table 5.3 Examples of estimated actual annual evapotranspiration for different land cover types from various reports

Land Cover Type	Annual evapotranspiration (mm y^{-1})	Reference
Tropics		
Tropical moist forest	1070	Coster (1937)
Tropical rain forest	1000-1500	Salati (1987)
Tropical rain forest	1481	Calder et al. (1986)
Evergreen dense tropical forest	1150	Huttel (1962)
Evergreen semi-deciduous forest	1168	Huttel (1962)
Woodland	1050	Malaisse (1973)
Eucalyptus plantation	1136	Banerjee (1972)
Temperate regions		
Forested land	445-489	Morton (1984)
Needle-leaf forest	550-660	Nonhebel (1987)
Deciduous forest	320-430	Nonhebel (1987)
Agricultural land	473-562	Morton (1984)

On account of the high net radiation income and the properties of forest canopies discussed above, forests are able to evaporate more water than other vegetative soil covers. Examples of measured evapotranspiration as presented in Table 5.3 illustrate that the actual water use may be different from the potential rates in Table 5.2, but the relative differences between forests and other land cover types are similar.

5.5 IMPLICATIONS OF LAND USE CHANGES FOR CLIMATE

The apportionment of absorbed solar radiation into sensible and latent heat is dominated by the availability of water at the earth's surface. Since the flux of latent heat associated with evaporative losses is dominant, greater vegetative coverage should be associated with increased latent heat loss and a reduction in surface temperature.

Deforestation generally causes increase in runoff, decrease in soil infiltration and decrease of the water use and this indicates, that even with sufficient rainfall, much of the water would not remain to evaporate back to the atmosphere. If there is a reduction of water returned to the atmosphere, this will result in a reduction in cloud cover, increase in net radiation and heat, which in turn would increase potential evapotranspiration, but decrease the actual evapotranspiration. If forest clearing leads to a reduction in water returning to the atmosphere, the effect will be determined by the scale of clearing, whereby the impact may be observable immediately downwind (precisely focused) or more regional within a more diffuse system.

The turbulence around the forest canopy accelerates the exchange of sensible heat and water vapour with the atmosphere. These mechanisms remove heat so efficiently, that despite being the site of positive radiation balance, forest canopies are cooler than cultivated or open land in the summertime (Henderson-Sellers, 1980). Denuded areas have reduced turbulence and will be warmer, a larger upward thermal radiation (reflection and emission), increased sensible and drastically reduced latent heat flux.

Variations in the amount of vapour condensing in the higher part of the troposphere may also influence climate on a global scale. During evapotranspiration solar energy is transformed into latent heat. This heat is subsequently released in the atmosphere where the water vapour condenses to form clouds. This energy is partly responsible for the circulation in the upper troposphere. On the other hand part of this vapour is transferred to higher latitudes where upon condensation energy is released. Hence, there is an energy transport from equatorial to polar zones. Deforestation and consequently reductions of the evapotranspiration rates may thus affect the atmospheric general circulation (Salati, 1987).

In the remainder of this section a number of studies of the climatic impact of deforestation using global circulation models will be discussed. More on this subject can be found in Henderson-Sellers (this volume).

Potter et al. (1975) used a two dimensional (zonal) atmospheric model to assess the impact of tropical deforestation. They assumed albedos of 0.07 and 0.25 of rain forest and cleared forest respectively. The chain of consequences predicted by the atmospheric model was: deforestation - increased surface albedo - reduced surface absorption of solar energy - surface cooling - reduced evapotranspiration and sensible heat flux from the surface - reduced convective activity and rainfall - weakened circulation, cooling in the middle and upper tropical troposphere - increased precipitation in the latitude bands 5 to 25°N and 5 to 25°S and a decrease in the Equator-pole temperature gradient - reduced meridional transport of heat and moisture out of equatorial regions - global cooling and a decrease in precipitation between 45 and 85°N and at 40 and 60°S.

Henderson-Sellers and Gornitz (1984) analyzed the effect of complete deforestation of the Amazon Basin using a 3-dimensional global climate model. They found no significant global temperature effect. Despite the fact, that the albedo increased, the temperature did not decrease. The cooling effect caused by the albedo increase was apparently balanced by a reduced evaporation and reduced cloud cover.

Dickinson and Henderson-Sellers (1988) included the land surface scheme (BATS) developed by Dickinson (1986) and Wilson et al. (1987) in a version of the NCAR-CCM to study the local effect of complete tropical deforestation of the South-American Amazon. All the tropical forest was replaced by a cover having characteristics of impoverished scrub-grassland, intended to be typical of the parts of the Amazon converted to cattle ranching. The major changes imposed by the deforestation were decreases in roughness length, in soil depth, in vegetation coverage, and in the sensitivity of the vegetation stomata to visible radiation, and increases in soil compaction, in surface runoff, in the albedos of vegetation and dry soil, and in the relative density of vegetation roots in the upper soil layer. Reduced mixing and less interception and evaporation from the canopy cause runoff to increase and surface temperatures to rise by 3-5 K. The periods of driest soil were extended in the model. Increased temperatures and drier soil could have detrimental impact on the survival of the remaining forest and on attempts at cultivation in deforested areas.

The above results should be viewed primarily as indications of possible change. It is difficult to compare modelling results. Furthermore, it should be noted, that the GCMs only consider the direct effects of deforestation and do not yet account for indirect effects such as increasing carbon dioxide concentration, increasing concentration of other greenhouse gases and atmospheric particulates.

5.6 CONCLUDING REMARKS

The conditions responsible for transpiration and wet canopy evaporation are so variable, both in space and time, that it is difficult to develop models which are generally applicable. Wet canopy evaporation is important in forests, especially in tropical rain forest environments, where the water recycling process may be essential in maintaining the fragile balance of the tropical rain forest ecosystem. And particularly these ecosystems are subject to massive deforestation practices.

Denudation of once forested land causes the ratio of sensible heat flux to latent heat flux to increase. The net result of this change in the energy balance is that water originally lost through evapotranspiration now has to drain superficially, thereby increasing the erosion hazard. A reduction of the water vapour flux into the atmosphere caused by deforestation has also a direct effect on the radiative balance both by reduced cloud cover (increase of incoming radiation) and by increased albedo. Added to this shift in the energy balance this means that a decline in the degree of coverage provided by vegetation will enhance aridification, it will imbalance local rainfall and influence the global energy balance.

In the last number of years scientists attribute an important microscale influence to forests and are reconsidering macro-scale effects. Present consensus tends to the opinion, that the overall effects of tropical deforestation are more likely to be observed through the loss of biological heritage, regional climate stability, soil degradation and erosion, reduced CO_2 uptake and water loss, rather than through modification of global climate. However, during recent years it is suggested that climatic changes in mid latitudes may be a consequence of anomalies in the tropics.

ACKNOWLEDGEMENTS

I wish to acknowledge H.A.R. De Bruin (Department of Meteorology, Wageningen Agricultural University, The Netherlands), I.R. Calder (Water Resources Department, Lilongwe, Malawi), R.E. Dickinson (NCAR, Boulder, USA), A.J. Dolman (Department of Physical Geography and Soil Science, University Groningen, The Netherlands), F.I. Morton (Wakefield, Quebec, Canada), J. Otterman (NASA GSFC, Greenbelt, USA), A. Pitman (Department of Geography, University of Liverpool, UK) and K. Szesztay (Budapest, Hungary) for the critical comments on earlier versions of this text. I also thank Sandrina Nonhebel (Department of Theoretical Production Ecology, Wageningen Agricultural University, The Netherlands) for critically reviewing the manuscript and for her suggestions for major improvements.

APPENDIX

Penman-Monteith equation. The latent heat flux ($M_R{}^*$) is commonly described as a combination of the energy balance and vapour and heat transport equations and on the basis of resistances (Penman-Monteith equation; Monteith, 1965):

$$M_{LH} = \frac{\Delta(M_R{}^* - M_G) + \rho\, C_p\, \delta(e)/r_a}{\Delta + \gamma\,(1 + r_c/r_a)} \qquad (5.3)$$

where: M_{LH} = latent heat flux density (W m^{-2})
$M_R{}^*$ = net radiation (W m^{-2})
M_G = soil heat flux (W m^{-2})
Δ = slope of saturation water vapour-temperature pressure curve (mbar K^{-1})
ρ = air density (kg m^{-3})
C_p = specific heat of air at constant pressure (J kg^{-1}K^{-1})
$\delta(e)$ = vapour pressure deficit (mbar)
r_a = aerodynamic resistance (s m^{-1})

r_c = canopy resistance (s m^{-1})

γ = psychometric constant (mbar K^{-1})

The resistance r_a depends on the surface roughness length and wind speed (see Bolle, this volume, equation 16.11). The canopy resistance r_c depends on many variables, such as vapour pressure deficit, temperature, radiation, carbon dioxide concentration, leaf area index and soil moisture status. The relationship between r_c and these environmental quantities varies from species to species and also depends on soil type. Mostly r_c is determined by using an independently measured M_R^* in the Penman-Monteith equation. Applying r_c's non-zero minimum value for situations in which the water supply in the root zone is optimal, equation 5.3 yields the potential transpiration.

The Penman-Monteith equation is a descriptive model and should not be used to predict evapotranspiration (McNaughton and Jarvis, 1983), since a number of interrelations exist:

1. When transpiration is reduced because of a decrease in the availability of water, the sensible heat flux from the surface increases in relation to the latent heat flux, i.e. the Bowen ratio (β; $\beta = M_{SH}/M_{LH}$) increases. On a basin scale this causes an increase in temperature and a decrease in the humidity of the overpassing air, thereby producing an increase in the transpiration estimated with the Penman-Monteith equation. Thus, evapotranspiration and vapour deficit are interrelated. (Bouchet, 1963; Morton, 1983, 1985).

2. When atmospheric vapour pressure deficits increase, the stomatal apertures of many plant species decrease in size and the stomatal resistance increases. With this mechanism plants try to keep transpiration independent of the vapour pressure deficit. A high vapour pressure deficit will result in a high transpiration estimate according to the Penman-Monteith equation, but at the same time causes a low transpiration caused by increased canopy resistance. Consequently forest transpiration is constrained within a comparatively narrow range (Roberts, 1983; McNaughton and Jarvis, 1983).

The Priestley-Taylor equation. Micrometeorological observations over well watered temperate arable crops show that evapotranspiration depends strongly on the available energy. In an attempt to provide a more convenient model to calculate the potential evapotranspiration, Priestley and Taylor (1972) proposed the following equation:

$$M_{LH} = \alpha \; \frac{\Delta}{(\Delta + \gamma)} \; (M_R^* - M_G) \tag{5.4}$$

where: α = coefficient

In conditions of non-limiting water supply the actual evapotranspiration equals the potential rate calculated according to equation 5.4 with α in the order of 1.2 to 1.3. Net radiation is well correlated with global radiation (except in winter time). On the basis of this correlation Makkink (1957) proposed a formula similar to the Priestley-Taylor method for well watered grassland:

$$M_{LH} = C \; \frac{\Delta}{(\Delta + \gamma)} \; M_{SH}^- \tag{5.5}$$

where: M_{SH}^- = downwelling shortwave radiation (W m^{-2})

C = constant (C = 0.65, De Bruin and Holtslag, 1987)

Wet canopy evaporation. For wet canopies the term r_c in equation 5.3 equals zero. Thus the evaporation of intercepted rainfall can be described in purely physical terms (Rutter et al., 1971):

$$M_{LH} = \frac{\Delta (M_R^*-M_G) + \rho \, Cp \, \delta(e)/r_a}{\Delta + \gamma} \qquad (5.6)$$

Models to predict wet canopy evaporation can be classified as practical (e.g. Gash, 1979; Calder and Newson, 1979; Mulder, 1985) and research models (e.g. Rutter et al., 1971). Mulder (1985) recognized the distribution of storms as a major determining factor for wet canopy evaporation (see also section 5.3). His model requires thrice daily observations of air temperature and relative humidity, daily means of wind run, daily totals of precipitation and the number of rainy hours and of bright sunshine. Vegetation parameters required are saturation storage capacity, free throughfall coefficient, zero plane displacement height and roughness length. Conditions influencing wet canopy evaporation are extremely variable and most models are not applicable on a routine basis and have local validity only.

Models of the Planetary Boundary Layer. The PBL can be considered as a box. Because heat and vapour are added to the box, the temperature and humidity must continuously rise. The rates of increase are given by:

$$M_{SH} = \rho \, C_p \, h \, (dT_m/dt) \qquad (5.7)$$
$$M_{LH}/L = \rho \, h \, (dq_m/dt) \qquad (5.8)$$

where: T_m = potential temperature (K)
     M_{SH} = sensible heat flux density (W m^{-2})
     L = latent heat of vaporization (J kg^{-1})
     t = time (sec.)
     h = height (m)
     q_m = specific humidity (g kg^{-1}).

From equations 5.7 and 5.8 the following relation is derived for variable height of the PBL:

$$M_{LH} = \alpha \, \frac{\Delta}{(\Delta + \lambda)} \, (M_R^*-M_G) \qquad (5.9)$$

which is identical to equation 5.4. Inputs into equation 5.9 are estimates of the surface fluxes M_{LH} and M_{SH}. The parameter α depends on many variables such as the vapour deficit, M_{SH} and M_R^*. De Bruin (1983) showed that $\alpha = 1.3$ for $r_c = 0$; $\alpha = 1$ for $r_c = 60$-90 sm-1; $\alpha < 1$ when $r_c > 100$ sm^{-1}. For some typical ranges of values of r_c is referred to Table 5.1. McNaughton and Spriggs (1986, quoted in Jarvis and McNaughton, 1986) presented similar results. In general a two-fold increase in r_c causes the evapotranspiration to decrease by only 10-20%. A description of this type of models can be found in McNaughton and Jarvis (1983) and De Bruin and Holtslag (1987).

CHAPTER 6

The Effect of Changing Land Cover on the Surface Energy Balance

A.F. BOUWMAN

International Soil Reference and Information Centre (ISRIC)
P.O. Box 353, 6700 AJ Wageningen, The Netherlands

ABSTRACT

The different longwave and shortwave components of the surface radiation balance are discussed. Denudation or partial deforestation cause an increase in the ratio of reflected to incident radiation at the earth's surface. Consequently net radiation will decrease and the ratio of sensible to latent heat flux (the Bowen ratio) will increase. Areas which in their original state are sources of latent heat, become foci for the generation of large amounts of sensible heat. This phenomenon has also been observed after partial deforestation. Particularly in areas of tropical rain forest the albedo effect of denudation is expected to have a great impact on local, regional and possibly global climate. Inclusion of this feature in global climate sensitivity models has received little attention so far.

6.1 INTRODUCTION

Until recently the land cover was assumed to be solely the result of climatic forcing with minimal feedback onto the climate itself. The interactions between land surface and the atmosphere however, are significant and complex. The land cover influences the atmosphere via radiation, transfer of momentum and transfer of sensible and latent heat. The albedo is one of the factors in the energy balance which determine the partitioning of the sun's incoming net radiation. The longwave terms in the radiation balance are the key factors in the greenhouse effect. Infrared (thermal) radiation is absorbed by atmospheric greenhouse gases.

Several studies have shown that denudation of once vegetated areas has an effect on the soil and the lower atmosphere as well as on the reflected radiation. On a regional or global scale the increased albedo caused by deforestation will have a cooling effect (Otterman, 1974; Schanda, 1986) which will result in decreased lifting of air necessary for cloud formation and precipitation. Other authors (Jackson and Idso, 1975) predicted that denudated surfaces should be warmer than vegetated ones. Bastiaanssen and Menenti (this volume) discuss the relationship between albedo and surface temperature, which is not similar for all surface reflectances.

Soils and the Greenhouse Effect. Edited by A.F. Bouwman
© 1990 John Wiley & Sons Ltd.

6.2 THE SURFACE ENERGY BALANCE

The processes at the earth's surface can be described by using the energy balance. This central equation sets boundary conditions to both the soil and the atmosphere subsystems. In a general form the surface energy balance can be written as:

$$M_R^* + M_{SH} + M_{LH} + M_G = 0 \qquad (6.1)$$

where: M_R^* = net radiation (net radiant flux density) (W m^{-2})
M_G = soil heat flux density (W m^{-2})
M_{SH} = sensible heat flux density (W m^{-2})
M_{LH} = latent heat flux density (W m^{-2}).

A frequently used expression is the dimensionless "Bowen ratio" (β), which is the ratio of sensible heat and latent heat: $\beta = M_{SH}/M_{LH}$. The sensible heat flux represents the heat transfer by dry ventilation and cooling of the surface by air passing over it. A strong feedback exists between the fluxes in the energy balance and surface properties. For example, the radiation balance of bare soils is affected by soil properties such as soil moisture content, organic matter content and soil temperature, since these variables influence surface albedo, emissivity and emittance respectively (see also Table 6.1 and Bastiaanssen and Menenti, this volume).

Sensible and latent energy (M_{LH}) are not lost from the surface atmosphere as a whole. Changes in their relative importance (changes in the Bowen ratio) can be significant for the hydrological cycle and therefore affect regional climates. The soil heat flux is traditionally measured with sensors buried just beneath the surface. This flux is dependent on the moisture condition and the type and density of the vegetative cover. In dry soils the soil heat flux is considerable (up to 30% to 50% of net radiation) while in grassland and forested soils this term is much smaller than net radiation.

6.3 RADIATION

6.3.1 Shortwave Radiation Terms

Global radiation (M_{SW}^-). M_{SW}^- is the major fraction of daytime incoming radiation. It is the shortwave radiant flux density (W m^{-2}) received at the surface resulting from the integration of radiance (W m^{-2} Sr^{-1}) over a solid angle 2πSr. M_{SW}^- can be estimated (not measured) from satellite data, whereby two different approaches exist: (i) Determination of statistical relationships between M_{SW}^- and cloudiness; (ii) Calculation of atmospheric transmittance by a radiative transfer model, whereby some of the required parameters such as cloudiness, albedo and planetary reflectance are obtained from satellites.

Net radiation (M_R^*). Net radiation is the sum of net shortwave and net longwave radiation:

$$M_R^* = (1-\rho)\, M_{SW}^- + M_{LW}^- - M_{LW}^+ \qquad (6.2)$$

where: ρ = albedo (-)
M_{LW}^+ = surface (longwave) upwelling radiation (W m^{-2})
M_{LW}^- = sky longwave downwelling radiation (W m^{-2})
M_{SW}^- = shortwave downwelling radiation (W m^{-2})

Emittance is a temperature dependent term (Bolzmann's law). Global radiation is the incoming shortwave radiation from the sun. The sun, a black body of 6000 K, emits radiation in the range of wavelengths of 0.2 to about 4 µm. The earth, a body with approximate temperature of about 300 K, emits radiation with wavelengths of 4 to 40 µm.

Hummel and Reck (1979) defined the surface albedo as the ratio of wavelength averaged solar radiation reflected by the earth's surface to that incident on it. An increased use of satellite data to monitor the albedo of arid lands has arisen from the dual importance of albedo as a potential indicator of arid land degradation and as a physical parameter with possible impacts on climate.

For a given surface and wavelength, the sum of reflectivity, absorptivity and transmissivity equals unity. For opaque bodies the sum of absorptivity and reflectivity equals unity since transmission is 0 (Ten Berge, 1986). The following formulas describe the above relations:

$$r = 1 - a \qquad (6.3)$$

The absorbed energy equals the amount emitted (Kirchhoff's law):

$$e = a \qquad (6.4)$$

where: r = reflectivity (dimensionless)
a = absorptivity (dimensionless)
e = emissivity (dimensionless).

Reflectivity depends on the wavelength of the incoming radiation and in general increases with wavelengths up to 1.2 µm (Coulson and Reynolds, 1971; Ten Berge, 1986). As surface reflectivity is also dependent on azimuth and zenith (incidence) angles, it will be clear, that the overall fraction of shortwave radiation reflected by the surface in reality is not a constant, but is related to atmospheric conditions and the position of the sun and thus depends on time of the day and the season.

6.3.2 Longwave Radiation Terms

The longwave radiation terms are essential in the greenhouse effect since they are absorbed by atmospheric greenhouse gases and aerosols and cause global warming. Radiation of wavelengths in the range of 4 to 40 µm is emitted by the earth. The quantity of emitted energy depends on the surface temperature. Longwave radiation is partly absorbed in the atmosphere by the so called greenhouse gases earlier listed in chapter 1. It is composed of a number of terms:

Sky radiation M_{LW}^-. Thermal sky radiation or longwave irradiance M_{LW}^- (W m^{-2}) constitutes an important term in the surface energy balance. In analogy to global radiation, the longwave irradiance is defined as an integral over azimuth, zenith angle and wavelength. In practice it is often taken as (Stefan Boltzmann's law):

$$M_{LW}^- = \varepsilon'_{sky} \; \sigma \; T_a^4 \qquad\qquad (6.5)$$

where: ε'_{sky} = apparent emissivity of the air (K^{-3}s^{-1}m^{-2})
$\quad\quad\;\; T_a$ = air temperature (K)
$\quad\quad\;\; \sigma$ = Stefan Boltzmann constant (5.67×10^{-8} W m^{-2}K^{-4})

The apparent emissivity concept implies that the atmosphere is considered as a grey body at the screen height temperature. One of the better known formulae to estimate the term ε'_{sky} is due to Brunt (1932), namely:

$$\varepsilon'_{sky} = a + b \sqrt{(e_{act})} \qquad\qquad (6.6)$$

where: e_{act} = actual vapour pressure (mbar)
$\quad\quad\;\; T_a$ = air temperature (K)
$\quad\quad\;\; a,b$ = coefficients to be determined from observational data

Brutsaert (1975) showed that the apparent emissivity under a clear sky can be empirically related to both vapour pressure and air temperature. In the presence of clouds the apparent clear sky emissivity has to be corrected for the fractional cloud cover and relative sunshine duration.

Surface emittance M_{LW}^+. The longwave radiation leaving the surface consists of the terms emittance and reflection. Over the whole wavelength interval the surface emittance can be calculated as:

$$M_{LW}^+ = \varepsilon \; \sigma \; T_s^4 \qquad\qquad (6.7)$$

where: ε = emissivity of the emitting body (K^{-3}s^{-1}m^{-2})
$\quad\quad\;\; T_s$ = surface temperature (K)

The surface emittance is particularly of interest in areas with bare soils. The emissivity ε (= the efficiency of emission of longwave radiation by a body) is a soil specific property, that ranges from 0.87 (sand) to 0.95 (material with higher organic matter content; Becker et al., 1985; 1987) to 0.97 (Menenti, 1984). The difference found between wet and dry soils usually amounts 0.02 to 0.04 (Ten Berge, 1986). Relatively few data are available on the emissivity of moisture conditions between dry and saturated (very wet) and the relation between emissivity, soil organic matter content and mineral composition. Differences in emissivity are small and hardly significant in the energy balance of bare soils, but they are of great importance in the interpretation of thermal infrared imagery. Differences in the ε have been reported to make cool, wet sand appear warmer on surface imagery than warm, dry sand.
        The reflection component of longwave radiation leaving the surface is:

$$(1-\varepsilon) \; M_{LW} \qquad\qquad (6.8)$$

The emissivity equals the absorptivity (Kirchhoff's law). The brightness temperature is always lower than the physical kinetic temperature of the emitting body (except when $\varepsilon = 1$). If the temperature is assumed to be equal to the brightness temperature, the real temperature is underestimated. The discrepancy increases with decreasing emissivities and with increasing surface temperatures (Thunnissen and Van Poelje, 1984; Schanda, 1986). Thus, the degree of underestimation is higher for bare soils than for vegetated surfaces.

Equations 6.7 and 6.8 can be combined in one equation:

$$M_{LW}^+ = \varepsilon\,\sigma\,T_s^4 + (1-\varepsilon)^* M_{LW}^- \qquad (6.9)$$

A surface which is partly covered by vegetation has a surface emittance (Thunnissen and van Poelje, 1984):

$$M_{LW}^+ = f_c \varepsilon_c \sigma T_c^4 + (1-f_c)\varepsilon_s \sigma T_s^4 + f_c(1-\varepsilon_c)M_{LW}^- + (1-f_c)(1-\varepsilon_s)M_{LW}^- \qquad (6.10)$$

where: f_c = vegetation coverage expressed as a fraction
  ε_s = soil emissivity
  ε_c = crop emissivity

6.4 REFLECTANCE PROPERTIES

6.4.1 Spectral Reflectance Characteristics of Bare Soils

A portion of 99% of the solar radiation is received in the spectral range 0.3-1.9 μm and 83% in the range 0.3-1.1 μm. Hence, a pyranometer sensing fluxes in the interval 0.3-2 μm provides representative values of reflected radiance. When a smaller band is selected, the relation between solar spectrum and soil spectral response must be taken into account, since in the visible and infra red spectrum the reflection is a consequence of scattering and has a definitive relation with incoming radiation.

The dielectric properties of a medium determine the dependence of reflection on wavelength, angle of incidence, etc. The dielectric properties in turn, are governed by the characteristics of the medium, such as soil moisture status and soil water retention characteristics, organic matter content, mineralogic composition. Other factors influencing the reflectivity are the surface roughness and texture.

For literature values of the reflectance of different bare soil surfaces the measuring technique (zenith angle) and averaging procedure need not be the same. As a result of the daily cycle of the angle of incidence, the albedo (ρ) will decrease in the morning and increase in the late afternoon. Furthermore ρ is dependent on wavelength of incident radiation. Values of ρ in the 0.4 to 1.1 μm spectral range (as in LANDSAT) is representative of a much wider spectral range of 0.2 to 3.5 μm. Knowledge of ρ-wavelength curves is helpful in classifying bare soil surfaces from LANDSAT-MSS data and in assessing soil moisture conditions. Broad band radiometers (0.5-1.1 μm) as carried in satellites allow a good determination of ρ. When dealing with multispectral scanners, one should account for the percentage of solar energy in each band to calculate ρ from the various single band reflectances.

The dielectric properties and reflectivity of soils are strongly related to soil moisture content. An increasing soil moisture suction or in other words, a decreasing free energy level of soil moisture, causes reflection to decrease in the optical and infra red wavelength range. Specific moisture absorption bands are 1.45 and 1.95 μm (Schanda, 1986). Soil organic matter also influences the reflectance (e.g. Baumgardner et al., 1970; Al-Abbas et al., 1972; Stoner et al., 1980; Latz et al., 1984; Stoner and Baumgardner, 1984; Menenti, 1984). An increase of organic matter content causes an increase of absorption (decrease of reflection) over the complete wavelength traject. Besides its proper reflectance characteristics, organic matter plays a role in soil structure as a binding agent. In that way it influences the surface roughness and -texture. Furthermore, organic matter may influence the soil moisture retention properties.

Other factors influencing the spectral reflectance characteristics of bare soil surfaces are the surface roughness, salt content, salt crusts, sealing of soil's surface, anorganic coatings on mineral parts, erosion class (Latz et al., 1984).

6.4.2 Spectral Reflectance Characteristics of Vegetation

The spectrum of a plant leaf can be divided into the following ranges (Mulders, 1987): (i) 0.4-0.7 μm: intense absorption of incident radiation by pigments in the plant leaves with major absorption bands of about 0.43-0.45 μm and 0.66-0.64 μm (Schanda, 1986); (ii) 0.7-1.3 μm: low absorption, high reflection; (iii) 1.3-2.6 μm: high absorption by water in the leaf; in this part of the spectrum the absorption of energy is related to moisture content and leaf thickness (maximum absorption at 1.45 and 1.95 μm). Although some land cover types have significantly different spectral response patterns, many tree species and many agricultural crops have spectral response curves, that are very similar (Hoffer, 1984). However, spectral differences between major cover types such as green vegetation, dry and dead vegetation are significant and distinct. The NOAA-AVHRR derived normalized difference vegetation index is a representation of the 0.4-0.7 and 0.7-1.3 μm areas. Therefore, this vegetation index cannot be used to discern between different species and vegetation types, but it can be useful for monitoring seasonal and annual patterns of change.

Visual colour is a strong indicator of pigmentation in vegetation. About 5% of the total light is still transmitted through 8 stacked leaves, and even in the reflected radiance a small difference may be recognized between stacks of 6 and 8 leaves. The status of pigmentation, inner structure and water content of a leaf can be observed almost independently. Recent developments show the ability to distinguish between vegetation species and to identify water stress (Schanda, 1986).

6.4.3 Seasonal Albedo Values

In the literature one finds mention of two types of definitions of albedo, i.e. the surface albedo and the planetary albedo (e.g. Bolle, this volume). The surface albedo is the desired quantity in climatological studies of the heat balance of the earth. Hummel and Reck (1979) developed a very useful albedo model based on land cover types, which was already mentioned in chapter 3. Their data illustrate the seasonal variation of albedo. Another seasonal global albedo data set is the one produced by Matthews (1985) using a 1x1° grid. Wilson and Henderson-Sellers (1986) also compiled data bases with a 1° resolution of soils and land cover for use in GCM's. Gutman (1988) developed a method to determine the monthly mean albedo of land surfaces using AVHRR data.

The albedo also shows daily variation. Pinker et al. (1980) report a mean albedo of 13% for tropical evergreen forest in Thailand, ranging from about 11% around midday to 18 to 19% in the early morning or late afternoon. The variation is largely suppressed on overcast days and the midday albedo is higher on overcast days than on clear days. Shuttleworth et al. (1984a) observed a mean albedo value of about 12.15±0.2% for the Amazon rain forest and less variation than reported by Pinker et al. (1980).

Variations in albedo can thus be caused by characteristics of the surface (soil, vegetation), by atmospheric conditions (cloudiness) and by the zenith angle. By using the concept of "normalized" albedo the atmoshperic influences and the effect of the zenith can be eliminated so as to enable analysis of soil and hydrological processes at the surface.

6.5 THE IMPACT OF CHANGING ALBEDO ON CLIMATE

Although it is difficult to consider climatic effects of albedo change separately, some of the literature on this subject will be reviewed below. Bolle (this volume) will also consider albedo effects using a simple model. Climatic impacts of deforestation through changing patterns of sensible and latent heat is further discussed in chapter 5.

The earth absorbs energy in the zones of small albedo depending on the incidence angle of the sun, thanks to the atmospheric transparency in this part of the spectrum. It loses energy rather uniformly over the globe due to the partial transparency in the thermal infrared. The extraordinary importance of water in the heat budget of the earth is obvious. For steep solar incidence, i.e. in the tropical region, the albedo of water is very low, hence absorption of the sun's radiation high. The polar ice caps with high albedo reflect the energy influx incident in the visible part of the spectrum. However, their lower albedo in the infrared causes effective radiation of thermal radiation. The strong dependence of emitted radiation on temperature (Boltzmann's law) balances the earth's temperature to a well defined mean and limits the ocean surface temperature to about 30°C (Schanda, 1986).

Sagan et al. (1979) report a global temperature change over the past several thousand years of -1K, caused primarily by desertification. The temperature decrease over the past 25 years was 0.2K in their estimate. Henderson-Sellers and Gornitz (1984) in modifying the estimates given by Sagan et al. (1979) considered the effect of complete deforestation of the Amazon basin. The $5x10^{12}$ m^2 of tropical moist forest was replaced by a grass vegetation cover. They arrived at maximum planetary albedo increase of between 0.00033 and 0.00064, corresponding to a global temperature decrease of 0.06 to 0.09 K, which is well below the larger period variability. The major difference with the estimates by Sagan et al. (1979) are reduced albedo changes. Henderson-Sellers and Gornitz (1984) concluded, that albedo changes induced by current levels of tropical deforestation appear to have a negligible effect on the global temperature. If there is such an effect, this will counterbalance the warming effects of the so called greenhouse gases discussed earlier in this paper.

Simply altering ground cover affects all the terms of the energy balance: the albedo and the Bowen ratio. Furthermore, the surface winds and run-off rate are influenced by denudation of soils. These changes in turn cause soil moisture, temperature and erosion rates to change. Where bare soils are at the surface the albedo depends on soil moisture. From the relationship between albedo and surface temperature reported by Bastiaanssen and Menenti (this volume) can be deduced that the value of the albedo of original and replacing land cover determines whether cooling or temperature increase will occur as a consequence of a variation in soil moisture content. Surface temperature is thus a result of many processes in the soil and in the atmosphere. The Bowen ratio, which has values of 0.3 to 0.6 for different vegetation covers, will change in extreme cases of devegetation to 2.0 or even 6.0 (see Table 5.2). This implies that areas that previously were sources of latent heat would generate large amounts of sensible heat (Salati and Vose, 1984).

Under conditions of unlimited moisture thermal emissions will be inversely related to the amount of vegetation. This relationship however, may be complicated by the effect of albedo, since high albedo results in low net radiation. Goward et al. (1985) concluded, that a satellite designed especially to observe the relation between vegetation index and thermal emission is needed. A decline in the surface density of vegetation will increase the surface albedo resulting in a net radiative loss, which gives rise to general drying over the area, thereby inhibiting or reducing the convection necessary for rain (Lo, 1987). This process will cause further vegetation declines through reduced precipitation (Sagan, et al., 1979). When high albedo soils are denuded, the resultant increase in albedo causes lower surface temperatures which in turn reduce the heat input in the lower atmosphere and reduce convective activity necessary for rainfall. The albedo of a bright desert sandy soil can be estimated to be 0.37, whereas the albedo of the same soil with appreciable vegetation cover would be 0.25 (Ottermann, 1974). Such large differences in albedo can have large environmental effects. Higher

albedos cause lower surface temperatures (Ottermann, 1974). Jackson and Idso (1975) however, reported an adverse relation.
Walker and Rowntree (1977) reported, that soil moisture and albedo provide a positive feedback to any tendency in aridity. It may be of importance to climatic stability both with respect to local rainfall and -because of the albedo influence of soil moisture- to the global heat balance.
Ghuman and Lal (1987) showed that maximum air temperature above the soil surface was always 2 to 5° lower under forest than in the open during daytime in a Nigerian rain forest. However, the forest at 1m was relatively warm at night. With the advance of the dry season the difference between the maximum air temperatures in the open and forest decreased. Some experiments (e.g. Ribeiro et al., 1982) show that in areas where vegetation cover is reduced, the relative humidity is reduced and the temperature is higher. Lal (1987) gives an extensive literature review on tropical forest microclimate and changes due to deforestation for various parts of the world.

6.6 CONCLUSIONS

The radiation terms which are particularly involved in the greenhouse warming are the longwave radiation terms discussed in 6.3.2. Net radiation consists of shortwave and longwave terms. The portion of global shortwave incoming radiation which is not reflected (i.e. 1 minus the albedo) is the shortwave fraction. Longwave terms are sky radiation, which depends on atmospheric vapour pressure and temperature, and surface emittance, depending on the surface temperature. The latter is determined by the albedo and incoming radiation. A decline in the surface density of vegetation or a complete denudation will increase the surface albedo and this will affect the net radiation directly by determining the shortwave component, and indirectly by determining the surface's temperature and longwave emittance. In short, net radiation will decrease and the ratio of sensible heat flux to latent heat flux will increase as a consequence of partial or complete denudation. The net result depends on the difference in albedo between the original and replacing surface cover. Areas which in their original state are sources of latent heat, become foci for the generation of large amounts of sensible heat. Especially in areas with tropical rain forest the albedo effect of denudation is expected to have a great impact of local, regional and possibly global climate. So far few attempts have been made to map albedo values. The references in the text consider albedo values of land cover types at 1° resolution. Inclusion of these data in global climate sensitivity models has received little attention so far.

ACKNOWLEDGEMENTS

The author acknowledges the critical comments on earlier versions of this text made by H.A.R. De Bruin (Department of Meteorology, Wageningen Agricultural University, The Netherlands), I.R. Calder (Water Resources Department, Lilongwe, Malawi), R.E. Dickinson (NCAR, Boulder, USA), A.J. Dolman (Department of Physical Geography and Soil Science, University Groningen, The Netherlands), F.I. Morton (Wakefield, Quebec, Canada), J. Otterman (NASA GSFC, Greenbelt, USA), A. Pitman (Department of Geography, University of Liverpool, UK) and K. Szesztay (Budapest, Hungary). Thanks are also due to Wim Bastiaanssen (Winand Staringcentre, Wageningen, The Netherlands) for critically reading the manuscript of this text.

CHAPTER 7

Remote Sensing Techniques
for Monitoring of Vegetation, and for Estimating Evapotranspiration and Phytomass Production

A.F. BOUWMAN

International Soil Reference and Information Centre (ISRIC)
P.O. Box 353, 6700 AJ Wageningen, The Netherlands

ABSTRACT

Aerial photography, multispectral scanning and radar sensing are important tools in environmental studies. For global and continental monitoring and change detection the NOAA-AVHRR is at present the most suitable sensor. Thanks to the repeat cycle of 1 day, the NOAA-AVHRR will give a cloud free image of any geographic area during any period of time. Analysis of time series of such data enable study of spatial changes and dynamics of vegetation and land use. Therefore, the NOAA-AVHRR sensor data are highly suitable for global monitoring of land cover. New AVHRR applications are being researched. The low spatial resolution AVHRR data can be supplemented with higher resolution MSS data or radar imagery when and where available.

Remote sensing of surface temperatures and correlation with meteorological measurements to estimate evapotranspiration has proved to be very accurate for agricultural crops on a large scale.

7.1 INTRODUCTION

Global and regional estimates of attributes such as the area of forests and the year by year changes in this area is basic information in for example studies of the global carbon cycle, but it is also essential with respect to changes in albedo, evaporation, run off rates and the area of agricultural land. The transition from forest to non forest cause large differences in albedo, that are easily detected using predominantly empirical remote sensing models.

For the determination of land cover changes different approaches are used. Strahler et al. (1986) divide remote sensing models into discrete and analogue (continuous) scene models and into deterministic and empirical models. Discrete models have elements with abrupt boundaries whereas in continuous models energy flows are taken to be continuous and there are no sharp or clear boundaries in the scene. Deterministic remote sensing formulates relations concerning real processes of energy and matter interaction (emissivity, scattering, absorption and properties related to thermodynamics). Empirical remote sensing models aim at associating sensor measurements with scene elements, typically in a

Soils and the Greenhouse Effect. Edited by A.F. Bouwman

statistical mode. In reality, methods that are basically deterministic are often formulated with empirical components (Strahler et al., 1986).

Active remote sensing techniques (e.g. lidar techniques) for measuring (trace) gas concentrations are available. However, for most gases this technique is not yet suitable, since the concentration gradients usually show a sharp drop at short distance above the emitting surface. Measurements usually are done at greater heights above the surface and not in the track where the gradient is.

7.2 REMOTE SENSING TECHNIQUES

7.2.1 General Aspects

Under clear sky conditions short wave reflectance measured from satellites will be only slightly different from values obtained on the ground (Menenti, 1984), but under cloudy conditions satellite measurements of reflectance in the visible part of the spectrum will be impossible.

For radiometrical observation of the surface it is important to use a range of wavelength where absorption and scattering by the atmosphere do not disturb. The desired information may be hidden by strong attenuation from amongst others atmospheric absorption and upwelling radiation created in the atmosphere itself (Schanda, 1986).

Long wave (thermal) radiation emitted by the surface can in principle be obtained by measuring the emittance in the infrared 8 to 14 μm spectral range. Use of radiometers installed in an airplane or a satellite is a suitable experimental technique allowing for the coverage of large areas.

For so-called active methods of remote sensing of the earth's surface (radar and lidar) it is obvious that the scatter behaviour of rough surfaces in general and characteristics of the radiation which is reflected back to the sensor in particular, are of fundamental interest. But also for observation of natural radiation of an object, the scatter and absorption properties of rough surfaces, the albedo and emissivity are fundamental. The utilization of surface scattering concepts are much more developed in the microwave than in the optical range of the spectrum. This is because radar systems are all weather day and night sensors, which measure backscattering of artificial radiation and allow sensitive discrimination between different surface properties.

7.2.2 Aerial Photography

In aerial photography four types of film are commonly used: black and white panchromatic, black and white infrared, colour and colour-infrared. Use of aerial photography offers the advantage of stereoscopic view which enhances recognition of objects and the interpretation for soil and vegetation mapping.

As a group, panchromatic films have the best resolution of any of the film types, which makes them useful for such measurements as heights of trees or diameters of crowns. But since panchromatic film is only sensible to the visible part of the spectrum (0.4-0.7 μm) it is not highly suitable for distinguishing between species. Black and white infrared films show less contrast, but can be used for multiband photography using different filters. In some applications a Wratten 89B filter (which filters all the visible wavelengths) is used to provide contrast between deciduous and coniferous forests. Wratten 25B filters sensitize films to both visible and reflected infrared wavelengths.

Table 7.1 Utility of different scales of aerial photography for vegetation mapping

Type/Scale	General Level of Discrimination.
Satellite imagery	Separation of extensive masses of evergreen vs. deciduous forest
1:25000-1:100000	Broad vegetation types, recognition by inferential process
1:10000-1:25000	Direct identification of major cover types and species occurring in pure stands
1:500 -1:10000	Identification of individual trees in pure stands
1:500 -1:2500	Identification of individual range plants and grassland types

Source: Hoffer (1984)

With colour films individual species of trees can be identified far better than with panchromatic film. Colour-infrared is sensitive to the visible and near infrared part of the spectrum (not thermal infrared) and can be used to detect spectral differences, which may be very small in the visible wavelengths but distinct in the near infrared. Another advantage of colour-infrared film (and also of black and white infrared film) is, that atmospheric penetration is better than for true colour film.

The type of film used and the scale depend on the degree of detail involved and the accuracy required. A summary of the scale of photography and the obtained degree of detail is presented in Table 7.1.

7.2.3 Multispectral Scanner Systems (MSS, TM)

In satellite-borne multispectral scanner systems the energy reflected or emitted from a small area on the earth's surface at a given moment, is reflected from a rotating or oscillating mirror through an optical system, which disperses the energy spectrally to an array of detectors. The motion of the mirror allows the energy along a scan line, which is perpendicular to the direction of flight, to be measured while the forward movement of the aircraft or spacecraft brings successive strips of terrain into view. The detectors simultaneously measure the energy in the different wavelength bands and the output signal from the detectors is amplified and recorded on magnetic tape. The latter feature makes this technique suitable for computer aided analysis techniques.

The spatial resolution of scanner systems is dependent on the characteristics of the scanner and its altitude. Usually the spatial resolution is not as good as can be obtained from photographs, but its spectral resolution is far better. Another aspect is the wide range of wavelengths that are accessible including part of the thermal region of the spectrum.

The multispectral scanner installed in the LANDSAT 1-3 collects data in 4 bands: 0.5-0.6 μm (green); 0.6-0.7 μm (red); 0.7-0.9 μm and 0.8-1.1 μm (reflected infrared). The data are handled in frames each covering a ground area of 185 x 185 km. A frame contains 2340 scan lines, each having 3236 picture elements (pixels). Each pixel represents an area of 0.46 ha. The thematic mapper in LANDSAT-4, launched in 1982 as part of a complete end-to-end highly automated earth monitoring system, obtains greater quantities of data since it has a higher spatial resolution (30 m, exc. band 6 with 120 m) and the thematic mapper multispectral scanning system (TM) has 7 bands. The wavelengths in each band are presented in Table 7.2.

Table 7.2 Wavelength ranges of the 7 bands of the Thematic Mapper (TM)

Band	Range (μm)	Description
1	0.45 - 0.50	Water body penetration, useful for coastal water mapping and differentiating soils from vegetation and deciduous from coniferous flora
2	0.51 - 0.60	Measurement of visible green reflectance of vegetation for plant vigor assessment
3	0.63 - 0.69	Chlorophyll absorption, important for vegetation discrimination
4	0.76 - 0.90	Determination of biomass content and delineation of water bodies
5	1.55 - 1.75	Indicative of vegetation moisture content and soil moisture and useful for differentiation of snow from clouds
6	10.40 -12.50	Thermal infrared band of use in vegetation stress analysis, soil moisture discrimination and thermal mapping
7	2.08 - 1.35	Potential for discriminating rock types and for hydrothermal mapping

Both manual interpretation (e.g. Crapper and Hynson, 1983) and computer aided analysis techniques have been used with LANDSAT imagery (Hoffer, 1984). LANDSAT data allow for delineation of major vegetation cover types and also disturbed forest lands can be reliably defined. However, brushland which develops after clear-cutting, is difficult to distinguish.

Use of LANDSAT data from different years offers potential for monitoring historical and actual deforestation in many critical areas of the world. Where available, coupling of LANDSAT data with NOAA-

AVHRR or radar imagery yields more information than can be obtained from either of the three separately. In this respect the repeat cycle of a satellite is important, especially in areas of frequent cloud cover. LANDSAT has a repeat cycle of 18 days. A few strategically placed days of poor viewing conditions can render a data set virtually useless. SPOT has a somewhat longer cycle of 26 days. The ERS-1 (Earth Resources Satellite, launch planned in 1990) will have a repeat cycle of 3 days while NOAA offers a daily data set.

7.2.4 Low spatial resolution scanners (AVHRR, CZCS)

The primary sensor for coarse resolution remote sensing is the advanced very high resolution radiometer on board the National Oceanographic and Atmospheric Administration's (NOAA) series of polar orbiting sun synchronous meteorological satellites (Townshend and Justice, 1986). The spectral range of NOAA-9 (launched in 1984) is 0.58-0.69 μm (band 1), 0.725-1.1 μm (band 2), 3.5-3.93 μm (band 3), 10.3-11.3 μm (band 4) and 11.5-12.5 μm (band 5). It provides low spatial resolution high radiometric resolution multispectral data for the entire surface of the earth on a daily basis (and at a low cost). It is planned for continued operation through the 1990's. The applications of AVHRR for environmental sciences are discussed in 7.3.4.

AVHRR offers local area coverage (LAC), global area coverage (GAC) and a global vegetation index data set (GVI). The LAC data have a resolution of 1.1 km (at nadir). As the name implies, these data are available for geographically limited areas on any one day. The GAC are generated from the 1.1 km resolution data on a daily basis for the whole globe. The GAC is obtained from LAC values by taking 4 out of 5 pixels in 1 out of 3 lines, which are averaged to represent GAC. The resulting resolution is approximately 4 km. This procedure is far from ideal, but in future more satisfying methods will be introduced.

The global vegetation index (GVI) data set is generated on receipt of the GAC data, whereby one GAC pixel is selected to represent the value of 4 pixels. The resulting resolution is about 8 km at the equator. The normalized difference vegetation index (NDVI) is calculated from the red and near infra red bands according to: NDVI = (band 2- band 1)/(band 2 + band 1). This ratio yields a measure of photosynthetic capacity such, that the higher the value of the ratio, the more photosynthetically active the cover type (Sellers, 1985). For specific periods of time (e.g. 7 days) a technique of maximum value compositing (MVC) is applied to correct for cloud contamination of the image. For specific geographic regions the highest NDVI-pixel is retained. The maximum value compositing technique is such, that near nadir pixels will usually be chosen, thereby reducing atmospheric attenuation and surface directional effects (Holben, 1986). On a continental broad scale there is close similarity between NDVI images obtained for different years in the same month. Sellers (1985) found a near linear relation between NDVI and IPAR (intercepted photosynthetically active radiation), but concluded that NDVI is an insensitive measure to estimate the leaf area index or biomass in the following cases: (i) when the leaf area index exceeds 2 or 3; (ii) when there are patches of bare soil in the sensor field of view; (iii) when there is an unknown quantity of dead material in the canopy; (3) when the leaf angle distribution is unknown and the solar elevation is high.

The NOAA-AVHRR (and also the Coastal Zone Color Scanner, CZCS, on board of the NIMBUS-7) do not suffer from LANDSAT's limitations with respect to the repeat cycle. NOAA and NIMBUS have cycles of 1/2 day and 6 days respectively. The sensors of AVHRR, LANDSAT and CZCS, with spectral responses in the visible and infra red regions of the energy spectrum, are found to respond comparably to incident radiation from agricultural targets (Cicone and Metzler, 1984). The principal variation in the signals of the 3 sensors is found to reside in two dimensions, that are highly correlated between sensors. These dimensions, called brightness and greenness, are related to target albedo and vegetative green leaf biomass.

The potential of AVHRR and CZCS for assessment of overall vegetation condition on a large area or small scale basis may exceed that of LANDSAT due to favourable temporal and data volume attributes. The spatial resolution of AVHRR and CZCS (1100 m and 825 m, respectively) however, do not favourably compare to MSS at 79 m resolution.

7.2.5 Radar systems (SLAR, SAR)

Radar operates in the microwave area of the electromagnetic spectrum. In radar radio signals are transmitted from a radar antenna, whereby the length of time required for the signal to travel to the target and back to the antenna, allows the distance of the target to be determined. The signals are pulses lasting only 10^{-7} seconds producing a range resolution of 15 m (range resolution is perpendicular to the flight-line). The along track resolution is proportional to the width of the beam of the microwave signal, which is inversely proportional to the length of the antenna.

The radar equation relating transmitted and received power shows a dependence of the received power on the fourth power of the distance. This is a severe limitation for satellite-borne remote sensing applications, which cannot always be overcome by increasing antenna aperture (Schanda, 1986). The antenna length can be artificially improved through the synthetic aperture systems (SAR), which have an along track resolution independent of the range, allowing for high resolution imagery of objects miles away. This is caused by the fact, that with increasing target distance the object remains in the beam of the antenna for a longer period of time. Thus, the effective length of the synthetic antenna is proportional to the range to the target, and the resolution is inversely proportional to the range. The result is a constant resolution.

One advantage of radar is the all-weather and day and night capability. Radar systems are usually side looking (SLAR; Side Looking Air-borne Radar), viewing the terrain from an oblique angle. For vegetation mapping differences between physiognomic classes can be enhanced; differences in moisture content also influence the radar signal reflection (reflection of certain bands of microwaves is influenced by the dielectric properties of the medium, which in turn are strongly related to moisture content). Radar patterns are marked by the presence of 'shadows' due to the reflection characteristics of the objects (facetted rocks, railways or even wiring). Areas behind tall terrain features facing the radar antenna often are in a radar 'shadow'. These objects do not return the radar signal and will appear black on the imagery. Mountainsides and slopes facing the radar antenna provide a much higher return than areas of similar cover types on flat terrain, thereby making the interpretation of radar imagery a difficult task. A basic rule for using radar imagery for change detection is radar look direction consistency for features having a predominant orientation (urban areas, cultivated areas), situated on non-level land or surfaces composed of non-isotropic scatterers. The interpreter needs to have a basic understanding of the nature of the terrain and of the nature of the interaction of radar energy and the surfaces imaged.

Changes such as recent clearing of forest areas are very distinct on radar imagery. Some wavelength bands penetrate vegetation which may be an important feature for the mapping of topographic features in forest areas. Synthetic aperture radar enables a high spatial resolution, e.g. 10m at a distance of 100km. Current radar systems usually operate in the K- and X-band of approximately 0.83 to 2.8 cm and 2.8 to 5.2 cm wavelength.

About 1200 LANDSAT scenes will be required to cover the tropical forest area of the earth. Many of these areas have never been imaged, because of the pervasive cloud cover. This gap can be filled with an active all-weather system and with coarse spatial resolution scanners (AVHRR) having short repeat cycles.

The registration of SAR imagery to maps and to LANDSAT images is not simple because of differences in the appearance of tie-points and the relief distortion caused by large off-nadir angles. Efforts are under way to solve these difficulties.

A number of unexplored possibilities of microwave sensing are (adapted from Billingsley, 1984): (1) measurement of stand density, which may show better on radar imagery than on LANDSAT-MSS imagery; (2) area delineation; (3) species differentiation; (4) tropical forest inventory; (5) tree height; (6) detection of tree stress and susceptibility to fire; (7) measurement and identification of interrelationships between soil type, soil moisture, surface roughness and vegetation cover.

7.3 APPLICATIONS OF REMOTE SENSING

7.3.1 Estimation of the soil moisture status

No operational satellite measures any aspect of the soil water balance directly at the present time. Geostationary satellites like METEOSAT can provide information about the precipitation and the soil moisture status in dryland areas. Information about the soil moisture condition can be derived from the thermal inertia and surface reflection (Milford, 1987). Thermal inertia is the ease by which the soil surface temperature is changed by heating from above. It is derived from measurements of the diurnal temperature and heat flux variations at the surface. The following equation is an example:

$$B = A \times DHC \qquad (7.1)$$

where: A   = amplitude of the surface temperature variation (K)
       B   = amplitude of the diurnal variation of the heat flux into the soil
       DHC = diurnal heat capacity

The amplitude A can be estimated from thermal infrared radiometer data, the heat flux is found as the residual in the surface energy balance (equation 3) or it may be estimated as a constant fraction of the net radiation M_R^* or of the total solar irradiation. With equation (7.1) the value of DHC can be calculated. From a run of DHC values of e.g. a year, the extreme values will give information on the dry conditions (minimum diurnal heat flux) and about field capacity (maximum diurnal heat flux). Only when coupled with ground data on soil moisture profiles, interpretations can be made. This method has only potential in dryland areas with a minimum of both clouds and vegetation and in these regions the method can yield a classification of the water content in the upper topsoil. Combined with information from other satellites (e.g. on changes in biomass) or with surface observations, the thermal inertia method might be improved (Milford, 1987). Surface reflection changes on drying or wetting and is of no use for operational monitoring, because the shallow surface layer, which is observed in remote sensing, dries out very quickly after rain.

Wetzel and Woodward (1987) used infrared surface temperature observations taken from the GOES-satellite (Geostationary Operational Environmental Satellites) to predict the soil moisture status. The morning surface temperature change turned out to be especially sensitive to the soil moisture status. The temperature change was linearly related to the square of the soil moisture deficit and to the remotely sensed (NOAA-AVHRR) vegetation index NDVI.

7.3.2 Estimation of evaporation

The evapo(transpi-)ration of water from vegetated surfaces is one of the most poorly understood aspects of the hydrological cycle. One reason for this is, that the measurement of evapotranspiration at a regional scale is very difficult. In recent years agricultural remote sensing has been mainly concerned with developing fundamental relationships for assessing plant condition and development on the basis of the emitted and reflected radiation from the plant canopy. Emitted thermal radiation from plant canopies has been related to evapotranspiration and plant water status (Monteith and Szeicz, 1962; Idso et al., 1977). Emitted thermal infrared and reflected visible and near infrared have been used to estimate total phytomass production (Asrar et al., 1985) and to asses evapotranspiration (Jackson et al., 1977; Nieuwenhuis et al., 1985; Reginato et al., 1985). A relation between evaporation and crop canopy temperature can be derived from the energy balance equation (6.1). Combining equations 6.1, 6.2 and 6.7 (chapter 6) the relation between evapotranspiration and crop temperature T_c can be found:

$$M_{LH} = \frac{T_a - T_c}{\rho\, C_p\, r_{ah}} + (1-\rho)\, M_{SW}^- + M_{LW}^- - \epsilon\sigma T_c^4 - M_G \qquad (7.2)$$

where: T_o = crop temperature (K)
T_a = air temperature (K)
(further clarifications in chapter 6, equations 6.1, 6.2)

T_c can be obtained by thermal infrared remote sensing. When T_a, r_{ah}, ρ, M_{SW}^-, M_{LW}^-, ϵ and M_G are known (or estimated) M_{LH} can be calculated. The turbulent diffusion resistance to heat transport r_{ah} depends on wind velocity, roughness length z_0 of the crop surface and atmospheric stability. In general T_a, M_{SW}^-, M_{LW}^- and the wind velocity may be taken to be constant over a regional area, implicating that standard meteorological measurements can be used. A major limitation of equation (7.2) is that calculating sensible heat flux from temperature differences between the air and plant leaves is only valid for uniformly evaporating surfaces. The formula will not yield satisfactory results in the case of partial cover, for example shrubs with roots reaching a water table and which transpire at potential rates, with large areas of bare soil between the shrubs. For non-uniform surface conditions it may not be possible to specify the factors required to compute sensible heat flux with sufficient accuracy to yield acceptable values of evaporation (Jackson et al., 1987).

The remote measurement of radiation temperature is strongly influenced by the atmosphere. But the absolute values of crop radiation temperatures are of little importance. Differences in radiation temperature are a practical tool for determining regional evapotranspiration (Nieuwenhuis et al., 1985). Jackson et al. (1977) proposed to relate the 24 hour evaporative flux M_{LH}^{24} (W m^{-2}) to 24 hour net radiation M_R^{*24} and the instantaneous temperature difference near midday $(T_c - T_a)^i$, i.e.:

$$M_{LH}^{24} = M_R^{24} - B\,(T_c - T_a)^i \qquad (7.3)$$

where: B = calibration constant (W m^{-2} K^{-1})

Equation (7.3) was developed for Phoenix, Arizona (USA). Nieuwenhuis et al. (1985) propose a simpler model for the Netherlands with intermittent cloudiness may occur frequently:

$$M_{LH}^{24} = M_{LH}pot^{24} - B' (T_c - T_c^*)^i \qquad (7.4)$$

where: $M_{LH}pot^{24}$ = potential 24-hour evapotranspiration (W m^{-2})
   T_c = temperature of a crop evaporating at the potential rate;
   B' = calibration constant (W m^{-2} K^{-1})

$M_{LH}pot^{24}$ can be calculated according to various methods (e.g. Monteith, 1973, quoted in Nieuwenhuis et al., 1985). The calibration constant is calculated on the basis of the assumption that the ratio of daily evaporation and instantaneous evaporation equals the ratio of daily and instantaneous net radiation and that wind speed is relatively constant over the daily period (Jackson et al., 1987). This allows for extrapolation of instantaneous to daily evaporation and yield satisfactory results.

Reginato et al. (1985) evaluated the evapotranspiration from cropped land by combining remotely sensed reflected solar radiation (using a multiband radiometer) and surface temperatures with ground meteorological data (incoming solar radiation, air temperature, wind speed and vapour pressure) and to calculate net radiation and sensible heat flux M_G as a fraction of net radiation ($M_G = [0.1-0.042h]M_{R.}$, h being crop canopy height after Clothier et al., 1986). They compared remotely sensed evaporation with measured water extraction rates and their results suggest, that ET maps of relatively large areas could be made using this method with data from airborne sensors. The extent of the area covered appeared to be limited by the distance over which air temperature and wind speed can be extrapolated. These results are confirmed by observations made by Desjardins et al. (1986) and Schuepp et al. (1987) who measured sensible heat flux densities using aircraft based eddy correlation techniques. In their experiments repeated passes in 4 distances were used to examine the variability of flux densities over areas of 20-500 km^2.

7.3.3 Estimation of Phytomass Production

Desjardins et al. (1987) showed that photosynthesis determined as CO_2 flux, measured with airborne sensors are related to vegetation indices (0.6-0.7 μm band/0.7-0.8 μm band; or red/near infrared) from LANDSAT D-MSS. Asrar et al. (1985) estimated phytomass production using the equation:

$$P = \sum_{i=1}^{n} e_c \times e_i \times e_s \times S \times C \qquad (7.5)$$

where: P = dry phytomass
   e_c = photochemical efficiency factor
   e_i = fraction absorbed PAR(photosynthetically active portion of solar radiation)
   e_s = fraction of energy in the PAR region of the electromagnetic spectrum

S = total incident solar radiation (J m^{-2} d^{-1})
C = crop stress index (ratio actual : potential evapotranspiration)

The factor C is a function of air and canopy temperature, wind speed and net radiation. In this way a combination of measured reflected visible and near infrared and emitted thermal radiation is used to estimate total above ground phytomass production. The results of this method showed, that at sparse canopies the yield estimates were too high due to improper partitioning of the solar energy but with dense canopies the method proved very accurate (Asrar et al., 1985).

7.3.4 Monitoring of Land Cover

The NOAA-AVHRR is the primary sensor for monitoring of land cover. Although its spatial resolution is low (1.1 km at nadir), there will mostly be a cloud-free coverage of any geographic area for a given period of time thanks to the repeat cycle of 1 day. Justice et al. (1985) showed, that the full resolution LAC images can be used successfully to delineate deforestation in Rondonia, Brazil. Individual fields however, could not be delineated. Nelson and Holben (1986) used AVHRR LAC data for delineating colonization clearings in Rondonia with Landsat MSS data as a ground reference. Available MSS data can provide the thresholds necessary for discrimination of the cleared areas from forest. Nelson and Holben also found, that GOES-VISSR imagery is of little value in this respect due to excessive data noise. Malingreau and Tucker (1988) selected the best NOAA-LAC images for each year and used channel 3 of NOAA (3.5-3.9 μm). Tentative classifications in forest – non forest categories were compared with maps, radar images, Landsat and Space Shuttle images and field checks and the extent of deforestation could be determined within Acre, Rondonia and Mato Grosso (Amazon Basin, Brasil).

AVHRR data for vegetation monitoring can be interpreted only on the basis of a thorough knowledge of the distribution of the surface cover, since the same normalized difference vegetation index (NDVI) may represent very different conditions for different vegetation communities (Townshend et al., 1986). If the NDVI is integrated over time, the results can be correlated with the total amount of biotic activity during the integrating period (Tucker et al., 1983; Tucker et al., 1985; Justice et al., 1986). A high value of the annual integrated NDVI corresponds to high net primary production (Sellers, 1985; Holben, 1986). In comparing the integrated NDVI image with the ecoclimatic map of East Africa (Pratt and Gwynne, 1977), Justice et al. (1986) found a good correspondence. Especially in semi-arid vegetation types with a considerable inter-annual variation controlled primarily by rainfall amount and distribution, the AVHRR data will give a good indication of vegetation conditions. AVHRR data can also be used to estimate the length of the growing season by assuming a threshold value for discriminating between presence and absence of photosynthesis and counting the number of days with NDVI values higher than the threshold (Justice et al., 1986). Hatfield et al. (1984) found a good correlation between NDVI and greenness (calculated as a function of the MSS bands 4,5,6 and 7) on the one hand and IPAR (Intercepted Photosynthetically Active Radiation) on the other hand. NDVI correlated significantly better with IPAR than greenness for all planting dates. At vegetation surface coverage of between 20 and 75% Huete et al. (1985) observed that greenness became strongly dependent upon soil background effect. This background is composed of both a soil spectral and a soil brightness effect. Normalization of soil background to a constant ratio or a perfect one-dimensional soil line only removed bare soil spectral influences and not the greater soil brightness influence.

Tucker et al. (1986) related the AVHRR derived NDVI to the seasonal atmospheric variation of the CO_2 concentration as measured at Point Barrow, Alaska and Mauna Loa, Hawaii. The monthly variations in atmospheric CO_2 concentrations and terrestrial NDVI show a good correlation. The authors concluded, that refining of the model with oceanic uptake and release, respiration and decomposition processes will greatly enhance the understanding of the global carbon cycle.

Gutman (1988) proposed that AVHRR data can be used to estimate surface albedo values. For that purpose AVHRR data were reorganized by 9 day repeat cycle, screened for clouds, averaged and atmospheric corrections were carried out.

7.4 CONCLUSIONS

A number of important factors influencing global climate, such as deforestation and desertification, can be remotely sensed. Accuracy in identifying changes in vegetation cover may be enhanced by proper selection of spectral bands. Narrower bands will be available in the future. LANDSAT-4 provides improved sensing, spatially and spectrally. For global monitoring the NOAA-AVHRR sensor data are highly suitable and new AVHRR applications are being researched. The low spatial resolution data of the AVHRR can be supplemented with high spatial resolution MSS data when and where available. SAR (synthetic aperture radar) promises all weather sensing, although operational spacecraft are still in the future and proper correlation with e.g. LANDSAT is in development. Deterministic remote sensing models to estimate biomass and evapotranspiration are being refined and have already proved their value for the environmental sciences. Calibration needs improvement in multi-temporal comparison.

For estimation of regional evaporation from agricultural land and grassland remote sensing techniques are available. However, for surfaces not uniformly evaporating (i.e. surfaces with partial coverage), and areas such as forested areas, where evaporation depends on more factors than net radiation alone, remote sensing techniques are as limited value as their basic theoretical equations.

As a tool for estimating stand density of vegetation, species differentiation and forest inventory, microwave sensing is very promising but needs further exploration.

ACKNOWLEDGEMENTS

Thanks are due to R. Escadafal (ORSTOM Teledetection Unit, Paris, France), J.U. Hielkema (FAO, Rome, Italy), R.D. Jackson (US Water Conservation Laboratory, USDA Agricultural Research Service, Phoenix, Arizona, USA), A.R.P. Janse (Wageningen, The Netherlands) and J.P. Malingreau (CEC Joint Research Centre, ISPRA, Italy) and M.A. Mulders (Department of Soil Science and Geology, Wageningen Agricultural University, Wageningen) for comments on earlier versuions of this text. Thanks are also due to John Pulles of ISRIC for critically reading the manuscript of this chapter.

References

Ajtay, G.L, P. Ketner and P. Duvigneaud (1979) Terrestrial primary production and Phytomass. In: Bolin, B., E.I. Degens, S. Kempe and P. Ketner (Eds.): The global carbon cycle, p 129-181. SCOPE Vol. 13. Wiley and Sons, New York.

Al Abbas, A.H., P.H. Swain and M.F. Baumgardner (1972) Relating organic matter and clay content to the multispectral radiance of soils. Soil Science 114:477-485.

Alexander, M. (1977) Introduction to soil microbiology (2nd edition). 467 p. Wiley and Sons, New York.

Allison, F.E. (1973) Soil organic matter and its role in crop production. Developments in Soil Science 3. 637 pp. Elsevier, Amsterdam.

Andersen, B. and L. Moseholm (1987) Effects on forests and other vegetation. In: H.M. Seip, Acid precipitation literature review, Nordisk Ministerrad, Kopenhagen, Miljorapport 1987:5.

Anderson, I.C. and J.S. Levine (1986) Relative rates of nitric oxide and nitrous oxide production by nitrifiers, denitrifiers and nitrate respirers. Applied Environmental Microbiology 51:938-945

Anderson, I.C. and Levine J.S. (1987) Simultaneous field measurements of nitric oxide and nitrous oxide. Journal of Geophysical Research 92:965-976.

Andreae, M.O. and D.S. Schimel (1989) Exchange of trace gases between terrestrial ecosystems and the atmosphere. Report of Dahlem workshop, February 19-24, 1989. 353 pp. Wiley and Sons, Chichester.

Andriesse, J.P. (1988) Nature and Management of tropical ped soils. 165 pp. FAO Soils Bulletin 59, FAO, Rome.

Armentano, T.V. (1980) Drainage of organic soils as a factor in the world carbon cycle. BioScience 30:825-830.

Armentano, T.V. and E.S. Menges (1986) Patterns of change in the carbon balance of organic soil-wetlands of the temperate zone. Journal of Ecology 74:755-774.

Armstrong, A.S.B. (1983) Nitrous oxide emissions from two sites in southern England during winter 1981/1982. Journal of the Science of Food and Agriculture 34:803-807.

Aselmann, I. and P.J. Crutzen (1989) Global distribution of natural freshwater wetlands and rice paddies: their net primary productivity, seasonality and possible methane emissions. Journal of Atmospheric Chemistry 8:307-358.

Asrar, G. and E.T. Kanemasu (1985) Estimation of total above ground phytomass production using remotely sensed data. Remote Sensing of Environment 17:211-220.

Austin, L.B., P.H. Schuepp and R.L. Desjardins (1987) The feasibility of using airborne CO_2 flux measurements for the imaging of the rate of biomass production. Agricultural and Forest Meteorology 39:13-23.

Banerjee, A.K. (1972) Evapotranspiration from a young Eucalyptus hybrid plantation of West Bengal. In: Proceedings and Technical papers Symposium man made forests in India (Society of Indian Foresters), quoted in: Unesco (1978).

Bartholomew, G.W. and Alexander, M. (1981) Soils as a sink for atmospheric carbon monoxide. Science 212:1389-1391.

Bartlett, K.B., D.S. Bartlett, R.C. Harriss and D.I. Sebacher (1987) Methane emissions along a salt marsh salinity gradient. Biogeochemistry 4:183-202.

Bartlett, K.B., P.M. Crill, D.I. Sebacher, R.C. Harriss, J.O. Wilson and J.M. Melack (1988) Methane flux from the central Amazonian floodplain. Journal of Geophysical Research 93:1571-1582

Bartlett, K.B., R.C. Harriss and D.I. Sebacher (1985) Methane flux from coastal salt marshes. Journal of Geophysical Research 90:5710-5720.

167

Baumgardner, M.F., S.J. Kristof, C.J. Johanson and A.L. Zachary (1970) Effects of organic matter on the multispectral properties of soils. Indiana Academy of Sciences Proceedings 79:413-422.

Baumgardner, M.F., L.F. Silva, L.L. Biehl and E.R. Stoner (1985) Reflectance properties of soils. Advances in Agronomy 38:1-44.

Baumgartner, A. (1965) Energetical bases for differential vaporization from forest and agricultural land. In: Proceedings International Symposium For. Hydr., p 381-389. London, Pergamon Press.

Baumgartner, A. and N. Kirchner (1980) Impacts due to deforestation. In: W. Bach, S. Pankrath and J. Williams (Eds.), Interactions of energy and climate, p 305-317. Reidel, Dordrecht.

Baumgartner, A. and E. Reichel (1975) The world water balance. Mean annual global, continental and maritime precipitation, evaporation and runoff. 179 pp. R. Oldenburg verlag, München, Wien.

Becker, F., F. Nerry, P. Ramanantsizehena and M.P. Stoll (1985). Emissivité dans l'infrarouge thermique: conditions d'une mesure par retrodiffusion et comparaison avec une méthode de mesure passive. In: Proceedings 3rd International Colloquium on Spectral Signatures of Objects in Remote Sensing, Les Arcs, France, 16-20 December 1985.

Becker, F., F. Nerry, P. Ramanantsizehena and M.P. Stoll (1987) Importance of a remote measurement of spectral infrared emissivities; presentation and validation of such a determination. Advanced Space Research 7:121-127.

Billingsley, F.C. (1984) Remote sensing for monitoring vegetation: an emphasis on satellites. In: G.M. Woodwell (Ed.), The role of terrestrial vegetation in the global carbon cycle. Measurement by remote sensing, p 161-180. SCOPE Vol. 23. Wiley and Sons, New York.

Bingemer, H.G. and P.J. Crutzen (1987) The production of methane from solid wastes. Journal of Geophysical Research 92:2181-2187.

Blackmer, A.M., S.G. Robbins and J.M. Bremner (1982) Diurnal variability in rate of emission of nitrous oxide from soils. Soil Science Society of America Journal 46:937-942.

Blake, D.R., Mayer, E.W., Tyler, S.C., Makide, Y., Montague, D.C. and F.S. Rowland (1982) Global increase in atmospheric methane concentrations between 1978 and 1980. Geophysical Research Letters 9:477-480.

Blake, D.R. and F.S. Rowland (1986) World wide increase in troposphere methane, 1978-1983. Journal of Atmoshperic Chemistry 4:43-62.

Bohn, H.L. (1976) Estimate of organic carbon in world soils II. Soil Science Society of America Journal 40:468-470.

Bolin, B. (1977) Changes of land biota and their importance for the carbon cycle. Science 196:613-615.

Bolin, B. (1986) How much CO2 will remain in the atmosphere ? The carbon cycle and projections for the future. In: Bolin, B., B.R. Döös, J. Jager, and R.A. Warric (Eds.), The greenhouse effect, climatic change and ecosystems, p 93-156. SCOPE Vol. 29. Wiley and Sons, New York.

Bolin, B., E.T. Degens, P. Duvigneaud and S. Kempe (1979) The global carbon cycle. In: Bolin, B., E.T. Degens, S. Kempe and P. Ketner (Eds.), The global carbon cycle, p 1-56. SCOPE Vol. 13. Wiley and Sons, New York.

Bolle, H.J., W. Seiler and B. Bolin (1986) Other greenhouse gases and aerosols. Assessing their role in atmospheric radiative transfer. In: Bolin, B., B.R. Döös, J. Jager and R.A. Warrick (Eds.), The greenhouse effect, climatic change and ecosystems, p 157-203. SCOPE Vol.29, Wiley and Sons, New York

Bont, J.A.M. de, K.K. Lee and D.F. Bouldin (1978) Bacterial oxidation of methane in a rice paddy. Ecological Bulletin 26:91-96.

Bormann, F.H., G.E. Likens, I.G. Siccana, R.S. Pierce and J.S. Eaton (1974) The export of nutrients and recovery of stable conditions following deforestation at Hubbard Brook. Ecological Monographs 44:255-277.

Bosch, J.M. and J.D. Hewlett (1982) A review of catchment experiments to determine the effect of vegetation changes on water yield and evapotranspiration. Journal of Hydrology 55:3-23.

Bouchet, R.J. (1963) Evapotranspiration reelle et potentielle, signification climatique. Proceedings IAHS 62:134-142.

Bouten, W., F.M. de Vre, J.M. Verstraten and J.J.H.M. Duysings (1984) Carbon dioxide in the soil atmosphere: simulation model parameter estimation from field conditions. IAHS Publication 150:23-30.

Bouwman, A.F. (1989a) The role of soils and land use in the Greenhouse Effect. Netherlands Journal of Agricultural Science 37:13-19.

Bouwman, A.F. (1989b) Modelling soil organic matter decomposition and rainfall erosion in two tropical soils after forest clearing for permanent agriculture. Accepted by Land Degradation & Rehabilitation.

Bouwman, A.F. (1989c) Land evaluation for rainfed Agriculture using global terrain information on a 1° longitude x 1° latitude grid. ISRIC Working Paper and Preprint 89/1. ISRIC, Wageningen.

Bowden, W.B. and F.W. Bormann (1986) Transport and loss of nitrous oxide in soil water after forest clear cutting. Science 233:867-869.

Brams, E.A. (1971) Continuous cultivation of West African soils: organic matter dimunition and effects of applied lime and phosphorous. Plant and Soil 35:401-414.

Breitenbeck, G.A., A.M. Blackmer and J.M. Bremner (1980) Effects of different nitrogen fertilizers on emission of nitrous oxide from soil. Geophysical Research Letters 7:85-88.

Bremner, J.M. and A.M. Blackmer (1978) Nitrous oxide: emission from soils during nitrification of fertilizer nitrogen. Science 199:295-296.

Bremner, J.M. and A.M. Blackmer (1981) Terrestrial nitrification as a source of atmospheric nitrous oxide. In: C.C. Delwiche (Ed.), Denitrification, Nitrification and atmospheric nitrous oxide, p 151-170. Wiley and Sons, New York.

Bremner, J.M., S.G. Robbins and A.M. Blackmer (1980) Seasonal variability in emission of nitrous oxide from soil. Geophysical Research Letters 7:641-644.

Brockman-Jerosch (1918) Die Einteilung der Pflantzengesellschaften nach ökologisch-physiognomischen Geschichtspunkten. Verlag Wilhelm Engelmann, Leipzig, 68 pp.

Broecker, W.S., T. Takahashi, H.J. Simpson and T.H. Peng (1979) Fate of fossil fuel carbon dioxide and the global carbon budget. Science 206:409-418.

Brown, S. and A.E. Lugo (1982) The storage and production of organic matter in tropical forests and their role in the global carbon cycle. Biotropica 14:161-187.

Brown, S. and A.E. Lugo (1984) Biomass of tropical forests: a new estimate based on forest volumes. Science 223:1290-1293.

Brunt, D. (1932) Notes on radiation in the atmosphere. Quarterly Journal of the Royal Meteorological Society 58:389-420.

Brutsaert, W.H. (1975) On a derivable formula for long-wave radiation from clear skies. Water Resources Research 11:742-744.

Bryan, M.L. and J. Clark (1984) Potentials for change detection using Seasat synthetic aperture radar data. Remote Sensing of Environment 16:107-124.

Budyko, M.I. (1958) Heat balance of the earth's surface. Transl. by M.A. Stevenson, Dept. of Commerce, Washington,D.C.

Burch, G.J., R.K. Bath, I.D. Moore and E.M. O'Loughlin (1987) Comparative hydrological behaviour of forested and cleared catchments in Southeastern Australia. Journal of Hydrology 90:19-42.

Buringh, P. (1979) Introduction to the study of soils in tropical and subtropical regions. 124 pp. Centre for Agricultural Publishing and Documentation. Wageningen, The Netherlands.

Buringh, P. (1984) Organic carbon in soils of the world. In: Woodwell, G.M. (Ed.)(1984), The role of terrestrial vegetation in the global carbon cycle. Measurement by remote sensing, p 91-109. SCOPE Vol. 23. Wiley and Sons, New York.

Burford, J.R., R.J. Dowdell and R. Crees (1981) Emission of nitrous oxide to the atmosphere from direct drilled and ploughed clay soils. Journal of the Science of Food and Agriculture 32:219-223.

Burton, M.A.S. (1987) Terrestrial ecosystems and biome types. A background for studying contaminants in global ecosystems. A Research Memorandum. Monitoring and Assessment Research Centre, King's College London, University of London, 42 pp.

Buyanovsky, G.A., G.H. Wagner and C.J. Gantzer (1986) Soil respiration in a winter wheat ecosystem. Soil Science Society of America Journal 50:338-344.

Calder, I.R. (1982) Forest evaporation, p 173-194. Proceedings Can. Hydr. Symposium 1981. National Research Council Canada, Ottawa, Ontario.

Calder, I.R. (1985) What are the limits on forest evaporation - comment. Journal of Hydrology 82:179-184.

Calder, I.R. (1986) What are the limits on forest evaporation - a further comment. Journal of Hydrology 89:33-36.

Calder, I.R. and M.D. Newson (1979) Land use and upland water resources in Brittain- a stategic look. Water Resources Bulletin 15:1628-1639.

Calder, I.R., I.R. Wright and D. Murdiyarso (1986) A study of evaporation from tropical rainforest - West Java. Journal of Hydrology 89:13-31.

CDIAC (1989) CDIAC communications summer 1989. Carbon Dioxide Informaiton analysis Center, Oak Ridge National Laboratory.

CEC (1985) Soil Map of the European Communities 1:1,000,000 (with acc. report). ECSC, EEC, EAEC, Brussels, Luxembourg.

Cates, R.L. and D.R. Keeney (1987a) Nitrous oxide emission from native and reestablished prairies in southern Wisconsin. The American Midland Naturalist 117:35-42.

Cates, R.L. and D.R. Keeney (1987b) Nitrous oxide production throughout the year from fertilized and manured maize fields. Journal of Environmental Quality 16:443-447.

Cicerone, R.J. and R.S. Oremland (1988) Biogeochemical aspects of atmospheric methane. Global Biogeochemical Cycles 2:299-327.

Cicerone, R.J., J.D. Shetter and C.C. Delwiche (1983) Seasonal variation of methane flux from a California rice paddy. Journal of Geophysical Research 88:11022-11024.

Cicone, R.C. and M.D. Metzler (1984) Comparison of Landsat MSS, Nimbus-7 CZCS and NOAA-7 AVHRR features for land use analysis. Remote Sensing of Environment 14:257-265.

Clark, W.C. (1986) Sustainable development of the biosphere: themes for a research program. In: W.C. Clark and R.E. Munn (Eds.), Sustainable development of the biosphere, p 5-48. IIASA, Laxenburg, Austria. Cambridge University Press.

Clements, F.E. (1916) Plant succession. An analysis of the development of vegetation. 512 pp. Carnegie Institute, Washington, D.C.

Clements, F.E. (1928) Plant succession and indicators. 453 pp. H.W. Wilson Co., New York, London.

Clothier, B.E., K.L. Clawson, P.J. Pinter Jr., M.S. Moran, R.J. Reginato and R.J. Jackson (1986) Estimation of soil heat flux from net radiation during the growth of alfalfa. Agricultural and Forest Meteorology 37:319-329.

C.O.L. (Commissie Onderzoek Luchtverontreiniging van de Vereniging Lucht Troposferische Chemie) (1987): Onderzoekprogramma t.b.v. de raad voor Milieu en Natuuronderzoek.

Cochran, V.L., L.F. Elliot and R.I. Papendick (1980) Nitrous oxide emissions fro ma fallow field fertilized with anhydrous ammonia. Soil Science Society of America Journal 45:307-310.

Colbourn, P. and I.W. Harper (1987) Denitrification in draied and undrained arable clay soil. Journal of Soil Science 38:531-539.

Colbourn, P., M.M. Iqbal and I.W. Harper (1984a) Estimation of the total gaseous nitrogen losses from clay soils under laboratory and field conditions. Journal of Soil Science 35:11-22.

Colbourn P., I.W. Harper and M.M. Iqbal (1984b) Denitrification losses from 15N labelled calcium fertilizer in a clay soil in the field. Journal of Soil Science 35:539-547.

Colbourn, P., J.C. Ryden and G.J. Dollard (1987) Emission of NOx from urine-treated pasture. Environmental Pollution 46:253-261.

Collins, N.M. and Wood, T.G. (1984) Termites and atmospheric gas production. Science 224:84-86.

Conrad, R. and W. Seiler (1982) Arid soils as a source of atmospheric carbon monoxide. Geophysical Research Letters 9:1353-1356.

Conrad, R., W. Seiler and G. Bunse (1983) Factors influencing the loss of fertilizer nitrogen in the atmosphere as N_2O. Journal of Geophysical Research 88:6709-6718.

Coster, C. (1937) De verdamping van verschillende vegetatie vormen op Java. Tectona 30:1-102. Quoted in Unesco (1978).

Coulson, K.L.and D.W. Reynolds (1971) The spectral reflectances of natural surfaces. Journal of Applied Meteorology 10:1285-1295.

Crapper, P.F. and K.C. Hynson (1983) Change detection using Landsat photographic imagery. Remote Sensing of Environment 13:291-300.

Crill, P.M., K.B. Bartlett, J.O. Wilson, D.I. Sebacher, R.C. Harriss, J. M. Melack, S. MacIntyre, L. Lesack and L. Smith-Morrill (1988): Tropospheric methane from an amazonian floodplain lake. Journal of Geophysical Research 93:1564-1570.

Crutzen, P.J. (1981) Atmospheric Chemical Processes of the oxides of nitrogen, including nitrous oxide. In: C.C. Delwiche (Ed.), Denitrification, nitrification and nitrous oxide, p 17-44. Wiley and Sons, New York.

Crutzen, P.J. (1983) Atmospheric interactions in homogeneous gas reactions of C, N and S containing compounds. In: Bolin, B. and R.B. Cook (Eds.): The major biogeochemical cycles and their interactions, p 67-112. SCOPE Vol. 21, Wiley and Sons, New York.

Crutzen, P.J. (1987) Role of the tropics in atmospheric chemistry. In: R.E. Dickinson (Ed.): The geophysiology of Amazonia, pp 107-130, Wiley and Sons, New York.

Crutzen, P.J., I. Aselmann and W. Seiler (1986) Methane production by domestic animals, wild ruminants, other herbivorous fauna and humans. Tellus 38B:271-284.

Crutzen, P.J. and L.T. Gidel (1983) A two dimensional model of the atmosphere 2: The tropospheric budgets of anthropogenic chlorocarbons, CO, CH4, CH3Cl and the effect of various NOx sources on tropospheric ozone. Journal of Geophysical Research 88:6641-6661.

Crutzen, P.J. and T.E. Graedel (1986) The role of atmospheric chemistry in environment-development interactions. In: W.C. Clark and R.E. Munn (Eds.), Sustainable development of the biosphere, p 213-250. IIASA, Laxenburg, Austria. Cambridge University Press.

Crutzen, P.J., L.E. Heidt, J.P. Krasnec, W.H. Pollock and W. Seiler (1979) Biomass burning as a source of atmospheric CO, H_2, N_2O, NO, CH_3Cl and COS. Nature 282:253-256.

Curran, P.J. (1986) Principles of remote sensing. Longman, London, New York.

Dam Kofoed, A. (1982) Humus in long term experiments in Denmark. In: Boels, D., D.B. Davies and A.E. Johnston (Eds.), Soil degradation, p 241-258. Balkema, Rotterdam.

Daniel, R.M., A.W. Limmer, K.W. Steele and I.M. Smith (1982) Anaerobic growth, nitrate reduction and denitrification in 46 Rhizobium Strains. Journal of Microbiology 28:1811-1815

De Bruin, H.A.R. (1983) A model for the Priestley-Taylor parameter α. Journal of Climate and Applied Meteorology 22:572-578.

De Bruin, H.A.R. (1987) From Penman to Makkink. In: J.C. Hooghart (Ed.), Evaporation and Weather, Proceedings and Information No. 39, p 5-31. TNO Committee on Hydrological Research, The Hague.

De Bruin, H.A.R. (1988) Evaporation in arid and semi-arid regions. In: I. Simmer (Ed.), Estimation of Groundwater Recharge, p 73-88. D. Reidel Publ. Comp., Dordrecht.

De Bruin, H.A.R. and A.A.M. Holtslag (1987) Evaporation and weather: interactions with the planetary boundary layer. In: J.C. Hooghart (Ed.), Evaporation and Weather, Proceedings and Information No. 39, p 63-83. TNO Committee on Hydrological Research, The Hague.

Delany, A.C., D.R. Fitzjerrald, D.H. Lenschow, R. Pearson, C.J. Wendel and B. Woodruff (1986) Direct measurements of nitrogen oxides and ozone fluxes over grasslands. Journal of Atmospheric Chemistry 4:429-444.

Delaune, R.D., C.J. Smith and W.H. Patrick Jr. (1983) Methane release from Gulf coast wetlands. Tellus 35B:8-15.

Delaune, R.D., C.J. Smith and W.H. Patrick Jr. (1986) Methane production in Mississipi River deltaic plain peat. Organic Geochemistry 9:193-197.

Denmead, O.T. (1983) Micrometeorological methods for measuring gaseous losses of nitrogen in the field. In: J.R. Freney and J.R. Simpson (Eds.), Gaseous loss of nitrogen from plant-soils systems, Developments in plant and soil science Vol. 9, p 133-157. Martinus Nijhoff/Dr. W. Junk Publishers.

Denmead, O.T. (1983) Micrometeorological methods for measuring gaseous losses of nitrogen in the field. In: J.R. Freney and J.R. Simpson (Eds.), Gaseous loss of nitrogen from plant soil systems, p 133-158. Martinus Nijhoff, The Hague.

Denmead, O.T., J.R. Freney and J.R. Simpson (1976) a closed ammonia cycle within a plant canopy. Soil Biology & Biochemistry 8:161-164.

Denmead, O.T., J.R. Freney and J.R. Simpson (1979a) Nitrous oxide emission during denitrification in a flooded field. Soil Science Society of America Journal 43:716-718.

Denmead, O.T., J.R. Freney and J.R. Simpson (1979b) Nitrous oxide emission from a grass sward. Soil Science Society of America Journal 43:726-728.

Dent, F.J. (1978) Guidelines for utilizing data bank profile field sheets. Prepared for the land capability Appraisal project at the Soil Research Institute, Bogor, Indonesia. Government of Indonesia, FAO.

Desjardins, R.L., E.J. Brach, J.I. MacPherson, P.H. Schuepp and L. Austin (1986) Regional measurements of evapotranspiration using aircraft mounted sensors. In: Proceedings ISLSCP Conference, Rome, Italy, 2-6 December 1985. p 381-386.

Desjardins, R., A. Mack, I. MacPherson and P. Schuepp (1987) Characterizing crop conditions using airborne CO_2 fluxes measurements and Landsat-D MSS Data. In: Preprint of Proceedings of 18th Conference on Agricultural and Forest Meteorology and 8th Conference on Biometeorology and Aerobiology, Sept. 14-18, 1987, p 97-100. American Meteorological Society, Boston, USA.

Detwiler, R.P. and C.A.S. Hall (1988) Tropical Forests and the Global Carbon Cycle. Science 239:42-47.

Detwiler, R.P., C.A.S. Hall and P. Bogdonow (1985) Land use change and carbon exchange in the tropics: II. Estimates for the entire region. Environmental Management 9:335-344.

Devol, A.H., J.E. Richey, W.A. Clark and S.L. Kiny (1988) Methane emissions to the troposphere from the Amazon floodplain. Journal of Geophysical Research 93:1583-1592.

Dickinson, R.E. (1980) Effects of deforestation on climate. In: V.H. Sutlive, N. Altshuler and M.D. Zamora, Blowing in the wind: deforestation and long range implications. Studies in Third World societies. Publ. no. 14. Dept.of Anthropology, College of William and Mary, Williamsburg, Virginia.

Dickinson, R.E. (1986) Evapotranspiration in global climate models. Paper prsented at the 26th COSPAR Meeting, July 2-10 1986, Toulouse, France.

Dickinson, R.E. and A. Henderson-Sellers (1988) Modelling tropical deforestation: a study of GCM land surface parameterizations. Quarterly Journal of the Royal Meteorological Society 114:439-462.

Diez, Th. and G. Bachtaler (1978) Auswirkungen unterschiedlicher Fruchtfolge, Dungung und Bodenbearbeitung auf den Humusgehalt der Boden. Landwirt. Forschung 55:368-377.

Dinel, H., S.P. Mathur, A. Brown and M. Lévesque (1988) A field study on the effect of depth on methane production in peatland waters: equipment and preliminary use. Journal of Ecology 76:1083-1091.

Dolman, A.J. (1988) Transpiration from an oak forest as predicted from porometer and weather data. accepted for publication. Journal of Hydrology 1988.

Dolman, A.J. and W.E. Oosterbaan (1986) Grondwatervoeding, interceptie en transpiratie can de Castricumse boslysimeters. H_2O (19)No.9:174-175.

Dolman, A.J., J.B. Stewart and J.D. Cooper (1988) Predicting forest transpiration from climatological data. Agricultural and Forest Meteorology 42:339-353

Dowdell, R.J. and K.A. Smith (1974) Field studies of the soil atmosphere II. Occurrence of nitrous oxide. Journal Soil Science 25:231-238.

Dowdell, R.J., J.R. Burford, R. Crees (1979) Losses of nitrous oxide dissolved in drainage water from agricultural land. Nature 278:342-343.

Duffy, L., I. Galbally and M. Elsworth (1988) Biogenic NO_x emissions in Latrobe valley. Clean Air 22/4:196-199.

Duvigneaud, P. (1972) Morale et écologie. La Pensée et les hommes 7:286-303.

Duxbury, J.M., D.R. Bouldin, R.E. Terry and R.L. Tate III (1983) Emissions of nitrous oxide from soils. Nature 298:462-464.

Edwards, K.A. and J.R. Blackie (1981) Results of East-African catchment experiments, 1958-1974. In: R. Lal and E.W. Russell (Eds.), Tropical Agricultural Hydrology, p 165-188. Wiley and Sons, New York.

Edwards, N.T. (1975) Effect of temperature and moisture on carbon dioxide evolution in a mixed deciduous forest floor. Soil Science Society of America Journal 39:361-365.

Edwards, N.T. and B.M. Ross-Todd (1983) Soil carbon dynamics in a mixed deciduous forest following clear-cutting with and without residue removal. Soil Science Society of America Journal 47:1014-1021.

Eggington, G.M. and K.A. Smith (1986a) Nitrous oxide emission from a grassland soil fertilized with slurry and calcium nitrate. Journal of Soil Science 1986,37:59-67.

Eggington, G.M. and K.A. Smith (1986b) Losses of nitrogen by denitrification from a grassland soil fertilized with cattle slurry and calcium nitrate. Journal of Soil Science 1986,37:69-80.

Ehhalt, D.H. (1988) How has the atmospheric concentration of CH_4 changed? In: F.S. Rowland and I.S.A. Isaksen, The Changing Atmosphere, pp 25-32, Wiley and Sons, Chichester.

Ehhalt, D.H. and U. Schmidt (1978) Sources and sinks of atmospheric methane. PAGEOPH 116:452-464.

Ehhalt, D.H., R.J. Zander and R.N. Lamontagne (1983) On the temporal increase of tropospheric CH4. Journal of Geophysical Research 88:8442-8446.

Ellenberg, H. (1973) Die Oekosysteme der Erde: Versuch einer Klassifikation der Oekosysteme nach funktionalen Gescichtspunkten. In: Ellenberg, H. (Ed.), Ökosystemforschung, p 235-265. Springer Verlag, New York, Heidelberg, Berlin.

Emanuel, W.R., H.H. Shugart and M.P. Stevenson (1985) Climatic change and the broad-scale distribution of ecosystem complexes. Climatic Change 7:29-43.

Endo, M., K. Minami and S. Fukushi (1986) Effects of interception of near ultraviolet radiation on nitrifier activity and nitrification process in a fertilized andosol under field conditions. Soil Science and Plant Nutrition 32:365-372.

Enquete-Kommission des 11. Deutschen Bundestags (1988) "Vorsorge zum Schutz der Erdatmosphäre", Schutz der Erdatmosphäre: eine internationale Herausforderung. Dt. Bundestag Referat Offentlichkeitsarbeit, Bonn.

Esser, G. (1987) Sensitivity of global carbon pools and fluxes to human and potential climatic impacts. Tellus 39B:245-260.

Farquhar, G.D., R. Wetselaar and P.M. Firth (1979) Ammonia volatilization from senescing leaves of maize. Science 203:1257-1258.

FAO (1949-1983) Production yearbook Vol.1-37. FAO Statistics series No.55, FAO Rome, Italy.

FAO (1978-1981) Reports of the Agro-Ecological Zones Project. World Soil Resources Project No48, Wol.1 - Methodology and Results for Africa, Vol.2 - Southwest Asia, Vol3 - South and Central America, Vol.4 - Southeast Asia. FAO, Rome.

FAO (1983) A physical resource base (map). Technical cooperation among developing countries. Main climatic and soil divisions in the developing world, scale 1:25,000,000. FAO, Rome.

FAO (1985-1988) Fertilizer yearbook, vol. 35-37. FAO Statistics series No.71, FAO, Rome, Italy.

FAO/UNEP (1981a) Tropical Forest Resources Assessment Project. Forest Resources of tropical Asia. FAO, Rome.

FAO/UNEP (1981b) Tropical Forest Resources Assessment Project. Forest Resources of tropical America. FAO, Rome.

FAO/UNEP (1981c) Tropical Forest Resources Assessment Project. Forest Resources of tropical Africa. FAO, Rome.

FAO/UNESCO (1971-1981) Soil Map of the World 1:5000000. Vol. I-X. FAO, Rome.

FAO/UNESCO (1988) Soil Map of the World. Revised Legend. World Resources Report 60. FAO, Rome.

Fearnside, P.M. (1987) Causes of deforestation in the Brasilian Amazon. In: R.E. Dickinson (Ed.), The Geophysiology of Amazonia. Vegetation and climate interactions, p 37-57. Wiley and Sons, New York.

Fillery, I.R.P. (1983) Biological Denitrification. In: Freney, J.R. and J.R. Simpson (Eds.), Gaseous loss of nitrogen from plant-soil systems. Developments in Soil Science Vol.9. Martinus Nijhoff/Dr.W.Junk Publishers, The Hague.

Fitzpatrick, E.A. (1983) Soils. Their formation, classification and distribution. 355 pp. Longman. London and New York.

Focht, D.D. (1974) The effect of temperature, pH and aeration on the production of nitrous oxide and gaseous nitrogen- a zero order kinetic model. Soil Science 118:173-179.

Folorunso, O.A. and D.E. Rolston (1984) Spatial variability of field measured denitrification gas fluxes. Soil Science Society of America Journal 48:1214-1219.

Fraser, P.J., P Hyson, R.A. Rasmussen and L.P. Steele (1986a) Methane, carbon monoxide and methylchloroform in the southern hemisphere. Journal of Atmospheric Chemistry 4:3-42.

Fraser, P.J., R.A. Rasmussen, J.W. Creffield, J.R. French and M.A.K. Khalil (1986b) Termites and global methane: another assessment. Journal of Atmospheric Chemistry 4:295-310.

Friedman, F. (1977) The amazon basin, another sahel. Science 197:7.

Friedman, H. (1986) The science of global change- an overview. In: T.F. Malone and J.G. Roederer (Eds.), Global Change, p 20-52. ICSU publ. Cambridge University Press.

Froment, A. (1972) Soil respiration in a mixed oak forest. OIKOS 23:273-277.

Galbally, I.E. (1985) The emission of nitrogen to the remote atmosphere: background paper. In: J.M. Galloway et al. (Eds.), The Biogeochemical cycling of sulfur and nitrogen in the remote atmosphere, p 27-53. D. Reidel Publishing Company, Dordrecht.

Galbally, I.E. and R.W. Gillett (1988) Processes regulating nitrogen compounds in the tropical atmosphere. In: H. Rodhe and R. Herrera, Acidification in tropical countries, p 73-116. SCOPE Vol. 36, Wiley and Sons, New York.

Galbally, I.E., J.R. Freney, W.A. Muirhead, J.R. Simpson, A.C.F. Trevitt and P.M. Chatte (1987) Emission of nitrogen oxides (NO_x) from a flooded soil fertilized with urea: relation to other nitrogen loss processes. Journal of Atmospheric Chemistry 5:343-365.

Galbally, I.E. and C.R. Roy (1978) Loss of fixed nitrogen by nitric oxide exhalation. Nature 275:734-735.

Garret, H.E. and G.S. Cox (1973) Carbon dioxide evolution from the floor of an oak-hickory forest. Soil Science Society of America Journal 37:641-644.

Gash, J.H.C. (1979) An analytical model of rainfall interception by forests. Quarterly Journal of the Royal Meteorological Society 105:43-55.

Gaussen, R. and F. Hädrich (1965) Atlas zur Bodenkunde. Bibliographisches Institut, Mannheim, West Germany.

Gentry, A.H. and J. Lopez Parodi (1980) Deforestation and increased flooding of the upper Amazon. Science 210:1354-1356.

Ghuman, B.S. and R. Lal (1987) Effects of partial clearing on microclimate in a humid tropical forest. Agricultural and Forest Meteorology 40:17-29.

Gildea and B. Moore (1985) FAO Global Soils Database. 1/2x1/2 degree Lou-Lat. University of New Hampshire, Science & Engineering Building, COmplex Systems Research Center, Durham, New Hampshire, USA.

Goodroad, L.L. and D.R. Keeney (1984) Nitrous oxide emission from forest, marsh and prairie ecosystems. Journal of Environmental Quality 13:448-452.

Goodroad, L.L. and D.R. Keeney (1985) Site of nitrous oxide production in field soils. Biology & Fertility of Soils 1:3-7.

Goudriaan, J. (1987) The biosphere as a driving force in the global carbon cycle. Netherlands Journal of Agricultural Science 35:177-187.

Goudriaan, J. (1988) Modelling biospheric control of carbon fluxes between atmosphere, ocean and land in view of climatic change. Manuscript draft, J. Goudriaan, Agricultural University, Wageningen, The Netherlands.

Goudriaan, J. and D. Ketner (1984) A simulation study for the global carbon cycle, including man's impact on the biosphere. Climatic change 6:167-192.

Goward, S.N., G.D. Cruickshanks and A.S. Hope (1985) Observed relation between thermal emission and reflected spectral radiance of a complex vegetated landscape. Remote Sensing of Environment 18:137-146.

Grainger, A. (1980) The state of the world's tropical forests. Ecologist 10:6-54.

Greenland, D.J. and P.H. Nye (1959) Increase in the carbon and nitrogen contents of tropical soils under natural fallows. Journal of Soil Science 10:284-299

Gregory, G.L., R.C. Harriss, R.W. Talbot, R.A. Rasmussen, M. Garstang, M.O. Andreae, R.R. Hinton, E.V. Browell, S.M. Beck, D.I. Sebacher, M.A.K. Khalil, R.J. Ferek and S.V. Harriss (1986) Air chemistry over the tropical forest of Guyana. Journal of Geophysical Research 91:8603-8612.

Groffman, P.M. (1985) Nitrification and Denitrification in conventional and no-tillage soils. Soil Science Society of America Journal 49:329-334.

Gutman, C. (1988) A sample method for estimating monthly mean albedo of land surfaces from AVHRR data. Journal of Applied Meteorology 27:973-988.

Haider, K., A. Mosier and O. Heinemeyer (1985) Phytotron experiments to evaluate the effect of growing plants on denitrification. Soil Science Society of America Journal 49:636-641.

Hager, H. (1975) Kohlendioxyd Konzentrationen, -Flusse und -Bilanzen in einem Fichtenholzwald. Wiss. Mitt. Meteor. Inst. München, Nr. 26. Quoted in: Baumgartner and Kirchner (1980).

Hall, C.A.S., R.P. Detwiler, P. Bogdonow and S. Underhill (1985) Land use change and carbon exchange in the tropics: I. Detailed estimates for Costa Rica, Panama, Peru and Bolivia. Environmental Management 9:313-334.

Hamilton, L.S. and P.N. King (1983) Tropical forested watersheds. Hydrologic and Soils Response to major uses or conversions. 168 pp. Westview Press, Boulder, Colorado.

Hao, W.M., D. Scharffe and P.J. Crutzen (1988) Production of N_2O, CH_4 and CO_2 from soils in the tropical savanna during the dry season. Journal of Atmospheric Chemistry 7:93-105.

Harriss, R.C., E. Gorham, D.I. Sebacher, K.B. Bartlett and P.A. Flebbe (1985) Methane flux from northern peatlands. Nature 315:652-653.

Harriss, R.C. and D.I. Sebacher (1981) Methane flux in forested freshwater swamps of the southeastern United States. Geophysical Research Letters 8:1002-1004.

Harriss, R.C., D.I. Sebacher and F.P. Day Jr. (1982) Methane flux in the great dismal swamp. Nature 297:673-674.

Harvey, H.R., R.D. Fallon and J.S. Patton (1989). Methanogenesis and microbial lipid synthesis in anoxic marsh sediments. Biogeochemistry 7:111-129.

Hatfield, J.L., G. Asrar and E.T. Kanemasu (1984) Intercepted photosynthetically active radiation estimated by spectral reflectance. Remote Sensing of Environment 14:65-75.

Hauck, R.D. (1986) Field measurement of denitrification - an overview. In: R.D. Hauck and R.V. Weaver (Eds.), Field measurement of dinitrogen fixation and denitrification, p 59-72. Soil Science Society of America Special Publ. No. 18, Madison, Wisconsin, USA.

Henderson-Sellers, A. (1980) The effects of land clearance and agricultural practices on climate. In: V.H. Sutlive, N. Altshuler and M.D. Zamora, Blowing in the wind: deforestation and long range implications. Studies in Third World Societies. Publication No. 14. Dept. of Anthropology, College of William and Mary, Williamsburg, Virginia.

Henderson-Sellers, A. (1987) Effects of land use on climate in the humid tropics. In: R.E. Dickinson (Ed.), The Geophysiology of Amazonia, p 463-493. Wiley Interscience Publ., Wiley and Sons, New York.

Henderson-Sellers, A., R.E. Dickinson and M.F. Wilson (1988) Tropical deforestation: important processes for climate models. Climatic Change 13:43-67.

Henderson-Sellers, A. and V. Gornitz (1984) Possible climatic impacts of land cover transformations, with particular emphasis on tropical deforestation. Climatic Change 6:231-256.

Henin, S. and M. Dupuis (1945) Essai de bilan de la matière organique du sol. Ann. Agron. 15:17-29.

Herwitz, S.R. (1985) Interception storage capacities of tropical rainforest canopies. Journal of Hydrology 77:237-252.

Hoffer, R.M. (1984) Remote sensing to measure distribution and structure of vegetation. In: G.M. Woodwell (Ed.), The role of terrestrial vegetation in the global carbon cycle. Measurement by remote sensing, p 131-159. SCOPE Vol.23. Wiley and Sons, New York.

Holben, B.N. (1986) Characteristics of minimum value composite images from temporal AVHRR data. International Journal of Remote Sensing 7:1417-1434.

Holdridge, L.R. (1967) Life Zone Ecology. Tropical Science Center, San Jose, Costa Rica, 206 pp.

Holzapfel-Pschorn, A., R. Conrad and W. Seiler (1986) Effects of vegetation on the emission of methane from submerged rice paddy soil. Plant and Soil 92:223-233.

Holzapfel-Pschorn, A. and W. Seiler (1986) Methane emission during a cultivation period from an Italian rice paddy. Journal of Geophysical Research :91:11803-11814.

Houghton, R.A., J.E. Hobbie, J.M. Melillo, B. Moore, B.J. Peterson, G.R. Shaver and G.M. Woodwell (1983) Changes in the carbon content of terrestrial biota and soils between 1860 and 1980: a net release of CO2 to the atmosphere. Ecological Monographs 53:235-262.

Houghton, R.A., R.D. Boone, J.M. Melillo, C.A. Palm, G.M. Woodwell, N. Myers, B. Moore III and D.L. Skole (1985) Net flux of carbon dioxide from tropical forests in 1980. Nature 316:617-620.

Houghton, R.A., R.D. Boone, J.R. Fruci, J.E. Hobbie, J.M. Melillo, C.A. Palm, B.J. peterson, G.R. Shaver, G.M. Woodwell, B. Moore, D.L. Skole and N. Myers (1987) The flux of carbon from terrestrial ecosystems to the atmosphere in 1980 due to changes in land use: geographical distribution of the global flux. Tellus 39B:122-139.

Huke, R.E. (1982) Rice area by type of culture: South, Southeast and East Asia. International Rice Research Intstitute, Los Baños, Laguna, Philippines, 32 pp.

Hueck, K. (1966) Die Wälder Südamerikas. Fischer Verlag, Stuttgart. 422pp.

Hueck, K. and P. Seibert (1972) Vegetationskarte von Südamerika. Gustav Fischer Verlag, Stuttgart.

Huete, A.R., R.D. Jackson and D.F. Post (1985) Spectral response of a plant canopy with different soil backgrounds. Remote Sensing of Environment 17:37-53.

Hummel, J. and R. Reck (1979) A global surface albedo model. Journal of Applied Meteorology 18:239-253.

Hutchinson, G.L. and A.R. Mosier (1979) Nitrous oxide emissions from an irrigated corn field. Science 205:1125-1127.

Hutchinson, G.L. and A.R. Mosier (1981) Improved soil cover method for field measurement of nitrous oxide flux. Soil Science Society of America Journal 45:311-316.

Hutchinson, G.L., A.R. Mosier and C.E. Andre (1982) Ammonia and Amine emissions from a large cattle feedlot. Journal of Environmental Quality 11:288-293.

Huttel, C. (1962) Estimation du bilan hydrique dans une foret sempervirente de basse Cote d'Ivoire. In: Radioisotopes in soil plant nutritiion studies, p 125-138. Vienne, AIEA, 1965. Quoted in Unesco (1978).

Idso, S.B., R.D. Jackson and R.J. Reginato (1977) Remote sensing of crop yields. Science 196:19-25.

Institut de la Carte Internationale de la Vegetation (1988) Catalogue 1988, Cartes, Livres. Université Paul Sabatier, 39 Allées Jules Guesde, 31062 Toulouse Cedex-France.

Jackson, P. (1983) The tragedy of our tropical rainforests. Ambio 12:252-254.

Jackson,R.D. and S.B. Idso (1975) Surface albedo, soils and desertification. Science 189:1012-1015.

Jackson, R.D., P.J. Pinter and R.J. Reginato (1985) Net radiation calculated from remote multispectral and ground station meteorological data. Agricultural and Forest Meteorology 35:153-164.

Jackson, R.D., M.S. Moran, L.W. Gay and L.H. Raymond (1987) Evaluating evaporation from field crops using airborne radiometry and ground-based meteorological data. Irrigation Science 8:81-90.

Jacobsen, P., W.H. Patrick Jr. and B.G. Williams (1981) Sulfide and methane formation in soils and sediments. Soil Science 132:279-287.

Janssen, B.H. (1984) a simple method for calculating decomposition and accumulation of 'young' soil organic matter. Plant and Soil 76:297-304.

Jarvis, P.G. and K.G. mcNaughton (1986) Stomatal control of transpiration, scaling up from leaf to region. In: Advances in Ecological Research 15:1-49. Academic Press London.

Jenkinson, D.S. (1977) Studies on the decomposition of plant material in soil. V. The effects of plant cover and soil type on the loss of carbon from ^{14}C labelled ryegrass decomposing under field conditions. Journal Soil Science 28:424-434.

Jenkinson, D.S. and A. Ayanaba (1977) Decomposition of carbon-14 labelled plant material under tropical conditions. Soil Science Society of America Journal 41:912-915.

Jenkinson, D.S. and J.H. Rayner (1977) The turnover of soil organic matter in some of the Rothamsted classical experiments. Soil Science 123:298-305.

Jenny, H. (1941) Factors of Soil formation. McGrawHill, New York.

Jenny, H., S.P. Gessel and F.T. Bingham (1949) Comparative study of decomposition rates of organic matter in temperate and tropical regions. Soil science 68:419-432.

Jenny, H., Salem, A.E. and J.R. Wallis (1968) Interplay of soil organic matter and soil fertility with state factors and soil properties. In: Study week on organic matter and soil fertility, p 5-36. Pontificea Academicae Scientinarum Scripta Varia 31. North Holland Publ. Co. Amsterdam, the Netherlands.

Johansson, C.L. (1984) Field measurements of emission of nitric oxide from fertilized and unfertilized forest soils in Sweden. Journal of Atmospheric Chemistry 1:429-442.

Johansson, C. and I.E. Galbally (1984) Production of nitric oxide in loam under aerobic and anaerobic conditions. Applied and Environmental Microbiology 47:1284-1289.

Johansson, C. and L. Granat (1984) Emission of NO from arable land. Tellus 36B:27-37.

Johansson, C., H. Rodhe and E. Sanhueza (1988) Emission of NO in a tropical savanna and a cloud forest during the dry season. Journal of Geophysical Research 93:7180-7192.

Johansson, C. and E. Sanhueza (1988) Emission of NO from savanna soils during rainy season. Journal of Geophysical Research 93:14193-14198.

Jones, M.J. (1972) The organic matter content of the savanna soils of West Africa. Journal of Soil Science 24:42-53.

Jong, E. de and H.J.V. Schappert (1972) Calculation of soil respiration and activity from CO_2 profiles in the soil. Soil Science 113:328-333.

Justice, C.O., B.N. Holben and M.D. Gwynne (1986) Monitoring East African vegetation using AVHRR data. International Journal of Remote Sensing. 7:1453-1475.

Justice, C.O., J.R.G. Townshend, B.N. Holben and C.J. Tucker (1985) Analysis of the phenology of global vegetation using meteorological satellite data. International Journal of Remote Sensing 6:1271-1318.

Kaplan, W. (1984) Sources and sinks of nitrous oxide. In: M.J. Klug and C.A. Reddy (Eds.), Current perspectives in microbial ecology, p 479-483. American Society of Meteorology.

Kaplan, W.A., S.C. Wofsy, M. Keller and J.M. de Costa (1988) Emission of NO and deposition of O_3 in a tropical forest system. Journal of Geophysical Research 93:1389-1395.

Keeney, D.R., I.R. Fillery, and G.P. Marx (1979) Effect of temperature on gaseous N products of denitrification in soil. Soil Science Society of America Journal 43:1124-1128.

Keepin, W., I. Mintzer and L. Kristoferson (1986) Emission of CO2 into the atmosphere. The rate of release of CO2 as a function of future energy developments. In: Bolin, B., B.R. Döös, J. Jager and R.A. Warrick (Eds.): The greenhouse effect, climatic change and ecosystems. p 35-91. SCOPE Vol.29. Wiley and Sons, New York.

Keller, M., W.A. Kaplan and S.C. Wofsy (1986) Emissions of N_2O, CH_4 and CO_2 from tropical forest soils. Journal of Geophysical Research 91:11791-11802.

Keller, M., W.A. Kaplan, S.C. Wofsy and J.M. Da Costa (1988) Emission of N_2O from tropical soils: response to fertilization with NH_4^+, NO_3^- and PO_4^{3-}. Journal of Geophysical Research 93:1600-1604.

Keller, M., S.C. Wofsy, T.J. Goreau, W.A. Kaplan, M.B. McElroy (1983) Production of nitrous oxide and consumption of methane by forest soils. Geophysical Research Letters 10:1156-1159.

Khalil, M.A.K. and R.A. Rasmussen (1983a) Increase and seasonal cycles of nitrous oxide in the earth's atmosphere. Tellus 35B:161-169.

Khalil, M.A.K. and R.A. Rasmussen (1983b) Sources, sinks and seasonal cycles of atmospheric methane. Journal of Geophysical Research 88:5131-5144.

Khalil, M.A.K. and R.A. Rasmussen (1984a) The global increase of carbon monoxide. In: V.D. Aneja (Ed.), Transactions APCA Specialty Conference Environmental Impact of Natural Emmissions, March 1984. p 403-414.

Khalil, M.A.K. and R.A. Rasmussen (1984b) Carbon monoxide in the earth's atmosphere: increasing trend. Science 224:54-56.

Khalil, M.A.K. and R.A. Rasmussen (1985) Causes of increasing atmospheric methane: depletion of hydroxyl radicals and the rise of emissions. Atmospheric Environment 19:397-407.

Kimura, M., H. Wada and Y. Takai (1984) Studies on the rhizosphere of paddy rice. IX. Microbial activities in the rhizosphere of paddy rice. Japanese Journal of Soil Science and Plant Nutrition 55:338-343 (In Japanese).

King, G.M. and W.J. Wiebe (1978) Methane release from soils of a Georgia solt marsh. Geochimica et Cosmochimica Acta 42:343-348.

Klemedtsson, L., B.H. Svensson and T. Rosswall (1987) Dinitrogen and nitrous oxide produced by denitrification and nitrification in soil with and without barley plants. Plant and Soil 99:303-319.

Kononova, M.M. (1975) Humus of virgin and cultivated soils. In: Gieseling, J.E.(Ed.): Soil components. Vol.I. Organic components. p 475-526. Springer Verlag, Berlin.

Köppen, W. (1936) Das Geographische System der Klimate. In: Köppen, W. and Geiger, G. (Eds.), Handbuch der Klimatologie, Vol. 1, Part C. Gebr. Borntraeger, Berlin.

Kortleven, J. (1963) Kwantitatieve aspecten van humusopbouw en humusafbraak. Verslagen Landbouwkundige Onderzoekingen 69-1. 109 pp. PUDOC. Wageningen, the Netherlands.

Kovacs, G. (1987) Estimation of average areal evapotranspiration - proposal to modify Morton's model based on the complementary character of actual and potential evapotranspiration. Journal of Hydrology 95:227-240.

Kovda, V.A., Ye.V. Lobova, and B.G. Rozanov (1967) Classification of the world's soils. Soviet Soil Science 851-863.

Kovda, V.A., V.E. Lobova, G.V. Dobrovolsky, J.M. Ivanv, B.G. Rozanov and N.A. Solomatina (1975) Soil Map of the World. 1:10,000,000 (In Russian and English). Institute for Geodesy and Cartography, Moscow.

Koyama, T. (1963) Gaseous metabolism in lake sediments and paddy soils and the production of atmospheric methane and hydrogen. Journal of Geophysical Research 68:3971-3973.

Küchler, A.W. (1965) Potential natural vegetation map at 1:7,500,000. Revised Edition. USDA Geological Survey, Sheet No. 90.

Lal, R. (1981) Deforestation of tropical rainforest and hydrological problems. In: R. Lal and E.D. Russell (Eds.), Tropical Agricultural Hydrology. Watershed Management and land use, p 131-151. Wiley Interscience Publication. Wiley and Sons, New York.

Lal, R. (1986) Tropical ecology and physical edaphology. Wiley Interscience Publication. Wiley and Sons, New York.

Lanly, J.P. (1982) Tropical forest resources. FAO Forestry Paper No. 30. 106 pp. FAO, Rome.

Latz, K., R.A. Weismiller, G.E. Van Scoyoc and M.F. Baumgardner (1984) Characteristic variations in spectral reflectance of selected eroded alfisols. Soil Science Society of America Journal 48:1130-1134.

Leemans, R. (1989) World Map ofHoldridge Life Zones (digital data set) 0.5° resolution map, presenting the 38 Holdridge life zones. IIASA, Laxenburg, Austria.

Letey, J., M. Valoras, D.D. Focht and J.C. Ryden (1981) Nitrous oxide production and reduction during denitrification as affected by redox potenital. Soil Science Society of America Journal 45:727-730.

Levine, J.S., T.R. Augustsson, I.C. Anderson, J.M. Hoell jr. and D.A. Brewer (1984) Tropospheric sources of NO$_x$: lightning and biology. Atmospheric Environment 18:1997-1804.

Lipschultz, F., O.C. Zafiriou, S.C. Wofsy, M.B. McElroy, F.W. Valois and S.W. Watson (1981) Production of NO and N$_2$O by soil nitrifying bacteria. Nature 294:641-643.

Livingston, G.P., P.M. Vitousek and P.A. Matson (1988) Nitrous oxide flux and nitrogen transformations across a landscape gradient in Amazonia. Journal of Geophysical Research 93:1593-1599.

Lo, C.P. (1987) Applied remote sensing. Longman, Essex.

Logan, J.A. (1983) Nitrogen oxides in the troposphere. Global and regional budgets. Journal of Geophysical Research 88:10785-10807.

Logan, J.A., M.J. Prather, S.C. Wofsy and M.B. McElroy (1981) Tropospheric chemistry: a global perspective. Journal of Geophysical Research 86:7210-7254.

Lugo, A.E., M.J. Sanchez and S. Brown (1986) Land use and organic carbon content of some subtropical soils. Plant and Soil 96:185-196.

Lull, H.W. and W.E. Sopper (1966) Factors that influence streamflow in the northeast. Water Resources Research 2:371-379.

Lundegard, H. (1927) Carbon dioxide evolution of soil and crop growth. Soil Science 23:417-454. Quoted in: Buyanovsky et al (1986).

MacPherson, J.I., R.L. Desjardins and P.H. Schuepp (1987) Gaseous exchange measurements using aircraft-mounted sensors. In: Preprint of Proceedings 6th Symposium on Meteorological Observations and Instrumentation, January 12-16, 1987. American Meteorological Society, Boston, Massachussets.

Makkink, G.F. (1957) Testing the Penman formula by means of lysimeters. Journal Inst. Water Engineering London. 11:277-288.

Malaisse, F. (1973) Contribution a l'etude de l'ecosysteme foret clair (miombo). Note 8. Le Projet Miombo. Ann. University Abidjan., E., Vol.6, No.2, 1973, pp 227-250. Quoted in Unesco (1978).

Malingreau, J.P. and C.J. Tucker (1987) The contribution of AVHRR data for measuring and understeating global processes: large scale deforestation in the Amazon Basin. Proceedings of IGARSS '87 Symposium, Ann Arbor, pp 443-448.

Malingreau, J.P. and C.J. Tucker (1988) Large scale deforestation in the Southeastern Amazon Basin of Brasil. Ambio 17:49-55.

Manabe, S. and R.T. Wetherald (1987) Large scale changes of soil wetness induced by an increase in atmospheric carbon dioxide. Journal of Atmospheric Science 44:1211-1235.

Mann, L.K. (1986) Changes in soil carbon storage after cultivation. Soil Science 142:279-288.

Manning, M.R., D.L. Lowe, W.H. Melhuish, R.J. Sparks, G. Wallace and C.A.M. Breuninkmeijer (1989). The use of radiocarbon measurements in atmoshperic studies. Radiocarbon 31 (in press).

Marks, P.L. and F.H. Bormann (1972) Revegetation and forest cutting: mechanism for return to steady state nutrient cycling. Science 176:914-915.

Martel, Y.A. and E.A. Paul (1974) The use of radiocarbon dating of organic matter in the study of soil genesis. Soil Science Society of America Journal 38:501-506.

Matthews, E. (1983) Global vegetation and land use: new high resolution data bases for climate studies. Journal of Climate and Applied Meteorology 22:474-487.

Matthews, E. (1984) Vegetation, land-use and seasonal albedo data sets. Documentation of archived data tape. NASA Technical memorandum 86107.

Matthews, E. (1985) Atlas of archived vegetation, land-use and seasonal albedo data sets. NASA Technical memorandum 86199.

Matthews, E. and I. Fung (1987) Methane emission from natural wetlands: Global distribution, area, and environmental characteristics of sources. Global Biogeochemical Cycles 1:61-86.

Matthews, E. and W.B. Rossow (1987) Regional and seasonal variations of surface reflectance from satellite observations at 0.6 μm. Journal of Climate and Applied Meteorology 26:170-202.

McElroy, M.B. and S.C. Wofsy (1986) Tropical forests, interaction with the atmosphere. In: G.T. Prance (Ed.), Tropical forests and World Atmosphere. AAAS symp. Westview Press, Boulder, Colorado.

McKenney, D.J., K.F. Shuttleworth and W.I. Findlay (1980) Nitrous oxide evolution rates from fertilized soils: effects of applied nitrogen. Canadian Journal of Soil Science 60:429-438.

McKenney, D.J., K.F. Shuttleworth, J.R. Vriesacker and W.I. Findlay (1982) Production and loss of nitric oxide from denitrification in anaerobic Brookston clay. Applied Environmental Microbiology 43:534-541.

Mc.Naughton, U.G. and P.G. Jarvis (1983) Predicting effects of vegetation changes on transpiration and evaporation. In: T.T. Kozlowski (Ed.), Water Deficits and Plant Growth, Vol. VII, p 1-47. Academy Press, New York.

Meentemeyer, V., E.O. Box, M. Folkoff and J. Gardner (1981) Climatic estimation of soil properties; soil pH, litter accumulation and soil organic content. Ecological Society of America Bulletin 62:104.

Menenti, M. (1984) Physical aspects and determination of evaporation in deserts, applying remote sensing techniques. 202 pp. PhD thesis and report No.10 of ILRI, Wageningen, The Netherlands.

Milford, J.R. (1987) Problems of deducing the soil water balance in dryland regions from METEOSAT data. Soil Use and Management 3:51-57

Minami, K.(1987) Emission of nitrous oxide (N_2O) from Agro-ecosystem. JARQ Vol.21:22-27.

Minami, K. and S. Fukushi (1983) Effects of phosphate and calcium carbonate application on emission of N_2O from soils under aerobic conditions. Soil Science and Plant Nutrition 29:517-524.

Minami, K. and S. Fukushi (1984) Methods for measuring N_2O flux from water surface and N_2O dissolved in water from agricultural land. Soil Science and Plant Nutrition 30:495-500.

Minami, K. and S. Fukushi (1986) Emission of nitrous oxide from a well aerated andosol treated with nitrite and hydroxylamine. Soil Science and Plant Nutrition 32:233-237.

Minami, K. and K. Yagi (1988) methode for measuring methane flux from rice paddies. Japanese Journal of Soil Science and Plant Nutrition 59:458-463. (In Japanese).

Minderman, G. (1968) Addition, decomposition and accumulation of organic matter in forests. Journal of Ecology 56:355-362.

Minderman, G. and J. Vulto (1973) Carbon dioxide production by tree roots and microbes. Pedobiologia Bd.18:337-343.

Mitsch, W.J. and J.G. Gosselink (1986) Wetlands. Van Nostrand Reinhold Company, NY.

Molion, L.C.B. (1987) Micrometeorology of an Amazonian rain Forest. In: R.E. Dickinson (Ed.), The Geophysiology of Amazonia. Vegetation and Climate Interactions, p 391-407. Wiley and Sons, New York.

Monteith, J.L. (1965) Evaporation and environment. Proceedings Symposium Society of Biology 19:205-234.

Monteith, J.L. and G. Szeics (1962) Radiative temperatures in the heat balance of natural surfaces. Quarterly Journal of the Royal Meteorological Society 88:496-507.

Monteith, J.L., G. Szeicz and K. Yabuki (1964) Crop photosynthesis and the flux of carbon dioxide below the canopy. Journal of Applied Ecology 6:321-337. Quoted in Buyanovsky et al (1986).

Mooney, H.A., P.M. Vitousek and P.A. Matson (1987) Exchange of materials between terrestrial ecosystems and the atmosphere. Science 238:926-932.

Moore, T.R. and R. Knowles (1987) Methane and Carbondioxide evolution from suarctic flux. Canadian Journal of Soil Science 67:77-81.

Moormann, F.R. (1981) The classification of "paddy soils" as related to soil taxonomy. In: Institute of Soil Science, Academia Sinica (Ed.): Proceedings of symposium on paddy soil, p 139-150. Science Press, Beijing, Springer Verlag. Berlin, Heidelberg, New York.

Moormann, F.R. and N. van Breemen (1978) Rice, Soil, Water, Land. International Rice Research Institute. Los Baños, Laguna, Philipines.

Morton, F.I. (1983) Operational estimates of areal evapotranspiration and their significance to the science and practice of hydrology. Journal of Hydrology 66:1-76.

Morton, F.I. (1984) What are the limits of forest evapotranspiration. Journal of Hydrology 74:373-398.

Morton, F.I. (1985) What are the limits on forest evaporation - reply. Journal of Hydrology 82:184-192.

Mosier, A.R. and G.L. Hutchinson (1981) Nitrous oxide emissions from cropped fields. Journal of Environmental Quality 10:169-173.

Mosier, A.R., G.L. Hutchinson, B.R. Sabey and J. Baxter (1982) Nitrous oxide emissions from barley plots treated with ammonium nitrate or sewage sludge. Journal of Environmental Quality 11:78-81.

Mosier, A.R. and W.J. Parton (1985) Denitrification in a shortgrass prairie: a modelling approach. In: D.E. Caldwell, J.A. Brierley and C.L. Brierley (Eds.), Planetary Ecology, p 441-451. Van Nostrand Reinhold Co., New York.

Mosier, A.R., M. Stillwell, W.J. Parton and R.G. Woodmansee (1981) Nitrous oxide emissions from a native shortgrass prairie. Soil Science Society of America Journal 45:617-619.

Mosier, A.R., W.D. Guenzi and E.E. Schweizer (1986) Soil losses of dinitrogen and nitrous oxide from irrigated crops in northeastern Colorado. Soil Science Society of America Journal 50:344-348.

Mueller-Dombois, D. (1984) Classification and mapping of plant communities: a review with emphasis on tropical vegetation. In: G.M. Woodwell (Ed.), The role of terrestrial vegetation in the global carbon cycle. Measurement by remote sensing, p 21-90. SCOPE Vol. 23, Wiley and Sons, New York.

Mulder, J.P.M. (1985) Simulating interception loss using standard meteorological data. In: B.A. Hutchison and B.B. Hicks (Eds.), The Forest Atmosphere Interaction, pp 177-196. D. Reidel Publ. Comp., Dordrecht.

Muzio and Kramlich (1988) Artifact in the measurement of N_2O of combustion sources. Geophysical Research Letters 15:1369-1372.

Mulders, M.A. (1987) Remote sensing in soil science. Elseviers publ. comp. Amsterdam, New York.

Mulvaney, R.L. and L.T. Kurtz (1984) Evolution of dinitrogen and nitrous oxide from nitrogen-15 fertilized soil cores subjected to wetting and drying cycles. Soil Science Society of America Journal 48:596-602.

Myers, N. (1980) Conversion of tropical moist forests. National Academy of Science, Washington, D.C., 205pp.

Myrold, D.D. and J.M. Tiedje (1985) Diffusional constraints on denitrification in soil. Soil Science Society of America Journal 49:651-657.

Nelson, R. and B. Holben (1986) Identifying deforestation in Brazil using multiresolution satellite data. International Journal of Remote Sensing 7:429-448.

Neue, H.H., and H.W. Scharpenseel (1984) Gaseous products of decomposition of organic matter in submerged soils. In: International Rice Research Institute, Organic Matter and Rice, p 311-328. IRRI, Los Baños, Laguna, Philipines.

Nieuwenhuis, G.J.A., E.H. Smidt and H.A.M. Thunnissen (1985) Estimation of regional evapotranspiration of arable crops from thermal infrared images. International Journal of Remote Sensing 6:1319-1334.

Nonhebel, S. (1987) Waterverbruik van Nederlandse Bossen: een modellen studie. Waterbeheer Natuur, Bos en Landschap rapport 7g, 47 pp + appendices, Studiecommissie Waterbeheer Natuur Bos en landschap.

Nordin, C.F. and R.H. Meade (1982) Deforestation and increased flooding of the upper Amazon. Science 215:426-427.

Nye, P.H., and D.J. Greenland (1960) The soil under shifting cultivation. Technical Communication N°51, 156 pp. Commonwealth Bureau of Soils, Harpenden.

Oades, J.M. (1988) The retention of organic matter in soils. Biogeochemistry 5:35-70.

O'Deen, W.A.and K.L. Porter (1986) Continuous flow system for collecting volatile ammonia and amines from senescing winter wheat. Agronomy Journal 78:746-749.

Odum, E.P. (1959) Fundamentals of Ecolgy, 2nd edition, W.B. Saunders Company, 546 pp.

O'Hara, G.W. and R.M. Daniel (1985) Rhizobial denitrification: a review. Soil Biology and Biochemistry 17:1-9.

Olson, J. (1963) Energy storage and the balance of producers and decomposers in ecological systems. Ecology 44:322-331

Olson, J.S., J.A. Watts and L.J. Allison (1983) Carbon in live vegetation of major world ecosystems. ORNL 5862. Environmental Sciences Division, Publ. No.1997. Oak Ridge National Laboratory, Oak Ridge, Tennessee. National Technical Information Service. U.S. Dept. Commerce.

Oremland, R.S. (1988) Biochemistry of methanogenic bacteria. In: Zehnder, A.J.B. (Ed.): Biology of anaerobic microorganisms, pp 641-706, Wiley and Sons, New York.

Oremland, R.M., L.M. Marsh and S. Polein (1982) Methane production and simultaneous sulfate reduction in anoxic marsh sediments. Nature 296:143-145.

Otterman, J. (1974) Baring high albedo soils by overgrazing: a hypothesized desertification mechanism. Science 186:531-533.

Palacpac, A.C. (1982) World Rice Statistics. International Rice research Institute, Dept. Agr. Economics, 152 pp.

Palm, C.A., R.A. Houghton and J.M. Melillo (1986) Atmospheric carbon dioxide from deforestation in Southeast Asia. Biotropica 18:177-188.

Parkin, T.B. (1987) Soil microsites as a source of denitrification variability. Soil Science Society of America Journal 51:1194-1199.

Parton, W.J, A.R. Mosier and D.S. Schimel (1988) Rates and pathways of nitrous oxide production in a shortgrass steppe. Biogeochemistry 6:45-48

Parton, W.J., D.S. Schimel, C.V. Cole and D.S. Ojima (1987) Analysis of factors controlling soil organic matter levels in Great Plains grasslands. Soil Science Society of America Journal 51:1173-1179.

Patrick, W.H. (1981) The role of inorganic redox systems in controlling reduction in paddy soils. In: Proceedings of Symposium on Paddy Soil, p 107-117. Science Press, Beijing, Springer Verlag, Berlin.

Paul, E.A. and J.A. Van Veen (1978) The use of tracers to determine the dynamic nature of organic matter. Paper presented at the 11th International Congress of Soil Science, June 19-27 1978, Edmonton, Canada.

Paulson, C.A. (1970) Mathematical representation of wind speed and temperature profiles in the unstable atmospheric surface layer. Journal of Applied Meteorology 9:857-861.

Pearce, A.J., L.K. Rowe and J.B. Stewart (1980) Nighttime wet canopy evaporation rates and the water balance of an evergreen forest. Water Resources Research 16:955-959.

Peck, A.J. and D.R. Williamson (1987) Effects of forest clearing on groundwater. Journal of Hydrology 94:47-65.

Penman, H.L. (1948) Natural evaporation from open water, bare soil and grass. Proceedings Royal Society of London Ser.A.193, p 120-145.

Pereira, H.C. (1973) Land use and water resources. Cambrdige University Press. 246 pp.

Perrier, A. (1980) Etude micro-climatique des rélation entre les propriété's de surface et les charactéristiques de l'air: application aux échanges regionaux. Meteorologie et Environment, EVRY (France), Octobre.

Persson, R. (1974) World forest resources. Royal College of Forestry, Stockholm No.17, 261pp.

Ponnamperuma, F.N. (1981) Some aspects of the physical chemistry of paddy soils. In: Proceedings of Symposium on Paddy Soil, p 59-94. Science Press, Beijing, Springer Verlag, Berlin.

Ponnamperuma, F.N. (1985) Chemical kinetics of wetland rice soils relative to soil fertility. In: Wetland Soils: Characterization, classification and utilization, p 71-89. Proceedings of Workshop IRRI 1984. IRRI, Los Baños.

Post, W.M., W.R. Emanuel, P.J. Zinke and A.G. Staugenberger (1982) Soil carbon pools and world life zones. Nature 298:156-159.

Poth, M. and D.D. Focht (1985) ^{15}N kinetic analysis of N_2O production by Nitrosomonas europaea: an examination of nitrifier denitrification. Applied Environmental Microbiology 49:1134-1141.

Potter, G.L., H.W. Ellsaessen, M.C. MacCracken and F.M. Luther (1975) Possible climatic effects of tropical deforestation. Nature 258:697-698.

Pratt, D.J. and M.D. Gwynne (1977) Rangeland Management and ecology in East Africa. Hodder and Stoughton, London.

Rakhmanov, V.V. (1970a) Dependence of streamflow upon the percentage of forest cover of catchments. pp 55-63, Proceedings FAO/USSR International Symposium on Forest Influences and Watershed Management, FAO, Rome.

Rakhmanov, V.V. (1970b) Effect of forests on runoff in the Upper Volga basin. pp 187-204, Proceedings FAO/USSR International Symposium on Forest Influences and Watershed Management, FAO, Rome.

Ramanathan, V., R.J. Cicerone, H.B. Singh and J.T. Kiehl (1985) Trace gas trends and their potential role in climatic change. Journal of Geophysical Research 90:5547-5566.

Rasmussen, R.A. and M.A.K. Khalil (1981) Atmospheric methane (CH_4): trends and seasonal cycles. Journal of Geophysical Research 86:9826-9832.

Rasmussen, R.A. and M.A.K. Khalil (1983) Global production of methane by termites. Nature 301:700-702.

Rasmussen, R.A. and M.A.K. Khalil (1986) Atmospheric trace gases: trends and distributions over the last decade. Science 232:1623-1624.

Reginato, R.J., R.D. Jackson and P.J. Pinter (1985) Evapotranspiration calculated from remote multispectral and ground station meteorological data. Remote Sensing of Environment 18: 75-89.

Reiners, W.A. (1968) Carbon dioxide evolution from the floor of three Minnesota forests. Ecology 49:471-483.

Reinke, J.J., D.C. Adriano and K.W. Mc Leod (1981) Effects of litter alteration on carbon dioxide evolution from a South Carolina pine forest floor. Soil Science Society of America Journal 45:620-632.

Ribeiro, M.M.G., E. Salati, N.A. Villa Nova, G.G. Demetrio (1982) Radicao solar disponivel em Manaus (Am) e su relacao com a duracao do brilho solar. Acta Amazon 12:339-346. Quoted in Salati (1987).

Rjabchakov, A.M. (1986) World Map on Actual Land Use at 1:15,000,000 (with 80 types). Legend translated in English at ISRIC. Faculty of Geography, Moscow State University.

Rockwood, A.A. and S.K. Cox (1978) Satellite inferred surface albedo over northwestern Africa. Journal of Atmospheric Science 35:513-522.

Rodhe, H., and R. Herrera (Eds.) (1988) Acidification in tropical countries. SCOPE Vol. 36, 405 p., Wiley and Sons, New York.

Rolston, D.E. (1981) Nitrous oxide and nitrogen gas production in fertilizer loss. In: C.C. Delwiche (Ed.), Denitrification, nitrification and atmospheric nitrous oxide. Wiley and Sons, New York.

Rolston, D.E., D.L. Hoffmann and D.A. Goldhamer (1976) Denitrification measured directly from nitrogen and nitrous gas fluxes. Soil Science Society of America Journal 40:259-266.

Rolston, D.E., D.L. Hoffman and D.W. Toy (1978) Field measurement of denitrification: I. Flux of N_2 and N_2O. Soil Science Society of America Journal 42:863-869.

Rotty, R.M. (1987) A look at 1983 CO_2 emissions from fossil fuels (with preliminary data for 1984). Tellus 39B:203-208.

Rubinoff, I. (1983) Strategy for preserving tropical rainforest. Ambio 12:255-258.

Russell, E.W. (1981) Role of watershed management for arable land use in the tropics. In: R. Lal and E.W. Russell (Eds.), Tropical Agricultural Hydrology, p 11-16. Wiley and Sons, New York.

Rutter, A.J., K.A. Kershaw, P.C. Robbins and A.J. Morton (1971) A predictive model of rainfall interception in forests. 1. Derivation of the model from observations in a plantation of Corsican Pine. Agricultural and Forest Meteorology 9:367-384.

Ryden, J.C. (1981) N_2O exchange between a grassland soil and the atmosphere. Nature 292:235-237.

Ryden, J.C. (1983) Denitrification loss from a grassland soil in the field receiving different rates of nitrogen as ammonium nitrate. Journal of Soil Science 1983:355-365.

Ryden, J.C., L.J. Lund and D.D. Focht (1978) Direct in-field measurement of nitrous oxide flux from soils. Soil Science Society of America Journal 42:731-737.

Ryden, J.C., L.J. Lund and D.D. Focht (1979) Direct measurement of denitrification loss from soils II. Developmentand application of field methods. Soil Science Society of America Journal 43:110-118.

Ryden, J.C. and L.J. Lund (1980) Nature and extent of directly measured denitrification losses from some irrigated crop production units. Soil Science Society of America Journal 44:505-511.

Sachse, G.W., R.C. Harriss, J. Fishman, G.F. Hill and D.R. Cahoon (1988) Carbon monoxide over the Amazon Basin during the 1985 dry season. Journal of Geophysical Research 93:1422-1430.

Sadovnikov, Yn.N.(1979) Changes in spectral reflectivity over the profiles of the major soil genetic types. Moscow University Soil Science Bulletin, 1979 35,1:34-39.

Sagan, C., B.T. Owen and J.D. Pollack (1979) Anthropogenic albedo changes and the earth's climate. Science 206:1363-1368.

Sahrawat, K.L. and D.R. Keeney (1986) Nitrous oxide emission from soils. Advances in Soil science Vol.4 pp 103-148. Springer Verlag, New York.

Sahrawat, K.L., D.R. Keeney and S.S. Adams (1985) Role of aerobic transformations in six acid climax forest soils and the effect of phosphorous and $CaCO_3$. Forest Science 31:680-684.

Salati, E. (1987) The forest and the hydrological cycle. In: R.E. Dickinson (Ed.), The Geophysiology of Amazonia, Vegetation and Climate Interactions, p 273-296. Wiley and Sons, New York.

Salati, E., T.E. Lovejoy and P.B. Vose (1983) Precipitation and water recycling in tropical rainforests. Environmentalist 3:67-74.

Salati, E. and P.B. Vose (1984) Amazon Basin: a system in equilibrium. Science 225:129-138.

Sanchez, P.A. (1976) Properties and management of soils in the tropics. Wiley Interscience Publication. Wiley and Sons, New York.

Schanda, E. (1986) Fysical fundamentals of remote sensing. Springer Verlag, Berlin, Heidelberg, New York, Toronto.

Scharpenseel, H.W. and H. Schiffmann (1977) Radiocarbon dating of soils, a review. Zeitschrift für Pflanzenernährung und Bodenkunde 140: 159-174.

Schimel, D.S., D.C. Coleman and K.A. Horton (1985a) Soil organic matter dynamics in paired rangeland and cropland toposequences in North Dakota. Geoderma 36:201-214.

Schimel, D.S., M.A. Stillwell and R.G. Woodmansee (1985b) Biogeochemistry of C, N and P in a soil catena of shortgrass steppe. Ecology 66:276-282.

Schlesinger, W.H. (1977) Carbon balance in terrestrial detritus. Annual Revue of Ecology and Systematics 8:51-81.

Schlesinger, W.H. (1982) Carbon storage in the caliche of arid soils. a case study from Arizona. Soil Science 133:247-255.

Schlesinger, W.H. (1984) Soil organic matter: a source of atmospheric CO_2. In: Woodwell, G.W. (Ed.): The role of terrestrial vegetation in the global carbon cycle, p 111-127. SCOPE Vol. 23. Wiley and Sons, New York.

Schlesinger, W.H. (1985) The formation of caliche in soils of the Mojave Desert, California. Geochimica et Cosmochimica 49:57-66.

Schlesinger, W.H. (1986) Changes in soil carbon storage and associated properties with disturbance and recovery. In: Trabalka, J.R. and D.E. Reichle (Eds.), The changing carbon cycle. A global analysis. p 194-220. Springer Verlag.

Schmidt, J., W. Seiler and R. Conrad (1988) Emission of nitrous oxide from temperate forest soils into the atmosphere. Journal of Atmospheric Chemistry 6:95-115.

Schmithüsen, J. (1968) 1:25 million vegetation maps of Europe, North Asia, South Asia, South West Asia, Australia, North Africa, Southern Africa, North America, Central America, South America (northern part), South America (southern part). In: GroBes Duden-Lexikon, Vol. 8, 321-346. Bibliographisches Institut A.G., Mannheim.

Schroeder, F.G. (1983) Die thermischen Vegetationszonen der Erde. Ein Beitrag zur Präzisierung der geobotanischen Terminologie. Mit einer Vegetationskarte. Sonderdrück aus Tuexenia, Mitteilungen der Floristisch-soziologischen Arbeitsgemeinschaft, Neue Serie, Band Nr.3, Tuexenia, Göttingen.

Schuepp, P.H., R.L. Desjardins, J.I. MacPherson, J. Boisvert and L.B. Austin (1987) Airborne determination of regional water use efficiency and evapotranspiration: present capabilities and initial field tests. Agricultural and Forest Meteorology 41:1-19.

Schütz, H., A. Holzapfel-Pschorn, R. Conrad, H. Rennenberg and W. Seiler (1989). A three years continuous record on the influence of daytime, season and ferilizer treatment on methane emission rates from an Italian rice paddy field. Submitted to Journal of Geophysical Research April 1989.

Sebacher, D.I., R.C. Harriss and K.B. Bartlett (1983) Methane flux across the air-water interface: air velocity effects. Tellus 35B:103-109.

Sebacher, D.I., R.C. Harriss and K.B. Bartlett (1985) Methane emissions to the atmosphere through aquatic plants. Journal of Environmental Quality 14:40-46.

Sebacher, D.I., R.C. Harriss, K.B. Bartlett, S.M. Sebacher, S.S. Grice (1986) Atmospheric methane sources: Alaskan tundra bogs, an alpine fen, and a subarctic boreal marsh. Tellus 38B:1-10.

Sedjo, R.A. (1989) Forests to offset the Greenhouse Effect. Journal of Forestry July 1989:12-15.

Seiler, W. (1984) Contribution of biological processes to the global budget of CH_4 in the atmosphere. In: M.J. Klug and C.A. Reddy (Eds.), Current perpectives in microbial ecology, p 468-477. American Society of Meteorology.

Seiler, W. and R. Conrad (1981) Field measurements of natural and fertilizer induced N_2O release rates from soils. Journal Air Poll. Contr. Ass. 31:767-772.

Seiler, W. and R. Conrad (1985) Exchange of atmospheric trace gases with anoxic and oxic tropical ecosystems. In: R. Dickinson (Ed.), Geophysiology of Amazonia, p 133-160. Wiley and Sons, New York.

Seiler, W. and R. Conrad (1987) Contribution of tropical ecosystems to the global budgets of trace gases, especially CH_4, H_2, CO and N_2O. In: R.E. Dickinson (Ed.), Geophysiology of Amazonia. Vegetation and Climate Interactions, p 133-160. Wiley and Sons, New York.

Seiler, W. and P.J. Crutzen (1980) Estimates of gross and net fluxes of carbon between the biosphere and the atmosphere from biomass burning. Climatic Change 2:207-247.

Seiler, W., A. Holzapfel-Pschorn, R. Conrad and D. Scharffe (1984) Methane emissions from rice paddies. Journal of Atmospheric Chemistry 1:241-268.

Sellers, P.J. (1985) Canopy reflectance, photosynthesis and transpiration. International Journal of Remote Sensing 6:1335-1372.

Sellers, P.J. (1987) Modeling effects of vegetation on climate. In: R.E. Dickinson (Ed.), The Geophysiology of Amazonia. Vegetation and climate interactions. p 297-339. Wiley and Sons, New York.

Sellers, P.J. and J.L. Dorman (1987) Testing the simple biosphere model (SiB) using point micrometeorological and biophysical data. Journal of Climate and Applied Meteorology 26:622-651.

Sextone, A.J., T.B. Parkin and J.M. Tiedje (1984) Temporal response of soil denitrification rates to rainfall and irrigation. Soil Science Society of America Journal 49:99-103.

Shields, J.A. and D.R. Coote (1988) SOTER procedures manual for small scale map and database compilation (for discussion). ISRIC Working Paper and Preprint 88/2. ISRIC, Wageningen.

Shuttleworth, W.J. and I.R. Calder (1979) Has the Priestley-Taylor equation any relevance to forest evaporation ? Journal of Applied Meteorology 18:639-646.

Shuttleworth, W.J., J.H.C. Gash, J.C.R. Lloyd, C.J. Moore, J. Roberts, A.de O. Marques Filho, G. Fisch, V. de Paula Silva Filho, L.C.B. Molion, L.D. de Abreu Sa, J.C.A. Nobre, O.M.R. Cabral, S.R. Patel, J.C. de Moraes (1984a) Observation of radiation exchanges above and below Amazonian forest. Quarterly Journal of the Royal Meteorological Society 110:1163-1169.

Shuttleworth, W.J., J.H.C. Gash, J.C.R. Lloyd, C.J. Moore, J. Roberts, A.de O. Marques Filho, G. Fisch, V. de Paula Silva Filho, L.C.B. Molion, L.D. de Abreu Sa, J.C.A. Nobre, O.M.R. Cabral, S.R. Patel, J.C. de Moraes (1984b) Eddy correlation measurements of energy partition for Amazonian forest. Quarterly Journal of the Royal Meteorological Society 110:1143-1162.

Shuttleworth, W.J., J.H.C. Gash, J.C.R. Lloyd, C.J. Moore, J. Roberts, A.de O. Marques Filho, G. Fisch, V. de Paula Silva Filho, L.C.B. Molion, L.D. de Abreu Sa, J.C.A. Nobre, O.M.R. Cabral, S.R. Patel, J.C. de Moraes (1985) Daily variations of temperature and humidity within and above Amazonian forest. Weather 40:102-108.

Simpson, J.R. and J.R. Freney (1986) Interacting processes of gaseous nitrogen loss from urea applied to flooded rice fields. Transactions of the XIIIth Congress of the International Society of Soil Science (ISSS), Hamburg, Vol. III, pp 968-969.

Simpson, J.R. and K.W. Steele (1983) Gaseous nitrogen exchanges in grazed pastures. In: J.R. Freney and J.R. Simpson (eds.), Gaseous losses of nitrogen from plant soil systems. Developments in Soil Science Vol.9. Martinus Nijhoff/Dr.W.Junk Publishers, The Hague.

Slemr, F., R. Conrad, and W. Seiler (1984) Nitrous oxide emissions from fertilized and unfertilized soils in a subtropical region (Andalusia, Spain). Journal of Atmospheric Chemistry 1:159-169.

Slemr, F. and W. Seiler (1984) Field measurement of NO and NO_2 emissions from fertilized and unfertilized soils. Journal of Atmospheric Chemistry 2:1-24.

Smith, C.J., M. Brandon and W.H. Patrick, Jr. (1982) Nitrous oxide emission following urea-N fertilization of wetland rice. Soil Science and Plant Nutrition 28:161-171.

Smith, C.J., R.D. Delaune and W.H. Patrick Jr. (1982) Carbon and nitrogen cycling in a spartina alterniflora salt marsh. In: J.R. Freney and I.E. Galbally (Eds.), Cycling of Carbon, nitrogen, sulphur and phosphorous in terrestrial and aquatic ecosystems. p. 97-103, Springer Verlag, New York.

Smith, C.J., R.D. Delaune and W.H. Patrick, Jr. (1983) Nitrous oxide emission from Gulf Coast Wetlands. Geochimica et Cosmochimica Acta 47:1805-1814.

Snedecor, G.W. and W.G. Cochran (1980) Statistical methods (7th edition). 507 p, the Iowa State University Press, Ames, Iowa.

Soderlund, R. and B.H. Svensson (1976) The global nitrogen cycle. In: B.H. Svensson and R. Soderlund (Eds.), Nitrogen, Phosphorous and Sulphur- Global cycles, p 23-73. SCOPE Vol.7, Ecological Bulletin, Stockholm.

Spycher, G., P. Sollins and S. Rose (1983) Carbon and nitrogen in the light fraction of a forest soil: vertical distribution and seasonal patterns. Soil Science 135:79-87.

Steele, L.P, P.J. Fraser, R.A. Rasmussen, M.A.K. Khalil, T.J. Conway, A.J. Crawford, R.U. Gamuron, K.A. Masaric and K.W. Thoning (1987) The global distribution of methane in the troposphere. Journal of Atmospheric Chemistry 5:127-171.

Stoner, E.R. and M.F. Baumgardner (1981) Characteristic variations in reflectance of surface soils. Soil Science Society of America Journal 45:1161-1165.

Stoner, E.R., M.F. Baumgardner, L.L. Biehl and B.F. Robinson (1980) Atlas of soil reflectance properties. Dept. Agron. Purdue University, West Lafayette, In. USA. Research Bulletin Agr. Exp. Station, Purdue University, 1980, No.962 75pp.

SSSA (1984) Glossary of Soil Science Terms. Publ. of the Soil Science Society of America, Madison WI 53711 USA.

Staatsbosbeheer (1986) Verslag van het vitaliteitsonderzoek 1986. Utrecht, afdeling bosontwikkeling, rapport 1986-21.

Staatsbosbeheer (1987) Effecten van luchtverontreiniging op terrestrische ecosystemen, met name bossen. Afd. Planologie en Milieu, Staatsbosbeheer.

Stanford, G. and S.J. Smith (1972) Nitrogen mineralization potential of soils. Soil Science Society of America Journal 36:465-472.

Steudler, P.A., R.D. Bowden, J.M. Melillo and J.D. Aber (1989) Influence of nitrogen fertilization on methane uptake in temperate forest soils. Nature 341:314-315.

Stevenson, F.J. (1986) Cycles of soil- carbon, nitrogen, phosphorous, sulphur, micronutrients. Wiley interscience publication. Wiley and Sons, New York.

Stevenson, F.J., R.M. Harrison, R. Wetselaar and R.A. Leeper (1970) Nitrosation of soil organic matter :III. Nature of gases produced by reaction of nitrite with lignins, humic substances and phenolic constituents under neutral and slightly acidic conditions. Soil Science Society of America Journal 34:430-435.

Stewart, J.B. (1988) Modelling dependence of surface conductance of Thetford forest on environmental conditions. Accepted for publication in Agricultural and Forest Meteorology 1988.

Stewart, J.B. and H.A.R. de Bruin (1985) Preliminary study of dependence of surface conductance of Thetford forest on environmental condition. In: B.A. Hutchison and B.B. Hicks (Eds.), The Forest Climate Interaction, pp 91-104. D. Reidel publ. Company, Dordrecht.

Strahler, A.H., C.E. Woodcock and J.A. Smith (1986) On the nature of models in remote sensing. Remote Sensing of Environment 20:121-139.

Svensson, B.H. (1976) Methane production in tundra peat. In: H.G. Schlegel, G. Gottschalk and M. Pfennig (Eds.), Microbial production and utilization of gases (H_2, CH_4, CO). p 135-139. E. Goltze, Göttingen.

Svensson, B.H. (1984) Different temperature optima in methane formation when enrichments from acid peat are supplemented with acetate or hydrogen. Applied Environmental Micriobiology 48:389-394.

Svensson, B.H. (1986) Methane as a part of the carbon mineralization in an acid tundra mire. In: F. Megusar and M. Gantar (Eds.), Perspectives in Microbial Ecology, Proceedings of teh IVth International Symposium on Microbial Ecology, p 611-616, Sloven Society for Microbiology, Ljubljana.

Svensson, B.H. and R. Rosswall (1984) In situ methane production from acid peat in plant communities with different moisture regimes in asubaretic mire. Oikos 43:341-350

Swarup, A. (1988) Influence of organic matter and flooding on the chemical and electrochemical properties of sodic soil and rice growth. Plant and Soil 106:135-141.

Szeicz, G., Endrodi, G. and S. Tajchman (1969) Aerodynamic and surface factors in evaporation. Water Resources Research 5:380-394.

Takai, Y. (1970) The mechanism of methane fermentation in flooded paddy soil. Soil Science and Plant Nutrition 10:238-244.

Ten Berge, H.F.M. (1986) Heat and water transfer at the bare soil surface. Aspects affecting thermal imagery. Proefschrift ter behaling v.d. graad v. doctor in de Landbouwwetenschappen. (PUDOC) Vakgroep Bodemkunde en Plantevoeding. Landbouwuniversiteit Wageningen.

Terry, R.E., R.L. Tate III and J.M. Duxbury (1981) The effect of flooding on nitrous oxide emissions from and organic soil. Soil Science 132:228-232.

Thompson, K. (1980) Forests and climate change in America: some early views. Climatic Change 3: 47-64.

Thorntwaite, C.W. (1948) An approach toward a rational classification of climate. Geographical Review 38:55-94.

Thunnissen, H.A.M. and H.A.C. van Poelje (1984) Remote Sensing studieproject Oost Gelderland. Deelrapport 3. Bepaling van regionale gewasverdamping met remote sensing. Projectteam Remote Sensing Studieproject. I.C.W. Wageningen.

Tiessen, H. and J.W.B. Stewart (1983) Particle size fractions and their use in studies of soil organic matter. II. Cultivation effects on organic matter composition in size fractions. Soil Science Society of America Journal 47:509-514.

Townshend, J.R.G. and C.O. Justice (1986) Analysis of the dynamics of African vegetation using the normalized difference vegetation index. International Journal of Remote Sensing 7: 1435-1445.

Tsutsuki, K. and F.N. Ponnamperuma (1987) Behaviour of anaerobic decomposition products in submerged soils. Effects of organic material amendment, soil properties and temperature. Soil Science and Plant Nutrition 33:13-33.

Tucker, C.J. (1980) A critical review of remote sensing and other methods for non-destructive estimation of standing crop biomass. Grass Forage Science 35:177-182.

Tucker, C.J., B.N. Holben, J.H. Elgin and J.E. McMurtrey (1981) Remote sensing of total dry matter accumulation in winter wheat. Remote Sensing of Environment 11:171-189.

Tucker, C.J., C.L. Vanpraet, E. Boerwinkel and A. Gaston (1983) Satellite remote sensing of total dry matter production in the Senegalese Sahel. Remote Sensing of Environment 13:461-474.

Tucker, C.J., C.L. Vanpraet, M.J. Sharman and G. van Itersum (1985) Satellite remote sensing of total herbaceous biomass production in the Senegalese Sahel: 1980-1984. Remote Sensing of Environment 17:233-249.

Tucker, C.J., I.Y. Fung, C.D. Keeling and R.H. Gammon (1986) Relationship between atmospheric CO_2 variations and a satellite-derived vegetation index. Nature 319: 195-199.

Udvardy, M.D.F. (1975) A classification of the biogeographical provinces of the world, IUCN Occasional Paper 18, 48 pp. IUCN, Gland, Switzerland.

Unesco (1973) International classification and mapping of vegetation. Unesco, Paris. 93 pp.

Unesco (1978) Tropical forest ecosystems: a state of the knowledge report. Unesco, Paris.

USDA (1971) Soils: Soil moisture regimes (map 1:50,000,000). Soil Geography Unit, USCS, Hyattsville, USA.

USDA (1972) Soils of the World. Distribution of Orders and Principal Suborders (map 1:50,000,000). Soil Geography Unit, SCS, Hyattsville, USA.

USDA (1975) Soil Taxonomy. A Basic System of Soil Classification for making and interpreting soil surveys. Agr. Handbook 436. Soil Conservation Service, U.S. Dept. of Agriculture.

Van Dam, A.J. and C.A. van Diepen (1982) The soils of the flat wetlands of the world, their distribution and their agricultural potential. 47 pp. International Soil Museum, Wageningen, The Netherlands.

Van der Linden, A.M.A., J.A. Van Veen and M.J. Frissel (1987) Modelling soil organic matter levels after long term applications of crop residues, farmyard and green manures. Plant and Soil 101:21-28.

Van Diepen, C.A. (1985) Wetland soils of the world, their characterization and distribution in the FAO-Unesco approach. In: Wetland soils: characterization, classification and utilization, p 361-374, Proceedings of Workshop IRRI, 1984. IRRI, Los Baños.

Van Veen, J.A. (1987) The use of simulation models of the turnover of soil organic matter: an intermediate report. Transactions XIII Congress ISSS Vol. VI pp 626-635.

Van Veen, J.A. and E.A. Paul (1981) Organic carbon dynamics in grassland soils. 1. Background information and computer simulation. Canadian Journal of Soil Science 61:185-201.

Van Veen, J.A., J.N. Ladd and M.J. Frissel (1984) Modelling C and N turnover through the microbial biomass in soil. Plant and Soil 76:257-274.

Van Veen, J.A., R. Merckx and S.C. van de Geijn (1987) Plant and soil related controls of the flow of carbon from roots through the soil microbial biomass. Submitted to Plant and Soil.

Vitousek, R.M. (1983) The effects of deforestation on air, soil and water. In: B. Bolin and R. Cook (Eds.), The major biogeochemical cycles and their interactions, p 223-245. SCOPE Vol.21. Wiley and Sons, New York.

Voroney, R.P., J.A. van Veen and E.A. Paul (1981) Organic C dynamics in grassland soils.2. Model validation and simulation of the long-term effects of cultivation and rainfall erosion. Canadian Journal of Soil Science 61:211-224.

Walker, J. and P.R. Rowntree (1977) The effect of soil moisture on circulation and rainfall in a tropical model. Quarterly Journal of the Royal Meteorological Society 435:29-46.

Walter, H. (1985) Vegetation of the earth. Translation by Joy Wieser, 274 pp. Springer Verlag, New York, Heidelberg, Berlin.

Watanabe, I (1984) Anaerobic decomposition of organic matter in flooded rice soils. In: International Rice Research Institute, Organic Matter and Rice. Los Baños, Laguna, Philippines.

Webster, C.P. and R.J. Dowdell (1982) Nitrous oxide emission from permanent grass swards. Journal of the Science Food and Agriculture 33:227-230.

Welch, R., T.R. Jordan and A.W. Thomas (1984) a photogrammetric technique for measuring soil erosion. Journal of Soil and Water Conservation 194:191-194.

Wetzel, P.J. and Jy-Tai Chang (1987) Concerning the relationship between evapotranspiration and soil moisture. Journal of Climate and Applied Meteorology 26:18-37.

Wetzel, P.J. and R.H. Woodward (1987) Soil moisture estimation using GOES-VISSR infrared data: a case study with a simple statistical method. Journal of Climate and Applied Meteorology 26:107-117.

Wahlen, M., N. Tanaka, R. Henry, B. Deck, J. Zeglen, J.S. Vogel, J. Southon, A. Shemesh, R. Fairbanks and W. Broecker (1989): Carbon-14 in methane sources and in atmospheric methane: the contribution from fossil carbon. Science 245:286-290.

Wahlen, M. and T. Yoshinari (1985) Oxigen isotope ratios in N_2O from different environments. Nature 313:780-782.

Whalen, S.C. and W.S. Reeburgh (1988) A methane flux time series for tundra environments. Global biogechemical Cycles 2,4:399-409.

Whittaker, R.H. (1970) Communities and Ecosystems. Current Concepts in Biology Series. The MacMillan Company, London, 162 pp; 2nd edition (1975), 287 pp.

Whittaker, R.H. and G.E. Likens (1975) The biosphere and man. In: Lieth, H. and R.H. Whittaker (Eds.), Primary production of the biosphere. Ecological Studies 14. Springer Verlag, Heidelberg, Berlin, New York.

Wiebe, W.J., R.R. Christian, J.A. Hansen, G. King, B. Sherr and G. Steyring (1981) Anaerobic respiration and fermentation. In: L.R. Pomeroy and R.G. Wiegert (Eds.) The ecology of a salt marsh, p.137-159, Springer Verlag, New York.

Wiersum, K.F. and P. Ketner (1984) Reforestation, a feasible contribution to reducing the atmospheric carbon dioxide content, In: P.O. Okken, R.J. Swart and S. Zwerver (Eds.) Climate and Energy, teh feasibility of controlling CO_2 emissions. Kluwer Publishing Company, Doordrecht.

Williams, E.J., D.D. Parrish and F.C. Fehsenfeld (1987) Determination of nitrogen oxide emissions from soils: results from a grassland site in Colorado, United States. Journal of Geographical Research 92:2173-2179.

Williams, R.J. and R.L. Crawford (1984) Methane production in Minnesota peatlands. Applied and Environmental Microbiology 47:1266-1271.

Wilson, M.F. and A. Henderson-Sellers (1985) A global archive of land cover and soils data for use in general circulation models. Journal of Climatology 5:119-143.

Wilson, M.F., A. Henderson-Sellers, R.E. Dickinson and P.J. Kennedy (1987) Sensitivity of the biosphere-atmosphere transfer scheme (BATS) to the inclusion of variable soil characteristics. Journal of Climate and Applied Meteorology 26:341-362.

Wilson, M.F., A. Henderson-Sellers, R.E. Dickinson and P.J. Kennedy (1987) Investigation of the sensitivity of the land surface parameterization of the NCAR Community Climate Model in regions of tundra vegetation. Journal of Climatology 7:319-344.

Witkamp, M. (1966) Rates of carbon dioxide evolution from the forest floor. Ecology 47:492-494.

Woodwell, G.M. (1978) The carbon dioxide question. Scientific American 238:34-43.

Woodwell, G.M. (1984) The carbon dioxide problem. In: Woodwell, G.M. (Ed.), The role of terrestrial vegetation in the global carbon cycle. Measurement by remote sensing. SCOPE Vol. 24. Wiley and Sons, New York.

Woodwell, G.M., R.H. Whittaker, W.A. Reiners, G.E. Likens, C.C. Delwiche and D.B. Botkin (1978) The biota and the world carbon budget. Science 199:141-146.

World Resources Institute (1986-1988) World Resources. An assessment of the resource base that supports the global economy (with data tables for 146 countries), World Resources Institute/International Institute for Environment and Development, Basic Books, New York.

Yoshida, T. (1978) Microbial metabolism in rice soils. In: International Rice Research Institute, Soils and Rice. Los Baños, Laguna, Philipines.

Yoshida, T. and M. Alexander (1970) Nitrous oxide formation by Nitrosomonas europaea and heterotrophic microorganisms. Soil Science Society of America Journal 34:880-882.

Young, A. (1976) Tropical Soils and Soil Survey. Cambridge University Press. Cambridge, London, New York, Melbourne.

Young, A. and A.C.S. Wright (1979) Rest period requirements of tropical and subtropical soils under annual crops. Consultants' Working Paper N°6. FAO, Rome.

Zablotowicz, R.M. and D.M. Focht (1979) Denitrificatie and anaerobic nitrate dependent acetylene reduct in cowpea rhizobium. Journal of General Microbiology 11:445-448.

Zimmermann, P.R., J.P. Greenberg, S.O. Wandiga and P.J. Crutzen (1982) Termites: a potentially large source of atmospheric methane, carbon dioxide and molecular hydrogen. Science 218:563-565.

Zobler, L. (1986) A world soil file for global climate modeling. NASA Technical Memorandum 87802, 32 pp.

Part III

Soils

CHAPTER 8

Soil Processes and Properties Involved in the Production of Greenhouse Gases, with Special Relevance to Soil Taxonomic Systems

N. VAN BREEMEN and T.C.J. FEIJTEL

Department of Soil Science and Geology, Agricultural University
P.O.Box 37, 6700 AA Wageningen, The Netherlands

ABSTRACT

In this paper the most important processes involved in the production or consumption by soils of the three major greenhouse gases, CO_2, N_2O and CH_4, are reviewed, and the soil properties determining the magnitude of these gas fluxes are related to criteria used in international soil taxonomic systems, in particular the FAO-UNESCO system.

The levels of soil organic carbon under steady state conditions and the net CO_2 sink term due to leaching of dissolved organic C and HCO_3^- can probably be estimated reasonably well by process-oriented simulation models using soil data such as incorporated in soil maps based on the FAO-UNESCO system. The present-day dramatic changes in land use over large areas in the tropics cause soils to change from (generally low-level) CO_2 sinks to high-level CO_2 sources. Such changes can be modelled too, but results may be difficult to validate properly. A case study of forested and reclaimed Ferralsols in Surinam is used to illustrate the application of simulation models for such a purpose. Provided geographic data on land use and changes in land use, in combination with data on soils, are sufficiently accurate, process oriented models should replace the more usual bookkeeping models to estimate changes in CO_2 fluxes from soils.

Soil data, together with other geographic data are very useful to indicate the poorly drained areas where methane emissions can be expected. Relatively little is known about environmental factors determining methane fluxes from poorly drained land. Therefore, process oriented models are not yet suitable to estimate methane emissions, and simple bookkeeping models are probably more appropriate.

Levels of nitrous oxide emissions seem to be determined mainly by management and environmental factors that cannot be extracted from soil taxonomic data. It may therefore be very difficult to estimate N_2O fluxes from large areas using only soil data.

8.1 INTRODUCTION

Soils play an important role in the production or consumption of the dominant greenhouse gases, CO_2, CO, CH_4 and N_2O. Table 8.1 shows the estimated ranges of global net annual emission of these gases, the

Soils and the Greenhouse Effect. Edited by A.F. Bouwman
© 1990 John Wiley & Sons Ltd.

present-day annual increase in these emissions, and the net emissions from soils.

Although estimates of the source- and sink terms vary considerably, soils appear to be important net sources of CO_2 (5-20% of total emissions), of CH_4 (about 30% of total emissions) and of N_2O (80-90% of total emissions), and act as a net sink of CO.

Table 8.1   Global net annual emission of CO_2, CO, CH_4 and N_2O, and soil-borne emissions in 1980. Derived mainly from literature data compiled by Bouwman, (this volume). Emissions are stated in units of Pg y^{-1} ($= 10^{15}g \, y^{-1}$)

	Total	Soil-borne	Annual increase (%) in total emission
CO_2	5.7-6.4*	0.2 to 0.9	0.5
CO	1.3 to 5.7	- 0.19 to -0.58	2 to 6
CH_4	0.40 to 0.64	0.1 to 0.24**	1.1
N_2O	0.007 to 0.021	0.006 to 0.019	0.25

*   Made up of fossil fuel combustion (5.2), volcanic emissions (0.3; Holland, 1978) and the soil-borne emissions
**   Excluding a soil-related sink term of 0.03

Most estimates of the emission of greenhouse gases from soils are rather crude. For instance, Palm et al. (1986) estimate CO_2 emissions from tropical soils using fixed values of pools of carbon in each of four types of virgin ecosystems (evergreen rain forest, seasonal forest, dry forest/woodland, plantation) and a certain fractional transformation of soil carbon to CO_2 (25 to 50%) in five years following clearing for any of five types of land use (undisturbed, permanent agriculture, two types of shifting cultivation and grasslands). Other bookkeeping models aimed at estimating CO_2 emissions for large land areas use similar categories (Melillo et al. 1988; Houghton et al., 1987; Detwiler and Hall, 1988). Such an approach may be acceptable, because the greatest errors are probably in the estimates of various land use categories, and of the temporal changes in land use. However, a large body of information on soils, e.g. the soil map of the world (FAO-UNESCO, 1971-1978, FAO, 1988), could be used to improve the estimates such as given in Table 8.1. Buringh (1984) has pioneered the use of such soil data for CO_2 emission rates by estimating the organic carbon pools in soil types corresponding to USDA (1975), both in forested and cultivated land. Much better estimates of gaseous emissions should be possible by combining knowledge on the fundamental soil processes involved in the exchange of greenhouse gases with such geographical information. An example of a process-based computer simulation model in a geographic framework is the CENTURY model by Parton et al. (1987), that describes the production and decomposition of soil organic matter over a large region (the North American Great Plains grasslands). Data derived from similar models could eventually become useful in policy and management decision making with regard to emission of greenhouse gases.

The aim of this paper is, therefore, to review the basic processes involved in soil-borne emission of greenhouse gases, and to identify key parameters in these processes that should be quantified for use in simulation models, and that can be related to soil taxonomic systems and to geographically oriented data such as soil maps. Since most information is available on CO_2, the emphasis in this paper will be placed on this abundant greenhouse gas. Also several examples will be given of model calculations of the CO_2 balance for different soil vegetation systems.

Shorter sections will be devoted to CH_4 and N_2O, and no further attention is given to CO.

8.2 CARBON DIOXIDE

8.2.1 Processes Involved in the Transfer of CO_2 between Soils and the Atmosphere

Figure 8.1 summarizes the main transfers of CO_2 between the atmosphere and various carbon pools in the soil. Each of these will be discussed below in relation to factors determining the rate of transfer. Finally the overall effect of the carbon cycle on the net CO_2 emission will be assessed for a number of cases, applying a computer simulation model.

The first process is the photosynthesis, the conversion of CO_2 to organic matter by plants under the influence of sunlight (Figure 8.1 a). Of all photosynthetically fixed CO_2 (gross primary production), 30 to 70% is respired again by the primary producers. In natural ecosystems the remaining net primary production (NPP) depends strongly on the climate and varies from virtually nil in arctic and some desert ecosystems, to 3×10^4 kg ha^{-1}y^{-1} in mature tropical rain forest. Part of the NPP is respired heterotrophically by herbivores and their predators, by parasites, and by decomposers before it reaches the soil. In forests, litter from leaves, and from stems plus branches supply most of the organic matter, but in marsh ecosystems roots may contribute up to 90% of the total organic matter input (Gosselink, 1984). Non-photosynthetic autotrophs, e.g. chemoautotrophic microorganisms, also supply organic matter to the soil, but their contribution is usually negligible. In systems with a high activity of burrowing decomposers such as earthworms, a large fraction of the litterfall is incorporated in the soil.

Under the influence of a large and complex community of soil organisms, most of the dead organic matter is respired to CO_2 and H_2O (Figure 8.b) over a period of months to years. Part of the freshly supplied organic C is stored as humified organic matter in the soil (Figure 8.c), which is respired too (Figure 8.d), albeit much more slowly than the fresh organic detritus. In addition to CO_2 from heterotrophic respiration, root respiration provides CO_2 to the soil (Figure 8.e). The combined effect of these biotic processes is that CO_2 is pumped from the atmosphere at a partial pressure of 0.03 KPa into the soil where its partial pressure is usually in the order of 0.1 to 10 KPa. Most of the soil CO_2 finds its way back to the atmosphere by diffusion, and by convection under the influence of temporal fluctuations in temperature and in soil water content (Figure 8.f). Part of the CO_2 dissolved in the soil solution according to Henry's law constant, dissociates to form HCO_3^- and H^+. If the H^+ is exchanged with cations at the negatively charged adsorption sites or bound in minerals (silicates, carbonates), the $CO_2 \rightleftharpoons HCO_3^-$ equilibrium is shifted to the right, so that CO_2-C can be removed by drainage as dissolved metal (usually Na^+, Ca^{2+} or Mg^{2+}) bicarbonate (Figure 8.1 g). Other sinks of atmospheric CO_2 are drainage of dissolved CO_2 and dissolved organic C (Figure 8.1 h,i).

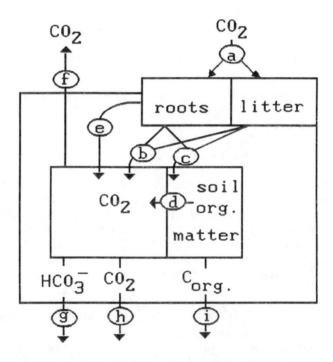

Figure 8.1 Main transfers of CO_2 between the atmosphere and various carbon pools in the soil. For explanation of a-i see text.

After establishment of a vegetation on bare land, the pool of soil organic matter increases to a steady state level, whereby the fluxes associated with c and d plus i in Figure 8.1 are equal. These steady levels depend mainly on climate (including the climatologicallydetermined vegetation cover), soil hydrology, soil texture, soil mineralogy, soil structure, and cultivation practices. The steady state soil organic carbon pools are in the order of 5 to 35 kg m^{-2}, and are reached over decades to centuries. In a climax vegetation the pool of dead organic carbon stored underground is in the same order to several times the amount of organic carbon stored in the living phytomass. During establishment of a climax vegetation and afterwards, as long as the steady state situation is maintained, a soil-vegetation system acts as a sink of atmospheric CO_2: a>f. This sink is formed initially by accumulation of living biomass and soil organic matter (c minus d) plus drainage of carbon originally derived from atmospheric CO_2 (g,h,i), and later only by drainage. In continuously wet conditions a steady state organic matter pool is never reached. Here decomposition is hampered by lack of oxygen and most fresh organic matter escapes from being respired (c>>b), while further heterotrophic respiration below the soil surface is virtually nill. As a result dead organic matter is accumulated continuously, forming peat deposits.

Soil-vegetation systems can turn from a CO_2 sink into a CO_2 source when conditions change so that the rate of respiration of soil organic matter exceeds the rate of its formation. Causes for such a shift can be a decreased supply of fresh organic matter to the soil (c), an increased rate of heterotrophic respiration of soil organic carbon (d) or any combination thereof so that d>c.

The supply of fresh organic matter to the soil may decrease as a result of a decrease in net primary productivity e.g. due to climatic change or due to forestry or agricultural activity. Decomposition of soil organic matter will be enhanced when heterotrophic respiration increases. This can be the result of increased aeration of the soil (e.g. after drainage), increased soil temperature (e.g. due to removal of vegetation cover) or increased availability of substrates (e.g. due to disturbing the soil by plowing). The present-day situation of soils acting globally as a net source of CO_2 is the result of human activities: drainage of wetlands and peat swamps, deforestation and reclamation of land for agriculture.

According to Silvola (1986), drainage of Finnish oligotrophic to mesotrophic forested peat mires by lowering the water table to 30-60 cm below the land surface caused a shift in CO_2-C emission from an estimated -25 g m^{-2}y^{-1} in virgin conditions to about +250 g m^{-2}y^{-1} after drainage. Mann (1986) selected 50 literature sources on C contents of uncultivated and cultivated soils, mainly about the Great Plains region in North America. After cultivation for several years to over 80 years, the upper 30 cm of the soil had lost on average 4 to 15% of their original soil carbon, amounting to 100 to 1500 g m^{-2}. The highest losses were observed in Mollisols, with Borolls losing on average 31% of their organic C in the upper 15 cm, amounting to 4900 g m^{-2}. Similarly, Voroney et al. (1981) reported a decrease of 36% in the organic C content of a Black Chernozemic soil (Udic Boroll) at a mid-slope position. However, depletion of organic C in the Ah horizon accounted for more than 90% of the total loss. The Ah horizon at the mid-slope position lost 58% of its organic C and 50% of its N in 70 years of cultivation (Voroney et al., 1981).

To know the source or sink term of soil-vegetation systems, i.e. the value of a-f in Figure 8.1, each of the below-ground fluxes indicated in the diagram mus be quantified. First, a qualitative enumeration will be given of the steering parameters of these fluxes that can be estimated on the basis of soil taxonomic systems, notably the FAO-UNESCO system. Next we will use a simulation model to show the order of magnitude of these below-ground fluxes, and the resulting net flux a-f for a number cases that can be more or less validated on the basis of available information.

8.2.2 Modelling the CO_2 Balance of Soils Using the FAO-UNESCO Soil Classification

In addition to data on climate and land use which primarily determine the NPP (and thus the inputs of fresh organic C to the soil) a number of soil factors need to be known in order to properly estimate parameters for simulation models of CO_2 transfer. Climate is not explicitly defined in the FAO-Unesco system except to separate arid and non-arid soils. While present-day climate can be estimated for many mapping units where soils derived their particular properties largely as a result of climate, this is not the case with the many azonal and intrazonal soils groups, and with soils developed under a different climate in the past. The problem can be circumvented by attributing an ecosystem type (e.g. according to Whittaker and Likens, 1975, or by Matthews, 1983) to each mapping unit of the FAO-UNESCO soil map of the world, as has been done by Bouwman (this volume).

The soil factors ideally include depth of the soil profile, and its textural make-up, the nature of the clay fraction (including the amount of allophane present), presence or absence of calcium carbonate, weatherability of the silicate minerals, and the soil drainage class. Clay minerals tend to protect soil organic matter by formation of humus-clay or humus-metal-ion-clay complexes, in part by catalyzing the oxidative polymerization of soluble polyphenols to humic substances (Wang et al., 1986). These complexes are highly resistant to decomposition (Martin and Haider, 1986). Moreover clay size particles may physically shield decomposable organic matter from microbial attack (Martin and Haider, 1986). The stabilizing effects are highest in allophane (amorphous or poorly crystalline aluminum silicates), and successively lower in smectites, illite and kaolinite (Martin and Haider, 1986; Boudot et al., 1986; Dalal and Mayer, 1986).

As discussed above, mineral weathering under the influence of CO_2 involves transfer of atmospheric CO_2 to aqueous HCO_3^-

An example of a model that we have used to estimate the fate of soil organic carbon and its role in the CO_2 balance of a particular soil-vegetation system will be outlined in 8.2.3.

Most of the textural and mineralogical characteristics influencing the C balance of soils can be deduced roughly from soil taxonomic data. For instance, the FAO-UNESCO system distinguishes three textural classes (fine, medium and coarse). The presence of calcium carbonate is implied in Calcisols, and all "calcaric" and "calcic" subgroups e.g. Calcaric Fluvisols, -Regosols, -Arenosols, -Cambisols, -Phaeozems, and Calcic Gleysols, -Vertisols, - Kastanozems. -Chernozems, and -Luvisols. Silicate weathering rates are potentially high (actual weathering rates may depend strongly on drainage rates) in most non-calcareous soils that are fertile and contain appreciable amounts of weatherable minerals. Examples are: most non-Dystric subgroups of Fluvisols, Gleysols, Leptosols, Andosols, Vertisols, Cambisols, and in Eutric Planosols, - Nitosols. Moderate weathering rates can be expected in a number of Dystric subgroups, and in most Arenosols, Luvisols and Lixisols. Low weathering rates can be expected in all highly acidic sandy soils such as Podzols and Ferralic Arenosols, and deeply weathered acidic soils common in the tropics, such as most Acrisols, Ferralsols and non-Eutric Planosols and -Nitosols. Clay mineralogy is often linked directly to the highest category soil units: allophanic in Andosols, dominantly smectitic in Vertisols, in Vertic Cambisols, and most Chernozems, dominantly kaolinitic in Acrisols, Nitosols, Ferralsols, and Plinthosols. A mixed clay mineralogy is probably typical for many Fluvisols, Gleysols, Luvisols and Lixisols. Poor internal soil drainage is typical for all soils with hydromorphic properties, which include all Gleysols, and all Gleyic subgroups. The concentration of dissolved organic carbon in the drainage water can be modelled as such, but can also be set at certain observed levels for different soil groups (e.g. high· in Podzols and Histosols, very low in most Ca- rich soils and tropical soils high in Fe and Al, and intermediate in most other soils of temperate zones.

8.2.3 Model Formulation and Assumptions

The model used (Feijtel and Meijer, 1989) simulates soil water and organic carbon dynamics for a soil with up to 4 layers or soil horizons, each characterized by thickness, texture, moisture content, permeability, and root percentage. Soil texture is the main governing variable of pore space, which determines moisture contents at saturation, field capacity, wilting point, and air-dry conditions, as well as the permeability for water at different moisture contents. The vertical flow through the soil is a function of precipitation, evaporation (for first soil layer only), transpiration (for total rooting depth), and intrinsic soil hydraulic properties (Feijtel and Meijer, 1989). Mean monthly temperature, precipitation and potential evapotranspiration are forcing variables imported from a data file or from another model.

The amount of infiltration equals canopy throughfall, which is calculated from rainfall minus canopy interception. The interception is estimated as a function of the leaf area index (LAI) (Feijtel and Meijer, 1989), or, if available, measured values can be used. NPP can in principle be estimated by the model from climatological data, but so far it is imported from input data files.

The hydrology of the system is governed by LAI, root distribution, and transpiration losses within each soil compartment. Combined with precipitation these effects ultimately determine soil moisture profiles and percolation through the profile.

The rate of decomposition of organic matter depends on a number of soil and climatic factors and also on the type of organic matter. Decomposition rates are usually reported in terms of a decomposition constant, k. The decay of organic material is described by the equation:

$$X_t/X_o = e^{-kt} \qquad (8.1)$$

where: X_o = amount of organic carbon at t=0;
       X_t = amount of organic carbon remaining at t=t;
       e = the base of natural logarithms.

The larger part of fresh detritus from plants and animals decays rapidly, with a decomposition constant ranging from 5 to 30 y^{-1}. A smaller fraction, generally rich in lignin, decays more slowly and is characterized by decomposition constants of less than 1 y^{-1}. Part of the decaying material is respired to CO_2, the remainder is incorporated or assimilated in microbial tissue. The fraction that is respired equals (1-a), where a is the assimilation coefficient, which ranges from 0.40 to 0.67 in most well-aerated soils. By assuming the presence of several fractions of organic matter differing in decomposition rate, the change with time of the organic carbon content of the soil can well be described (see e.g. Jenkinson and Rayner, 1977; Janssen, 1984; Van der Linden et al., 1987; Parton et al., 1987).

The fate of litter and soil organic carbon is modelled for distinct time intervals: the situation at the end of the time interval forms the starting position for calculations on the next interval. During each interval primary material is partly converted into secondary compounds in all fractions. Some of the decomposable secondary material is degraded subsequently to tertiary components and so on.

The types of organic carbon distinguished in our model are: (i) Fresh organic material, consisting of root and crop residues (farm land) or leaves, branches and stems (forest), (ii) Roots, root exudates and dead root cells discarded by the plant, (iii) Soil organic carbon, consisting of dead organic soil material and of living biomass, and (iv) Soluble organic carbon.

We have modelled C decomposition according to the following scheme:

Plant litter → Soluble org. C → Soil org. C → CO_2
     ↓   ↓
    CO_2 CO_2
(evolved into   (evolved into soil layers)
the atmosphere)

Roots and root exudates are assumed to decompose directly to CO_2 and soil organic carbon, without soluble intermediates:

Roots + exudates → Soil organic C → CO_2
       ↓   (evolved into soil CO_2 layers)
    CO_2
(evolved into soil layers)

Above-ground plant litter is added in monthly time steps according to a seasonally determined fall rate function. The fresh organic material is considered to consist of three fractions (leaves, branches and stems), decomposing each at its own rate. The assumption that all above-ground litter passes through a solubilization step is a simplification which implicitly involves extra downward transport by burrowing soil fauna such as earthworms.

For each time increment the amount added, degraded and remanent in each fraction (leaves, branches, and stems) are calculated. As indicated above, all plant litter falling on the soil surface was assumed to degrade into in a gaseous CO_2 pool and a soluble pool, according to a partitioning constant p. Once in solution, all fractions are assumed to decompose at the same rate. Within each time step fresh organic C on the soil surface is partly decomposed to CO_2, the remainder is converted to soluble organic carbon and a solid rest fraction. The rest fraction consists of litter, green mass, and wooden parts, which can be degraded in subsequent time steps. This may lead to the formation of an organic litter layer (forest floor) on top of the mineral soil, consisting of increasingly recalcitrant material until a steady-state composition is reached. The soluble organic carbon formed within each time-step will be on its turn partly degraded into CO_2 and soil organic matter in the next time step.

The order of magnitude of decomposition rates under optimal moisture and temperature conditions were taken from various literature sources, and calibrated for specific cases. Decomposition virtually stops at 0°C, its rate increases to a maximum between 20 and 40°C, and decreases at higher values to stop above 50-60°C. Soil moisture is optimal for decomposition at pF-values of 1 to 2.5 and decomposition rates decrease under wetter and dryer conditions (Van der Linden et al., 1987) (Figure 8.2). Actual decomposition rates were estimated by adjusting the optimal rates for each time-step for in-situ temperature and moisture regimes according to Figure 8.2.

In the model there are two sources of soil organic C: (i) precipitation and decomposition of soluble organic material and (ii) by decomposition of roots. The equations are:

$$dSOC_s(t)/dt = a\ k_s\ (SolOC)_t\ theta_t - k_h\ SOC_s(t) \qquad (8.2)$$

where: $SOC_s(t)$ = Soil organic C or humus from soluble organic C (kg m^{-2}) at time t,
  a = assimilation factor (fraction of assimilated carbon stored in living cells, between 0 and 1)
  k_s = respiration rate of soluble organic C (y^{-1})
  SolOC = soluble organic C concentration (kg m^{-3}) (determined by percolation input and drainage output during previous time step)
  theta = average water volume during time step (m^3 m^{-3})
  k_h = decomposition rate of soil organic C (y^{-1})

$$dSOC_r(t)/dt = a'\ k_e\ E_t - k_h\ SOC_r(t) \qquad (8.3)$$

where: $SOC_r(t)$ = Soil organic C from roots (kg m^{-2})
  a' = assimilation factor
  k_e = consumption rate of exudates (y-1)
  E_t = amount of root exudates per month (kg m^{-2})
  k_h = decomposition rate of soil organic C (y^{-1})

Total soil organic carbon (SOC_T) is equal to the sum of SOC_s and SOC_r.

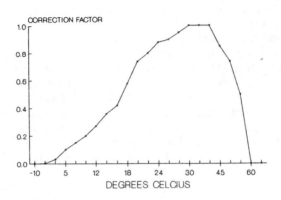

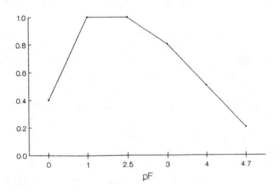

Figure 8.2   Correction factors (expressed as fraction of the value under optimal conditions) for decomposition rates as a function of temperature and of pF (negative logarithm of pressure potential in mbar) of soil water.

The production of CO_2 in the soil layers is coupled to the degradation of (SolOC), roots, SOC_s and $SOCr$ as follows:

$$dCO_2/dt = (1-a)\ k_s\ (SolOC)_t\ theta_t + (1-a')\ k_e\ E_t + k_h\ (SOC_T)_{(t)}\quad (8.4)$$

Note that in the third term no assimilation factor appears: decomposition of humus is assumed to yield only CO_2, no microbial C, so the factor **a** has been set at 0.

8.2.4 Model Calibration and Scenario Analysis

The development of soil organic carbon pools to steady state levels were simulated for an undisturbed tropical forest ecosystem in Surinam, as described by Poels (1987). We assumed in our model that primary production was governed by temperature, radiation, and supply of water, although in fact, production was affected by nutrient availability (Poels, 1987). Primary production was closely linked to water availability with LAI's ranging from 5 to 7. The soil is a Haplortox (USDA, 1975) or Ferralsol (FAO, 1988) with a sandy clay loam texture throughout. Mean monthly temperature varied from 26 to 31°C, and mean monthly evapotranspiration from 125 to 190 mm.

Actual monthly rainfall data for a three-year period, and three-year mean monthly potential evapotranspiration data were used to determine the moisture regime of a sandy clay loam (Figure 8.3), assuming that the leaf area index varied seasonally between 5 and 7.

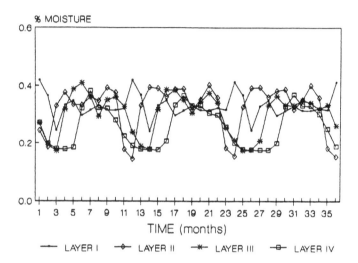

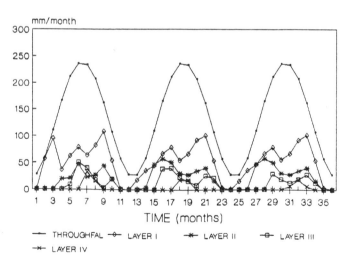

Figure 8.3   Mean monthly soil moisture content (volume fraction, %) and monthly percolation (mm) for each of four successive soil layers (0-20, 20-40, 40-60 and 60-80 cm depth) over three years for undisturbed primary forest on a Ferralsol in Surinam.

Moisture and temperature regimes for the litter layer were assumed to be similar to those for the first soil layer. Poels (1987) determined litter fall of leaves, branches, and stems as 11.2, 4.63, and 4.43 ton/ha.yr respectively. Simulation of steady state amounts of each of these fractions in the forest floor, resulted in a good agreement with the forest floor composition as determined by Poels (1987) (Figure 8.4). Steady state amounts were reached in about 50 years.

Table 8.2 summarizes the calibration settings and simulation results for the generation of the forest floor, soil organic carbon and the root mass. Kinetic rate constants reported in Table 8.2 are for optimal moisture and temperature conditions, and were adjusted for sub-optimal conditions according to Figure 8.2. Simulated soil organic carbon profiles corresponded well with measured contents for each layer (Table 8.2). Measured root production was 4.16 ton $ha^{-1}y^{-1}$, and the root mass was about 108 ton ha^{-1} over a total depth of 150 cm. Root production in our simulation was set at 4.0 ton $ha^{-1}y^{-1}$ with a root distribution of 30, 30, 20, and 20% over the four layers. The simulated root mass in the upper 80 cm of the profile was 65.7 ton ha^{-1} (Table 8.2), of the same order as reported by Jordan (1985) for the upper 1 m of soil in tropical forest ecosystems.

Table 8.2 Degradation rate constants for above-ground litter (k_l), roots (k_r), soluble organic C (k_s) and soil organic C (k_h) at different depths in the soil. The value of the assimilation factor (0.425) is typical for a sandy clay loam, and was used for degradation of root material and soluble organics at all depths. The partition coefficient setting the fraction of above-ground plant litter that is mineralized to CO_2 was taken as 0.6. Quantities of organic matter are expressed as mass or mass fraction of C

Litter degradation k_l (y^{-1})	Leaves 2.8		Branches 0.96	Stems 0.22
depth (cm)	0-20	20-40	40-60	60-80
k_r (y^{-1})	0.10	0.09	0.08	0.07
k_s (y^{-1})	25	23	22	20
k_h (y^{-1})	0.055	0.035	0.027	0.027

Model Results:				
Forest floor ton ha^{-1}:	Leaves 2.54		Branches 3.26	Stems 14.54
Depth (cm):	0-20	20-40	40-60	60-80
Root Mass (ton ha^{-1}):	8.27	10.58	7.0	6.99
Soil Org. C. (%) simulated:	1.536	0.847	0.464	0.296
measured: (Poels 1987)	1.53	0.88	0.51	0.30

Total CO_2 production accounted for about 370 g C m^{-2} y^{-1}. Steady state soil CO_2-production ranges from 223 g C m^{-2} y^{-1} in the upper layer to about 31 g m^{-2} y^{-1} in the lower soil layer. About 30 g C m^{-2} y^{-1} was lost in the form of DOC and H_2CO_{3aq}. Seasonal fluctuation of CO_2 production in each of the four soil layers is shown in Figure 8.5.

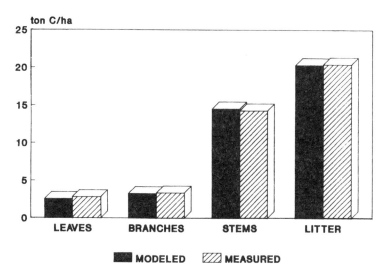

Figure 8.4 Steady state composition of the forest floor under undisturbed primary forest on a Ferralsol in Surinam according to the model, and as measured by Poels (1987).

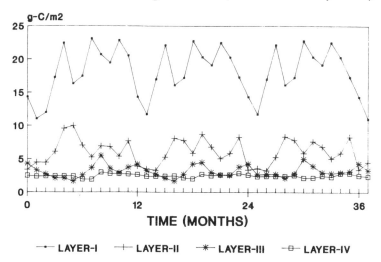

Figure 8.5 The seasonal variation in CO_2 production (modelled) in successive deeper soil layers (0-20, 20-40, 40-60 and 60-80 cm) under undisturbed primary forest on a Ferralsol in Surinam.

Next we investigated changes in soil carbon as a result of changing several environmental conditions, but all within the same climatological setting. In most cases decomposition rate constants were the same as calibrated earlier for the undisturbed forest.

8.2.5 Effect of Clear Cutting and Cultivation

The effect of clear cutting of the forest followed by continuous cultivation of maize was simulated. Based on data reported by the Centre of Agricultural Research, University of Surinam (Celos, 1980, 1981, 1982, 1983) the production of maize (grain) ranged from 2000 to 3000 kg ha^{-1}. Maize was planted twice a year with growing seasons corresponding to the rainy periods in March–July and November–February). The leaf area index increased from 2 after seeding to 5 at harvest, and roots were concentrated in the upper layers (fractions of total root mass were 0.4, 0.4, 0.1, 0.1 in the successive layers). Annual litter and root production were 5 ton ha^{-1} (70% leaves, 25% "branches" and 5% "stem") and 1 ton ha^{-1}, respectively. Simulation with kinetic parameters identical to those reported in Table 8.2 resulted in a significant decrease of organic carbon content in all soil layers (Figure 8.6). The content of organic carbon in the upper layer dropped from 1.536% under forest to about 0.411% at steady state under maize, while in the fourth layer the carbon content decreased from 0.296% to 0.048% Organic carbon contents decreased dramatically in the first 10 to 15 years after deforestation, especially in the upper layers.

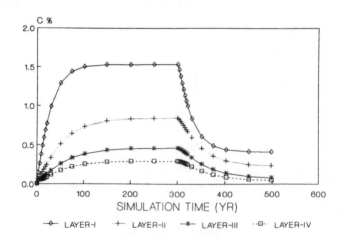

Figure 8.6 Changes in soil organic carbon at different depth (Layers I, II, III and IV correspond to resp. 0-20, 20-40, 40-60 and 60-80 cm) in a Ferralsol under undisturbed tropical forest (0-300 y), and after deforestation and continuous maize cultivation (300-500 y).

Measured organic carbon contents for the same soil type after 10 to 15 years of cultivation following deforestation were around 1.2, 0.8, 0.4, and 0.3% for layers 1, 2, 3, and 4 respectively (Poels, pers.comm.). Simulated values for the same layers after 12 years of cultivation were respectively 1.215, 0.753, 0.424, and 0.276% (Table 8.3). To test the model the organic carbon profile was also simulated for a similar maize crop, but starting from a bare soil substrate with only 0.1% carbon throughout the profile (Table 8.3). Steady state soil organic C levels reached after 300 years are identical to those reached after clear cut of the forest.

Table 8.3 Soil organic carbon levels (mass fraction, %) in a Ferralsol under continuous maize (two crops per year) as a function of depth and time after clear cut of a primary forest, and after cultivation on bare soil with 0.05% organic C initially

Simulation Time (y)	Clear Cut				Bare Soil			
	L-1	L-2	L-3	L-4	L-1	L-2	L-3	L-4
0	1.536	0.847	0.458	0.293	0.05	0.05	0.05	0.05
6	1.383	0.806	0.448	0.289	0.101	0.058	0.048	0.045
12	1.215	0.753	0.424	0.276	0.154	0.072	0.048	0.043
18	1.067	0.699	0.397	0.260	0.200	0.086	0.049	0.040
51	0.608	0.459	0.257	0.168	0.345	0.157	0.057	0.039
150	0.415	0.251	0.097	0.059	0.408	0.222	0.069	0.042
300	0.410	0.230	0.070	0.042	0.410	0.230	0.070	0.042

8.2.6 Effect of Texture

Texture influences organic carbon dynamics in the soil by its impact on the hydrology (and thus the decomposition rates in as much they are affected by soil moisture, cf Figure 8.2), and by the effect of clay and silt on microbial kinetics and assimilation of organic carbon into microbial biomass. The following empirical relationships between soil texture and degradation and assimilation coefficients were derived from Parton et al. (1987):

$$\text{k-effective} = \text{k-hypoth} (1 - 0.75 \text{ T}) \qquad (8.5)$$

$$\text{a-effective} = \text{a-hypoth} [1-(0.85 - 0.68 \text{ T})] \qquad (8.6)$$

where T represents the mass fraction of the silt plus clay fraction, k-hypothetical and a-hypothetical represent the optimum decomposition rate constant and assimilation factor. Equation (8.5) shows that the decay rate of organic carbon (from dissolved and root organic C) decreases as the silt plus clay content increases. Equation (8.6) indicates that assimilation into microbial mass becomes more effective as the silt plus clay content increases.

To illustrate effects of differences in texture organic C levels for the subsoil were simulated at three different textures (Table 8.4). Organic carbon contents after 150 years increased by 36 and 52%

respectively at 40-60 and 60-80 cm depth due to the change from sandy clay loam to clay loam. Substituting sandy loam for silty clay loam caused an even greater increase in C content (see Table 8.4). If only hydrological effects resulting from a change from sandy clay loam to clay loam are considered, and the effects on decomposition constants according to equations 1 and 2 are neglected, organic C contents increased only by 8.7 and 9.5% respectively in layers three and four after 150 years. These results indicate that microbial degradation parameters affect the carbon profile to a much greater extent than hydrological effects.

Table 8.4 Effect of clay+silt on the steady state organic carbon levels (mass fraction %) in the subsoil of a Ferralsol under primary forest in Surinam. The degradation rate constants (in y^{-1}) of soluble organic matter (k_s), of root litter (k_r) and the microbial assimilation coefficient a were adjusted from the calibrated values for sandy clay loam (cf Table 8.2) using Parton's (1987) relationships. Clay + silt mass fractions (T in equations 1 and 2) were 0.4, 0.65 and 0.90 for sandy clay loam, clay loam, and silty clay loam respectively

| | ----- 40-60 cm layer ----- | | | | ----- 60-80 cm layer ----- | | | |
	k_s	k_r	a	org. C	k_s	k_r	a	org. C
Sandy clay loam	22	0.08	0.43	0.069	20	0.07	0.43	0.042
Clay loam	16	0.06	0.59	0.094	15	0.05	0.59	0.064
Silty clay loam	10	0.05	0.76	0.099	9.3	0.04	0.76	0.076

Actual effects of textural differences on soil organic carbon can be shown from measured data on surface soils occuring in the same area about one decade after deforestation and reclamation for maize. A sandy clay loam surface soil had 1.33% organic C, while these values were 1.2% and 0.46% respectively for a sandy loam and a sand (Celos, 1981, 1982, 1983).

8.2.7 Effects of Ploughing

Ploughing increases the rate of decomposition of soil organic carbon by improving aeration, and by increasing the access to soil organic carbon by soil fauna and flora due to increased disruption of soil clods, exposing physically protected organic material. Increased decomposition rates of soil organic C as a result of disruption can be estimated from observed changes in soil organic C after reclamation (e.g. Brams, 1971; Voroney et al., 1981; Mann, 1986; Lugo et al., 1986; Dalal and Mayer, 1986) or by fitting these changes using varying decomposition rates in simulation models.

Effects of ploughing on organic carbon levels were simulated by increasing the degradation constants of the first layer by 10, 30, and 50% (Figure 8.7). Simulation was started at steady state levels of organic carbon for an undisturbed forest. After clear cutting, cultivation started with an above-ground crop litter production of 5000 kg ha^{-1}y^{-1} and a root production of 1000 kg ha^{-1}y^{-1}. Results suggested that after 150 years of cropping organic carbon decreased respectively by 9, 24, and 34% in the first layer. Table 8.5 suggests that decreased tillage intensity (rotovating- ploughing- no-tillage) indeed decreases the rate of decomposition of soil organic carbon after deforestation, but in view of spatial variability of the experimental plots, these data need further statistical analysis before firm conclusions can be drawn.

Table 8.5 Organic C profiles (mass fraction, %) in Ferrasols in Surinam reclaimed from forest and cultivated with mais with three different tillage methods

Depth (cm)	Ploughing	Rotovating	No-tillage
0 -10	0.94	1.09	1.25
10-20	1.06	1.04	1.21
20-30	1.09	0.75	0.82
30-40	0.64	0.48	0.64
40-50	0.38	0.33	0.44

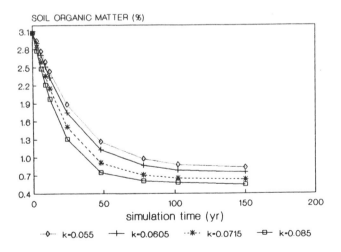

Figure 8.7 Effect of increased rates of decomposition (k, varying from 0.055 to 0.086 y^{-1}) of organic matter, as can be expected from regular ploughing, on the change of organic C contents of the surface soil of a Ferralsol after deforestation and continuous maize cultivation.

8.2.8 Effect of Weathering

Weathering of calcium carbonate is a soil forming process typical for all calcareous soils in areas where rainfall exceeds evapotranspiration. Van Breemen and Protz (1988) reported that concentrations of dissolved calcium carbonate in the soil solution calculated from the weathering rates of $CaCO_3$ and the rate of water drainage are in the range expected for the equilibrium with calcite. Over a wide range of conditions the dissolution of calcite appears to be high enough to maintain equilibrium concentrations in water percolating calcareous soils and rocks. Consequently, the rate of calcium carbonate weathering could be predicted from the solubility of $CaCO_3$ and the percolation rate. Aqueous CO_2 will act as proton donor in the dissolution of calcite :

$$CaCO_3 + CO_2 \text{ (g)} + H_2O \rightleftharpoons Ca^{2+} + 2\ HCO_3^-$$

At equilibrium, the solubility of calcite follows from the equilibrium constant for this reaction:

$$K_{eq} = [Ca^{2+}].[HCO_3^-]^2 / P_{CO2} \qquad (8.7)$$

In a $CaCO_3$-CO_2-H_2O system the concentration of HCO_3^- is a function of temperature and P_{CO2} only (because $[Ca^{2+}]$ can be expressed in terms of $[HCO_3^-]$, according to $2[Ca^{2+}]=[HCO_3^-]$), and can thus be calculated from the equilibrium constant K_{eq}. The partial pressure of CO_2 was calculated from the amount produced within one month and the amount dissolved according to:

$$pCO_2 = CO_{2tot}/(1 + K_{CO2}) \qquad (8.8)$$

where K_{CO2} stands for Henry's constant. Temperature corrections were made as described by Feijtel and Meijer (1989).

Simulation of calcite weathering in a sandy clay loam with 1% calcite (but otherwise under conditions similar to those in Surinam) cropped with maize gave a removal rate of about 6.8 g C $m^{-2}y^{-1}$ of atmospheric CO_2 as HCO_3^- assuming a crop litter production of 5000 kg $ha^{-1}y^{-1}$. Under a forest vegetation with a litter production of 20000 kg $ha^{-1}y^{-1}$, the calculated removal rate was about twice as high (13.9 g C $m^{-2}y^{-1}$), in spite of slightly lower drainage rates of water (about 120 mm under forest and 150 mm under maize). A doubling of HCO_3^- leaching (from calcareous dune sand) in spite of decreased water drainage was also observed in lysimeters when comparing forested to unvegetated sites (Minderman and Leeflang, 1968). From data summarized by van Breemen and Protz (1988), removal of atmospheric CO_2 by dissolution of calcium carbonates in calcareous terrain is in the order of 4 to 30 g C $m^{-2}y^{-1}$.

In non-calcareous areas the drainage flux of HCO_3^- is determined by the rate of silicate weathering, which depends strongly on the types and quantities of minerals present in the soil and in the bedrock. The CO_2 sink term due to weathering is generally much lower in areas with silicate rock than in calcareous areas. Lowest rates are observed in acidic soils on rocks low in easily weatherable minerals (e.g. in Hubbard Brook watershed, N.H., USA: 0.003 g C $m^{-2}y^{-1}$; data from Likens et al., 1977). In areas with easily weatherable silicate rock values may range from 0.6 g C $m^{-2}y^{-1}$ (e.g. in a forested watershed underlain by pelitic schist in Maryland, USA; data from Cleaves et al., 1970) to as high as 12 g C $m^{-2}y^{-1}$ (at very high precipitation in a glacial environment in the Cascade Mountains, Washington: data from Reynolds and Johnson, 1971).

Net removal of atmospheric CO_2 by drainage of water with dissolved CO_2 present in excess of that dissolved in equilibrium with the atmospheric CO_2 pressure (generally 1 to 30 mg CO_2-C per litre) may be considerable too: in the order of 0.1 to 10 g CO_2-C $m^{-2}y^{-1}$. Part of this may become involved in weathering reactions in aquifers (and as such is accounted for in silicate and carbonate weathering), and the rest is largely undone by degassing when ground water reenters open streams.

Export of dissolved organic carbon from soils is another potential sink of atmospheric CO_2. Concentrations of dissolved organic carbon in soil solution below the root zone are generally in the order of 2 to 10 mg C per litre. The high values are typical for Podzols and Histosols, the low values for Ca-rich soils and tropical soils high in Fe and Al, and intermediate values are common in most other temperate-climate soils. At annual drainage rates between 100 and 1000 mm, these concentrations imply that 0.2 to 10 g $m^{-2}y^{-1}$ of atmospheric CO_2-C is removed by drainage of organic C from soils. Part of this organic carbon is retained as solid humic material in deeper strata, part may be oxidized to CO_2 in oxygenated aquifers, and part is respired to CO_2 after the ground water has surfaced in streams.

World-wide, the effect of carbonate and silicate weathering in soils and rocks on the CO_2 balance is appreciable. Weathering causes an annual removal of about 0.16 and 0.27 Pg of CO_2-C from the atmosphere in the form of dissolved HCO_3^- (Holland, 1978), compared to an estimated 0.2 to 0.9 Pg of CO_2-C nowadays emitted from soils as a result of human activities. This is partly undone by new formation of minerals in the oceans, leaving a net removal of about 0.1 Pg of CO_2-C per year. Nevertheless, the comparison shows that the role of weathering in the CO_2 balance of soils cannot be neglected, as has been done by most researchers dealing with this question.

8.2.9 Discussion

Figure 8.8 is helpful in wrapping up the discussion on the role of soils in the CO_2 balance of the atmosphere. It shows the net effect of afforestation and deforestation followed by cultivation on the net flux of atmospheric CO_2 into or out of the Ferralsol from Surinam, discussed in the previous paragraphs. Starting with a soil with 0.05% C throughout, the build-up of soil organic C under the influence of a forest makes the soil a strong sink for atmospheric CO_2.

As steady state conditions are approached, the role of soil organic carbon in removing atmospheric CO_2 diminishes, and (except in peatland) eventually ceases. When the external conditions change so that soil organic material undergoes net decomposition, e.g. as a result of deforestation and reclamation for agriculture, the soil becomes a source of CO_2, until a steady state with respect to organic matter (now at a lower level) is reached again. Under steady state conditions with respect to organic carbon the soil is usually a sink of atmospheric CO_2, due to formation of HCO_3^-, followed by leaching, and due to drainage of organic C. These effects are too small to be shown on the scale of Figure 8.8: in the Surinam Ferralsol, simulated CO_2 losses due to weathering, drainage of dissolved organic C, and drainage of dissolved CO_2 were in the order of 1.5 g C $m^{-2}y^{-1}$ under maize, and 4 g C $m^{-2}y^{-1}$ under forest. Higher weathering rates in case calcite or easily weatherable silicates were present would cause CO_2 losses in the order of 5 to 30 g C $m^{-2}y^{-1}$.

Simulation runs similar to those shown here could be carried out for other soil types with other vegetation types. Using soil data from taxonomic systems and making assumptions about various parameters needed for the simulations is probably a feasible approach to estimate the steady state carbon levels and the steady state CO_2 sink term due to weathering typical for a particular soil-vegetation system. Our present preoccupation with this subject, however, is the result of dramatic changes taking place when the steady state situation is disturbed as a result of sudden changes in environmental conditions. However, accurate prediction of peaks in CO_2 production caused by deforestation, as shown in Figure 8.8, may be much more difficult. Because small differences in degradation kinetics may have relatively large effects on such CO_2 production peaks it is particularly important to calibrate and validate such simulations for a large number of specific conditions. Much research still needs to be done to provide the data necessary for such an exercise.

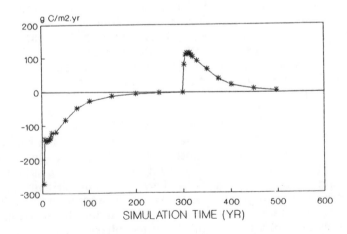

Figure 8.8 Net effect of afforestation (0-300 y) and deforestation followed by cultivation (300-500 y) on the net flux of atmospheric CO_2 into or out of a Ferralsol in Surinam. At year 0 the soil has 0.05% organic C at all depths.

8.3 METHANE

Production and consumption of methane in soils are both microbial processes. Methane is formed in anaerobic soils when methanogenic bacteria take part in the decomposition of organic matter. On the other hand methane can be oxidized in aerated soils or soil horizons under the influence of methane oxidizing bacteria. Most peats and poorly drained or artificially flooded mineral soils serve as a source of methane, while well drained soils serve as a sink. Cicerone and Oremland (1988) recently reviewed the subject of biogeochemical aspects of atmospheric methane, and we will mainly refer to their paper for details about methanogenesis.

Figure 8.9 summarizes the processes of exchange of methane between soils and the atmosphere. Methane forming bacteria, or so-called methanogens, are strictly anaerobic, so in addition to metabolizable carbon, absence of oxygen (O_2) is a requirement for their activity. The requirement that O_2 is absent is usually fulfilled when the soil is saturated with water. Because diffusion of gases in water is about 10^4 times slower than in air, the exchange of gases virtually stops when the soil is waterlogged. In water-saturated soil with sufficient metabolizable organic matter, O_2 is consumed more rapidly than it can be replenished by diffusion. All O_2 trapped during water saturation usually disappears within a few hours to days, and the soil becomes anaerobic, except in a layer of a few mm thickness at the soil- air (or soil - aerated water-) interface where sufficient O_2 can penetrate. After flooding of an initially aerobic soil, a thermodynamic suite of other oxidants takes over the role of O_2 as electron acceptor in the respiration of organic matter. First any nitrate present is reduced, next MnIV,III- and FeIII oxides are partially reduced to MnII and FeII, followed by reduction of sulfate to sulfide. Only after the disappearance of sulfate will fermentation to methane take place.

Methanogens are active in almost any anoxic ecosystem, and they can withstand extremes in temperature, salinity and pH. Because they can metabolize only a limited number of substrates, methanogens depend on other microbes for producing these from organic matter. They tend to be out-competed by sulfate reducers, and are usually inactive in the presence of sulfate. This probably explains the gradient in methane production along a salinity gradient in marine swamps, with little or no methane production close to the sea, and higher production inland (DeLaune et al., 1983; Bartlett et al., 1987). In the thin oxidized zone on top of anaerobic soil, bacterial methane oxidation may remove a large part of bacterially formed CH_4, particularly where the methane evolves by diffusion, rather than by ebullition. Emission of methane through aerial channels present in marsh plants (de Bont et al., 1978; Sebacher et al., 1985) could also circumvent oxidation by methanotrophs.

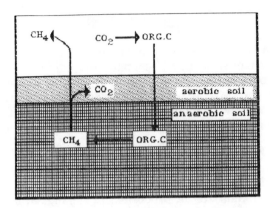

Figure 8.9   Processes involved in the exchange of methane between soils and the atmosphere.

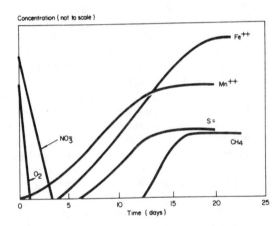

Figure 8.10 Sequence of chemical changes in an initially aerobic soil after waterlogging (taken from Patrick and Reddy, 1978).

The natural soil environments where methane is produced include fresh and salt water (mangrove) swamps, fresh and salt water marshes, bogs, fens, waterlogged tundras, and shallow open water. Apart from the above-cited lower emission rates in (sulfate-rich) saline environment, little is known about systematic differences in emission rates, which vary widely among the various environments. For instance Harris et al. (1985) observed no differences between ombrotrophic (pH 3-4) and minerotrophic (pH 6-7) peats in CH_4 production, while Svensson and Rosswall (1984) found higher methane production in minerotrophic peats. Although high temperatures stimulate methane formation, annual methane emissions estimated from field measurements are in the same range (1-200 g $m^{-2}y^{-1}$) for tropical, cool temperate, and subarctic environments (Bouwman, this volume). The environments considered so far are natural, and it is unlikely that they have contributed to the increase in atmospheric CH_4 in the past decades. If soils indeed contribute to the increased levels of atmospheric methane, wetland rice fields are the most likely candidates, because they are the only wetland environments that have increased strongly over the past decades. Seiler and Conrad (1987) estimate that the CH_4 emission from rice paddies may have increased from 0.075 Pg y^{-1} in 1950 to 0.117 Pg y^{-1} in 1980. Although the production of methane from wetland rice soils is well documented (De Bont et al., 1978; Kimura et al., 1984; Seiler, 1984) very little is known about specific factors influencing the rate of methane formation in wetland rice soils.

Methane oxidizing bacteria are probably very important in limiting the flux of methane to the atmosphere from reduced soil zones, and they could provide a sink for atmospheric methane in well-drained soils. The amount of methane removed from the atmosphere by soils is probably very small (Keller et al., 1986; Table 1). Most methanothrophic bacteria require CH_4 concentrations for their growth that are appreciably higher than atmospheric concentrations, although one species with very high affinity for CH_4 has been isolated (Jörgensen and Degn, 1983).

Methane emitting soils can be identified in part from the World Soil Map. They encompass all peat soils (Histosols), and most seasonally wet soils, which include all Gleysols, and Gleyic subgroups. However, in the wetter climates and in subarctic areas, water saturated conditions may prevail in a small-scale pattern in depressions in otherwise well-drained areas, which cannot be indicated on a global scale. Wetland rice is grown on a wide variety of soils, but predominantly on Gleysols, Fluvisols and Luvisols.

With the present limited knowledge on the relationship between soil and climatic factors and methane emission within the group of waterlogged soils, it is probably not possible at present to attempt the use of process-oriented simulation models for estimating CH_4 emissions. However, soil inventory data should be useful in making such estimates by assigning reasonable values of methane emission rates obtained from field measurements in different soils types. The most comprehensive effort to date in this direction is that by Matthews and Fung (1987) for methane emissions from natural wetlands. They combined and integrated three independent global sources, compiled to digital 1°resolution data bases: (i) the UNESCO vegetation data base, (ii) the FAO-UNESCO soil map, and (iii) the Operational Navigation Charts.

8.4 NITROUS OXIDE

Of the three greenhouse gases considered here, nitrous oxide shows the highest emission from soils relative to other sources. Because nitrous oxide emissions from soils were reviewed recently (Sahrawat and Keeney, 1986), we will only summarize the main aspects of the process here. N_2O can be formed both during oxidation of NH_3 (nitrification) and during reduction of NO_3^- (denitrification). Because oxidized as well as reduced soil conditions can give rise to nitrous oxide emissions, the process can be expected to occur in a wide variety of soils.

In well aerated soils, nitrification after fertilization with urea, anhydrous NH_3, and ammonium sulfate increases N_2O emission compared to non-treated or $Ca(NO_3)_2$-amended controls. After fertilizer doses in the order of 100 to 250 kg N ha^{-1}, maximum losses of up to 20 kg N ha^{-1} over a period of 3 to 5 months were observed. Usually, however, fertilizer induced N_2O losses on well drained soils are less, and in the order of 0.1 to 1% of the applied nitrogen (Bouwman, this volume). N_2O production increases with increasing temperature, increasing water content to field capacity, and increasing pH between 4.7 to 6.7. These effects are mainly due to the microbial nature of the nitrification process, which is performed mainly by autotrophic bacteria with an optimum under near-neutral conditions. The strong diurnal, seasonal and spatial variation of N_2O emission are also related to the microbial nature of nitrification.

Acid forest soils (pH $CaCl_2$ 3.9-4.4) also produced appreciable N_2O (6-60 mg N kg^{-1} dry soil) upon aerobic incubation, but liming doubled or tripled the emission. Duxbury et al. (1982) reported very high N_2O emissions from drained and cultivated organic soils (7-165 kg N ha^{-1} y^{-1}, compared to cropland on mineral soils (1-4 kg N ha^{-1}y^{-1}). Probably, strong mineralization and nitrification of organic N in these peat soils is responsible for these extreme levels. This is corroborated by results by Mosier et al. (1982) and Cates and Keeney (1987), who observed relatively high N_2O losses (in the order of 0.8 t0 2% of the applied N, cf. Bouwman this volume) after adding manure.

Microbial denitrification is a respiration process in the absence of oxygen whereby nitrate is used as electron acceptor in the oxidation of organic substances. The overall process of complete denitrification can be written as:

$$4NO_3^- + 5CH_2O + 4H^+ \rightarrow 2N_2 + 5CO_2 + 7H_2O,$$

where CH_2O stands for the organic energy source. However, N_2 is not the only N product of denitrification. Denitrification takes place in the sequence: $2NO_3^- > 2NO_2^- > 2NO > N_2O > N_2$, with N_2O and N_2 being by far the most important N products. Ratios of N_2O/N_2 vary greatly, and increase with NO_3^-, pO_2, decreasing pH (reduction of N_2O to N_2 is inhibited at pH<6), and decrease with increasing C availability. Alternate oxidation/reduction strongly increases N_2O emission (Letey et al., 1981; Smith and Patrick, 1983), apparently because the complete denitrification sequence is interrupted by regular supply of limited amounts of oxygen. Continuous flooding on the other hand, gives mainly denitrification to N_2 with little N_2O production. Therefore, N_2O emissions are probably low from flooded rice fields (cf. Table 4.17 in Bouwman, this volume). However, while farmers aim at keeping rice fields flooded continuously, in actual field situations limited water supply from rain or irrigation water often causes temporary aeration of the surface soil (Moormann and Van Breemen, 1978). Under these conditions, both nitrification and denitrification following nitrification may cause much higher N_2O emissions than observed under ideal experimental conditions.

Whereas much work has been done on the effect of fertilizer application on N_2O relatively little is known about N_2O emission from uncultivated soils. Field measurements by Keller et al. (1986) indicated that well-drained soils (including Oxisols and Ultisols) under tropical forests may emit appreciable amounts of atmospheric N_2O and may contribute about 40% of the current global N_2O emissions. Emissions from temperate forests seem to be much lower. However, Bowden and Bormann (1986) found that whole tree harvesting in New Hampshire (USA) hardwood forests increased N_2O losses (up to 7 kg N $ha^{-1}y^{-1}$) from Haplorthods, presumably by increased nitrification as well as increased denitrification. Part of the loss did not take place from the soil, but by degassing of dissolved N_2O from streams.

N_2O can be emitted from a wide variety of soils, whereby soil characteristics useful in soil taxonomic systems generally play a subordinate role. Moreover, anthropogenic factors (fertilization, drainage, irrigation, deforestation) have an overwhelming effect on the N_2O emission rate from a given piece of land. Therefore, soil taxonomic data will probably be of limited use in estimating global N_2O emission rates from budget models, and still less so from process-oriented models.

8.5 CONCLUSIONS

We feel that levels of soil organic carbon at steady state and the net CO_2 sink term due to leaching of dissolved organic C and HCO_3^- can be estimated reasonably well by process-oriented simulation models using soil data from the FAO-UNESCO system plus climatic and land use data. Even so, much work remains to be done in translating FAO-UNESCO and other data to values for model parameters. Extensive data on soil organic matter profiles can be used to validate the model results.

Our preoccupation with soils as a source of greenhouse gases, however, is directly linked to the present-day dramatic changes in land use over large areas in the tropics cause soils. As a result of these shifts, soils change from (generally low-level) CO_2 sinks to (high-level) CO_2 sources. The effect of such changes on the CO_2 balance can be modelled too, but results will be difficult to validate. Nevertheless, with good geographic data on land use and changes in land use, in combination with data on soils, process-oriented models should replace the more usual bookkeeping models to estimate changes in CO_2 fluxes from soils.

Soil data, together with other geographic data (e.g. on wetland rice cultivation) are indispensable to indicate the poorly drained areas where methane emissions can be expected. However, because relatively little is known about environmental factors determining methane fluxes from poorly drained land, process-oriented models are not yet suitable to estimate methane emissions, and simple bookkeeping models are probably more appropriate.

Levels of nitrous oxide emissions seem to be determined mainly by management en environmental factors that cannot be extracted from soil taxonomic data. Therefore, it may be very difficult to estimate N_2O fluxes from large areas using soil data.

In conclusion, therefore, process-oriented simulation models for soil-borne emissions of the three greenhouse gases considered here using soil taxonomic data are most promising for CO_2 (for which soil emissions are of minor importance relative to other sources) and least promising in case of N_2O (for which soils are the dominant source).

ACKNOWLEDGEMENTS

The authors thank Dr R.L.H. Poels and Dr B.H. Janssen of the Agricultural University, Wageningen, Dr J.A. van Veen of ITAL Wageningen, for reviewing earlier drafts of this paper.

REFERENCES

Bartlett, K.B., D.S. Bartlett, R.C. Harriss and D.I. Sebacher (1987) Methane emissions along a salt marsh salinity gradient. Biogeochemistry 4:183-202.

Boudot, J.P., B.A. Bel Hadj, T. Chone and B.A.B. Hadj (1986) Carbon mineralization in Andosols and aluminium-rich highland soils. Soil Biology and Biochemistry 18:457-461.

Bouwman, A.F. (1989) International conference "Soils and the greenhouse effect", Background paper. Int Soil Reference and Information Centre, Wageningen, the Netherlands, August 1988, 143 p.

Bowden, W.B. and F.W. Bormann (1986) Transport and loss of nitrous oxide in soil water after forest clear cutting. Science 233:867-869.

Brams, E.A. (1971) Continuous cultivation of West African soils: organic matter diminution and effects of applied lime and phosphorous. Plant and Soil 35:401-414.

Buringh, P. (1984) Organic carbon in soils of the world. In: Woodwell, G.M. (Ed.)- (1984), The role of terrestrial vegetation in the global carbon cycle. Measurement by remote sensing, p 91-109. SCOPE Vol. 23. Wiley and Sons, New York.

Cates, R.L. and D.R. Keeney (1987) Nitrous oxide production throughout the year from fertilized and manured maize fields. Journal of Environmental Quality 16:443-447.

Celos (1980, 1981, 1982, 1983) Annual Reports of the Centre of Agricultural Research. University of Surinam. Paramaribo, Surinam.

Cicerone, R.J. and R.S. Oremland (1988) Biochemical aspects of atmospheric methane. Global Biochemical Cycles 2:299-327.

Cleaves, E.T., A.E. Godfrey and O.P. Bricker (1970) Geochemical balance of a small watershed and its geomorphic implications. Geological Society America Bulletin 81:3015-3032.

Dalal, R.C. and R.J. Mayer (1986) Long-term trends in fertility of soils under continuous cultivation and cereal cropping in southern Queensland. II. Total organic carbon and its rate of loss from the soil profile. Australian Journal of Soil Research 24:281-292.

De Bont, J.A.M., .K.K. Lee and D.F. Bouldin (1978) Bacterial oxidation of methane in a rice paddy. Ecological Bulletin 26:91-96.

Detwiler, R.P. and C.A.S. Hall (1988) Tropical Forest and the Global Carbon Cycle. Science 239:42-47.

DeLaune, R.D., C.J. Smith and W.H. Patrick (1983) Methane release from Gulf coast wetlands. Tellus 35B:8-15.

Duchaufour, Ph (1973) Actions des cations sur les processus d'humification. Science du Sol 3:151-163.

Duxbury, K.A., D.R. Bouldin, R.E. Terry and R.L. Tate III (1983) Emissions of nitrous oxide from soils. Nature 298:462-464.

Feytel, T.C.J and E.L. Meyer (1989) Simulation of Soil forming processes. Internal course manual. Dept. Soil Science and Geology, Agricultural University Wageningen. The Netherlands.

FAO-UNESCO (1971-1978) Soil Map of the World 1:5000000. Vol. I-X. FAO. Rome.

FAO-UNESCO (1988) Revised legend of the FAO-Unesco Soil map of the World. World Soil Resources Report N`60. FAO. Rome, 109 p.

Gosselink, J.G. (1984) The ecology of delta marshes of coastal Louisiana: A community profile. U.S. Fish and Widl. Serv., Office of Biological Services, Washington D.C. FWS/OBS.

Harris, R.C., E. Gorham, D.I. Sebacher, K.B. Bartlett and P.A. Flebbe (1985) Methane flux from northern peatlands. Nature 315:652-653.

Holland, H.D. (1978) The chemistry of the atmosphere and oceans. Wiley and Sons, N.Y., 351 p.

Houghton, R.A., R.D. Boone, J.R. Fruci, J.E. Hobbie, J.M. Melillo, C.A. Palm, B.J. Peterson, G.R. Shaver, G.M. Woodwell, B. Moore, D.L. Skole and N. Meyers (1987) The flux of carbon from terrestrial ecosystems to the atmosphere in 1980 due to changes in land use: geographical distribution of the global flux. Tellus 39B:122-139.

Janssen, B.H. (1984) A simple method for calculating decomposition and accumulation of 'young' soil organic matter. Plant and Soil 76:297-304.

Jenkinson, D.S. and J.H. Rayner (1977) The turnover of soil organic matter in some of the Rothamsted classical experiments. Soil Science 123:298-305.

Jordan, F. (1985) Nutrient cycling in tropical forest ecosystems. Wiley and Sons. 190 p.

Jörgensen, L. and H. Degn (1983) Mass spectrometric measurements of methane and oxygen utilization by methanotrophic bacteria. FEMS Microbiol.Lett. 20:331-336.

Keller, M., W.A. Kaplan and S.C. Wofsy (1986) Emission of H_2O, CH_4 and CO_2 from tropical forest soils. Journal of Geophysical Research 91:11791-11802.

Kimura, M., H. Wada and Y. Takai (1984) Studies on the rhizosphere of paddy rice. IX. Microbial activities in the rhizosphere of paddy rice. Japanese Journal of Soil Science and Plant Nutrition 55:338-343 (In Japanese).

Likens, G.E., F.H. Bormann, R.S. Pierce, J.S. Eaton and N.M. Johnson (1977) Biogeochemistry of a forested ecosystem. Springer Verlag, N.Y., 135 p.

Lugo, A.E., M.J. Sanchez and S. Brown (1986) Land use and organic carbon content of some subtropical soils. Plant and Soil 96:185-196.

Letey, J., M. Valoras, D.D. Focht and J.C. Ryden (1981) Nitrous oxide production and reduction during denitrification as affected by redox potential. Soil Science Society of American Journal 45:727-730.

Mann, L.K. (1986) Changes in soil carbon storage after cultivation. Soil Science 142:279-288.

Matthews, E. (1983) Global vegetation and land use: new high resolution data bases for climate studies. Journal of Climate and Applied Meteorology 22:474-487.

Matthews, E. and I. Fung (1987) Methane emission from natural wetlands: global distribution, area, and environmental characteristics of sources. Global Biochemical Cycles 1:61-86.

Martin, J.P. and K.Haider (1986) Influence of mineral colloids on turnover rates of soil organic carbon. p. 283-304 in: P.M. Huang and M. Schnitzer (eds) Interactions of soil minerals with natural organics and microbes. SSSA Special Publication 17, Soil Sci. Soc. Amer., Madison, Wisconsin, USA, 606 p.

Melillo, J.M., J.R. Fruci, R.A. Houghton, B. Moore III and D.L. Skole (1988) Land-use change in the Soviet Union between 1850 and 1980: causes of a net release of CO_2 to the atmosphere. Tellus 40B:116-128.

Minderman, G. and K.W.F. Leeflang (1968) The amounts of drainage water and solutes from lysimeters. Plant and Soil 28:61-80.

Mosier, A.R., G.L. Hutchinson, B.R. Sabey and J. Baxter (1982) Nitrous oxide emissions from barley plots treated with ammonium nitrate or sewage sludge. Journal of Environmental Quality 11:78-81.

Moormann, F.R. and N. van Breemen (1978) Rice, Soil, Water, Land. International Rice Research Institute. Philippines. 185 pp.

Odum, E.P. (1971) Fundamentals of Ecology, 3rd Ed. Saunders Company, Philadelphia, 574 p.

Palm, C.A., R.A. Houghton and J.M. Melillo (1986) Atmospheric carbon dioxide from deforestation in Southeast Asia. Biotropica 18:177-188.

Parton, W.J., D.S. Schimel, C.V. Cole and D.S. Ojima (1987) Analysis of factors controlling soil organic matter levels in Great Plains grasslands. Soil Science Society of America Journal 51:1173-1179.

Patrick Jr., W.H., and C.N. Reddy (1978) Chemical changes in rice soils, p.361-379 in: Soils and Rice, Int. Rice Research Institute, Los Baños, Philippines.

Poels, R.L.H. (1987) Soil, Water and nutrients in a forest ecosystem in S᠎rinam. Agricultural University Wageningen. 253 pp.

Reynolds, R.C. and N.M. Johnson (1972) Chemical weathering in the temperate glacial environment of the northern cascade mountains. Geochimica Cosmochimica Acta 36:537-554.

Sahrawat, K.L. and D.R. Keeney (1986) Nitrous oxide emission from soils. Advances in Soil Science, Vol. 4:103-148. Springer Verlag, New York.

Sebacher, D.I., R.C. Harriss, and K.B. Bartlett (1985) Methane emission to the atmosphere through aquatic plants. Journal of Environmental Quality 14:40-46.

Seiler, W. and P.J. Crutzen (1980) Estimates of gross and net fluxes of carbon between the biosphere and the atmosphere from biomass burning. Climatic Change 2:207-247.

Seiler, W. and R. Conrad (1987) Contribution of tropical ecosystems to the global budgets of trace gases, especially CH_4, H_2, CO and N_2O. p 133-160 in R.E. Dickinson (Ed.), Geophysiology of Amazonia. Vegetation and Climate Interactions. Wiley and Sons. New York.

Silvola, J. (1986) Carbon dioxide dynamics in mires reclaimed for forestry in eastern Finland. Annales Botanici Fennici 23:59-67.

Smith C.J. and W.H. Patrick, Ir. (1983) Nitrous oxide emission as affected by alternate anaerobic and aerobic conditions from soil suspensions enriched with ammonium sulfate. Soil Biology and Biochemestry 15:693-697.

Svensson, B.H. and T. Rosswall (1984) In-situ methane production from acid peat in plant communities with different moisture regimes in a subarctic mire. Oikos 43:341-350.

USDA (1975) Soil Taxonomy. US Government printing office. Washington DC. 754 pp.

Van Breemen, N and R. Protz. 1988. Rates of calcium carbonate removal from soils, Canadian Journal of Soil Science 68, 449-454.

Van der Linden, A.M.A., J.A. van Veen and M.J. Frissel (1987) Modelling soil organic matter levels after long term applications of crop residues, farmyard and green manures. Plant & Soil 101:21-28.

Voroney R.P, J.A. Veen and E.A. Paul (1981) Organic C dynamics in grassland soils. 2. Model validation and simulation of the long-term effects of cultivation and rainfall erosion. Canadian Journal of Soil Science 61:211-224.

Wang, T.S.C., P.M. Huang, Chang-Hung Chou and Jen-Hsuan Chen (1986) The role of soil minerals in the abiotic polymerization of phenolic compounds and formation of humic substances. p. 251-281 in: P.M. Huang and M.Schnitzer (eds) Interactions of soil minerals with natural organics and microbes. SSSA Special Publication 17, Soil Sci. Soc. Amer., Madison, Wisconsin, USA, 606 p.

Whittaker R.H. and G.E. Likens (1975) The biosphere and man. In: Lieth, H. and R.H. Whittaker (Eds.): Primary production of the biosphere. Ecological Studies 14. Springer Verlag, Heidelberg, Berlin, New York.

CHAPTER 9

Geographic Quantification of Soils and Changes in their Properties

W.G. SOMBROEK

ISRIC, Wageningen
P.O. Box 353, 6700 AJ Wageningen, the Netherlands

ABSTRACT

The most comprehensive global-scale soil map of the world is the one of FAO/Unesco at scale 1:5.000.000, published between 1971 and 1981. It contains a wealth of information, but being soil-classification oriented it does not include all of the soil and terrain attributes, in quantified form, that may be needed for Global Change studies.

For a number of regions the map is out-of-date; more accurate data on soil geographic patterns have been assembled by national institutions since the time of active data compilation by FAO.

Soil attribute-oriented global-scale maps and tabular data bases do not yet exist. A start has been made through the SOTER project of the International Society of Soil Science, but it will take 10 to 20 years before such a Global Soil and Terrain Database with digitized maps at scale/resolution of 1:1.000.000 and computer-stored tabular data, is completed. For an early overview one could however extract a provisional data base from the existing FAO documentation on soils and agro-ecological zoning, with the incorporation of the data of a number of new national soil maps published at 1:1.000.000 scale.

Geo-referenced information on changes in soil and terrain conditions is incomplete. A global overview, at an average scale of 1:10.000.000, of the present status of human induced soil degradation in its various forms will be available in 1990, through the current GLASOD project of UNEP/ISRIC.

A provisional quantification of worldwide degradation hazards was proposed by FAO/UNEP/Unesco in 1979. This was executed, at 1:5.000.000 scale, for Northern Africa and the Middle East only. A more definite quantification of the hazards, encompassing the whole world, awaits completion of the SOTER project.

Associated geo-referenced data bases for other aspects of the land, such as surface climate, hydrology, vegetation and land use, animal life, and human population pressure are needed. For some of these, categorical systems of classification and legend structures still have to be developed.

Soils and the Greenhouse Effect. Edited by A.F. Bouwman
© 1990 John Wiley & Sons Ltd.

9.1 INTRODUCTION

In the context of a Global Change research programme the world's soil mantle may be seen to constitute a special facet or subsystem: the pedosphere.

Soils are rooting media and sources of bio-production and therefore *key components of life-support systems*. They are also *sinks and sources* of biogeochemical elements and membranes or filters for pollutants. Soils are however also organised and structured natural entities in their own right, constituting *a mosaic of small reactors* where other spheres interrelate and form something new and unique. The end products are relatively stable, and form *blocks-of-memory* of past atmospheric-hydrologic-lithospheric-biologic interactions. At the same time these archives are liable to change, gradually or abruptly and partly irreversibly, under human influence of differing type and intensity.

The present-day contribution of the world soils and their cover ("terrestrial biota") to greenhouse gas production varies from relatively modest in the case of CO_2 (30%), to predominant for CH_4 (70%) and to nearly exclusive in the case of N_2O (cf. Bouwman, 1989). The soil condition also affects albedo and evaporation. In the likelihood of a Greenhouse Effect the world's soils therefore deserve careful attention.

As every farmer, road builder or nature conservationist knows, soils vary from place to place and the pattern can be very intricate. At detailed level, relationships with climatic conditions, geology, geomorphology, hydrology, land cover type and past human influence are usually quite clear, and the soil properties, or attributes, can be fully quantified. At regional and global level, however, one needs generalizations and abstractions. Already from a scientific point-of-view the description, laboratory analysis, classification and mapping becomes problematic, because soils form a continuum in space and time, and are neither dead matter nor fully living bodies with self-reproducing capacity (though living, soils have no sex).

On the utilitarian side, generalized but still quantified data on soil attributes have to serve many purposes: guidelines for agricultural production, for civil engineering works, for environmental management planning, and now also for an assessment of their actual and potential contribution to a greenhouse warming. No wonder that different schools of grouping and mapping of soils have evolved, often linked to cultural traditions and even national pride. Most of the schools are pedogenetically oriented, and take the subsurface-subsoil features as starting point, because of their supposedly more stable character than the surface and topsoil features.

When discussing soils and soil changes in relation to global climatic change, these topsoil features, and the associated changes in land cover (vegetation, crops) become however much more interesting.

9.2 SOIL CLASSIFICATION-ORIENTED GLOBAL-SCALE SOIL MAPS

9.2.1 FAO/Unesco

The largest-scale, i.e. the highest-resolution map of the soils at global level is the FAO/Unesco Soil Map of the World at scale 1:5.000.000. It was compiled by the World Soil Resources Office of FAO with the active participation of the International Society of Soil Science (ISSS), and published by Unesco in ten volumes (FAO, 1971-1981). It consists of 18 partly overlapping sheets with accompanying explanatory texts and tables in 9 volumes, and a separate volume with the details of the legend criteria. The map is mainly a compilation of a multitude of national soil maps available at the time, of differing detail, accuracy and classification system. To achieve uniformity, many regional correlation meetings and field expeditions were organised. It resulted in a worldwide applicable legend, with names partly derived from existing classification systems, partly newly coined, that are easily pronounced and translated into major languages (English, French, Spanish and Russian versions have been printed).

The Legend (FAO, 1974) can be regarded as a two-level soil classification system, based on soil genetic principles but with quantitative limits between the 106 units. Main criteria used are the so-called "diagnostic (soil) horizons". Additional differentiating criteria are non-horizon related "diagnostic properties".

Because of the limitations of scale, most cartographic units, or series-of-polygons of identical content, are composed of a predominant soil classification unit as well as associated soils, and inclusions. Land features not directly related to these units, but rather to the mapping unit as a whole, are indicated as "phases". The main diagnostic horizons, properties and phases are summarised in Table 9.1.

Also the topsoil textural class of the predominant soil unit is indicated (coarse, medium, or fine textured), as well as the average slope of the mapping unit as a whole (level to gently undulating, rolling to hilly, or steeply dissected to mountainous). Areas of "non-soil" are indicated as miscellaneous land units: dunes or shifting sands, glaciers and snow caps, salt flats, rock debris, and desert detritus. Boundaries of permafrost or intermittent permafrost are indicated separately, but no other soil moisture and soil temperature characteristics are used, except implicitly in hydromorphic and desert soils. To estimate moisture and temperature characteristics, FAO devised special maps on "agro-ecologic zoning" (FAO, 1978-1981) which can be used as overlays. These zones are in fact agro-climatic regions, based on overhead climatic conditions, and expressed in "length-of-growing periods". In this context, mention should still be made of the 1:25M world map of Unesco (1977) on the distribution of arid regions showing a) the degree

Table 9.1 Soil attributes as defined in the FAO/Unesco Legend of 1974 (not exhaustive, and limits deleted; between brackets the revisions of 1988)

1. DIAGNOSTIC SOIL HORIZONS

a) H, A or E horizons: topsoils

histic:	peaty	(histic)
mollic:	mineral soil material, rich in organic matter, base-saturated	(histic)
umbric:	mineral soil material, rich in organic matter, acid	(umbric)
ochric:	mineral soil material, low in organic matter	(ochric)
albic:	bleached subsurface layer	(albic)
-o-	thick man-made surface layer	(fimic)

b) B horizons: subsoils

argillic:	illuvial clay-accumulated, acid or non-acid	(argic)
-o-	distinct textural increase from A to B horizon	(argic)
natric:	clay- and sodium accumulated	(natric)
cambic:	slightly weathered soil material	(cambic)
spodic:	accumulation of iron, aluminium and/or organic matter	(spodic)
oxic:	strongly weathered, without differential clay accumulation	(ferralic)
calcic:	accumulation of calcium carbonates, non-cemented	(calcic)
gypsic:	accumulation of gypsum, non-cemented	(gypsic)
sulfuric:	accumulation of sulfuric acid	(sulfuric)

2. DIAGNOSTIC SOIL PROPERTIES

abrupt textural change:	claypan	(abrupt textural change)
albic material:	bleached	-o-
amorphous material-dominated:	allophanic	(andic properties)
ferralic properties:	low-activity clay-minerals dominated	(ferralic properties)
ferric properties:	with non-indurated concentrations of non-active iron oxide	(ferric properties)
thin iron pan:		(placic phase)
hydromorphic properties:	oxigen-deficient, reduction colours	
	ground-water induced	(gleyic properties)
	surface-water induced	(stagnic properties)
takyric features:	polygonal surface cracks	(takyric phase)
gilgai:	microrelief due to swell-shrink clays	(gilgai phase)
plinthite:	reddish mottled clay, irreversibly hardening on exposure	(plinthite)
vertic properties:	strong vertical cracking	(vertic properties)
sulfidic materials:	waterlogged, with sulfides	(sulfidic materials)
permafrost:	continuously frozen layer	(permafrost)
high organic matter content throughout:		(strongly humic)
-o-	clayey, with active iron oxides	(nitic properties)
-o-	carbonate-rich material	(calcareous)
-o-	calcareous throughout	(calcaric)
-o-	strongly saline throughout	(salic properties)
-o-	gypsum rich material	(gypsiferous)
-o-	regularly replenished sediments	(fluvic properties)
-o-	extremely weathered material	(geric properties)

Table 9.1 (continued)

3. DIAGNOSTIC SOIL PHASES

stony:	stony materials at the surface	(rudic)
lithic:	coherent hard rock at shallow depth	(lithic)
petric:	oxidic concretions in the soil	(skeletic)
petrocalcic:	continuously cemented calcium-rich layer	(petrocalcic horizon)
petrogypsic:	continuously cemented gypsum-rich layer	(petrogypsic horizon)
petroferric:	continuously cemented iron-rich layer	(petroferric)
phreatic:	groundwater table at depth	(phreatic)
fragipan:	high-density brittle layer	(fragipan)
duripan:	silica-cemented layer	(duripan)
saline:	layer with medium to high salt content	(salic)
sodic:	layer with medium to high sodium content	(sodic properties)
-o-	desert surface features	(yermic)
-o-	man-induced surface water hydromorphism	(anthraquic)
-o-	microrelief due to frost-heaving	(gelundic)
-o-	temporary submerging or flooding	(inundic)

of bioclimatic aridity, b) temperature regimes, and c) length-of-drought periods.

The FAO/Unesco soil legend and the published maps have contributed to the development of a number of national soil classification systems such as those of Kenya and Mexico. It also greatly stimulated systematic soil cartographic work in many countries, often resulting in country maps at 1:1M scale. Sometimes radically different pictures of the soil pattern emerged - with the Amazon region as a glaring example. In a way the FAO/Unesco effort has been defeated by its own success at stimulating soil geographic research at national and regional levels.

FAO and other entities such as ISRIC have therefore long pleaded to update the existing maps with the newly generated spatial information. For some regions this updating has indeed been recorded at the FAO office in Rome, but funds for a systematic worldwide updating and subsequent publication programme have most unfortunately not been made available.

Not only updating at the same scale of 1:5M was suggested, but also a more detailed map presentation, at scale 1:1M or so. The latter was effectuated for Western Europe only, thanks to recent financial support from EEC and from ISSS in the case of Austria and Switzerland (CEC, 1985; ISSS, 1986). Its legend was already somewhat more detailed than the one used for the 1:5M map. Recently a systematic revision of the world-level FAO/Unesco legend was completed (FAO, 1988), for use at the same 1:5M scale or larger, i.e. higher-resolution scales such as 1:1M. It contains 28 major soil groupings, subdivided into 153 soil units, with suggestions for third-level subdivisions. This Revised Legend has already been used for 1:1M aggregation of the soil conditions for irrigation development purposes in the Nile-basin countries of North-eastern Africa by ISRIC (Hakkeling and Endale, 1988), the data now being digitized at FAO-Rome. Digitizing of all 1:5M map sheets, either by polygon vectoring or by gridding, has been undertaken by several organizations, such as UNEP-GRID in Geneva (see the Conference Background document, section II of this volume, for details).

Several generalisations of the 1:5M map have been made, for instance for FAO's Physical Resources Base map at scale 1:25M, where soils and agro-climatic conditions are delineated in conjunction (FAO, 1983). Some of the units of the FAO/Unesco Soil Legend have been combined on the latter map, either because of close geographic association or because of a degree of similarity in properties. These combinations are given and described in the Conference Background paper (Section II of this volume), because the 1:25M generalisation and its digitizing may be of convenient use for preliminary Global Change modelling.

9.2.2 World Soil Maps Produced by National Agencies

USA. The Soil Conservation Service of the US Department of Agriculture has for many years maintained a World Soil Geographic Unit in Lanham/Washington, where soil data from all parts of the world were annotated on 1:1M Operational Navigation Charts (ONC). These were never published but negatives of the annotated sheets were graciously supplied to ISRIC. The Unit used an early US system of soil classification (Baldwin, Kellogg and Thorp, 1938), now officially declared obsolete and replaced by the US "Soil Taxonomy" system (Soil Survey Staff, 1975). This new system, intended for world-wide use, is a detailed quantitative system of five categoric levels, which is moreover regularly being amended through a system of international committee consultations and correlation meetings. It includes criteria on soil moisture and soil temperature regimes, on "soil" slope and on soil mineralogy. For the highest categories, a tentative correlation with the units of the FAO/Unesco Legend is given in Section II of this volume.

The aim to prepare and publish world soil maps on the basis of the Soil Taxonomy system has thus far been realised at very low resolutions only, such as the 1:50M map of USDA/SCS of 1972. A tentative plan to inventorize the soil resources of all Third-World countries at 1:1M scale using Soil Taxonomy, through the USAID/USDA Soil Management Support Services project, has recently been abandoned.

The very precise soil morphometric criteria would facilitate modelling but at the same time have thwarted compilation at smaller resolution. The identification and the agricultural relevance of the soil moisture and soil temperature regimes as defined have remained a bone of contention. The moisture regimes, with suggested subdivisions, have been published separately by Cornell University (Van Wambeke, 1981, 1982, 1985 and in CEC, 1985). It did however not use actual measurements in the soils themselves, but relied on the computational model of Newhall for derivation of the moisture regimes from atmospheric climatic data. It should moreover be noted that the Newhall model applies the Thornthwaite formula for calculation of the potential evapotranspiration rather than the Penman formula as FAO and Unesco used. As a consequence there are rather strong spatial differences between the Cornell maps and the FAO and Unesco ones mentioned above, especially as regards aridic regions (Sombroek, 1987).

USSR. Soil scientists of the USSR can be proud of a long history of soil science in general, and of soil geography in particular. Soil genetic principles are of primordial importance in their classification systems, quantified morphometric criteria being sparsely used. The system of Kovda et al. (1967) served as a basis for the compilation of a world soil map at scale 1:10M. It was published at about the same time as the FAO/Unesco map (Kovda et al. 1975), in both Russian and English. It derives its data mainly from the same sources as FAO, but gives less details on the soil association/complex per mapping unit/polygon. The main grouping of the legend units is on primary vegetation type and climate type. Topographic conditions and surface textures are not given, except where extremes occur (mountains, sand areas). Stony and skeletal soils are indicated separately.

Somewhat more quantified criteria were used for a generalized world soil map at 1:15M, published by Glasovskaya and Friedland (1986; in Russian only).

It is not known whether the two maps concerned have been digitized by some national or international institution.

France. Francophone soil scientists have basically the same genetic approach as the USSR, although of late much effort has been given to underpin this with quantified and mutually exclusive morphometric criteria (Segalen et al., 1979/1984). The system that is used formally is however the one of the French "Commission de Pédologie et de Cartographie des Sols" (CPCS, 1975). At the occasion of the golden anniversary of the French Association for Soil Studies in 1984, a world map of the "grandes tendances" of soils at scale 1:10M, applying CPCS criteria, was shown. This map was not printed for general use.

Others. A number of national soil classification systems have been developed, or are being developed, that are geared towards the specific conditions of the country or region concerned (West Germany, England, Australia, Brazil, Canada, South Africa, China, Poland, etc.).

Each of these systems has its own merits, and has often resulted in low-resolution national maps such as the 1:5M soil map of Brazil of 1980, but by the nature of these systems no attempts were made to produce world maps. Correlation of all these systems and the maps produced with them, with a view to improve world soil mapping, is difficult and time-consuming. For all of them, however, reference material is present at ISRIC as well as at the Soils Branch of FAO in Rome.

Elsevier-ISRIC's wall chart on "Soils of the World" (1987) gives a tentative correlation between some main systems.

9.3 SOIL ATTRIBUTE-ORIENTED GLOBAL-SCALE MAPS AND TABULAR DATABASES

As expounded in a preceding conference paper (Van Breemen), and voiced repeatedly by IGBP Panels and Working Groups, there is a need for global geographically referenced information on a number of soil and terrain attributes that are thought to have direct influence on gas fluxes, albedo, evapo(transpi)ration and other aspects of the world hydrological cycle. These are: surface roughness, surface sealing/crusting, surface colour in relation to reflectance; amount, type, stability and vertical distribution of organic matter; soil biological activity; surface run off, soil moisture storage and transmittal properties; drainage and flooding conditions; and soil temperature regimes, but also relatively simple attributes such as bulk density per layer/horizon. The latter are also needed to express soil characteristics and properties as commonly given on a weight basis in units-of-volume as used by modellers of climate and plant growth.

9.3.1 Geo-referenced Soil Attribute Data Derived from Classification-oriented ("pedologic") World Maps

Many attributes can be derived from a judicious analysis of the criteria used in the more quantitative morphometrically oriented soil classification systems such as the FAO/Unesco Legend and the US Soil Taxonomy. Since for the latter no world map exists, geo-referencing can only be done with the FAO/Unesco maps - accepting the locally out-dated character of the information. The two USSR world maps are less suitable because of the scarcity of quantitative limits between the soil units that are distinguished.

In particular the combination of the FAO/Unesco map and the FAO agro-ecological zoning map provides many clues for semi-quantification of soil attributes that are required for Global Change studies, and their subsequent digitizing and tabular computer storage - even though the FAO-approach is oriented on crops rather than on ecosystem processes. Such an exercise would gain much value if new national or regional maps, based fully or largely on the FAO/Unesco Legend criteria but with often larger resolution (EEC-countries, Brazil, Mexico, Kenya, Ethiopia, Canada, etc), are incorporated.

Unfortunately a programme of geo-referenced extraction and machine processing of soil and terrain attributes on the basis of the world map has not yet been formulated, let alone funded. At regional level it is however being undertaken for the EEC countries. Examples are the current soil moisture storage studies by Verheye (1989) and by Groenendijk (see elsewhere in this volume). Mention should also be made of the world map at 1:5M scale of the wetlands and the soils with hydromorphic properties as produced for the polders-of-the-world symposium, based on the FAO map but for Latin America already updated by Van Dam and Van Diepen (1982). The comprehensive computerized and geo-referenced study on the Land Resources of tropical South America (Cochrane et.al., 1985) makes also use of more recent national maps than the ones used for the published FAO/Unesco documentation on the region.

An example of the compilation of soil-attribute digital maps using US Soil Taxonomy mapping is the set prepared for Indiana State by Valenzuela and Baumgardner, as recorded in Baumgardner and Oldeman (1986).

There is much merit in a concerted and concentrated effort, by relevant international organisations and countries, to compile soil-attribute global maps on the basis of the FAO/Unesco documentation, especially if some updating for critical areas is executed. It could result in the availability of much-wanted data within a year or two, and funding requirements would not be excessive.

9.3.2 Geo-referenced Soil and Terrain Attribute Databases by Direct Compilation

In the longer run, attribute-oriented global databases on soil and terrain conditions will have to be made available, in addition to global pedologically-oriented maps by conventional cartography. Soil classification systems and pedological maps have a scientific value of their own, just as plant taxonomy and vegetation mapping, but they can never be as utilitarian as to satisfy the needs of the many groups of users of soils and landscape information (Dudal, 1986; Bouma, 1988). For many purposes the 1:5M global soil map of FAO/Unesco is anyhow too general, and 1:1M or 1:0.5M mapping of the whole world by conventional cartography would be excessively expensive - even if a wholehearted agreement and cooperation between institutions of different "schools" of classification and legend construction would emerge (an ISSS Working Group is currently devising an International Reference Base for soil classification).

Fortunately, the advance of digitizing techniques and computerized data storage offers revolutionary possibilities, not previously available, for comprehensive geo-referenced soil and terrain attribute databases, with an adaptable level of spatial resolution and inbuilt updating capacity. Such a direct, digital database would in fact side-step the problems of attempting to interpolate soil properties from soil maps constructed with differing legend structure and differing soil taxonomic systems (Sombroek, 1986). An ad-hoc Working Group of the International Society of Soil Science met on this subject in January 1986 at the ISRIC premises in Wageningen (Baumgardner and Oldeman, 1986). The Workshop resulted in a proposal for a world soils and terrain digital database at an average scale of 1:1M, acronymed SOTER, to be completed in a 10 to 20 years period. A number of pilot areas were identified, and UNEP provided funding for a first pilot area in South America (portions of Argentina, Brazil and Uruguay). ISRIC was asked to coordinate the initial SOTER activities. Since then a procedures manual for small scale map and database compilation was prepared (Shields and Coote, 1988), drawing much on experience gained at a national level such as in Canada. The manual was revised in 1989 to take into account experiences in the first SOTER pilot area. A database structure using a Relational Data Base Management System was developed for data storage and an international panel assessed various GIS systems suitable for SOTER.

The essence of the approach is to screen all existing data in an area, whether or not registered on official soil maps, and where necessary to complement them with Remote Sensing indications. The data are then rearranged for the database, going from landform and terrain component features (with 9 attributes at polygon level and 19 terrain component parameters) to soil layer attributes. For all soil attributes – 52 in total – quantitative class limits are used. Baumgardner (this volume), in his paper on the details of the SOTER project, gives the full listing of all attributes and class codes for the polygon file, the terrain component file and the soil layer file; suggestions for additional attributes and their classes in relation to Global Change studies are most welcome! Those attributes that can not yet reliably be quantified from available data, or from limited additional field- and laboratory measurements, are flagged as such, for amendment at a later stage. Right now, algorithms and computer programmes are being developed for the use of such a comprehensive database for a number of purposes, for instance the risks of several forms of soil degradation. In due course the data system will be incorporated in UNEP's GRID system.

The pilot areas of the SOTER programme are strategically located, often in frontier areas of two or more countries with differing approaches to soil and terrain classification and mapping. The need to converge towards a common approach makes the initial database production slow, but it is expected that, once such a common approach has been agreed upon, further work on a SOTER database within a country can largely be done by national organizations – as already exemplified by a newly initiated programme of the EMBRAPA organization in Brazil. Details on progress of the SOTER project are given by Baumgardner (this volume).

9.4 GEO-REFERENCED INFORMATION ON CHANGES IN SOIL AND TERRAIN CONDITIONS.

9.4.1 Changes which Have Occurred in the Past

To be able to put the present atmospheric greenhouse conditions and any imminent changes in the right perspective, global changes of the past have to be studied, as reference material. This may hold in particular for the soil factor: soils generally respond to climatic change somewhat slower than vegetation does, but they show probably a more true record of the regional climate than pollen, which would seem to show considerable variation over a short time range. Paleosols, defined as soils which were formed in a landscape of the past, have been identified in various environments, from polar to tropical deserts, and in a number of lithological sequences (coastal sands, tills, and loess mantles in particular). It has been suggested (Yaalon, 1989) to start with the preparation of "Paleo-pedomaps" at scale 1:15M for certain slices of time, representing both colder and warmer periods of the past.

Concurrently, a geo-referencing may take place of major human-induced soil changes on the past, such as salinization of major floodplains and the erosion of major landscapes through the mismanagement of the land by early civilizations (see for instance Stamp, 1961).

9.4.2 Changes Occurring at Present

Much publicity is being given to changes in soil and terrain conditions occurring right now, under the influence of increased human activities: acidification, erosion, soil structure decline, deforestation, desertification, salinization, soil biological degradation, etc. There is however little systematic information on the extent and seriousness of these changes worldwide in the form of maps. Even at the national level, such mapping is often of an incidental nature only. It is true that for one particular form of degradation, viz desertification (or better said "aridification"), FAO and Unesco prepared in 1977 a provisional world assessment at scale 1:25M. Its criterion of vulnerability-of-land-to-desertification can be taken to approximate the present changes. The map shows "surfaces subject to sand movement", "strong or rocky surfaces subjected to areal stripping by deflation and sheetwash", "alluvial or-residual surfaces subject to stripping of topsoil and accelerated run-off, gully erosion on slopes and/or sheet erosion or deposition on sand flats", and "surfaces subject to salinization or alkalization".

Soon after completion of the Soil Map of the World, FAO, UNEP and Unesco started a three-years project for the global assessment of actual and potential soil degradation. This was to be based on the compilation of existing data and on the interpretation of environmental factors influencing the extent and the intensity of soil degradation. A provisional methodology was established, and maps at scale 1:5M for

North Africa and the Middle East were prepared (FAO-UNEP-Unesco, 1979). One set of maps relates to the "present degradation rate" and the "present state of soil". It draws from the 1:5M soil map and gives, in a provisional form:

a) the present rate of water erosion, wind erosion, salinization and sodication, each in four classes of intensity (none to slight, moderate, high, very high);
b) areas of shallow soil (between 10 and 50 cm, less than 10 cm), of light texture (medium but not stony, coarse but not stony), and of high salinity/sodicity (4-15 mmho, more than 15 mmho electrical conductivity; 6-15%, more than 15% exchangeable sodium);
c) miscellaneous land units that can be considered, in part, to be the result of past, largely natural degradation processes (dunes, salt flats, rock debris or rock outcrops).

This effort for North Africa and the Middle East was not followed-up by a worldwide programme, not only because of the excessive costs of map compilation and printing at 1:5M scale, but also because it was realised that the database from which the assessment had to be derived was not sufficiently accurate for a number of regions, and some doubts about the validity of the proposed methodology.

This situation induced UNEP, about a decade later, to call for a small-scale/low resolution world map of the status of human-induced soil degradation, mainly for awareness purposes. A contract to this effect was awarded to ISRIC end 1987. This project, acronymed GLASOD (for Global Assessment of Soil Degradation) will produce a world map at an average scale of 1:10M, and the data aggregation is mainly through an expert system approach - relying on appraisal by experienced natural-resources geographers that have broad overview knowledge of a particular region or subcontinent. Guidelines for such general assessment have been developed (Oldeman, 1988). They itemise the various forms of soil or land degradation, the method to estimate their extent within delineable physiographic units, their degree of severity, and their speed ("recent-past average rate"). The resulting world map and matrix tables per polygon, to be ready in the course of 1990, will be included in the UNEP-GRID system. Quantification of land degradation at relatively high resolution (1:1M) is to be obtained for window areas only, coinciding with the SOTER pilot areas.

In principle it will be possible to carry out a repeat inventory every 10 or 20 years, as a means to monitor the advance of soil and land degradation (c.q. its halting or improvement in certain areas) over decades to come.

Details on the GLASOD project are given in a poster presentation at the Conference (Oldeman et al., this volume).

9.4.3 Changes to be Expected in the Future: Hazards or Risks

The hazard of further soil and land degradation can in principle be assessed on the basis of a modelling programme that takes into account relevant soil constraints, water and wind action, on-going salinization processes, air-borne depositions etc. This should be coupled with expected direct human influence (increased arable cropping, animal husbandry, irrigation/drainage development etc.), and preferably be verified with experimental results in key locations such as standardized erodibility and erosivity field tests (Lal, 1988).

An early estimate on one aspect, viz. desertification hazard, is given on the FAO/Unesco 1:25M map of 1977. This was followed by a more systematic evaluation of desertification hazards, prepared by FAO for UNEP in the period 1979 to 1984. It gives maps of "component analyses" for the African continent at 1:25M scale (soil constraints, water action, wind action, salinization, animal pressure, population pressure), accompanied by an Explanatory Note (FAO, 1984). Included is a "window" map at 1:5M for a part of Northeastern Africa, and one at 1:10M on soil elements used in assessing desertification and degradation for the Indian subcontinent.

A more comprehensive effort to quantify soil degradation hazards was given in the aforementioned FAO project for a provisional methodology for soil degradation assessment (FAO-UNEP-Unesco, 1979) which produced, as a second set of maps, a 1:5M "provisional map of soil degradation risks" for Northern Africa and the Middle East. It combines ongoing dominant soil degradation processes and their intensity (water erosion, wind erosion, salinization and sodication) with secondary soil degradation risks and their intensity (chemical degradation through leaching or toxicity; physical degradation through rainfall or bad agricultural practises or through waterlogging and irrigation, and biological degradation).

Again, this effort was not followed by a worldwide programme, for the same reasons as mentioned under 4.2. Since the seventies the seriousness of the problem of worldwide changes in soil conditions through human interference has however become more apparent than ever. The Changes are not limited to the Third World countries on which the early FAO-Unesco-UNEP efforts were concentrated. Also in zones with mediterranean, temperate or cold climates the hazards for further human-induced changes in soil and terrain conditions are apparent: water erosion in mountainous areas; air-borne industrial pollution; acidification and toxification by over-fertilizing and bio-industrial activities; soil organic matter destruction by artificial drainage; soil compactation by use of heavy agricultural machinery; land dishevelment by mining, or transformation of arable land into asphalted or concrete surfaces - a special form of desertification! In the tropics and subtropics the expansion and intensification of paddy rice cultivation, with new cultivars requiring high fertilizer input, is now causing concern, as well as monocultures of export crops such as cassava (Thailand) and soybeans (Southern Brazil). All these changes will have some influence on the soils contribution to the greenhouse function of the atmosphere. Therefore, worldwide and much more quantified information on soil and land degradation hazards in its manyfold forms is needed. It should make use of machine-processable quantified soil and terrain attribute data[1]. The SOTER approach can provide this in principle, and a quantified soil degradation hazard assessment for its pilot areas is foreseen through the establishment of degradation subfiles (see the second part of the Manual of Shields and Coote, 1989 for details). As mentioned before, it will take a number of years before a comprehensive global database will be completed.

Within a relatively short time, however, a geo-referenced soil degradation hazard picture could be obtained for some regions: for North America by combining the Canadian generalised soil-landscape mapping at 1:1M with a USDA/SCS 1:1M Soil Taxonomy based mapping of the USA and with the 1:1M set of maps on the natural resources of Mexico (SPP, 1981); for Europe by combining the 1:1M FAO-based soil map of the EEC, the 1:1M Soil Taxonomy-based mapping of the European part of the USSR, and national soil maps of other Eastern European countries. The question is: is any funding agency prepared to foot the bill?

9.5 REQUIRED ASSOCIATED LAND DATABASES

Soil is only one aspect of the "land" to be considered at assessing its contribution, now and in the future, to the atmospheric greenhouse functioning. Some other aspects are briefly mentioned below (see Young, 1987, for a more comprehensive review).

9.5.1 Climate

Geo-referenced data on surface temperatures, rainfall, evapo-(transpi)ration etc. can relatively easily be obtained through applying and expanding FAO's agro-ecological zoning concepts, and the CLICOM programme of WMO. Hopefully, it can be combined with IIASA's gridded world map of "Holdridge life zones" as prepared by Leemans (1989), which is a climate data set derived from about 8000 weather stations (see section II of this volume for details). It should be mentioned that the SOTER programme has a separate climate file.

---

[1]Such a database will also be required to assess positive or negative feedbacks of a greenhouse warming and associated sea-level rise on soil conditions and the latitudinal or altitudinal shifting of agricultural belts and biomes - as to be discussed at a UNEP/ISSS sponsored Conference in february 1990 in Nairobi.

9.5.2 Hydrology

The spatial pattern of the terrestrial hydrological conditions is poorly inventorized. Deep groundwater reserves and their quality have been put on maps for Europe and parts of Latin America (International Association of Hydrogeologists, in cooperation with Unesco, 1983), as well as the discharge of the major rivers of the world and the reservoir storage of large dams (Unesco, 1978). Shallow groundwater reserves and their condition, surface run-off, and land flooding are little quantified. Also in view of the importance of wetlands to the greenhouse functioning, there is a need for a classification system and legend construction - separate from those for soils - for the different types of flooding or submergence of land, whether natural or artificial. Depth and length of flooding or submergence during the growing season; chemical characterization of the water; sediment content; physical damaging force of flooding, etc. need to be quantified (see also Van Diepen, 1985). A first effort was recently made by Aselmann and Crutzen (1989). It may be elaborated by the Working Group on the Geomorphology of River and Coastal Plains of the International Geographical Union (IGU), which once had an abortive project for a geomorphological map of flood-affected areas and drainage basins of the world at 1:5M scale. Mention should be made of an inundation analysis for Bangladesh at scale 1:750.000 carried out by FAO (Brammer, 1985), for ideas on some criteria to be used at a worldwide mapping effort on terrestrial hydrological conditions.

9.5.3 Vegetation and Land Use

The classification and geo-referenced inventory of vegetation and land use will be discussed in Matthews (this volume). On natural vegetation one particular biome may however be singled out here. FAO's Forestry Department carried out a first inventory of the extent of tropical rain-forest (Lanly, 1982). All available geo-referenced data were compiled on draft maps at 1:1M scale, but the data were then tabularised per administrative unit (country, state, province) and the draft maps subsequently destroyed! FAO is now starting on a new project of assessing the extent of tropical deforestation since 1980, and it is sincerely hoped that this time the draft maps will remain available, for ultimate publication and for digitizing purposes.

The various IGBP Panels and Working Groups give at present much attention to land cover and the proposed classification scheme is purposely left relatively coarse and robust viz. the scheme for the USA (Anderson, 1976). The extent of undeveloped land still primarily shaped by the forces of nature ("wilderness" land) is however only one-third of the global land surface (McCloskey and Spalding, 1989). The other two-third is modified by human settlement, in differing patterns and degrees of intensity. The types of land use vary enormously, also in relation to the greenhouse phenomenon. Some examples:

(i)   it may make a big difference for CH_4 production whether wetland rice growing is being done throughout the year (double or triple cropping) such as in South-east Asia, or is restricted to a single yearly crop due to rainfall and/or temperature constraints such as in North-east Asia. In large mechanized rice farms of the subtropics such as those of Southern France/Italy, Southern Brazil/Uruguay and the USA the single cropping is even of an intermittent type to ensure effective weed control. Also the depth of the watertable varies much: from shallow rainfed lowland (0-25 cm) to deep water (100-600 cm) rice cultivation types (see also Aselmann and Crutzen, 1989, and this Volume).

(ii) Shifting cultivation with long fallow periods will contrast substantially with semi-permanent small-holders settling in tropical deforested areas as regards the amount of atmospheric CO_2 that will be taken up by secondary forest growth.

(iii) Plantation cropping of oil palm will have a much higher leaf area index than plantation cropping of black pepper with its bare open spaces in between rows, inducing a lot of reflectance.

Global geo-referenced information on land use is available through the "World Atlas of Agriculture" prepared under the aegis of the International Association of Agricultural Economists (Medici et al., 1969). It gives land utilization maps at scale 1:2.5M or 1:5M (depending on the region) and accompanying smaller-scale relief maps. This atlas was however published twenty years ago and has a rather coarse legend.

A recent world map on land use is the one at 1:15M prepared by the Faculty of Geography of Moscow State University (Rjabchakov, 1986; in Russian only, but legend translated for use by ISRIC by Targulian).

There is a definite need for a new, computer-aided effort of world mapping of vegetation and land utilization types at scales 1:1M or 1:2.5M, taking into account the many changes that have occurred in the last 20-30 years, and giving details on the land- and water management practises involved. For this purpose a multicategorical system of land utilization types may have to be developed. For mapping purposes it can be combined with a vegetation classification system such as the one developed by Unesco (1973), yielding a legend that accommodates percentages classes of land use and national vegetation wherever the pattern is very detailed. Examples are the scheme elaborated by Malingreau et al. for Indonesia (Wood and Dent, 1983) and the approach of the "Institut de la Carte Internationale de la Végétation (ICIV) of Toulouse, France and Pondichéry, India (Blasco, this volume).

Some international entity should take the lead, may-be FAO, with the scientific support of an appropriate Working Group of the International Geographical Union. A first scheme can then be tested at the forthcoming IGBP Land Cover Change Pilot Study.

9.5.4 Animals

Certain groups of animals are known to produce substantial amounts of methane. For herbivores a geo-referenced global picture can be obtained relatively easily, because census data on cattle, sheep etc., and on wildlife herbivores and their foraging habits are quite abundant. For termites, another source of methane, such mapping is much more difficult. Not only may the CH_4 production vary between individual species, but also their frequency per physiographic soil unit and vegetation/land use type is poorly known. Above-ground termite structures are often an adaptation to extreme conditions of moisture (poor soil drainage) and temperature (above-canopy ventilation) rather than a sure sign of high population per unit area of land. With favourable below-ground moisture and temperature conditions termite colonies tend to remain hidden within the soil.

9.6 CONCLUSION

There is a clear need for a concentrated effort for the collection of geo-referenced and machine-readable information on quantified soil and land attributes, as a prerequisite for an accurate assessment of the present day contribution of soils and their cover to the greenhouse system, as well as for an estimation of the influence of any future changes in soil and land cover conditions on a surmised global warming. Provisionally, FAO/Unesco's soil map of the world and its updating can serve that purpose. In the longer run, however, a fully quantified global soil and terrain attribute data base will be required, to be linked with data bases on surface climatic conditions, on land cover and land use, and hydrology. Major landforms (physiographic land units) can possibly function as binding, key-entry elements.

REFERENCES

Anderson, J.R., E.E. Hardy, J.T. Roach and R.E. Witmer (1976) A Land Use and Land Cover Classification System for Use with Remote Sensor Data. Geological Survey Professional Paper 964. US Govnt. Printing Office, Washington D.C.

Aselmann, I. and P.J. Crutzen (1989) Freshwater Wetlands; global distribution of natural wetlands and rice paddies, their net primary productivity, seasonality and possible methane emissions. Max Planck Inst. L. Chemistry, Mainz, FRG. (accepted for publication by the Journal of Atmospheric Chemistry).

Baldwin, M., C.F. Kellogg and J. Thorp (1938) Soil Classification. In: Soils and Man. 979-1001. USDA Yearbook of Agriculture 1938. Washington DC.

Baumgardner, M.F. and L.R. Oldeman (1986) Proceedings of an International Workshop on the Structure of a Digital International Soil Resources Map annex Database. SOTER Report no. 1. ISSS/ISRIC, Wageningen.

Bouma, J. (1988) When the Mapping is Over, Then What? In: Proceedings of the International Interactive Workshop on Soil Resources, their inventory, analysis and interpretation for use in the 1990's. Educational Development System, Univ. of Minnesota, St. Paul, Minnesota, USA.

Bouwman, A.F. (1989) The Role of Soils and Land Use in the Greenhouse Effect. Netherlands Journal of Agricultural Sciences 37:13-19.

Brammer, H. (1985) Land Resources Appraisal of Bangladesh. Report 4, vol. 2. Ministry of Agriculture of Bangladesh and FAO/UNDP Project BGD/81/035. FAO, Rome.

CEC (1985) Soil Map of the European Communities 1:1.000.000. Nine sheets and Explanatory Booklet. Office For Official Publications of the European Communities, Luxembourg.

Cochrane, T.T., L.F. Sanchez, L.G. de Azevedo and J.A. Porras (1985) Land in Tropical America. A guide to climate, landscapes and soils for agronomists in Amazonia, the Andean Piedmont, Central Brazil and Orinoco. vol. 1-3. CIAT, Cali-Colombia and EMBRAPA-CPAC, Planaltina, DF. Brazil.

CPCS (1957, reprint 1977) Classification des Sols. Docum. Lab. Géol.-pédol. ENSA, Grignon, France.

Dudal, R. (1986) The Role of Pedology in Meeting the Increasing Demands on Soils. In: Transactions XIII Congress of the International Society of Soil Science. Vol. 1:80-96. ISSS, Wageningen.

Elsevier/ISRIC (1987) Soils of the World, wall chart 85x135 cm. Elsevier, Amsterdam.

FAO (1974) FAO/Unesco Soil Map of the World. 1:5.000.000. Volume I. Legend. Unesco, Paris.

FAO (1978-1981) FAO/Unesco Soil Map of the World 1:5.000.000. Volumes II-X. Maps per (Sub)continent and explanatory texts. Unesco, Paris.

FAO (1978-1981) Report on the Agro-ecological Zones Project. Vols. I-IV. World Soil Resources Report 48/1-4. FAO, Rome.

FAO (1983) A physical Resource Base; main climatic and soil divisions in the developing world, scale 1:25.000.000 (map only). FAO, Rome.

FAO (1984) Map of Desertification Hazards, Africa 1:25.000.000 and Explanatory Note, prepared for UNEP. FAO, Rome.

FAO (1988) Soil Map of the World, Revised Legend (prepared by FAO, Unesco and ISRIC). World Soil Resources Report 60. FAO, Rome.

FAO/UNEP/Unesco (1979) A Provisional Methodology for Soil Degradation Assessment. Maps and Explanatory Note. FAO, Rome.

FAO/Unesco (1977) Desertification Map of the World, 1:25.000.000 and Explanatory Note. UNEP, Nairobi.

Glasovskaya, M.A. and V.M. Friedland (1986) Soil Map of the World at 1:15.000.000 Geogr. Faculty of Moscow State University and Dokutchaev Soil Institute, Moscow (in Russian).

Hakkeling, R.T.A. and D.M. Endale (1988) Soils of Eastern and Northeastern Africa at 1:1 Million Scale and their Irrigation Suitability. Consultancy Mission Report 88/1, ISRIC, Wageningen.

ISSS (1986) Soil Map of Middle Europe 1:1.000.000; one sheet and explanatory text. CEC, Brussels and ISRIC, Wageningen.

Kovda, V.A., E.V. Lobova and B.G. Rozanov (1967) Classification of the World's Soils. Pochvovedeniye [Soviet Soil Science]. 4 and 7 (in Russian).

Kovda, V.A., E.V. Lobova, G.V. Dobrovolsky, J.M. Ivanov, B.G. Rozanov and N.A. Solomatina (1975) Soil Map of the World 1:10.000.000. Institute for Geodesy and Cartography, Moscow (in Russian and English).

Lal, R. (ed.) (1988) Soil Erosion Research Methods. pp. 244. Soil and Water Conservation Society, Ankony, Iowa and ISSS, Wageningen.

Lanly, J.P. (1982) Tropical Forest Resources. FAO Forestry Paper 30. pp. 106. FAO, Rome.

Leemans, L. (1989) World map of Holdridge Life Zones, IIASA, Laxenburg, Austria.

Medici, C., C. Vanzetti, J.R. Anderson et al. (1969) World Atlas of Agriculture. Istituto Geografico De Agostino, Novara, Italy.

McCloskey, J.M. and H. Spalding (1989) A Reconnaissance Level Inventory of the Amount of Wilderness Remaining in the World. Ambio 18(4):221-227.

Oldeman, L.R. (ed.) (1988) Guidelines for General Assessment of the Status of Human-induced Soil Degradation. Working Paper and preprint 88/4. ISRIC, Wageningen.

Rjabchakov, A.M. (1986) [World Map of Actual Land Use at 1:15.000.000] Faculty of Geography, Moscow State University, Moscow (in Russian).

Segalen, P., R. Fauck, M. Lamouroux, A. Perraud, P. Quantin, P. Roederer and J. Vieillefon (1979/1984) Projet de Classification des Sols. ORSTOM, Paris (english version "Project of Soil Classification" published in 1984 as Technical Report no. 7 by ISRIC, Wageningen).

Shields, J.A. and D.R. Coote (1988, 1989) SOTER Procedures Manual for Small-scale Map and Database Compilation, and Procedures for Interpretation of Soil Degradation Status and Risks. ISRIC Working Paper en Preprint 88/2, 89/3, Wageningen.

Soil Survey Staff (1975) Soil Taxonomy: a basic system of soil classification for making and interpreting soil surveys. Agric Handbook 436, USDA-Soil Conservation Service, Washington DC.

Sombroek, W.G. (1986) Establishment of an International Soil and Land Resources Information base. In: Baumgardner and Oldeman, 1986 (see above). 118-124.

Sombroek, W.G. (1987) Aridisols of the World, Occurrence and Potential. Proceedings Fourth International Soil Correlation Meeting, oct. 1987, Lubbock, Texas, U.S.A. (in press), and ISRIC Working Paper and Preprint 87/2.

SPP (1981) Cartas geologicas, Cartas edafologicas, Cartas de Humedad en el Suelo y Cartas de uso del suelo y vegetación 1:1.000.000. [eight sheets each]. Dirección General de Geografia del Territorio Nacional, Secretaria de Programación y Presupuesto dos Estados Unidos Mexicanos. Mexico City.

Stamp, L.D. (ed.) (1961) A history of Land use in Arid Regions. Arid zone Research 17. Unesco, Paris.

Unesco (1973) International classification and mapping of vegetation. Ecology and Conservation 6. Unesco, Paris.

Unesco (1977) Map of the World Distribution of Arid Regions, with Explanatory Note. MAB Technical Notes 7. Unesco, Paris.

Unesco (1978) World Water Balance and Water Resources of the Earth. Studies and Reports 25. Unesco, Paris

Unesco (1983) International legend for Hydrogeological Maps. SC 84/WS/7. Unesco, Paris.

USDA/SCS (1972) Soils of the World; distribution of Orders and principal Suborders (map 1:50.000.000). Soil Geography Unit, Hyattsville, USA.

Van Dam, A.J. and C.A. van Diepen (1982) The Soils of the flat Wetlands of the World; their distribution and their agricultural potential. ISRIC, Wageningen.

Van Diepen, C.A. (1985) Wetland Soils of the World, their characterisation and distribution in the FAO-Unesco approach. In: Wetland soils: characterisation, classification and utilization. Proceedings of a workshop at IRRI, 1984. 361-374. IRRI, Los Baños.

Van Wambeke, A. (1981) Soil Moisture and Temperature Regimes. South America. SMSS Technical Monograph 2. Cornell University, New York.

Van Wambeke, A. (1982) Soil Moisture and Temperature Regimes, Africa. SMSS Technical Monograph 3. Cornell University, New York.

Van Wambeke, A. (1985) Calculated Soil Moisture and Temperature Regimes of Asia. SMSS Technical Monograph 9. Cornell University, New York.

Verheye, W.H. (1989) Le régime hydrique des sols d'Europe, basé sur des données pédologiques et climatologiques courantes; principes et approche méthodologique. Science du Sol 27(2):117-130.

Wood, S.R. and F.J. Dent (1983) LECS, a Land Evaluation Computer System User Manual. Min. of Agriculture, Govnt. of Indonesia, and FAO, Rome (AGOF/INS/78/006 Manuals 5 and 6).
Yaalon, D. (1989) The Relevance of Soils and Paleosols in interpreting past and Ongoing Climatic Changes. ISSS Bulletin no. 75. Wageningen, The Netherlands.

Part IV

Sources and Sinks
of
Greenhouse Gases

CHAPTER 10

Modelling Global Terrestrial Sources and Sinks of CO_2 with Special Reference to Soil Organic Matter

G. ESSER

General Ecology Group
Biology/Chemistry Department of the University
Barbarastrasse 11, D-4500 Osnabrück, F.R.G.

ABSTRACT

The Osnabrück Biosphere Model, a global grid-based carbon balance model of the terrestrial biosphere, was used to calculate the influences of (i) the rising atmospheric CO_2 level, (ii) the temperature change due to the greenhouse effect, (iii) the clearing of forests and other natural vegetation, on the development of the biospheric pools, the soil organic carbon, and the atmospheric CO_2. The calculations were made for two periods: (i) The period 1860-1980 using historical data for emissions of fossil carbon and clearings together with a geographical information system which includes data on climate, soil type, vegetation, land use; (ii) The period 1980-year of doubling of preindustrial CO_2. For this model run scenarios were used for assumed future developments of the data necessary to drive the model functions.

10.1 INTRODUCTION

The annual carbon turnover through the terrestrial biosphere amounts to about 45 Gt (45×10^9 t), which is 7% of the atmospheric carbon pool. This large atmospheric pool may be influenced by the terrestrial biosphere, if the flux atmosphere-biosphere, which is the net primary production, and the flux back into the atmosphere by depletion of organic material, are of different size. This normally occurs during the course of a year, since production and depletion of biomass dominate in different seasons causing a phase-difference between the time functions of both fluxes. As a consequence, the atmospheric CO_2 concentration shows a seasonal amplitude of ca. 6 ppm at Mauna Loa and up to 15 ppm at northern subpolar latitudes (Point Barrow, Alaska). These seasonal fluctuations are levelled out on a medium-term scale.

On the other hand both principal fluxes may be unbalanced over a period of several years, which commonly is the result of different weather conditions from year to year. These fluctuations show up in

Soils and the Greenhouse Effect. Edited by A.F. Bouwman

seasonally adjusted time-series of the atmospheric CO_2 concentration (Keeling, 1983).

Climate effective are above all 'irreversible' carbon transfers to the atmosphere caused by long-term changes of the terrestrial pools. If these changes are the direct effect of human impacts (clearing of tropical forests, land use changes) then the related mechanisms are quite well known, although the availability of global data sets is often poor. If indirect effects of anthropogenic impacts are to be considered (CO_2 fertilization, climate change and land use effects on depletion rates of organic carbon) the mechanisms are presently not sufficiently understood to be able to develop suitable models.

Therefore, current carbon balance models of the terrestrial biosphere must be improved in both fields. This requires reliable data bases (geographical information systems) and sound representation of indirect processes (Esser et al., 1989). This is particularly true for processes concerning soil organic matter and litter. If any long-term changes of these pools exist, either caused by direct anthropogenic impacts or as a consequence of climatic change, a strong impact on atmospheric CO_2 may result.

In this paper an attempt is made to quantify the main terrestrial carbon pools and fluxes with special reference to soil-related processes, using a global regionalized model. These processes include the accumulation and depletion of the labile and stable fractions of soil organic carbon and litter, the influences of land use changes and of probable climatic changes.

10.2 MODELLING THE CARBON BUDGET OF THE TERRESTRIAL BIOSPHERE

The model used is Version 3 of the Osnabrück Biosphere Model (OBM) which was, in comparison to previous versions, improved in certain aspects:

- CO_2 fertilization effect
- data base for land use changes up to 1980
- land use scenario for the period after 1980
- new submodel for soil organic carbon

10.2.1 The Principal Model Features

The OBM was developed as an instrument to investigate the carbon balance of the terrestrial biosphere, the impacts of the rising atmospheric CO_2 level and climatic changes. It had to meet, therefore, the following requirements: (i) to include the major carbon fluxes and pools of the terrestrial biosphere; (ii) to quantify the fluxes by means of equations, which reflect relationships with the environment; (iii) to make the environmental variables in the equations valid throughout the whole of the terrestrial biosphere; (iv) to include any important indirect effects which may influence the global carbon budget.

The reasons which require the regionalization of the model were: (i) non-linearity of the model's equations; (ii) arbitrary human influences which act on different regional natural vegetation complexes; (iii) yields in agriculture depending on regional different socio-economic rather than ecological influences; (iv) complex regional patterns of change in the independent variables used by the model.

Regionalization was realized on a 2.5 degrees grid of longitude and latitude as a compromise between data availability and resolution needs on the one hand, and computing time on the other hand. Since the grid size of 2.5 degrees was too coarse for a sound representation of vegetation and agriculture, and to a certain degree of soils, extensive sub-grid processing of data was necessary.

On the grid element level, the OBM is one-dimensional. As a consequence, the set of model equations is applied once per time step to each of the 2433 grid elements, using an individual set of the driving variables for a grid element and time step. Time series of input data are thus required for each grid element.

Model structure. Figure 10.1 gives an overview of the model structure. In Table 10.1 a summary is given of the way the principal variables are computed.

Litter and soil organic carbon submodel. For each grid element of the model, depletion coefficients for woody and herbaceous fresh litter and for lignin-compounds are calculated from the variables *mean annual temperature* and *average annual precipitation*. It is assumed that lignin-compounds have depletion coefficients which amount to 1% of the coefficients of fresh herbaceous litter. The lignin content of wood is assumed to be 30%, that of herbaceous material 11%. Thus three fractions of decomposing material are distinguished in each grid element: herbaceous fresh litter, woody fresh litter, and lignin-compounds which are assumed to contribute to soil organic matter.

The function $k_i = f\,(T,Pp,i)$ with k_i being the depletion coefficient for the respective material, T and Pp being the climatic variables and i the litter fraction, as discussed in Esser et al. (1982) and Esser and Lieth (1989).

The pools of the litter and soil organic carbon fractions in a grid element are balanced with the litter production and depletion fluxes. The litter production flux is a balance of productivity and biomass change while the depletion flux is assumed to be proportional to the respective pool with k_i being the coefficient of proportionality. The complex of differential equations is integrated by use of a numerical library routine.

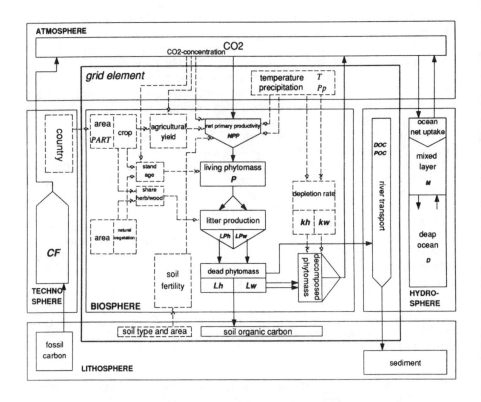

Figure 10.1 Simplified structure of the Osnabrück Biosphere Model. The global grid size is 2.5 degrees latitude and longitude. Carbon pools are represented by rectangles, carbon fluxes by pentagons, driving forces by dashed rectangles, mass relations by solid arrows, control relations by dashed arrows.

Geographical information system. The OBM reads the input data, which are necessary as driving functions, from a geographical information system which includes statistical data as well as digitized global thematic maps. It is updated continuously whenever new data sets become available. It includes the following sets of data:

- Climate: Mean annual temperature and average annual precipitation. Sources: NCAR data tapes including the WMO standard net (NCAR), the World Atlas of Climate Diagrams (Walter and Lieth, 1960 ff.), and a data collection with climatic zones maps by Müller (1982).
- Soils: 106 soil units according to the Soil Map of the World (FAO Unesco, 1974 ff.).
- Potential natural vegetation: 172 vegetation units are included globally (Schmithüsen, 1976).
- Land use: World Atlas of Agriculture (Instituto Geographico de Agostini, 1969, 1971, 1973), FAO Production Yearbooks (FAO-Unesco 1980 ff.), FAO Agro-Ecological Zones Project Results (FAO-Unesco, 1978 ff.).

Table 10.1 Principal biospheric variables of the Osnabrück Biosphere Model, their computation, and references for further information

Biospheric variable	Calculated from	References
Net primary productivity	temperature, precipitation, soil fertility, agricultural yield, conversion factors yield → productivity, land use areas, CO$_2$-fertilization	(Esser, 1987, 1989a, b)
Land use changes 1860-1981	statistical data on country basis remote sensing	(Esser, 1989c), (Richards et al., 1983)
Land use changes after 1981	scenarios considering: land use density in 1970, natural productivity, soil fertility, increase of of agricultural areas 1950-1980; options: preservation of natural vegetation, minimum-fertility for economic uses	(Esser, 1989c)
Cleared phytomass	land use changes, natural phytomass, crop phytomass	(Esser, 1987)
Soil fertility	empirical function for major 37 out of 106 soil units of Soil Map (FAO, 1974 ff.)	(Esser, 1984) (Esser et al., 1982)
Conversion factor Yield → productivity	empirical factor for major field crops	(Aselmann and Lieth, 1983) (Esser, 1989b)
CO$_2$-fertilization	atmosph. CO$_2$-conc., soil fertility	(Esser, 1989b)
Phytomass	net primary productivity, mean stand age	(Esser, 1984, 1987)
Litter production	net primary productivity, phytomass change	(Esser, 1987)
Litter pool	litter production minus depletion	(Esser and Lieth, 1989) (Esser et al., 1982)
Litter depletion	depletion coefficient, litter pool	(Esser, 1989b) (Esser and Lieth, 1989) (Esser et al., 1982)
Litter depletion coefficient	temperature, precipitation, material depleted	(Esser, 1989b) (Esser and Lieth, 1989) (Esser et al., 1982)
Soil organic carbon production	litter production, lignin content	this paper
Soil organic carbon	production minus depletion	this paper
Soil organic carbon	temperature, precipitation, soil	this paper
depletion	organic carbon pool	
Leaching of dissolved and particulate org. C	precipitation	(Esser and Kohlmaier, 1989)
Leaching of minerals due to land use changes	soil type, precipitation, element concentration in phytomass, cleared amount of phytomass	(Esser, Lieth and Clüsener Godt, 1989)
Atmosphere	balanced by: fossil emissions, ocean exchange, net primary production, depletion fluxes of litter and soil organic carbon, burnt biomass	
Ocean	box diffusion ocean 1 mixed layer, 43 deep sea boxes	(Oeschger et al., 1975)

- Land use changes: Factor-array for 121 countries for each year 1860-1981, based on data published by Richards et al. (1983), and evaluation of 934 Landsat images for the years 1972-1982 carried out by Esser and Lieth (1986).
- Crop yields: FAO Production Yearbooks (FAO-Unesco, 1980 ff.).
- Anthropogenic emissions from fossil sources compiled by Keeling (1973) and Marland and Rotty (1983).
- Share factors herbaceous/woody, stand ages, soil fertility factors, mineral content of the phytomass are based on our database DATAVW, which includes the relevant literature data. The data sets were completed using the method of 'ranking' (Esser, 1986).

10.3 RESULTS

The results of the OBM Version 3 differ from those of earlier versions (Esser, 1987) with respect to the fluxes productivity, litter production, forest clearing and others, while the total biospheric balance since 1860 is only slightly changed. The reason is that the fertilization effect of CO_2 has been reduced due to inclusion of soil fertility limitations, while on the other hand land use changes and cleared areas were reduced in the new regionalized land use database. In former versions of the model land use changes depended upon the development of the global population and did not consider different developments in individual countries. In Version 3, the global agricultural area increases by 10.0 × 10^6 km^2 in the period 1860-1980 (Esser, 1989c), while in former versions the increase was 12.5 × 10^6 km^2. Moreover, the increase is now quasi-linear, while it was exponential in earlier versions. All fluxes and pools are influenced by these modifications due to dynamic feedback.

10.3.1 Development 1860-1980

Net primary productivity. The only input flux of carbon into the terrestrial biosphere is the net primary productivity (NPP). As shown in Table 10.3, this flux has been almost constant since 1860. It amounted 44 Gty^{-1} up till 1930, while it has been slightly increasing to 46 Gty^{-1} further to 1980.

As agricultural land use increased in this period, a growing part of the NPP is contributed by field crops and plantations. In many countries the productivity of crops is much lower than the NPP of the replaced natural vegetation (see Table 10.2). Therefore the total global NPP should also have decreased. However, this is not observed, since the CO_2 fertilization effect hides the reduction. The mean productivity of all land not under agricultural use including deserts has increased from 330×10^6gC m^{-2}y^{-1} in 1860 to 346×10^6gC m^{-2}y^{-1} in 1980, although about 10 Mill. km^2 of productive land (ca. 8% of the global land area) were converted to agriculture in that period.

If the CO_2 fertilization is excluded in the model, the importance of this effect can be easily demonstrated. The global NPP decreases in that case from 48 Gty^{-1} in 1860 to 45 Gty^{-1} in 1980. The CO_2 fertilization effect is thus responsible for an additional NPP of about 5 Gty^{-1} of carbon or 10% (Table 10.3, column (6)).

Litter and soil organic carbon. The litter depletion and soil organic carbon depletion fluxes return carbon to the atmosphere. The source pools are the litter (dead but not yet decomposed plant material) and the soil organic carbon (in the model the long-lived fraction which stems from the lignin compounds of the litter). Since the global NPP and thus the production of litter are not changed, the pools and the depletion fluxes are also unchanged in model runs with standard climate. This is clearly pointed out in Table 10.3, columns (7) and (8).

This result contradicts papers which expect a large source from the decomposition of soil organic carbon after land use changes (collection of papers and discussion in Bouwman (this volume)). Those papers calculate the CO$_2$ emissions by multiplication of the losses after clearing of natural vegetation with the areas affected, but commonly ignore the CO$_2$ fertilization effect. If again a model run is carried out without considering CO$_2$ fertilization, net losses of 5 Gt C from litter and 26 Gt C from soil organic carbon occur in the period 1860-1980. The mean annual losses between 1970 and 1980 were 0.6 Gty^{-1} of carbon. This corresponds well with estimates given by Schlesinger (1984).

This stresses once more the importance of the CO$_2$ fertilization effect. As far as we know at present, it is probably so large, that well known source functions of C are compensated. Ignoring this effect will result in unrealistic values for atmospheric CO$_2$ in 1860 (261 ppm as calculated with the OBM).

Table 10.2 Comparison of the productivity of areas covered with natural vegetation and agricultural crops for some tropical and extratropical countries. The productivity of the agricultural crops were calculated from the yields by use of conversion factors given by Aselmann and Lieth (1983). Productivity are stated in units of g of dry matter (for carbon multiply by 0.45)

Country	Agricultural productivity (g$\times$m^{-2}xy^{-1})	Natural productivity (g$\times$m^{-2}xy^{-1})	Relation agric./natural
Zaire	180	1960	0.10
Kenya	350	1300	0.13
Niger	150	890	0.17
Kampuchea	310	1800	0.17
Bolivia	280	1500	0.19
Brazil	310	1620	0.19
Spain	510	750	0.68
FR Germany	1130	1190	0.95
Belgium, Luxembourg	1290	1210	1.07

Table 10.3 Development of the global sums of the major annual or accumulated carbon fluxes of the terrestrial biosphere and the ocean in the period 1860-1980. Results of a model run with standard climate (no changes). Fluxes are stated in units of Gt or Gty^{-1} of C, respectively)

Year	Fluxes (Gt)							
	Accumulation fluxes (since 1860)				Annual fluxes			
	Fossil Source	Ocean	Clearings	Fertil. effect	NPP	Litter Depl.	SOC	burnt Biomass
(1)	(2)	(3)	(4)	(5)	(6)	(7)	(8)	(9)
1860	-0	0	-0	0	44	-33	-11	-0.0
1870	-1	1	-7	2	44	-33	-11	-0.3
1880	-3	3	-15	5	44	-34	-11	-0.3
1890	-6	4	-22	8	44	-33	-11	-0.1
1900	-10	8	-30	12	44	-34	-11	-0.3
1910	-17	12	-37	15	44	-34	-11	-0.2
1920	-26	17	-44	19	44	-34	-11	-0.2
1930	-36	23	-54	25	44	-34	-11	-0.2
1940	-47	30	-63	31	45	-34	-11	-0.2
1950	-61	37	-73	38	45	-34	-11	-0.3
1960	-82	47	-81	47	45	-33	-11	-0.2
1970	-115	59	-89	58	45	-34	-11	-0.2
1980	-163	76	-96	73	46	-34	-11	-0.1

The well known global temperature rise since 1860 may have had some influence on the depletion of litter and soil organic carbon. A global uniform rise of the mean annual temperature was implemented into the OBM as a function of the atmospheric CO_2 concentration, whereby doubling of the present CO_2 level induces a temperature change of +3.5°C. For the period 1860-1980 the temperature rise was 0.8°C. In the model this change does not affect NPP. Moreover, in the model, vegetation does not migrate as climatic zones move. The climatic change therefore affects decomposition processes only. Net losses of 12 Gt C from the soil organic carbon pool and 0.5 Gt from litter were calculated by the OBM under these conditions.

Burnt biomass. During tropical deforestation, part of the cleared phytomass is subject to burning, while the major share of phytomass remains unburnt and is subject to natural decomposition. Accordingly, the OBM considers the clearcut phytomass as litter production, except for 50% of the herbaceous and 30% of the woody material, which are converted to CO_2 immediately. This 'burnt material'-flux amounts 0.1-0.3 Gty^{-1} without any significant change since 1860 (Table 10.3, column (9)).

Table 10.4 Development of the global sums of the major
carbon pools of the terrestrial biosphere and the atmosphere
in the period 1860-1980. Results of a model run with
standard climate (no changes). The Pools are stated in units
of Gt C, the atmosphere in ppm (vol.)

| Year | Pools | | | | |
| | Atmosphere (ppm) | Phytomass nat. | agri. (Gt) | Litter | SOC |
(1)	(2)	(3)	(4)	(5)	(6)
1860	285.0	668	1.6	91	1536
1870	286.7	663	1.7	92	1537
1880	288.6	658	1.9	92	1538
1890	290.7	654	2.1	92	1539
1900	293.1	650	2.2	92	1539
1910	296.1	646	2.4	92	1538
1920	299.4	643	2.5	92	1538
1930	303.2	639	2.6	92	1538
1940	307.0	636	2.8	92	1537
1950	311.6	633	2.9	92	1536
1960	317.6	634	3.0	92	1535
1970	326.5	637	3.1	92	1533
1980	338.8	644	3.2	92	1531

The current OBM does not include burning in 'natural' environments
such as savannas, grasslands and forests. Related charcoal production is
not included, therefore. Assuming that herbaceous vegetation is burnt
annually, fire frequency in coniferous forests is once in 20-100 years,
(depending on the precipitation regime), suggest that the global charcoal
formation since 1860 has been as much as 8 Gt in grasslands and
savannas and 1.5 Gt in coniferous forests. This amount is considerable
and further efforts should be made to refine the relevant processes in
the OBM.

Clearings and CO₂ fertilization. The carbon loss due to clearings is 96
Gt (Table 10.3, column (4)). The land use changes after clearing cause
additional emissions from litter and soil organic carbon of 5 and 26 Gt
respectively. The total source due to clearing is thus probably 127 Gt.
The fertilization effect as assumed in the OBM compensates for almost
the entire source. The remaining *net* sources are -1 Gt from litter and
5 Gt from soil organic carbon (Table 10.4, columns (5) and (6)). The net
source from phytomass is 23 Gt (Table 10.4, columns (3) and (4), and
Table 10.3, column (4) plus (5)). The reason for this great net source is
that the phytomass is calculated from NPP and stand age. The influence
of NPP is compensated by CO₂ fertilization, while the stand age is
definitely reduced from 10-300 years to 0.6-1.0 years at conversion of
forests to field crops.

A strong negative feedback follows from the fertilization effect:
NPP is enhanced, intensifying the input fluxes into the pools phytomass,

litter, and soil organic carbon. The growing pools are retarding the atmospheric CO_2 increase. The terrestrial biosphere thus behaves similarly as the ocean. The clearing of natural vegetation counteracts this process. Up to the 1960s the fertilization effect fell behind the emissions from cleared forest. In the 1970s, CO_2 fertilization began to overcompensate the clearings, turning the terrestrial biosphere into a small sink (ca. 0.5 Gt C y^{-1}).

Atmosphere and ocean. In a model run with standard climate, the box diffusion ocean absorbed 76 Gt carbon in the period 1860-1980, while 114 Gt C remain in the atmosphere (Tables 3 and 4). The atmospheric CO_2 level increases from 285 ppm to 338.8 ppm. The modelled increase in the period of the Mauna Loa records 1958-1980 is 22.6 ppm. The corresponding increase measured at Mauna Loa value is 23.2 ppm (Keeling, 1982). The model does not reproduce the year by year deviations of the seasonally corrected Mauna Loa values. The value of 285 ppm for 1860 is close to the ice core measurements for that time which suggest 287 ±3 ppm (Oeschger, 1986).

Overall balances 1860-1980. In Table 10.5, the results of three model runs with different scenarios for the period 1860-1980 are summarized. Some results were already discussed in detail in the past sections. The scenarios include the following assumptions:

- Standard climatic data set as mentioned in section 10.2.1, no climate variability.
- Increase of the mean annual temperature in each grid element depending on the atmospheric CO_2 concentration to yield +3.5°C for double CO_2. The change is similar for all grid elements. Only

Table 10.5 Net changes of global carbon pools for three scenarios for the period 1860-1980. Standard climate means the data set mentioned in 10.2.1, temperature increase relates to a change of the mean annual temperature of each grid element coupled to the atmospheric CO_2 to yield +3.5°C for double CO_2 (+0.8°C in 1980), no CO_2 fertilization assumes that NPP is uninfluenced by the atmospheric CO_2 level. Figures are net changes in Gt and Gty^{-1}, respectively

	Standard climate	Temperature increase	No CO_2 fertil.
CO_2 (1860) ppm	285	283	261
Phytomass	- 23	- 20	- 83
Litter	+ 1	- 0.5	- 5
Soil organic C	- 5	- 12	- 26
Ocean	+ 76	+ 78	+112
Atmosphere	+114	+117	+164
Fossil source		-163	

depletion coefficients of litter and soil organic carbon are assumed to be sensitive to temperature changes, while the NPP is not influenced.
- No CO_2 fertilization effect exists.

A disadvantage of dynamic models like the OBM is the fact that results are frequently somewhat peculiar when one effect is influencing all model pools and fluxes due to the dynamic feedbacks. An example is the temperature increase scenario in Table 10.5. Although the change affects the depletion of organic matter only, in the model results even the phytomass pool is influenced.

The most realistic scenario is probably the one which considers a global rise of the mean temperature of 0.8°C. In comparison, the greenhouse warming estimated for the past 100 years by climate models is 0.5-1.0°C. The observed warming is 0.7 ± 0.2°C (Hansen, et al. 1988). The CO_2 concentration of 283 ppm in 1860 is somewhat low compared with recent measurements in ice cores, but some sinks exist which are not considered in the model run: charcoal, leaching of organic compounds to rivers (particulate and dissolved material).

The atmospheric concentration of 261 ppm in 1860 obtained by not considering CO_2 fertilization is probably unrealistic, unless large carbon sinks have been overlooked.

10.3.2 Presumable Future Development after the Year 1980

The model runs for historic periods are, although numerous assumptions had to be included in the model, based on known effects and the results are subject to verification. The conditions to be expected in the coming centuries never existed in man's history and may hardly be foreseen at present. Scenarios are based on extrapolations of presently observable mechanisms and probably will be of little use for the prediction of the behaviour of vegetation in a 2 × CO_2 world.

Besides that, it is the author's conviction that it is the duty of the scientific community to put together even poor knowledge for making prognoses as a basis for political decisions, since the alternative is not a *better*, prognosis but *no* prognosis *at all*. Most relevant here is the trend of the development of carbon pools in future and the probable year of doubling of atmospheric CO_2. The results given by the OBM are listed in Table 10.6 for some obvious scenarios.

Model run 152 in comparison to 148 confirms the well known fact that the growth-rate of emissions from fossil sources is an important element. Similarly important is the rate of clearing of natural vegetation (run 149). In land use scenario 3 about 45 × 10⁶ km² of land is cleared up till the year 2123, while in scenario 6 about 43 × 10⁶ km² is cleared up till the year 2181 (run 148). Since all other assumptions are similar in both scenarios, the faster clearing rate of run 149 has reduced the time needed to reach doubling of CO_2 by 58 years. As side effect, the ocean absorbed 187 Gt C less and the fossil fuel emissions were 295 Gt C less in this shorter time period in run 149.

Neglecting the CO_2 fertilization in scenario 151 causes the $2 \times CO_2$ level to be reached 41 years earlier.

The effect of the predicted greenhouse warming was low in model run 150, but the suspected strong feedback relations between ocean and atmosphere are not included in the OBM. Therefore, we are presently planning large scale model coupling under the responsibility of the Max-Planck Institute for Meteorology in Hamburg (FRG), which will bring together the OBM with General Circulation Models and sophisticated Ocean Models to investigate this important feature.

Table 10.6   Results of five scenarios considering combinations of probable future developments for climate, land use, and consumption of fossil fuels. The pool changes and fossil emissions refer to the period 1980 to year of $2 \times CO_2$. All figures are stated in Gt carbon

Model run	148	149	150	151	152
Climate[1]	fixed	fixed	ΔT	fixed	fixed
Land use[2]	6	3	6	6	6
CO_2 fertil.[3]	+	+	+	-	+
Energy[4]	0.1	0.1	0.1	0.1	0.5
Year of $2 \times CO_2$	2181	2123	2173	2140	2091
Δ atmosphere			+490		
Δ ocean	+714	+527	+711	+673	+454
Δ phytomass	-159	-308	-157	-213	-110
Δ litter	-8	-7	-11	-18	+2
Δ soil org. C	-86	-46	-121	-190	-10
Fossil emissions	-951	-656	-912	-742	-825

[1]   fixed: Standard data set, no change
      ΔT: temperature increasing with CO_2 up to $+3.5°C$ at $2 \times CO_2$
[2]   3: progressive clearing of total natural vegetation up to year 2400
      6: progressive clearing until 50% of area of natural vegetation cleared in each grid element (reached 2300)
[3]   + CO_2 fertilization effect considered
      - not considered
[4]   0.1: 0.1% annual increase of fossil emissions relative to 1981
      0.5: 0.5%

10.4   CONCLUSIONS

In the model runs carried out for this paper, the Osnabrück Biosphere Model was coupled with very simple ocean and atmosphere models. Although the quantitative results were certainly influenced by this fact, some preliminary conclusions may be drawn:

- The CO_2 fertilization effect is very important for the behaviour of the vegetation in the past and as well for the prediction of future developments. The functions to calculate this effect consider the CO_2 concentration of the atmosphere and limitations by soil, but should be modified to include the probable feedback with the hydrological cycle. Since transpiration is suppressed at higher CO_2 levels, the fertilization effect may even be enhanced, especially in arid zones.

- An improved database for land use changes is indispensable. The estimates of clearings which have taken place since the beginning of the 80's must be improved by remote sensing. Scenarios for predicting future land use development should also be improved.

- The very simple assumption that the lignin fraction of litter contributes to the soil organic carbon is sufficient to predict the probable present soil organic carbon pool. For a better regionalization the nitrogen budget should be included at least.

- Soil organic carbon losses after clearing are sufficiently predictable considering the reduced net primary productivity of field crops. It is not necessary to assume changed depletion.

- The effects of land use changes and clearings on soil organic carbon and litter are, on a global scale, compensated by the CO_2 fertilization effect. Phytomass in contrast is reduced, since in addition to the reduction of the productivity, which is compensated by CO_2 fertilizing, there is also a strong influence of stand age.

A final conclusion is that we are presently not in a position to validate model results with independent data. There are plenty of estimates for vegetation parameters like phytomass, productivity, and others (see Esser, 1987), but all are based on the same limited set of data which is also used for the calibration of model functions. A model may be tested by use of the time series of atmospheric CO_2 concentrations such as the Mauna Loa records. But if this is done with annual means, model errors may compensate. Therefore, the annual cycle should be preferred since then the biospheric fluxes will not be in phase and may be validated by use of the seasonal and latitudinal CO_2 concentrations. OBM is at present being modified into a seasonal model.

REFERENCES

Aselmann, I. and H. Lieth (1983) The implementation of agricultural productivity into existing global models of primary productivity. In: Degens, Kempe, Soliman (eds.) Transport of carbon and minerals in major world rivers, Part 2, Mitt. Geolog. Paläontolog. Inst. Univ. Hamburg, SCOPE/UNEP Sonderband 55:107-118.

Esser, G. (1984) The significance of biospheric carbon pools and fluxes for the atmospheric CO_2: A proposed model structure. Progress in Biometeorology 3:253-294.

Esser, G. (1986) The carbon budget of the biosphere-structure and preliminary results of the Osnabrück Biosphere Model (in German with extended English summary). Veröff. Naturf. Ges. zu Emden von 1814, New Series Vol. 7. 160 pp. and 27 Figures.

Esser, G. (1987) Sensitivity of global carbon pools and fluxes to human and potential climatic impacts. Tellus 39:245-260.

Esser, G. (1989a) Global implications of climate impacts on production and decomposition in grasslands and coniferous forests. Invited paper for a SCOPE workshop in Woods Hole, Ma., April 1989, on 'Ecosystem response to climate change:

The effects of climate change on production and decomposition in coniferous forests and grasslands'; to be published in a special issue of Oikos.

Esser, G. (1989b) Geographically referenced models of net primary productivity and litter decay. Invited paper for a SCOPE workshop in Woods Hole, Ma., April 1989, on 'Ecosystem response to climate change: The effects of climate change on production and decomposition in coniferous forests and grasslands'; to be published in a special issue of Oikos.

Esser, G. (1989c) Global land use changes from 1860 to 1980 and future projections to 2500. Ecological Modelling 44:307-316.

Esser, G., I. Aselmann and H. Lieth (1982) Modelling the Carbon Reservoir in the System Compartment 'Litter'. Mitt. Geolog.-Paläontolog. Inst. Univ. Hamburg. 52:39-58. SCOPE/UNEP Sonderband.

Esser, G. and G.H. Kohlmaier (1989) Modelling terrestrial sources of nitrogen, phosphorus, sulfur, and organic carbon to rivers. In: Kempe (ed.), Biogeochemistry of Major World Rivers, chapter 14, SCOPE Report.

Esser, G. and H. Lieth (1986) Evaluation of climate relevant land surface characteristics from remote sensing. Proc. ISLSCP Conference, Rome, 2-6 Dec. 1985, ESA SP-248, May 1986, Rome.

Esser, G. and H. Lieth (1989) Decomposition in tropical rain forests compared with other parts of the world. In: Lieth, Werger, (eds.), Tropical Rain Forest Ecosystems. Ecosystems of the World Vol. 14B, 571-580, Elsevier Science Publ. Amsterdam.

Esser, G., H. Lieth and M. Clüsener Godt (1989) Assessment of P, K, Ca dynamics during land use changes. In: Ittekkot et al. (Eds.), Facets of Modern Biogeochemistry, Chapter 10, 102-115, Springer Verlag, Berlin, Heidelberg.

Esser, G., A. Spitzy and B. Zeitzschel (1989) The carbon cycle. In: Arbeitsgemeinschaft der Großforschungseinrichtungen (ed.), Physical foundations of present climate models (in German), AGF Bonn-Bad Godesberg.

FAO-Unesco (1974 ff) Soil Map of the World. Vol. I-X, Paris.

FAO-Unesco, (1978 ff) Report on the Agro-Ecological Zones Project. World Soil Resources Report 48,/1 ff. Food and Agricultural Organization of the U. N., Rome.

FAO-Unesco, (1980 ff) Production yearbooks, Vol. 33 ff. FAO Statistics Series No. 28, ff. Food and Agricultural Organization of the U. N., Rome.

Hansen, J., I. Fung, A. Lacis, D. Rind, S. Lebedeff, R. Ruedy, G. Russell and P. Stone (1988) Global climate changes as forecast by Goddard Institute for Space Studies three-dimensional model. J. Geophys. Research 93:9341-9364.

Instituto Geographico de Agostini (1969, 1971, 1973) World Atlas of Agriculture. Novara, Italy.

Keeling, C.D. (1973) Industrial production of carbon dioxide from fossil fuels and limestone. Tellus 25:174-197.

Keeling, C.D. (1982) Measurements of the concentration of carbon dioxide at Mauna Loa Observatory, Hawaii. In: Clark (ed.), Carbon Dioxide Review 1982, pp. 377-385, Oxford.

Keeling, Ch.D. (1983) The global carbon cycle: What we know and what we could know from atmospheric, biospheric, and oceanic observations. Proceedings: Carbon Dioxide Research Conference: Carbon Dioxide, Science and Consensus. USDE CONF-820970, Washington, D.C. 20545.

Marland, G. and R.M. Rotty (1983) Carbon dioxide emissions from fossil fuels: A procedure for estimation and results for 1950-1981. Report DOE/NBB-0036 for U.S. Dept. of Energy.

Müller, M.J. (1982) Selected climatic data for a global set of standard stations for vegetation science. In: Lieth (ed.), Tasks for vegetation science, Vol. 5, Dr. W. Junk Publ., The Hague, Boston, London.

NCAR, National Center for Atmospheric Research, Data tape documentations TD-9645 and TD-9618. Boulder, Co.

Oeschger, H. (1986) Investigation of climate and environmental systems by analysis of ice cores (in German). Ann. Meteorol. 23:1-3.

Oeschger, H., U. Siegenthaler, U. Schotterer and A. Gugelmann (1975) A box diffusion model to study the carbon dioxide exchange in nature. Tellus 27:168-192.

Richards, J.F., J.S. Olson and R.M. Rotty (1983) Development of a data base for carbon dioxide releases resulting from conversion of land to agricultural uses. Institute for Energy Analysis, Oak Ridge Ass. Universities, ORAU/IEA-82-10(M); ORNL/TM-8801.

Schlesinger, W.H. (1984) Soil organic matter: a source of atmospheric CO_2. In: Woodwell, G. W. (ed.), The role of terrestrial vegetation in the global carbon cycle. SCOPE Vol. 23:111-127, Wiley and Sons, New York.

Schmithüsen, J. (1976) Atlas for Biogeography. Meyers Grosser Physischer Weltatlas 3 Bibl. Inst. Mannheim, Wien Zürich.

Walter, H. and H. Lieth (1960 ff) World Atlas of Climate Diagrams. Gustav Fischer Verlag Jena, GDR.

CHAPTER 11

Biotic Sources of Nitrous Oxide (N₂O) in the Context of the Global Budget of Nitrous Oxide

M.M. UMAROV

Dept. of Soil Science, Moscow State University,
119899, Moscow, USSR

ABSTRACT

Microorganisms are the main biotic sources of N_2O in terrestrial and aquatic ecosystems. There are three pathways of N_2O formation - denitrification, autotrophic and heterotrophic nitrification, and dissimilatory reduction of nitrate to ammonia.

Denitrification is carried out by numerous of bacteria in anaerobic and aerobic conditions. The content of organic matter plays an important role in total activity of this process. Hence denitrification proceeds more active in plant rhizosphere where there are favourable conditions for denitrifying bacteria. Nitrate concentration, temperature, pH have secondary meaning for denitrification.

Autotrophic nitrification is carried out by highly specialized bacteria and proceeds in specific ecological conditions - pH nearly neutral, sufficient aeration, lack of organic matter, low concentration of ammonia. So the role of autotrophic nitrification in N_2O formation in the biosphere is probably insignificant. In contrast to autotrophic nitrification, heterotrophic nitrification is carried out by numerous procaryotic and eucaryotic microorganisms. The activity of this process is generally about 10^3 - 10^4 times less than autotrophic nitrification, but heterotrophic nitrification is very important due to the enormous ecological versatility of the microorganisms involved and the great number of nitrogen compounds which they can oxidize.

The formation of nitrous oxide in natural conditions is caused by the capability of all organisms for oxidation and reduction of nitrogen in different mineral and organic compounds. Some of these processes have not been examined yet. Between them the process of "N-oxygenation" is well known because it leads to the formation of toxigenic, mutagenous and carcinogenic products. One of these products is nitrous oxide, a narcotic substance, commonly called "laughing gas". The main pathways of its formation are closely connected with such products of N-oxygenation as hydroxilamine, nitrate and nitrite.

Microorganisms are able to produce only some intermediates of N-oxygenation: hydroxilamino-, nitro- and nitrosoderivatives. There are few examples of gaseous oxide formation by animals. For instance, the catalase of liver and some pro-oxidases are able to reduce nitrate to nitrite and subsequently to N_2O only in presence of H_2O_2 and Mn^{+2} (Hlavica, 1982).

Microorganisms are capable of oxidation and reduction of all nitrogen compounds producing not only hydroxilamine, but also to nitrate,nitrite and nitrous oxide. Formation and consumption of N_2O are carried out by large group of heterotrophic and autotrophic

Soils and the Greenhouse Effect. Edited by A.F. Bouwman
© 1990 John Wiley & Sons Ltd.

microorganisms. Most of them are bacteria, but there are some eucaryotic organisms, especially fungi, which are able to produce N_2O.

11.1 DENITRIFICATION

The term denitrification is used to describe the utilization by microorganisms of nitrate and nitrite as ultimate accepters of electrons when oxidizing the different organic compounds (C_{org}) leading to the formation of gaseous products including N_2O:

$$C_{org} + NO_3 \text{ ---> } (CH_2O)_n + CO_2 + N_2O$$

This reaction scheme indicates that denitrification proceeds actively in media with the excess of available organic substances.

Up till now it was suggested that the utilization of nitrogen as terminal acceptors of electrons is repressed by oxygen. But recently the possibility of "aerobic" denitrification was demonstrated (Kucera et al., 1987). Earlier it was found that 10% O_2 did not inhibit the reduction of nitrate to N_2O (Betlach and Tiedje, 1981) and the high denitrification level with the formation of N_2O is observed in well-aerated soils but only after a short period of anaerobic conditions (O'Hara et al., 1983).

The physiological group of denitrifying bacteria includes many species, mainly saprophytes, which can oxidize organic substances using free oxygen. In the absence of oxygen they use bound oxygen from nitrates. It was considered earlier that only a small specialized group of denitrifying microorganisms is able to carry out this process. But at present it is recognized that all procaryotic microorganisms are capable of denitrification. It was proved by direct enumeration of denitrifying bacteria: more than 65% of these bacteria can carry out this process (Umarov, 1986).

This statement is also supported by the fact that the same bacteria can conduct two opposite processes - nitrogen fixation and denitrification. Particularly, some strains of Azotobacter chroococcum, Azospirillum brasilense and Az. lipoferum, Rhodopseudomonas sphaeroides and different strains of Rhizobium, that are well-known as active nitrogen-fixers, are also capable of denitrification. Such microorganisms have nitrogenase and nitrate reductase, that have close functional dependence on the base of common Mo-co factor.

There is particularly a close correlation between the activity of nitrate reductase and the main enzymes of N_2-fixation - glutamine synthetase and glutamate synthase (Lang, Golwano, 1988). Even more, nitrogenase can often perform as nitrate reductase (Vaughn, Burgess, 1988).

Nitrogen fixation is known to be a fundamental property of procaryotic cell. As N_2-fixation and denitrification are closely connected, finally we can assume that all procaryotic microorganisms are capable for denitrification. Therefore it is impossible to give a complete list of denitrifying bacteria. New data about new groups of such microorganisms are constantly appearing, for example, the group of Archaebacteria.

Thus, we can make a first general conclusion: there is constantly a great number of denitrifying bacteria in the biosphere; it includes all procaryotic populations of soils and water reservoirs. Their efficiency is dependent on environmental conditions, the most significant of them is the content of available organic matter.

Being heterotrophes, denitrifying bacteria are able to oxidize a great number of organic compounds. It was demonstrated that N_2O evolution from soils is correlated with the content of soluble carbon and is increased by the addition of organic matter (Umarov, 1986). The energy of organic matter oxidation is utilized more efficiently than the energy obtained by denitrification. When microorganisms turn to "nitrate respiration", they are "obliged" to utilize about 5 times more organic substance and nitrate than in ordinary oxidation (Kurakova, Umarov, 1985).

In soils, denitrification proceeds most active in rhizosphere, where is enough root exudates and root debris form a good substrate. The total mass of root exudates and root debris from actively growing plants is 35-60% of total photosynthetic production (Panikov et al., 1988). The roots stimulate denitrification indirectly, by utilizing O_2 and producing CO_2. Thus, in rhizosphere there are 500 times as much denitrifying bacteria as in fallow (Vedenina, Lebedinski, 1984).

Hence, there are favourable conditions for denitrification in the rhizosphere: more available organic matter, low partial oxygen pressure, and more abundant population of bacteria.

It has been established by means of ^{13}N and gas chromatographic methods that a high level of N_2O evolution occurs in the rhizosphere of different plant species (Casella et al., 1984). For instance, in the rhizosphere of barley the losses of nitrous oxide reach 90% and only 5% in fallow. In the rhizosphere of rice denitrification proceeds 14 times more intensive than in fallow.

Other ecological factors, such as nitrate concentration, level of aeration, temperature and pH have secondary significance. They affect only the denitrifying potential, but the actual activity of the process is only affected in the presence of sufficient organic matter (Umarov, 1986).

11.2 NITRIFICATION

The formation of N_2O as the result of N-oxygenation is one of microbiological process widely spread in the biosphere.

During autotrophic nitrification, which is conducted by a highly specialized group of bacteria, ammonia ions serve as electron donors (energetic substrate). The N_2O formation during autotrophic nitrification was established more than 50 years ago and now there are 4 genera of bacteria known to form nitrous oxide - Nitromonas, Nitrosolobus,

Nitrosococcus, and Nitrospora. They conduct the first phase of nitrification, and N_2O is formed as a result of the process:

$$NH_4^+ \text{ ---> } NH_2OH \text{ ---> } NOH \text{ ---> } N_2O + NO_2^-$$

Bacteria, oxidizing NO_2^- to NO_3^- (second phase to autotrophic nitrification) are not able to form N_2O (Vedenina and Lebedinski, 1984).

Autotrophic nitrification proceeds in specific ecological conditions: pH's nearly neutral, good aeration, lack of organic matter, low concentration of ammonia. So the role of autotrophic nitrification in the common process of N_2O formation is assumed to be insignificant.

Heterotrophic nitrification is more widely spread in nature. It is defined as co-oxidation of ammonia during mineralization of different organic substances. So heterotrophic nitrification has no energetic functions. As well as NH_4^+, heterotrophic nitrifiers can oxidize not only NH_4^+, but also aminonitrogen, hydroxilamine and hydroximates (Sorokin, 1989).

In contrast to autotrophic nitrification, heterotrophic nitrification can be carried out by a great variety of microorganisms - procaryotes and eucaryotes. The most active are microscopic fungi Aspergillus and Penicillium, bacteria Achromobacter, Arthrobacter, Corynebacterium, Flavobacterium, Nocardia, Pseudomonas, Vibrio, Xantomonas etc. This list is constantly extending.

Recently the possibility of N_2O formation during heterotrophic nitrification was disclaimed. Investigations using ^{13}N isotope and inhibitor analyses have shown that N_2O is formed by different organisms mentioned above (Malinovski, Ottow, 1985).

The main factor influencing the activity of these microorganisms is the C/N ratio. For heterotrophic nitrification the C/N ratio must be less than 10. For active heterotrophic nitrification, the content of nitrogen must exceed the amount of nitrogen necessary for normal growth.

Heterotrophic nitrification proceeds rapidly in neutral or slightly acid soils. Some ions (Fe^{+2}, Fe^{+3}, Cu^{+2}) stimulate this process (Sorokin, 1989).

The properties of heterotrophic nitrification are not investigated yet in appreciable degree. A few data are available on the physiological nature of heterotrophic nitrification. Earlier it was considered that this process is not of great importance because its activity is $10^3 - 10^4$ times less than autotrophic nitrification. The importance of heterotrophic nitrification in biosphere is now confirmed by the enormous ecological versatility and great amount of nitrogen sources that can be oxidized in this process. For instance, the contribution of heterotrophic nitrification to total nitrification was 64-90% in acid forest soils (Sorokin, 1989).

Heterotrophic nitrification is not changed much by acidification of soils after acid rains, while autotrophic nitrification is inhibited under acid conditions. The pollution of environment by pesticides (herbicides) also increases the role of heterotrophic nitrification while it causes inhibition of autotrophic nitrification.

Heterotrophic nitrifiers are able to oxidize hydroxilamine and oximes. These products constantly presenting in soils and water are the main intermediates of oxidizing and reducing reactions in the nitrogen cycle of a great number of microorganisms.

Thus, we can make the second general conclusion: heterotrophic nitrification plays a great role in biosphere, which is not considered yet.

11.3 DISSIMILATORY FORMATION OF AMMONIA

Besides denitrification, the dissimilatory formation of ammonia can also take place in the oxygen deficient conditions in many micro habitats. This process also leads to the evolution of N_2O (Vedenina, Lebedinski, 1984).

The ecological meaning of dissimilatory reduction of nitrate to ammonia as one of biotic sources of N_2O cannot be estimated yet because of its unknown physiology and biochemistry.

As in denitrification nitrate and nitrite are the main ultimate electron acceptors. However, the enzyme system responsible for dissimilatory ammonia formation is different for the two processes (Kaspar and Tiedje, 1981). The main agents of dissimilatory formation of ammonia are different bacteria: Achromobacter, Aerobacter, Bacillus, Campylobacer, Clostridium, Desulfovibrio, Esherichia, Erwinia, Klebsiella, Serratia, Vibrio and others.

Eucaryotes (fungi and yeasts) are also able to form N_2O, but during assimilatory reduction of nitrate and nitrite. Very little is known about the mechanism of this process. Nevertheless it seems to be important that reduction of nitrate and nitrite to ammonia can be carried out by large groups of microorganisms. This confirms its role as one of biotic sources of N_2O in nature.

For dissimilatory ammonia formation the C/N ratio is very important. C/N ratios higher than 10 are necessary for this process. Besides this, velocity of the process increases with increasing of anaerobiosis.

Thus, the dissimilatory ammonia formation with nitrous oxide evolution proceeds in different ecological habitats with low O_2 concentration and abundance of organic matter.

REFERENCES

Betlach, M.P. and J.M. Tiedje (1981) Kinetic explanation for accumulation of nitrite, nitric oxide and nitrous oxide during bacterial denitrification. Applied Environmental Microbiology 42:1074-1084.

Casella, S., C. Leporini and M. Nuti (1984) Nitrous oxide production by nitrogen-fixing fast-growing rhizobia. Microbiological Ecology 10:107-114.

Hlavica, P. (1982) Biological oxidation of nitrogen in organic compounds and disposition of N-oxidized products. Crit. Rev. Biochemistry 12:39-101.

Kaspar, H.F. and Tiedje, J.M. (1981) Dissimilatory redaction of nitrate and nitrite in the bovine rumen: nitrous oxide production and effect of acetylene. Applied Environmental Microbiology 41:705-709.

Kucera, I., L. Kozak and K.V. Dada (1987) Aerobic dissimilatory redaction of nitrate by cells of Paracoccus denitrificans: the role of nitrite oxide. Biochem. Biophys. Acta 894:121-126.

Kurakova, N.G. and M.M. Umarov (1985) The significance of denitrification in nitrogen budget of soils. Agrochemistry 5:118-129.

Lang, P. and M. Golwano (1988) Effect of nitrate on carbon metabolism and nitrogen fixation in lupine root nodules. In: H. Bothe, F. De Bruijn and W.E. Newton (Eds.) Nitrogen fixation hundred years after, p. 562. Gustav Fisher, Stuttgart - N.Y.

Malinovski, P. and C.G. Ottow (1985) Okologische Bedingungender "Denitrifikation" bei Pilzen. Landwirtsch. Microbiol. 38:115-121.

O'Hara, J.W., R.M. Daniel and K.W. Steele (1983) Effect of oxygen on the synthesis, activity and breakdown of the Rhizobium denitrification system. Journal General Microbiology 129:2405-2413.

Panikov, N.S., A. Yu. Gorbenko and D.G. Zviagintsev (1988) Oscillatory pattern of microbial growth dynamics in the soil and its nature. Bull. Moscow State Univ., ser. Soil Science 1:34-41.

Sorokin, Yu.D. (1989) Heterotrophic nitrification of Alcaligenes bacteria. Microbiology 58:9-14.

Umarov, M.M. (1986) Associative nitrogen fixation. Moscow Univ. Publish., Moscow.

Vaughn, I.Y. and B. Burgess (1988) Nitrogenase reactivity toward nitrate and nitrite. In: H.Bothe, F. De Bruijn and W.E.Newton (Eds.) Nitrogen fixation hundred years after, p. 138.Gustav Fisher, Stuttgart - N.Y.

Vedenina, I.J. and A.V. Lebedinski (1984) Transformation of nitrous oxide during denitrification, dissimilatory formation of ammonia and nitrification. Advances of Microbiology (in Russian) 19:135-165.

CHAPTER 12

Soil and Land Use Related Sources and Sinks of Methane (CH₄) in the Context of the Global Methane Budget

H. SCHÜTZ, W. SEILER, and H. RENNENBERG

Fraunhofer-Institut für Atmosphärische Umweltforschung
Kreuzeckbahnstr. 19, D-8100 Garmisch-Partenkirchen, F.R.G.

ABSTRACT

Soils represent the most important source of atmospheric methane. This greenhouse gas, with an atmospheric residence time of approximately 10 years, is produced predominantly by microbial degradation of organic carbon in rice paddies (100 ± 50 Tg y^{-1}), natural wetlands (100 ± 50 Tg y^{-1}) and landfills (50 ± 20 Tg y^{-1}). Together, these sources account for approximately half of the total CH_4 emission of approximately 496 ± 251 Tg y^{-1}. Non-soil related sources of atmospheric methane include the emissions by ruminants (85 ± 15 Tg y^{-1}), biomass burning (80 ± 20 Tg y^{-1}), leakages related to natural gas consumption, coal mining and others, which total approximately 233 ± 60 Tg y^{-1}.

Under aerobic conditions, soils act also as a sink of atmospheric methane. The deposition of CH_4 has been studied only sporadically and in only a few soil habitats. Based on the few data on CH_4 deposition, the global sink strength may be as high as 23 to 56 Tg, which is relatively small compared with the oxidation of CH_4 by OH radicals in the atmosphere. As the role of large areas of potential CH_4 deposition (e.g. savannas, deserts) is unknown, this figure is, however, speculative. Because of the temporal increase of CH_4 mixing ratios in the atmosphere, the CH_4 sink strength may increase with time.

12.1 INTRODUCTION

With the exception of carbon dioxide, methane is the most abundant atmospheric carbon species. Its present tropospheric mixing ratio is in the order of 1.7 to 1.8 ppmv and it is increasing with time at a present rate of approximately 0.8 to 1.0% per year. Compared with the pre-industrial values of approximately 0.6 to 0.7 ppmv, the CH_4 mixing ratio has more than doubled within a period of less than 150 years.

Despite its relatively low concentration, atmospheric methane is of particular significance for our environment. Methane is an important, climatically relevant substance and accounts for approximately 20% of the supposed greenhouse warming of 0.7°C during the last 100 years.

Soils and the Greenhouse Effect. Edited by A.F. Bouwman
© 1990 John Wiley & Sons Ltd.

This effect is strengthened by the fact that the oxidation of methane by OH reactions at high NO_x-concentrations leads to the formation of ozone in the troposphere which also is climatically relevant. Furthermore, the tropospheric ozone concentration determines the oxidation potential of the troposphere and thus has a significant influence on the distribution and abundance of other trace constituents of environmental relevance.

Methane is produced exclusively in the troposphere. It is transported into the stratosphere, however, where it is oxidized, forming water vapour that contributes to the creation of the polar clouds that have a significant impact on the formation of the Antarctic ozone hole.

Because of the importance of methane for the earth's climate and the chemistry of the atmosphere, the budget and cycle of atmospheric methane, in particular their perturbation by human activities, have received considerable attention. During the last 10 years or so, studies of the atmospheric cycle of methane have been intensified. Nevertheless, there are still considerable gaps in our knowledge which need further attention.

It is generally accepted that the most dominant CH_4 sources are related to land surfaces, which account for approximately 150 million km². Two thirds of the total land area is located in the northern hemisphere and only one third in the southern hemisphere. According to Bouwman (1988), the land area can be subdivided into 14 ecosystem types, as listed with their total land areas in Table 12.1. The ecosystems with the largest surface areas (deserts, savannah and woodland/shrubland) are those with relatively small precipitation and consequently dry soils.

Although the ocean surface area exceeds the land surface area by more than a factor of two, the emission of methane from oceans (and lakes) does not play a significant role in the atmospheric CH_4 budget. Fluxes of CH_4 from these sources account for approximately 15 ± 12 Tg y^{-1}, which is less than 5% of the total budget (Seiler, 1984; Bolle et al., 1986; Cicerone and Oremland, 1988; and others).

Significant CH_4 production and, hence, emission into the atmosphere occurs in ecosystems where anaerobic soil conditions prevail. This condition is found in waterlogged soils (e.g. in natural wetlands, swamps, marshes, and irrigated and rainfed rice paddies). The sediments of these ecosystems are generally anaerobic and thus provide excellent habitats for methanogenic bacteria.

In contrast to waterlogged anoxic soils, well-aerated and dry soils can act as sinks of atmospheric methane. Obviously, methane is oxidized by methanotrophic and/or ammonium-oxidizing bacteria.

In the following, the individual sources and sinks of atmospheric methane are summarized and their contribution to the atmospheric methane budget is discussed.

Table 12.1: Global soil and land use.

Global area: 510×10^6 km^2
Global land area: 149×10^6 km^2 (100 NH, 49 SH)
Global ocean area: 361×10^6 km^2 (155 NH, 206 SH)

Ecosystem	Area ($\times 10^6$ km^2)[*]
1. tropical rainforest	7.11
2. tropical seasonal forest	7.105
3. temperate evergreen forest	7.306
4. temperate deciduous forest	6.834
5. boreal (taiga)	7.013
6. woodland/shrubland	7.173
7. savanna	10.695
8. tropical grassland	2.115
9. temperate grassland	10.467
10. desert/semi-desert scrub	12.001
11. extreme desert	12.575
12. cultivated land	15.776
(rice paddies	1.45)
13. swamp/marsh	2.101
14. tundra/alpine	6.947
15. miscellaneous	15.210
Total	130.428

relevant CH$_4$ production in ecosystems 12, 13 and 14.
CH$_4$ deposition in ecosystems 1,2,3,4,5,7,12,13.
[*] Source: Bouwman (1988).

12.2 SOIL AND LAND USE RELATED SOURCES OF METHANE

As noted above the most important natural source of atmospheric methane is the microbial mineralization of organic matter under strictly anaerobic conditions. These conditions are generally observed in sediments of waterlogged soils (such as swamps, rice paddies, marshes, landfills) and during the enteric fermentation in the digestive tract of ruminants and other herbivores. The contribution of these and other individual sources to the atmospheric CH$_4$ budget is described below.

12.2.1 Rice Paddies

An area of 1.45 million km^2 is used for rice cultivation, which is approximately 10% of the global cropland. Nevertheless, the CH$_4$ emission from rice paddies is one of the most dominant individual sources of atmospheric methane. Approximately 90% of the global rice paddy area, and thus of the CH$_4$ emission from this source, is located in Asia, where rice represents the most important crop. Water saturation

272   *Soils and the Greenhouse Effect*

leading to continuously anoxic soil conditions occurs during most of the vegetation period in irrigated and probably also rainfed rice paddies, which together cover approximately 86% of the whole rice growing area (Table 12.2).

Table 12.2 Rice area allocation for Asia[a]

Country	Dry-land	Deep-water	Irrigated			Rainfed			Total
			wet	dry	total	shallow 0-30 cm	intermediate 30-100 cm	total	total
Southeast-Asia[b]	4642	1686	7685	4100	11785	12396	4238	16634	34747
South Asia[c]	6951	3604	12569	3513	16082	17979	7349	25328	51965
Korea	112	-	1620	-	1620	249	-	249	1981
Japan				2254					2254
China	606	-	23986	9690	33676	1880	-	1880	36162
Grand Total	12311	5290	45860	17303	65417	32504	11587	44091	127109
% of total	9.7	4.2	36.1	13.6	51.5	25.6	9.1	34.7	100

[a] Source: Huke (1982).
[b] including Burma, Thailand, Vietnam, Kampuchea, Laos, Malaysia, Indonesia, Philippines.
[c] including India, Bangladesh, Pakistan, Sri Lanka, Nepal, Bhutan.

The influence of CH_4 produced in rice paddies on the concentration and distribution of methane in the atmosphere was first discussed by Koyama (1964). On the basis of laboratory experiments using soil samples taken from rice paddies in Japan, he calculated the CH_4 emission from this source into the atmosphere to be in the order of 190 Tg (Tg = 1012 g) per year in the early 1960s. Extrapolated to the middle of the 1970s, Ehhalt and Schmidt (1968) even postulated a source strength of 280 Tg per year, which would account for more than 50% of the total atmospheric CH_4 production.

The first in situ measurements of CH_4 emissions from rice paddies were obtained in 1980 in California by Cicerone and Shetter (1981), who estimated the global CH_4 emission to be approximately 59 Tg per year. A similar value (35 to 59 Tg y^{-1}) was reported by Seiler et al. (1984a) using results obtained from field measurements in rice paddies in Spain. Both measuring programs included either only a few measurements distributed sporadically during the rice vegetation period or did not cover a whole vegetation period, so that the given total CH_4 source strength was highly uncertain. In addition, the measurements in Spain were influenced by penetration of water from the Mediterranean

Sea into the ground water of the paddy field, which may have reduced methanogenesis.

The first measurements covering a complete rice vegetation period, obtained by Holzapfel-Pschorn and Seiler (1986) in rice paddies in Italy, indicated a much higher source strength, varying between 70 and 170 Tg y^{-1}. These measurements were repeated during three additional vegetation periods in the following years. The data indicated an average CH_4 flux rate of approximately 12 ± 6 mg $m^{-2}h^{-1}$. Using the average distribution of rice paddies and considering the actual soil temperatures in these regions, Schütz et al. (1989) estimated a global CH_4 source strength of 100 ± 50 Tg per year.

CH_4 emission rates from rice paddies in the Far East became available only recently (Wang et al., in preparation). These measurements were obtained in rice paddies near Hangzhou in the People's Republic of China; approximately one third of the annually harvested paddy area in Asia is in PR China (Table 12.2). Data are available for a measuring period of 1.5 years covering three vegetation periods. Though not yet fully analyzed, these data can be used for a preliminary estimate, which is in the order of 100 Tg per year. This figure agrees reasonably well with the observations in Italy. Nevertheless, this estimate remains uncertain, since the influence of climate, soil type, plant variety, field management, type and application forms of fertilizers on CH_4 emission rates have still not been studied satisfactorily.

Most interestingly, the intensive field campaigns in Italy and the People's Republic of China show significant diurnal and seasonal variations, which also seem to be dependent on climate and field management. In Italy, maximum daily CH_4 emission rates were observed during the late afternoon when soil temperatures at approximately 5 cm depth had reached maximum values. The positive correlation between soil temperature and CH_4 emission rates was observed during almost the entire vegetation period. On a seasonal average, the CH_4 emission rates increased by a factor of two for a temperature increase of 5°C (Q_{10}=4). In contrast to the observations in Italy, the CH_4 emission rates in the Chinese rice paddies exhibited two maxima during the early rice vegetation period, i.e. between the end of April and the end of June. During this vegetation period, maximum CH_4 emission rates occurred at noon and during the night (see Figure 12.1A). Surprisingly, the CH_4 emission rates showed the maximum value only in the night during the late vegetation period, i.e., between the beginning of August and the beginning of September (Figure 12.1B).

In addition to the high diurnal variations, a well expressed seasonal variation of CH_4 emission rates was found in Italy with typically two maxima of the daily average. The first maximum was in May shortly after flooding during tillering of the rice. This maximum was caused by the degradation of organic material (e.g. rice straw)

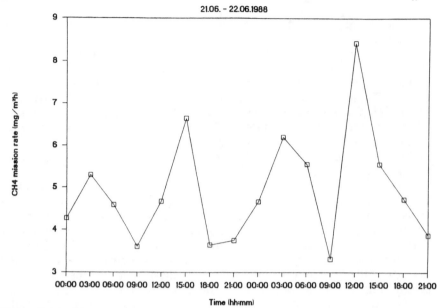

Figure 12.1A   Diurnal variations of CH_4 emission rates (June, 21-22) in an unfertilized rice paddy during the early vegetation period (May to July) of 1988 in Hangzhou, Zhejiang province, China.

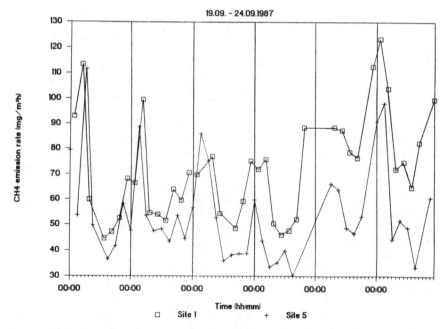

Figure 12.1B   Diurnal variations of CH_4 emission rates (September, 19-24) in an unfertilized rice paddy during the late vegetation period (August to October) of 1987 in Hangzhou, Zhejiang province, China.

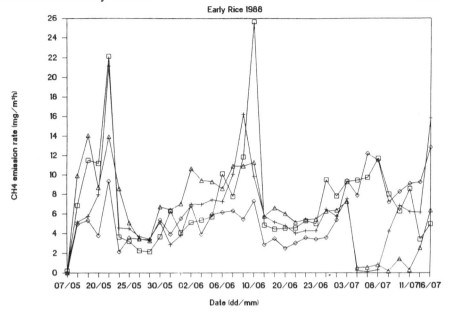

Figure 12.2A   Seasonal variations of daily average CH_4 emission rates in an unfertilized rice paddy (squares), as well as in rice paddies fertilized with either KCl (crosses), organic manure plus KCl (diamonds), or organic manure (triangles), during the early rice period of 1988 in Hangzhou, Zhejiang province, China.

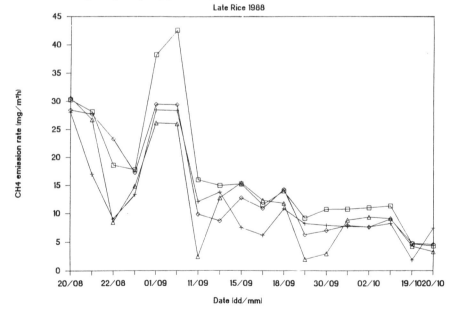

Figure 12.2B   Seasonal variations of daily average CH_4 emission rates in an unfertilized rice paddy (squares), as well as in rice paddies fertilized with either K_2SO_4 (crosses), organic manure plus K_2SO_4 (diamonds), or organic manure (triangles), during the late rice period of 1988 in Hangzhou, Zhejiang province, China.

present in the soil at flooding. The second maximum occurred in July during the reproductive phase of the rice plants and could be attributed to the stimulation of methanogenesis caused by the excretion by the rice plants of organic exudates. During two vegetation periods in Italy, a third maximum of the CH_4 emission rates was found in late August, which was probably caused by the degradation of dying plant materials and plant litter.

Seasonal variations of CH_4 emission rates were also found in China, showing three maxima during the early rice period, the first one at tillering, the second during the reproductive phase, and the third at flowering (Figure 12.2A). High average CH_4 emission rates were found at the very beginning of the late rice period and also during tillering (Figure 12.2B).

The influence of the application of mineral fertilizer on CH_4 emission rates was first studied by Cicerone and Shetter (1981), who observed a five fold increase of CH_4 emission after fertilization with ammonium sulfate. Extended field experiments in Italy indicated that the influence of fertilizers on CH_4 emission rates is rather complicated and needs further intensive study. In general, CH_4 emission rates were found to be strongly dependent on the type, rate and application mode of fertilizer. Application of organic fertilizers (e.g., straw) enhanced the rate of CH_4 emission by a factor of two whereas fertilization with deeply-incorporated urea (200 kg N ha^{-1}) or ammonium sulfate decreased the CH_4 emission rates by approximately half, which contrasts with the observations reported by Cicerone and Shetter (1981). The preliminary data obtained from field experiments in Chinese rice paddies apparently do not indicate a significant dependence of CH_4 emission rates on the mineral fertilizers (e.g., KCl, K_2SO_4) and organic manure (animal excrement, rape seed cake) used locally.

12.2.2 Natural Wetlands

Another type of ecosystem with strongly anaerobic sediments are the natural wetlands, which are even more complex than rice paddies with respect to their global distribution, water saturation, soil types, vegetation and nutrient input. Recent detailed global land survey studies (Matthews and Fung, 1987; Aselmann and Crutzen, 1989) have evaluated the individual distribution of the main categories of wetlands, i.e., bogs, fens, marshes, swamps, floodplains and shallow lakes. Using rates measured in wetlands of subtropical and temperate regions (Baker-Blocker et al., 1977; Harriss and Sebacher, 1981; Harriss et al., 1982; Sebacher et al., 1983; Burke et al., 1988; Harriss et al., 1988), tropical regions (Bartlett et al., 1988; Devol et al., 1988; Thathy et al., 1988; Crill et al., 1988b), and boreal regions (Clymo and Reddaway, 1971; Svensson, 1980; Svensson and Rosswall, 1984; Sebacher et al., 1986; Whalen and Reeburgh, 1988; Harriss et al., 1985; Crill et al., 1988a), the average annual CH_4 emissions from these regions can be estimated. For temperate and subtropical wetlands, estimated annual emissions amount to 26 Tg CH_4 y-1 (Aselmann and Crutzen, 1989). For the tropics, the emissions vary from 17 Tg y-1 (Bartlett et al., 1988) to 23 Tg y-1 (Aselmann and Crutzen, 1989). Boreal wetlands are a source of 67 Tg y-1 (Matthews and Fung, 1987) to 70-90 Tg y-1 (Crill et al., 1988a). These estimates may be highly questionable, mainly because of the high spatial and temporal variations of the CH_4 emissions within the individual wetland ecosystems. The overall average CH_4 emission from wetlands has been estimated to be approximately 110 Tg y^{-1} (Matthews and Fung, 1987) and 42 to 154 Tg y^{-1} (Table 3; Aselmann and Crutzen, 1989). These values are higher than the rates of 25 to 70 Tg y^{-1} reported earlier by Bolle et al. (1986).

Shallow freshwater lakes have been included in natural wetlands. The corresponding source strength for atmospheric methane was estimated to be approximately 2 Tg per year (Aselmann and Crutzen, 1989). CH_4 emission from deep water parts of freshwater lakes is most unlikely to contribute significantly to atmospheric methane because by far most of the CH_4 produced in the anoxic sediment is oxidized in the oxic sediment surface and in the oxic parts of the water body.

In contrast to the freshwater sediments, CH_4 emission rates from marine sediments were found to be rather low, probably because of the inhibition of methanogenesis by high sulfate and/or salt content. Thus CH_4 emission from saltwater marshes is negligible in the total global CH_4 budget (approximately 0.34 Tg y^{-1}, Table 3, Bartlett et al., 1985).

12.2.3 Animals and Humans

Methane is produced by enteric fermentation in the digestive tract of ruminants, other herbivorous fauna and humans. The total contribution of this source to atmospheric methane is estimated to range between 70 and 100 Tg y^{-1} (Table 12.3) with approximately 74% coming from cattle and another 8 to 9% from buffaloes and sheep. The remainder comes from camels, mules and asses, pigs and horses. Human CH_4 production is probably less than 1 Tg y^{-1} (Crutzen et al., 1986). This estimate is based on a detailed survey of the global distribution of individual populations and on the ratios of biomass consumption and its relative conversion to CH_4. The values for the CH_4 yield from the gross energy intake vary in the relatively narrow range of 3 to 10% for cattle and sheep, and 0.25 to 3% for other animals, indicating that this source estimate has a high degree of certainty. These CH_4 production rates have been obtained under "laboratory" conditions only, however, and nothing is known about CH_4 production "in situ".

The production of methane by termites is still a matter of controversy. The only field measurements of CH_4 emissions from termite nests were reported by Seiler et al. (1984b). In this study, variations of CH_4 fluxes relative to carbon digestion were found to be dependent on the genera and species of termites and varied over two orders of magnitude. The global contribution of termites to atmospheric methane was estimated to be only approximately 2 to 5 Tg y^{-1}. This is

Table 12.3 Global soil and land use related CH_4 sources and sinks.

Ecosystem/type	Area $10^6 ha^2$	CH_4 flux rates (- = deposition) $\mu g\ m\text{-}2h^{-1}$	Prod. period Days	CH_4 emission or CH_4 deposition(-) Tg y^{-1}
1. trop. rainforest	711	- 6 - -24 [h]	365[a]	-0.43 - -17.1
2. trop.seas. forest	710	-10 - -21 [a, h]	365[a]	-0.71 - -1.5
3. temp.evergr.forest	731	-10 - -160 [a, i, k]	200[a]	-0.35 - -5.6
4. temp.decid. forest	683	-10 - -160 [a, i, k]	200[a]	-0.33 - -5.3
5. boreal (taiga)	701	-10 - -160 [a]	120[a]	-0.34 - -5.4
6. woodland/shrubl.	717	-52 [l]	365	- 3.7
7. savanna	1069	-52 [l]	365	- 5.6
8. tropical grassland	211	?	-	?
9. temp. grassland	1047	?	-	?
10. desert/semi- desert scrub	1200	-52 [l]	365	- 6.2
11. extreme desert	1257	-52 [l]	365	- 6.5
12. cropland	1578	?	-	?
12.a. rice paddies	145	1.3 - 40.7 × 10^3	75-300[b]	100 ± 50[b]
13. swamp/marsh/ floodplain	210	0.4 - 24.6 × 10^3	122-247[c]	53 (35-84)[c]
14. tundra/alpine[m]	695	?	-	?
14.a. bogs/fens	241c	0.1 - 15.01 × 10^3	169-178[c]	25 (7-70)[c]
15. miscellaneous	-	-	-	-
15.a. lakes	012c	0.2 - 24.38 × 10^3	365[c]	2 (1-4)[c]
15.b. digestive tract	-	see text	365	70 - 100[d]
15.c. marine sedim.	038e	0.1 - 22.9 10^3	365	0.34[e]
15.d. landfills	-	see text	365	30 - 70[f]
15.e. natural gas reservoirs	-	see text	365	65 - 75[g]
15.f. biomass burn.	-	see text	365	55 - 100[g]
Total of sources				313 - 653
Total of sinks				23 - 56

[a] Steudler, personal communication; [b] Schütz et al., 1989; [c] Aselmann and Crutzen, 1989; [d] Crutzen et al., 1986; [e] Bartlett et al., 1985; [f] Bingemer and Crutzen, 1987; [g] Bolle et al., 1986; [h] Keller et al., 1986; [i] Keller et al., 1983; [k] Harriss et al, 1982; [l] Seiler et al., 1984b; [m] tundra included in 14.a.

considerably lower than the figure of 150 Tg y^{-1} reported by Zimmerman et al. (1982) and the figure of 50 Tg y^{-1} reported by Rasmussen and Khalil (1983), but it is closer to the figure of less than 15 Tg y^{-1} reported by Fraser et al. (1986). Clearly, more detailed field studies are needed to improve our knowledge of the importance of termite species in different ecosystems, population sizes, organic carbon digestion and its relative conversion to CH_4 under field conditions. Also, no data on possible CH_4 production by insects others than termites are available, but studies on, for example, cockroaches and dung beetles should be considered (Crutzen et al., 1986).

12.2.4 Landfills

CH_4 production in landfills has been estimated to be in the range of 30 to 70 Tg y^{-1} (Table 3; Bingemer and Crutzen, 1987). This estimate is based on a detailed survey of global waste production, its carbon content and assuming that at 35°C 80% of the degradable organic carbon fraction is converted to biogas containing approximately 50% CH_4 by volume. The resulting CH_4 import from landfills into the atmosphere is, however, largely unknown because only very few data on emission rates are available. In addition, Bingemer and Crutzen (1987) assumed that the CH_4 produced in the landfills is released completely into the atmosphere. Because considerable oxidation of CH_4 may take place in the upper aerobic layers, especially under dry conditions, the CH_4 emission from this source as reported by Bingemer and Crutzen (1987) may have been overestimated. Furthermore, little is known about CH_4 emission from landfills in developing countries, which definitely consist of a mixture of materials different from that in industrial countries; these landfills are expected to grow rapidly because of increasing population and urbanization. Consequently, methane release rates from landfills may contribute significantly to a future increase of atmospheric CH_4 mixing ratios. Clearly, this source needs further attention.

12.2.5 Biomass Burning

Another source is the CH_4 produced by biomass burning, which is estimated to range between 55 and 100 Tg y^{-1} (Table 3, Bolle et al., 1986). This estimate is based largely on a study by Crutzen et al. (1979). These authors gathered data on global areas of biomass burning and measured, among other things, the CH_4 yield by volume relative to that of CO_2. Including data from another study, they arrived at a range of 1.0 to 2.2% CH_4 of CO_2 (v/v) with an average value of 1.6%. These data were obtained from forest fires in a temperate region and from burning of agricultural wastes. Most biomass is burned in the tropics in the dry season, however (e.g., by fires in savanna and bushland, or for deforestation). Data from this potentially large source of atmospheric methane are lacking. The variability of CH_4 produced from burning biomass may be attributed to different materials burned, which may cause different ratios of CH_4 relative to CO_2 (Cicerone and Oremland, 1988).

12.2.6 Coal Mining and Natural Gas and Oil Exploration and Distribution

Estimates of CH_4 production from coal mining and natural gas and oil exploration and distribution require information on the gas loss rates, which is scarce. The estimate of 65 to 75 Tg y^{-1} (Table 12.3) for CH_4 emission from natural gas reservoirs (Bolle et al., 1986) is close to an updated estimate of 50 to 95 Tg y^{-1} (Cicerone and Oremland, 1988). The estimate of CH_4 production from coal mining (35 Tg y^{-1}, Bolle et al.,

1986) is based largely on an updated figure of global coal production, the resulting coal field gas production and its volume content of CH_4 given by Koyama (1964). Gas loss rates from natural gas exploration and distribution were assumed to range between 2 and 4% (average 2.5% for 1980), but the latter figures are poorly documented. In addition, emission of natural gas from oil exploration and recovery, and from venting and incomplete flaring at oil and gas wells, and losses caused by explosive events should be investigated (Cicerone and Oremland, 1988).

12.3. SOILS AS SINKS FOR ATMOSPHERIC METHANE

In contrast to waterlogged soils, aerated and dry soils generally act as sinks for atmospheric methane. The uptake of CH_4 by soils was first demonstrated by Seiler et al. (1984b) by measurements in a subtropical broad-leafed-type savannah in South Africa. Their measurements showed a decomposition rate of CH_4 at the soil surface of 52 μg m^{-2}h^{-1} during the rainy season. Speculating that these rates are representative for other land surfaces in the subtropics, (e.g., savannas, woodland/shrubland and deserts), the total annual consumption of CH_4 in subtropical soils was estimated to be approximately 21 Tg. Since CH_4 production appears to be negatively correlated to the soil moisture content and the data were obtained during the rainy season, it is speculated that the CH_4 uptake by soils in the dry savannah regions may even exceed the value of 21 Tg y^{-1}.

It is assumed that the flux of atmospheric CH_4 to the soil surface is a net result of simultaneous production and decomposition of CH_4 in soils. In fact, experiments using atmospheres with very low initial CH_4 mixing ratios show a net flux of CH_4 from the soil into the atmosphere until the CH_4 mixing ratio within the experimental system stabilizes at an equilibrium value, which then remains constant with time. Similarly, a net flux of CH_4 from the atmosphere to the soil surface is observed when CH_4 mixing ratios higher than the equilibrium value were used. Provided that the environmental parameters such as soil temperature and soil moisture remain constant, both experiments resulted in the same equilibrium value which was in the order of 1.0 to 1.2 ppmv in the case of the experiments in South-Africa (Seiler et al., 1984b). The existence of an equilibrium value in connection with the observed overall trend of atmospheric methane means that the CH_4 uptake by soils may increase with the rise of the atmospheric CH_4 mixing ratios.

Production of CH_4 may occur in the so-called microniches of soils where anaerobic environmental conditions are observed and where methanogenic microorganisms may be active. CH_4 decomposition may be due to activities of methanotrophic microorganisms and/or ammonium-oxidizing microorganisms which are widespread in soils.

More recently, CH_4 uptake has also been observed in forest soils of tropical, temperate and boreal regions. Average CH_4 decomposition rates in tropical forest were found to be 6 - 24 μg m^{-2}h^{-1} (Keller et al., 1986;

Steudler et al., 1989). CH_4 decomposition rates for temperate and boreal forest soils ranged between 10 and 160 μg m^{-2}h^{-1} (Steudler et al., 1989) and, thus, were well within values reported by Keller et al. (1983) and Harriss et al. (1982). From these data the CH_4 uptake by tropical, temperate and boreal forest soils may vary between 2 and 35 Tg CH_4 per year.

Adding the CH_4 uptake by subtropical soils, estimated to be 21 Tg per year, the total CH_4 sink strength by soils in forests, savannah, woodland/shrubland, and desert may account for 23 to 56 Tg per year (Table 12.3; ecosystems 1 to 7 and 10 to 11). Compared with the total turnover of the atmospheric CH_4 cycle of approximately 500 Tg per year this sink strength represents 7 to 8% of the total CH_4 decomposition. This figure is highly uncertain because of severe lack of information on the dependency of the CH_4 uptake by soils on a variety of parameters such as soil type, soil temperature, soil moisture. Almost nothing is known on the influence of the major croplands on the CH_4 decomposition which cover an area of approximately 14 million km^2.

12.4. CONCLUDING REMARKS

It was demonstrated that soil and land use related processes represent the most dominant sources of atmospheric CH_4. This does not apply for the destruction of atmospheric methane which predominantly is due to the photochemical decomposition by reactions with OH (Table 12.4).

As demonstrated by Bolle et al. (1986) the source strengths of the individual land related sources have increased with time, mainly due to anthropogenic activities. Based on statistical data, these authors estimated the temporal increase of the total CH_4 emission into the atmosphere between 1940 and 1980 to be on the order of approximately 50% relatively to the value observed in 1940. This agrees reasonably well with the overall increase of the atmospheric CH_4 mixing ratios of 0.8 to 1.0% per year (Table 12.4) observed during the last 10 to 15 years. An exception are the CH_4 emissions from swamps/marshes which have decreased with time because of land change and, thus, decreasing surface area of swamps on a global scale.

Most interestingly is the positive correlation between the atmospheric CH_4 mixing ratios and the human world population as shown in Figure 12.3. This remarkable good positive correlation is indicative for the strong influence of human activities on the atmospheric CH_4 cycle, predominantly by changing the CH_4 source strengths of the major CH_4 emissions, e.g. by ruminants, rice paddies, biomass burning, usage of natural gas and others.

The increase of the atmospheric CH_4 burden may also be caused by a decrease of the OH concentrations and thus, declining CH_4 sink strengths of the photochemical CH_4 destruction. Decreases of OH radicals are proposed to be the result of increasing anthropogenic emissions of CO which strongly influences the abundance of OH. Bolle

Table 12.4 Atmospheric CH_4 burden, CH_4 residence time, and total of global sources and sinks for atmospheric methane

Atmospheric CH_4 burden	4800 Tg[a]
Atmospheric residence time	8.1 - 11.8 years[a]
Annual rate of increase	0.8 - 1.0%[a]
Sources: Rice paddies	100 ± 50 Tg
Natural wetlands	100 ± 50 Tg
Digestive tract	85 ± 15 Tg
Landfills	50 ± 20 Tg
Natural gas reservoirs	70 ± 5 Tg
Biomass burning	78 ± 32 Tg
Total of sources	318 - 673[b] Tg
Average	495.5 Tg
Sinks: tropospheric oxidation by OH (85% of sources[a])	421 Tg
stratospheric oxidation by OH	60[a, c] Tg
soil uptake	40 Tg
Total of sinks	521 Tg

[a] Cicerone and Oremland, 1988; [b] including 5-20 Tg y[-1] for oceans (Cicerone and Oremland, 1988); [c] Bolle et al., 1986.

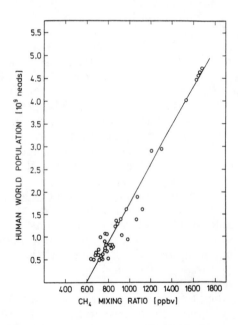

Figure 12.3 Correlation between the increasing world's human population and the CH_4 mixing ratios obtained from Arctic and Antarctic ice cores.

et al. (1986) based on data of the global CO distribution calculated a possible upward trend of CO of 0.6 to 1.0% per year during the last 10 years which would have caused an average reduction of the tropospheric OH abundance of only 0.2% per year. This reduction would not be sufficient to explain the increase of atmospheric CH_4 abundance of 1.0% per year. This conclusion agrees with recent estimates by Cicerone and Oremland (1988) who calculated that the possible reduction of the tropospheric OH concentration may account for a decrease of the CH_4 sink strength of approximately 0.5 to 3.0 Tg y^{-1} which, however, is considerably lower than the 48 Tg y^{-1} which is necessary to explain the observed increase of the atmospheric CH_4 burden.

Although considerable efforts have been made in the past to study the individual sources and sinks, considerable gaps still exist in our knowledge of the atmospheric CH_4 cycle. Further internationally coordinated studies are urgently needed to provide the data basis which is necessary for allowing prediction of the future potential trends of atmospheric methane and its impact on our environment.

REFERENCES

Aselmann, I. and P.J. Crutzen (1989): Freshwater wetlands: Global distribution of natural wetlands and rice paddies, their net primary productivity, seasonality and possible methane emissions. Journal of Atmospheric Chemistry 8:307-358.

Baker-Blocker, A., T.M. Donahue and K.H. Mancy (1977): Methane flux from wetland areas. Tellus 29:245-250.

Bartlett, K.B., R.C. Harriss and D.J. Sebacher (1985): Methane flux from coastal salt marshes. Journal of Geophysical Research 90:5710-5720.

Bartlett, K.B., P.M. Crill, D.I. Sebacher, R.C. Harriss, J.O. Wilson and J.M. Melack (1988): Methane flux from the central Amazonian floodplain. Journal of Geophysical Research 93:1571-1582.

Bingemer, H.G. and P.J. Crutzen (1987): The production of methane from solid wastes. Journal of Geophysical Research 92:2181-2187.

Bolle, H.-J., W. Seiler and B. Bolin (1986): Other greenhouse gases and aerosols, assessing their role for atmospheric radiative transfer. In: B. Bolin, B.R. Döös, J. Jäger and R.A. Warrick (eds.), The greenhouse effect, climatic change, and ecosystems, pp. 157-203. Chichester, New York, Brisbane, Toronto, Singapore: Wiley and Sons.

Bouwman, A.F. (1988): International conference on: "Soils and the Greenhouse effect". Draft background paper. Working Paper and Preprint 88/3, International Soil Reference and Information Centre, Wageningen.

Burke, R.A.Jr., T.R. Barber and W.M. Sackett (1988): Methane flux and stable hydrogen and carbon isotope composition of sedimentary methane from the Florida Everglades. Global Biogeochemical Cycles 2:329-340.

Cicerone, R.J. and R.S. Oremland (1988): Biogeochemical aspects of atmospheric methane. Global Biogeochemical Cycles 2:299-327.

Cicerone, R.J. and J.D. Shetter (1981): Sources of atmospheric methane:measurements in rice paddies and a discussion. Journal of Geophysical Research 86:7203-7209.

Cicerone, R.J., J.D. Shetter and C.C. Delwiche (1983): Seasonal variation of methane flux from a California rice paddy. Journal of Geophysical Research 88:11.022-11024.

Clymo, R.S. and E.J.F. Reddaway (1971): Production of Sphagnum (bog-moss) and peat accumulation. Hydrobiologia 12:181-192.

Crill, P.M., K.B. Bartlett, R.C. Harriss, E. Gorham, E.S. Verry, D.I. Sebacher, L. Madzar and W. Sanner (1988a. Methane flux from Minnesota peatlands. Global Biogeochemical Cycles 2:371-384.

Crill, P.M., K.B. Bartlett, J.O. Wilson, D.I. Sebacher, R.C. Harriss, J.M. Melack, S. MacIntyre, L. Lesack and L. Smith-Morrill (1988b. Tropospheric methane from an Amazonian floodplain lake. Journal of Geophysical Research 93:1564-1570.

Crutzen, P.J., L.E. Heidt, J.P. Krasnec, W.H. Pollock and W. Seiler (1979): Biomass burning as a source of atmospheric gases CO, H2, N2O, NO, CH3Cl and COS. Nature 282:253-256.

Crutzen, P.J., I. Aselmann and W. Seiler (1986): Methane production by domestic animals, wild ruminants, other herbivorous fauna, and humans. Tellus 38B:271-284.

Devol, A.H., J.E. Richey, W.A. Clark, S.L. King and L.A. Martinelli (1988): Methane emissions to the troposphere from the Amazon floodplain. Journal of Geophysical Research 93:1583-1592.

Ehhalt, D.H. and U. Schmidt (1978): Sources and sinks of atmospheric methane. Pageoph. 116:452-464.

Fraser, P.J., R.A. Rasmussen, J.W. Creffield, J.R. French and M.A.K. Khalil (1986): Termites and global methane - another assessment. Journal of Atmospheric Chemistry 4:295-310.

Harriss, R.C. and D.J. Sebacher (1981): Methane flux in forested freshwater swamps of the southeastern United States. Geophysical Research Letter 8:1002-1004.

Harriss, R.C., D.J. Sebacher, K.B. Bartlett, D.S. Bartlett and P.M. Crill (1988): Sources of atmospheric methane in the South Florida environment. Global Biogeochemical Cycles 2:231-243.

Harriss, R.C., D.J. Sebacher and F.P. Day (1982): Methane flux in the Great Dismal Swamp. Nature 297:673-674.

Harriss, R.C., E. Gorham, D.J. Sebacher, K.B. Bartlett and P.A. Flebbe (1985): Methane flux from northern peatlands. Nature 315:652-654.

Holzapfel-Pschorn, A. and W. Seiler (1986): Methane emission during a vegetation period from an Italian rice paddy. Journal of Geophysical Research 91:11803-11814.

Huke, R.E. (1982): Rice area by type of culture: South, Southeast, and East Asia. International Rice Research Institute, Los Baños, Laguna, Philippines.

Keller, M., T.J. Goreau, S.C. Wofsy, W.A. Kaplan and M.B. McElroy (1983): Production of nitrous oxide and consumption of methane by forest soils. Geophysical Research Letter 10:1156-1159.

Keller, M., W.A. Kaplan and S.C. Wofsy (1986): Emissions of N2O, CH4 and CO2 from tropical forest soils. Journal of Geophysical Research 91:11.791-11.802.

Koyama, T. (1964): Biogeochemical studies on lake sediments and paddy soils and the production of hydrogen and methane. In: T. Miyaka and T. Koyama (eds.), Recent Researches in the Fields of Hydrosphere, Atmosphere, and Geochemistry, pp. 143-177, Maruzen, Tokyo, 1964.

Matthews, E. and I. Fung (1987): Methane emission from natural wetlands: global distribution, area, and environmental characteristics of sources. Global Biogeochemical Cycles 1:61-86.

Rasmussen, R.A. and M.A.K. Khalil (1983): Global production of methane by termites. Nature 301:700-702.

Schütz, H., A. Holzapfel-Pschorn, R. Conrad, H. Rennenberg and W. Seiler (1989): A 3 years continuous record on the influence of daytime, season, and fertilizer treatment on methane emission rates from an Italian rice paddy. Journal of Geophysical Research in press.

Sebacher, D.J., R.C. Harriss and K.B. Bartlett (1983): Methane flux across the air-water interface: air velocity effects. Tellus 35B:103-109.

Sebacher, D.J., R.C. Harriss, K.B. Bartlett, S.M. Sebacher and S.S. Grice (1986): Atmospheric methane sources: Alaskan tundra bogs, an alpine fen, and a subarctic boreal marsh. Tellus 38B:1-10.

Seiler, W. (1984): Contribution of biological processes to the global budget of CH_4 in the atmosphere. In: M.J. Klug and C.A. Reddy (eds.), Current Perspectives in Microbial Ecology, pp. 468-477. Washington D.C.: American Society of Microbiology.

Seiler, W. and R. Conrad (1987): Contribution of tropical ecosystems to the global budget of trace gases, especially CH_4, H2, CO, and N2O. In: R.E. Dickinson (ed.), The Geophysiology of Amazonia, pp. 133-160, Wiley and Sons, New York.

Seiler, W., A. Holzapfel-Pschorn, R. Conrad and D. Scharffe (1984a): Methane emission from rice paddies. Journal of Atmospheric Chemistry 1:241-268.

Seiler, W., R. Conrad and D. Scharffe (1984b): Field studies of methane emission from termite nests into the atmosphere and measurements of methane uptake by tropical soils. Journal of Atmospheric Chemistry 1:171-186.

Steudler, P.A., R.D. Bowden, J.M. Melillo and J.D. Aber (1989) Influence of nitrogen fertilization on methane uptake in temperate forest soils. Nature 341:314-316.

Svensson, B.H. (1980): Carbon dioxide and methane fluxes from the ombrotrophic parts of a subarctic mire. In: M. Sonesson (ed.), Ecology of a subarctic mire, pp. 235-250. Ecol. Bulletin (Stockholm) 30.

Svensson, B.H. and T. Rosswall (1984): In-situ methane production from acid peat in plant communities with different moisture regimes in a subarctic mire. Oikos 43:341-350.

Thathy, J.P., R. Delmas, B. Cros, A. Marenco, J. Servant and M. Labat (1988): Methane emissions from flooded forest in Central Africa. Eos 69:1066.

Wang, M.X., A. Dai, R.X. Shen, H.B. Wu, H. Schütz, H. Rennenberg and W. Seiler, CH_4 emission from a Chinese rice paddy field. in preparation.

Whalen, S.C. and W.S. Reeburgh (1988): A methane flux time-series for tundra environments. Global Biogeochemical Cycles 2:399-409.

Zimmerman, P.R., J.P. Greenberg, S.O. Wandiga and P.J. Crutzen (1982): Termite: A potentially large source of atmospheric methane, carbon dioxide and molecular hydrogen. Science 218:563-565.

Part V

Methods

CHAPTER 13

Gas Flux Measurement Techniques with Special Reference to Techniques Suitable for Measurements over Large Ecologically Uniform Areas

A.R. MOSIER

USDA-ARS, P.O. Box E, Ft. Collins, CO, USA

ABSTRACT

The soil/atmosphere flux of many trace gases have been measured by soil chamber techniques while micrometeorological methods have been less generally used. Micrometeorological methods are conceptually ideal for measuring trace gas emissions over large ecologically uniform areas. This is because the general techniques, eddy correlation and flux-radient, require large uniform terrain to function correctly. The techniques have not been extensively used to measure N_2O and CH_4 flux, however, because analytical methods that respond rapidly enough or are sensitive enough to quantify these gases were not available. Soil chamber methods have been used to measure fluxes of these gases and such studies generally infer that these measurements represent the larger area from the which the samples were taken. As development of sensitive, fast response chemical detectors permit, eddy correlation flux measurement will undoubtly complement chamber methods to measure trace gas flux.

This paper provides a brief, simple description of techniques that have been or could be used to measure trace gas flux from terrestrial ecosystems. Three examples of how different techniques have been used to measure the flux of N_2O or CH_4 from very different but relatively uniform terrestrial ecosystems to the atmosphere are presented.

13.1 INTRODUCTION

According to Baldocchi et al. (1988), most studies of biosphere/ atmosphere exchange of trace gases in the ecological literature have relied on chamber techniques. Micrometerological techniques provide an alternative means for measuring exchanges of chemicals between the biosphere and the atmosphere which one would intuititively consider superior to the chamber techniques. The micrometeorological methods integrate large surface areas thereby reducing the spatial variability problems inherent in chamber methods. Chambers also alter the immediate environment of the site of soil/atmosphere gas exchange by

Soils and the Greenhouse Effect. Edited by A.F. Bouwman
© 1990 John Wiley & Sons Ltd.

interfering with the normal air turbulence, changing temperature, altering gas concentrations, or altering solar radiation.

Micrometerological methods eliminate these problems, but present an array of problems themselves. One of the major limitations to the use of micrometerological methods is the lack of chemical detectors for N_2O and CH_4 that have sufficiently rapid response time and sensitivity to permit the more adaptable eddy correlation methods to be used. The recently developed tunable diode laser techniques may soon alleviate this problem. Until these techniques are generally available, however, chamber flux measurements will continue to be the principal method of determining trace gas emissions from the soil to the atmosphere.

Fowler and Duyzer (1989) suggest that the design of field trace gas flux studies should integrate micrometerological measurements with independent measurements of exchange rates using chambers. This combination provides an important contribution to quality assurance in the measurements and an opportunity to study the small scale while obtaining fluxes representative of a much larger scale.

Chambers are still the method of necessity but micrometerological methods should soon be widely used to measure CH_4 and N_2O fluxes. I'll briefly describe both techniques along with some of their advantages and limitations. Three examples of how different techniques have been used to measure the flux of N_2O or CH_4 from relatively uniform terrestrial ecosystems to the atmosphere are also presented.

13.2 CHAMBER METHODS

Basic design, theory, and limitations of chambers used to collect gases emitted from the soil have been discussed by a number of authors (Conrad et al., 1983; Denmead, 1979, 1987; Hutchinson and Mosier, 1981; Johansson, 1989; Jury et al., 1982; Matthias et al., 1978; Mosier and Heinemeyer; 1985; Sebacher and Harriss, 1982) so a detailed discussion will not be presented here. Some version of two basic chamber designs have been used to estimate fluxes of CO_2, CH_4, NO, N_2O, H_2S, and other biogenic sulphur emissions from soil and water surfaces. Both of the common chamber types enclose a distinct volume of air above a known area of soil or water surface and prevent or control mixing of the emanating gas with the external atmosphere. The two chamber types are those with forced flow-through air circulation, designated as "open soil chambers," and those with closed-loop air circulation or no forced air circulation, designated as "closed soil chambers" (Mosier, 1989).

13.2.1 Closed Chambers

Several different designs of closed chambers have been used to measure N_2O and CH_4 fluxes (Conrad et al., 1983; Denmead, 1979; Duxbury et al., 1982; Hutchinson and Mosier, 1981; Matthias et al, 1980; Sebacher and Harris, 1982). Gas flux from the soil using closed chambers can be measured by periodically collecting gas samples from the chamber and determining the change in concentration of the gas with time. Typical advantages of closed covers are that very small fluxes can be measured, no electrical supply is needed, the covers need to be in place for only a short time so that disturbance of the site due to the cover is limited, the chambers are simple to construct, and, since they are easy to install and to remove, they provide the opportunity to measure different locations at different times with the same equipment.

Problems commonly attributed to closed soil covers include: 1) concentrations of gas in the enclosure head space can build up to levels where they inhibit normal gas diffusion. This problem can be limited by using short collection periods and correction equations (Jury et al., 1982; Hutchinson and Mosier, 1981). 2) Closed covers either eliminate or alter the atmospheric pressure fluctuations which normally are found at the soil surface due to the natural turbulence of air movement. An appropriately designed vent allows pressure equilibration in and outside the chamber (Hutchinson and Mosier, 1981). 3) Chamber disturbs the soil air boundry layer. 4) Pressure changes in the soil can be caused by inserting the chamber into the soil. This problem may be overcome by installing collars in the soil that are normally open to the atmosphere and to seal the cover to the collar when the chamber is used (Seiler and Conrad, 1981; Duxbury et al., 1982). Alternatively, after initially inserting the chambers into the soil the chambers may be removed for a few minutes to allow dissipation of any gas released during the disturbance and then replaced (Livingston et al., 1988). 5) Temperature changes in the soil and atmosphere under the chamber can occur. Temperature differences within and outside the chamber can be reduced by insulating the chamber and covering it with reflective material and by short gas collection periods.

13.2.2 Open Chambers

Open soil covers such as those used by Denmead (1979), Ryden et al. (1978), Sebacher and Harriss (1982) and Steudler and Peterson (1985) are coupled to the atmosphere by an air inlet through which outside air is continuously drawn into the cover and forced to flow over the enclosed soil surface. The gas flux from the soil surface can be calculated from concentration difference between incoming and outgoing air, flow rate, and area covered by the open soil cover. The main advantage of open chambers covers is that they maintain environmental conditions closer to those of the uncovered field. This implies that open systems are more applicable for a continuous long-term monitoring of gas flux. Open chambers are, however, sensitive to pressure deficits inside the chamber caused by the induced air flow, which may in turn cause artificially high fluxes. (Kanemasu et al, 1974). This can be readily overcome by insuring that the size of the inlet gas orifices are large compared to size of outlet (Denmead, 1979). An additional consideration in open cover systems is the time required for gas concentration in the soil and the chamber air to adjust to new equilibrium values. Measurements assume an equilibrium flux between soil atmosphere and chamber atmosphere so estimates will be erroneous during the equilibration period (Denmead, 1979). Alterations of solar radiation inside the chamber is unavoidable but can be minimized (Schutz and Seiler, 1989).

13.2.3 Spatial Variability

Site spatial variability is undoubtedly the greatest problem in using chamber techniques to estimate a given gas flux from a field or ecosystem (Folorunso and Rolston, 1984). Coefficients of variation (CV) of N_2O or CH_4 between measuring points within a "uniform" site location typically range between 50 and 100% (Whalen and Reeburgh, 1988; Mosier and Hutchinson, 1981) Matthias et al. (1980) found that within a 100 m2 area of an Iowa field that N_2O emissions had CV's ranging from 31 to 168 %. Duxbury et al. (1982) found CV for daily N_2O flux to range from 0 to 224% for mineral soils in New York and organic soils in Florida. They observed no relationship between flux magnitude and CV.

13.3 MICROMETEOROLOGICAL METHODS

The basic concept of micrometeorological approaches for measuring trace gas flux to or from the soil surface is that gas transport is accomplished by the eddying motion of the atmosphere which displaces parcels of air from one level to another (Denmead, 1983). Transport of a gas through the free atmosphere to within a mm or so of the absorbing or emitting surface is then provided by turbulent diffusion in which the displacement of individual eddies is the basic transport process. Very close to the surface, typically over distances <1 mm, turbulence is damped by viscous forces, and transport over these very short distances relies on molecular diffusion. For the distance scales over which measurements are practical, turbulent transport is therefore the dominant mechanism, and in the simplest of the micrometerological methods, the flux may be measured by sensing the concentrations and velocities of components of the turbulence (Fowler and Duyzer, 1989).

With the exception of the mass balance method (briefly described below) micrometerological methods are based on the assumption that the flux to or from the surface is identical to the vertical flux measured at the reference level some distance above the surface. These two quantities may differ as a consequence of three processes: i) chemical reaction within the air column between the measurement level and the surface, ii) changes in concentration with time and therefore changes in the storage of the trace gas within the air column, and iii) horizontal gradients in air concentration leading to advection (Fowler and Duyzer, 1989).

These methods also require extensive uniform surface areas, a flux to the surface which is uniform throughout the upwind area influencing the sample point (fetch), and constant atmospheric conditions during each measurement period. In flat, homogeneous terrain the flux measured at the chosen sampling points above the surface provides the average vertical flux over the upwind fetch. The sample points must be in the height range in which the vertical flux is constant with height. This constant flux layer generally extends vertically to about 0.5% of the upwind uniform fetch, and is therefore about 1 m deep at the downwind edge of a 200 m uniform field (Monteith, 1973). It is evident that experimental areas need to be quite large and uniform before these techniques can be used in accord with theory.

Two general micrometerological techniques are used to measure trace gas flux density; eddy correlation and flux-gradient (gradient diffusion). I'll provide a brief verbal description of these methods, derived primarily from Baldocchi et al. (1988); Denmead (1983); and Fowler and Duyzer (1989).

13.3.1 Flux-Gradient Method

Flux-gradient theory assumes that turbulent transfer of a gas is analogous to molecular diffusion. But unlike molecular diffusion which results from the random motion of molecules, eddy diffusion results from the movement of parcels of air from one level to another. Consequently, eddy diffusitivities are usually several orders of magnitude greater than molecular diffusion. Their actual magnitudes are determined by wind speed, height above the surface (plant canopy, soil, water), the aerodynamic roughness of the surface and the vertical temperature gradient. The turbulent flux is then proportional to the product of the mean vertical mixing ratio gradient of the gas and an eddy diffusivity. The vertical transport of a gas towards the surface is the product of the transfer coefficient (eddy diffusivity) for the gas and its vertical concentration gradient in air in the constant flux layer. If the gradient in concentration decreases towards the surface the gas concentration gradient is negative by convention and the flux is towards the surface, and vice versa.

Gradients with height of wind velocity, air temperature and concentrations of trace gases, measured within the constant flux layer, provide the basic data for several methods of measuring vertical fluxes over extensive surfaces. Two of the more popular techniques for computing vertical fluxes are the aerodynamic and the Bowen-ratio (energy balance) methods.

13.3.2 Aerodynamic method

The aerodynamic method is based on the relationship between the momentum flux equation and the wind speed gradient. This technique requires the measurement of windspeed at two or more heights and the concentration gradient of the gas at these heights. The trace gas flux is influenced by vertical air density gradients due to water vapor and heat fluxes, and stability corrections must be made (Fowler and Duyzer, 1989).

13.3.3 Bowen Ratio (Energy Balance) Method

This technique does not require wind velocity profile measurements and is based on the energy balance at the surface. The incoming net radiation is partitioned between sensible heat flux, latent heat flux, and the soil heat flux. The ratio of sensible heat flux/latent heat flux is the Bowen ratio. The energy balance methods require measurements of vertical gradients of gas concentration, temperature and humidity and these provide estimates of the fluxes of sensible heat, water vapor and the trace gas without the complication and uncertainty introduced by stability corrections necessary in the aerodynamic method. The method also permits calculation of water evaporation rate (Denmead, 1983).

The largest drawback, and most important one for trace gas fluxes, is the substantial net radiation fluxes required by energy balance methods. In cloudy, night or winter conditions the available net radiation is frequently too small to permit satisfactory flux estimates (Fowler and Duyzer, 1989). A further nighttime problem is that the condensation of dew on radiation instruments leads to erroneous measurements. The employment of both aerodynamic and energy balance methods would seen advisable at any time, but particularly so when the diurnal pattern of trace gas loss is being investigated (Denmead, 1983).

13.3.4 Eddy Correlation Approach

Using the eddy correlation technique a trace gas flux density (vertical transport of a gas past a point in the atmosphere) is obtained by correlating the instantaneous vertical wind speed at a point with the instantaneous concentration of that gas. In the natural environment the eddies which are important in the transport process occur with frequencies extending up to 5 or 10 Hz. Therefore a rapid response detector is required (Denmead, 1983).

If a fast response detector is available, eddy correlation methods offer many advantages. In particular they require a minimum number of assumptions about the nature of the transport process. The method requires real time, continuous measurement of wind speed and gas concentration at only one height above the surface. The technique can also be used inside plant canopies and at night.

The method does present some problems, however. The method requires sophisticated recording and computing facilities to cope with the rapid data acquisition rates and large storage capacity. This problem becomes less dramatic almost daily, however, with the rapid advances that are being made in computer technology. Another problem for which special consideration must be made is the correction of the trace gas density required due to water vapor and heat transfer. These corrections become quite large for some gases (Denmead, 1987).

Application of both eddy correlation and flux-gradient approaches are limited to situations in which the air analyzed has passed over a homogeneous exchange surface for a long distance so that profiles of gas concentration in the air are in equilibrium with the local rates of exchange. The methods also require that horizontal concentration gradients are negligible.

13.4 ALTERNATIVE MICROMETEROLOGICAL APPROACHES

13.4.1 Eddy Accumulation

The eddy accumulation technique has been proposed as a possible means of measuring the flux of constituents for which no fast response sensor is available, N_2O for example. The technique involves the collection of upward and downward transported material in two separate containers at rates proportional to the vertical wind speed. The flux is evaluated as the difference in concentration accumulated over the sampling period. Unfortunately the practical difficulties of measuring small concentration differences in the accumulators and sampling the air according to vertical wind velocity have proven difficult (Baldocchi et al., 1988).

13.4.2 Mass Balance Technique

The mass balance method has been used to measure fluxes of ammonia from small fields (Denmead et al., 1977; Wilson et al., 1983). Gas flux density is related to the horizontal distance from the upwind edge of the field and the top of the air layer influenced by the emission of the gas. The method assumes that the mean horizontal turbulent flux is much smaller than the mean, horizontal advective flux. The top of the air layer influenced by the emission of the gas is a function of stability and surface roughness but can generally be simply estimated (Denmead, 1983). Denmead (1983) recommended that concentrations of the gas and wind speeds be measured at five levels or more above the surface. The horizontal distance from the upwind edge must be known precisely. To minimize the effect of changing wind direction on this horizontal distance it is recommended that experiments be conducted in a circular plot with the instrument array in the center (Baldocchi et al., 1988).

13.5 EXAMPLES OF METHANE AND NITROUS OXIDE GAS FLUX MEASUREMENTS

13.5.1 Example 1

Trace gas flux from soil systems, agricultural and "undisturbed", are characteristically spatially and temporally variable. Proper measurement of gas fluxes to account for both of these sources of variability require large numbers of chambers and frequent measurement or frequent measurements by micrometeorological methods. An approach to the problem of estimating spatial variability and monitoring temporal variability is an automated closed chamber technique. Conrad et al. (1983) used such a system to measure N_2O flux from terrestrial systems. A modification of this technique was used by Schutz et al. (1988, 1989) to measure CH_4 flux from flooded rice fields.

Schutz and Seiler (1989) summarize rice (Oryza sative L.) field studies in Italy and China where the automated closed chamber technique was used to observe temporal variations of CH_4 emission rates as well as the influence of N fertilizer application on CH_4 emissions. This system allowed the semi-continuous determination of CH_4 flux rates at 16 individual field sites using closed chambers made of Plexiglas. The boxes were covered with a removable lid which was opened and closed by a pneumatic pressure cylinder. In the field, the gas collection boxes were mounted on stainless steel frames fixed in the ground prior to field flooding. These frames remained in place during the entire vegetative period. The height of a frame is adjusted so that approximately 5 cm of the box is below the water surface. In this way the air volume inside the box is separated from the ambient atmosphere, but the water beneath the box can exchange with the surrounding water body, thus avoiding temperature increases in the flooded soil. The system consists of two completely separate units, each with a gas chromatograph (GC) connected to a dual-channel integrator. Magnetic valves controlling the air flow in the system are operated by a programable micro-computer which also stores the raw data output of the GC-system. Air samples from inside the individual closed boxes are taken periodically and analyzed for CH_4 by flame ionization detector (FID)-GC. The system allows automatic determination of CH_4 emission rates at each field plot 8 times per 24 h period.

Using the semicontinuous system, Schutz et al. (1988) found high diurnal and seasonal variations in CH_4 flux. The system also allowed a comparative study of flux rates in fertilized and unfertilized rice fields. Fertilization with ammonium sulfate or urea depressed CH_4 emission by up to 50% while fertilization with rice straw enhanced emissions up to two fold. From the results of these studies Schutz and Seiler (1989) suggest that rice paddies are likely to be the most important individual source for global atmospheric CH_4.

13.5.2 Example 2

Nitrous oxide flux measurements were made in a 120 ha maize (Zea mays L.) field located near Berthoud, Colorado by closed cover and aerodynamic techniques (Hutchinson and Mosier, 1979; Mosier and Hutchinson, 1981). The field site was selected based on its suitability for micrometerological measurement of vertical N_2O flux density. The site was flat, and fetch across the crop in the direction of prevailing winds was about 500m with no obstructions for at least twice that distance. The principal soil type in the maize field was Nunn clay loam (fine, montmorillonitic, mesic Aridic Argiustoll) which was fertilized with 200 kg NH_3-N ha^{-1} in early June, about 6 weeks after planting.

Nitrous oxide flux measurements were made periodically from mid-May through mid-September by both methods. Closed chambers similar to those described by Hutchinson and Mosier (1981) were used to estimate N_2O vertical flux density. Quadruplicate measurements were made at new locations within a small uniform area of the field each sampling date, with two chambers placed in the maize row and two in the irrigation furrow between rows. After 0, 15, 30 and 60 minutes, 30 ml air samples were withdrawn from the jars by 60-ml polypropylene syringes fitted with vacuum-tight stop-cocks and transported to the laboratory for same-day analysis by electron capture detector (ECD)-GC (Mosier and Mack, 1980). Nitrous oxide flux was computed from the concentration increase and corrected for the reduction in soil N_2O concentration gradient with time as the gas accumulated (Hutchinson and Mosier, 1981).

Data obtained by the soil cover method were compared periodically with simultaneous measurements of vertical N_2O flux density determined by the aerodynamic technique described by Thom (1975). Estimates of eddy diffusivity required by the method were obtained from a momentum balance. Vertical wind speed profiles were measured with cup anemometers and temperature profiles were determined with linear thermistors mounted in radiation shields patterned after those described by Lourence and Pruitt (1969). Nitrous oxide concentration profiles were determined by using a modified constant rate syringe pump to simultaneously fill eight 60-ml polypropylene syringes fitted with vacuum-tight stop-cocks each with air from a different sampling height. Air samples, accumulated over a 1-h period, were returned to the laboratory for same-day analysis by ECD-GC. Vertical profiles of wind, temperature and N_2O were determined in the first few meters of free air stream above the maize canopy.

The high spatial variability of N_2O emissions from the soil enhance the attractiveness of micrometerological techniques for N_2O flux determinations, since they provide a spatially integrated vertical flux estimate. Unfortunately, the aerodynamic technique could be used only during periods of high flux (only four days) because only then were the differences in N_2O concentration between sampling heights greater then the minimum detectable difference of the ECD-GC (1 to 2 ppb). The N_2O flux density estimated by the aerodynamic method always fell within the range of the four individual flux estimates made by the soil cover method. The following data are an example of the variability of the data from soil covers and its comparison to the micrometerological determination of N_2O flux made over the same time period: Emissions measured by the chamber technique averaged 258 ± 163 ng N m^{-2} sec^{-1}. Half the samples were collected within the maize plant row and averaged 373 ±20 while the other half of the samples that were collected between the maize plant rows averaged 143 ± 30 ng N m^{-2} sec^{-1}. The N_2O flux measured by the aerodynamic method was 350 ng N m^{-2} sec^{-1}.

13.5.3 Example 3

The last example comes from a study where CH_4 fluxes from tundra environments were estimated (Whalen and Reeburgh, 1988). These measurements were made at a set of permanent sites chosen to represent components typical of large areas of arctic tundra. Seasonal measurements of net methane flux were made in Eriophorum tussocks, intertussock depressions, moss-covered areas and Carex stands near Fairbanks, Alaska.

Net CH_4 fluxes were determined using static chambers based on the design of Conrad et al. (1983). The chambers consisted of three parts: a permanent aluminum base, a topless Plexiglas vertical section that could be stacked to vary volumes, and a Plexiglas cover. A watertight channel was located on each base and at the top of each vertical section. This channel was filled with water and provided a seal between the chamber components. Some covers were equipped with thermistors to monitor chamber temperature during experiments. The covers and vertical sections were removed from the bases after each flux measurement. Gas samples were drawn by syringe from each chamber immediately after installation and at two approximately equally spaced time intervals. Sampling intervals and chamber volumes were adjusted to minimize the time needed for a ≥ 0.3 ppm by volume methane increase in the chamber and to maintain linearity in concentration versus time plots. Minimum and maximum time intervals were 0.2 and 24 h, respectively. Typical summer sampling intervals were 0.25 to 0.5 h; winter sampling intervals were 12 to 24 h. Methane was analyzed using FID-GC. Methane fluxes were calculated using the rate of methane increase determined by linear least squares fits of chamber methane concentration versus time, base areas, chamber volumes and the molar volume of methane at ambient temperature.

Methane fluxes showed high diel, seasonal, intra site, and between site variability. *Eriophorum* tussocks and *Carex* dominated CH_4 release to the atmosphere, with mean annual net CH_4 fluxes of 8.05 ± 2.5 g methane m^{-2} and 4.88 ± 0.73 g methane m^{-2}, respectively. Methane fluxes from the moss sites and intertussock depressions were much lower (0.47 ± 0.16 and 0.62 ± 0.28 g CH_4 m^{-2} y^{-1}). Over 90% of the mean annual CH_4 flux from the *Eriphorum*, intertussock depressions, and *Carex* sites occurred between thaw and freeze-up. Some 40% of the mean annual

CH_4 flux from the moss sites occurred during the winter. Composite CH_4 fluxes for tussock tundra and *Carex*-dominated wet meadow tundra environments were produced by weighting measured components fluxes according to areal coverage. Tussock and wet meadow tundra accounted for an an estimated global CH_4 emission of 19-33 Tg y^{-1}.

This study provides an example of where micrometeorological methods would simplify integrated methane gas flux measurements from the tundra ecosystem. The site is ideal for such methods since the topography is relatively flat and the vegetation is close to the ground. The variability present within the ecosystem is small scale variability in which the Whalen and Reeburgh (1988) made individual site allocations. Micrometeorological methods would provide a spatially integrated gas flux measurement across the small scale topographic variations. If, however, determining the variability of CH_4 flux within and between the different sites is part of the objective of the study then chamber methods can best accomplish them.

13.6 RECENT TECHNOLOGICAL ADVANCES

Measurements of N_2O and CH_4 flux by micrometerological or other methods which integrate spatially over large surface areas are almost nonexistent, because detectors possessing sufficient speed or sensitivity have not been available. Recent development of tunable diode laser technology has produced a detector for CH_4 which has permitted eddy correlation measurement of CH_4 flux (R.C. Harriss, 1989). When such technology becomes readily available, rapid advances in our knowledge of gas fluxes from both large uniform ecological systems and surface rough, heterogeneous systems will be possible by the general research community.

Another analytical system that is currently being tested to measure trace gas concentrations on the ecosystem scale is the long-path fourier-transform infrared spectroscopy (FTIR) (Gosz et al., 1988). The FTIR is capable of measuring the concentration of multiple trace gases simultaneously over spatial scales up to 1 km. Detections limits of about 1 uL/kL for many gases (CH_4, N_2O, CO_2, O_3 for example) make the detection method potentially useful to study spatial patterns of biological processes, fluxes between ecosystems and scale-dependent processes as well as gas fluxes from homogeneous ecosystems. The instrument can respond as rapidly as the second range therefore permits short time temporal change measurements. The technique is limited to gas concentration measurements until the methodology for making the necessary accompanying measurements to estimate gas flux is perfected. Research is underway to extend the FTIR measurements to estimate gaseous fluxes for spatial scales much larger than presently possible (Gosz et al., 1988).

REFERENCES

Baldocchi, E.D., B.B. Hicks and T.P. Meyers (1988) Measuring biosphere-atmosphere exchanges of biologically related gases with micrometerological methods. Ecology 69:1331-1340.

Conrad, R., W. Seiler and G. Bunse (1983) Factors influencing the loss of fertilizer nitrogen into the atmosphere as N_2O. Journal of Geophysical Research 88:6709-6718.

Denmead, O.T. (1987) Notes on measuring trace gas fluxes with enclosures. In: Workshop on Measurement of Surface Exchange and Flux Divergence of Chemical Species in the Global Atmosphere. New York, NY. Columbia University.

Denmead, O.T. (1983) Micrometerological methods for measuring gaseous losses of nitrogen in the field. In: J.R. Freney and J.R. Simpson (Eds.) Gaseous Loss of Nitrogen from Plant-Soil Systems. pp. 133-157. Martinus Nijhoff/Dr.W. Junk Publishers, The Hague, Netherlands.

Denmead, O.T. (1979) Chamber systems for measuring nitrous oxide emissions from soils in the field. Soil Science Society of America Journal 43:89-95.

Denmead, O.T., J.R. Simpson and J.R. Freney (1977) A direct field measurement of ammonia emission after injection of anhydrous ammonia. Soil Science Society of America Journal 41:1001-1004.

Duxbury, J.M., D.R. Bouldin, R.E. Terry and R.L. Tate (1982) Emissions of nitrous oxide from soils. Nature 298:462-464.

Folorunso, O.A. and D.E. Rolston (1984) Spatial variability of field measured denitrification gas fluxes. Soil Science Society of America Journal 48:1214-1219.

Fowler, D. and J. Duyzer (1989) Micrometerological techniques for the measurement of trace gas exchange. In: M.O. Andreae and D.S. Schimel (Eds.), Exchange of Trace Gases Between Terrestrial Ecosystems and the Atmosphere. Dahlem Konferenzen. Chichester: John Wiley & Sons Ltd., (In Press).

Gosz, J.R., C.N. Dahm and P.G. Risser (1988) Long-path FTIR measurement of atmospheric trace gas concentrations. Ecology 69:1326-1330.

Harriss, R.C. (1989) The arctic Boundary layer expedition (ABLE-3A). EOS 70:282.

Hutchinson, G.L. and A.R. Mosier (1981) Improved soil cover method for field measurement of nitrous oxide fluxes. Soil Science Society of America Journal 45:311-316.

Hutchinson, G.L. and A.R. Mosier (1979) Nitrous oxide emissions from an irrigated cornfield. Science 205:1225-1226.

Johansson, C. (1989) Fluxes of NOx above soil and vegetation. In: M.O. Andreae and D.S. Schimel (Eds.), Exchange of Trace Gases Between Terrestrial Ecosystems and the Atmosphere. Dahlem Konferenzen. Chichester:John Wiley & Sons Ltd., (In Press).

Jury, W.A., J. Letey and T. Collins (1982) Analysis of chamber methods used for measuring nitrous oxide production in the field. Soil Science Society of America Journal 46:250-256.

Kanemasu, E.T., W.L. Powers and J.W. Sij (1974) Field chamber measurements of CO_2 flux from soil surfaces. Soil Science 118:233-237.

Livingston, G.P., P.M. Vitousek and P.A. Matson (1988) Nitrous oxide flux and nitrogen transformations across a landscape gradient in Amazonia. Journal of Geophysical Research 93:1593-1599.

Lourence, F.J. and W.O. Pruitt (1969) A psychrometer system for micrometerology profile determination. Journal of Applied Meterology 8:492-498.

Matthias, A.D., A.M. Blackmer and J.M. Bremner (1980) A simple chamber technique for field measurement of emission of nitrous oxide from soil. Journal of Environmental Quality 9:251-256.

Matthias, A.D., D.N. Yarger and R.S. Weinbeck (1978) A numerical evaluation of chamber methods for determining gas fluxes. Geophysical Research Letters 5:765-768.

Monteith, J.L. (1973) Principles of Environmental Physics. E. Arnold Press.

Mosier, A.R. (1989) Chamber and isotope techniques. In: M.O. Andreae and D.S. Schimel (Eds.), Exchange of Trace Gases Between Terrestrial Ecosystems and the Atmosphere. Dahlem Konferenzen. Chichester:John Wiley & Sons Ltd., (In Press).

Mosier, A.R. and O. Heinemeyer (1985) Current methods used to estimate N_2O and N2 emissions from field soils. In: H.I. Golterman (Ed.) Denitrification in the Nitrogen Cycle. pp. 79-99. Plenum Publishing Corp.

Mosier, A.R. and G.L. Hutchinson. 1981. Nitrous oxide emissions from cropped fields. Journal of Environmental Quality 10:169-173.

Mosier, A.R. and L. Mack (1980) Gas chromatographic system for precise, rapid analysis of N_2O. Soil Science Society of America Journal 44:1121-1123.

Ryden, J.C., L.J. Lund and D.D. Focht (1978) Direct in-field measurements of nitrous oxide flux from soil. Soil Science Society of America Journal 42:731-738.

Schutz, H. and W. Seiler (1989) CH_4 flux measurements: Methods and results. In: M.O. Andreae and D.S. Schimel (Eds.) Exchange of Trace Gases Between Terrestrial Ecosystems and the Atmosphere. Dahlem Konferenzen. Chichester: John Wiley & Sons Ltd. (In Press).

Schutz, H., A. Holzapfel-Pschorn, A.R. Conrad, H. Rennenberg and W. Seiler (1988) A three year continuous study on the influence of daytime, season, and fertilizer treatment on methane emission rates from an Italian rice paddy field. Journal of Geophysical Research (In Press).

Sebacher, D.J. and R.C. Harriss (1982) A system for measuring methane fluxes from inland and coastal wetland environments. Journal of Environmental Quality 11:34-37.

Seiler, W.A. and R. Conrad (1981) Field measurements of natural and fertilizer-induced N_2O release rates from soils. Journal Air Pollution Control Association 31:767-772.

Steudler, P.A. and B.J. Peterson (1985) Annual cycle of gaseous sulfur emissions from a New England Spartina alternifolra marsh. Atmospheric Environment 19:1411-1416.

Thom, A.S. (1975) Momentum, mass and heat exchange of plant communities. In: J.L. Monteith (Ed.) Vegetation and the Atmosphere. Vol. 1: Principles. pp 57-109. Academic Press Inc., New York.

Whalen, S.C. and W.S. Reeburgh (1988) A methane flux time series for tundra environments. Global Biogeochemical Cycles 2:399-409.

Wilson, J.D., V.R. Catchpoole, O.T. Denmead and G.W. Thurtell (1983) Verification of a simple micrometerological method for estimating the rate of gaseous mass transfer from the ground to the atmosphere. Agricultural Meterology 29:183-189.

CHAPTER 14

Analysis of Vegetation Changes Using Satellite Data

F. BLASCO and F. ACHARD

Centre National de la Recherche Scientifique
Institute of the International Map of the Vegetation
31062 Toulouse Cedex, France

ABSTRACT

A critical study of the actual capabilities of satellites for the monitoring of tropical vegetation reveals that these tools are not yet operational. Monitoring capabilities are improving thanks to a combination of conventional tools and satellite data (NOAA, LANDSAT, SPOT). Present vegetation mapping techniques have a scale of time incompatible with the scale of events to be monitored.

The probable annual rate of deforestation in the tropics is of about 200,000 km². Out of the remaining 8 million km² of humid forests, 30 to 50% are already more or less deeply transformed by man.

14.1 INTRODUCTION

Large scale and almost irreversible destructions of forest habitats have recently created a major ecological issue, for a number of reasons: (i) the actual rate of deforestation is exceedingly high; (ii) it is concentrated in a few focal points (Amazon basin, Indonesia, West Africa); (iii) we do not have at present adequate tools for a monitoring of forest depletion at a global scale; (iv) the exact implications of forest destruction for atmospheric budgets of CO_2, CH_4 or N_2O are not sufficiently known to date.

According to recent estimates, the main greenhouse gas is carbon dioxide which contributes about 40% to the total greenhouse warming (for comparison: CFCs contribute 10 to 20%). Most of the atmospheric carbon dioxide is obviously released by burning of fossil fuels, but at least 20% of the total emission stems from deforestation in tropical countries. Some authors (Woodwell et al., 1983) estimates that the annual release of carbon from forests and soils exceeded the release from fossil fuels until about AD 1960 (Figure 14.1).

Forests are important stores of carbon. The carbon stocks in tropical forests may amount to 150 t C ha^{-1} in vegetation and in addition 120 t C ha^{-1} in soils. Regrowing secondary forests or actively growing planted

Soils and the Greenhouse Effect. Edited by A.F. Bouwman
© 1990 John Wiley & Sons Ltd.

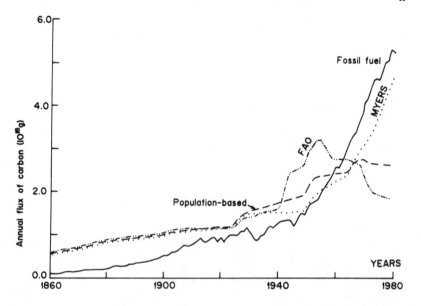

Figure 14.1 Annual and global carbon released from biota and soils (from a synthesis given by Woodwell et al., 1983).

trees in tropical landscapes may accumulate considerable amounts of carbon.

The aim of this paper is to analyze the capabilities and limitations of modern remote sensing techniques for an evaluation of vegetation changes in tropical countries.

14.2 AREAL EXTENT OF TROPICAL FORESTS AND DEFORESTATION RATES

According to official figures published by FAO (1981) and UNESCO (1981) the remaining closed forests in the tropical world covered about 10 million km^2 in the eighties.

Although the evergreen and semi-evergreen closed forests (see figure 14.2) occur in some 60 countries, three countries have theoretically more than 50% of the world's reserves (1980): Brazil (3.5 million km^2 of closed forests, which is approximately 35% of the world's area of rain forest), Indonesia (1.1 million km^2 or 11%), Zaire (1 million km^2 being 10%).

In those places where detailed vegetation maps are available (e.g. at 1:1,000,000 scale) these maps show that the areal extent of dense humid forests has generally been overestimated.

The case of Indonesia is interesting (Guppy, 1984). Out of a total forest area of approximately 1,100,000 km^2 (11% of earth's reserves), in

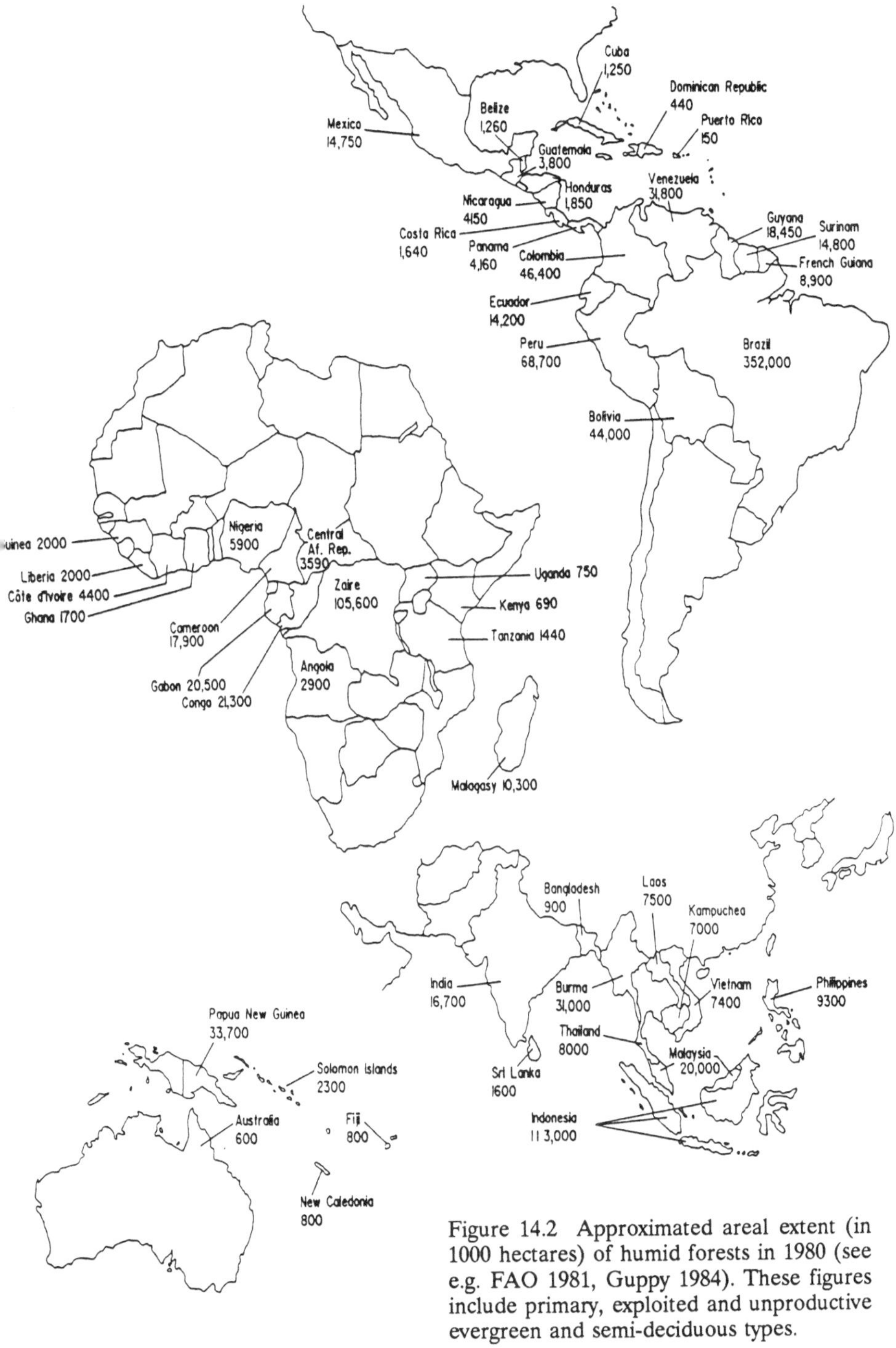

Figure 14.2 Approximated areal extent (in 1000 hectares) of humid forests in 1980 (see e.g. FAO 1981, Guppy 1984). These figures include primary, exploited and unproductive evergreen and semi-deciduous types.

1984 about 750,000 km² where already either exploited or degraded and unproductive.

In the absence of measured (objective) data the estimations reported by UNESCO (1981) presented in Table 14.1 may be illustrative.

Table 14.1 Total area of forest estimated for the end of the 19th century, the mid 1970's and the projected area for the end of the 20th century

	Tropical America	Tropical Africa	Tropical Asia and Australia	Total
End of 19th Century	803	362	435	1600
1976	506	175	254	935
End of 20th Century	300?	140?	160?	600?

Provisional conclusions are the following:

- The conventional methods to estimate the actual rates of deforestation in tropical countries are not adequate, and this explains the great diversity of reported global deforestation rates which range from 71,000 km²y⁻¹ (FAO, 1981) to 200,000 km²y⁻¹ (Myers, 1986).
- A combination of conventional tools and satellite data may improve our monitoring capabilities. A major limitation in this respect is, that it took almost 10 years to obtain a complete Landsat MSS coverage for the island of Sumatra (total area of 473,000 km²). The climate of Sumatra is very humid and under such conditions the probability to obtain a cloud-free image en a certain date is low and, in addition, the repeat cycle of Landsat of 16 days is rather long.
- The production of a vegetation map in tropical countries usually requires 4 to 10 years, depending on size, number of technicians available and existing data. Some examples of recently produced vegetation maps and the duration of each project are presented in Table 14.2.

Table 14.2 Vegetation maps for a number of areas produced by the Institute of the International Map of the Vegetation, their scale, areal coverage and time used for their production. Data from Blasco (1988)

Country	Scale	N° of sheets	Area	Years required
Kampuchea	1:1,000,000 scale	1 sheet	(180,000 km²)	4 years
Sumatra	1:1,000,000 scale	3 sheets	(473,000 km²)	6 years
Madagascar	1:1,000,000 scale	3 sheets	(587,000 km²)	5 years
Cameroon	1:500,000 scale	8 sheets	(474,000 km²)	10 years
South America	1:5,000,000 scale	2 sheets	(17.8 million km²)	5 years

- Hence, the existing vegetation mapping techniques are still time consuming process. Moreover, vegetation maps represent a static image at a certain moment. Deforestation processes and magnitude are not represented on such maps.
- Our cross checking investigations lead to the assumption that remaining humid tropical forests in the world in 1989 cover about 8 million km², out of which 30 to 50% are already more or less deeply transformed by man.

14.3 THE USE OF SATELLITES FOR VEGETATION STUDIES

To date, the actual capacity of satellite products does not allow the monitoring of forest resources in the tropical belt.

According to the characteristics of each satellite or sensor we can distinguish three main levels of perception (Table 14.3).

Table 14.3 Theoretical applications of remote sensing data

Level of discrimination	Scale	Satellite or sensor
Global or regional: discrimination forest - non forest	1:2,000,000 or smaller	NOAA (AVHRR)
National: distinction of vegetation types (physiognomy and phenology)	1:250,000 to 1:1,000,000	LANDSAT-MSS IRS1
Local: accurate inventories, management purposes, floods control, etc...	1:50,000 to 1:100,000	LANDSAT-TM SPOT-HRV

14.3.1 Studies at global or regional scales

In moist tropical regions the separation between forests and non-forests is always a difficult task, in a single spectral band or using combinations of spectral bands.

- In the Amazon Basin where contacts between forest and non-forest units are usually clear (clearing for agriculture and rangeland development), it is attempted to base their separation upon the analysis of thermal data. This may be possible because the canopy of virgin forests is usually cooler than neighbouring deforested areas. These features have been locally identified in Southern Brazil using NOAA-AVHRR (Advanced Very High Resolution Radiometer) channel 3 data (3.5 - 3.9 μm), but to date these remote sensing data are not self sufficient and have to be confirmed by other sources of information, such as vegetation maps, shuttle images and field sampling (Malingreau et al., 1989). The estimated deforested area in Southern Brazil was 90,000 km² in 1985.

- In West Africa data provided by different satellites were combined, i.e., NOAA, SPOT and LANDSAT-TM (Achard, 1989). The principal characteristics of the different types of data used are high acquisition frequency (every 9 days at nadir) and low resolution (1 km) for the NOAA-AVHRR data and high spatial resolution (20 or 30 m) and low frequency (every 26 or 16 days at nadir) for SPOT and LANDSAT-TM data.

The Normalized Difference Vegetation Index (NDVI = {Near Infrared - Red} / {Near Infrared + Red}) and the Surface Temperature (ST) have been computed for the major vegetation classes. Surface Temperature (ST) has been calculated by a "split window" method to obtain the emission temperatures from bands 4 and 5. The existing relationships between the NDVI and vegetation characteristics is based on the fact that this ratio is directly related to the amount of photosynthetically active radiation (PAR) absorbed in the observed vegetation canopy.

At a global scale it appears that the discrimination of savannas and forests is possible at least during the dry season, whereas the discrimination of crops is always difficult. Recent estimates of deforestation are lacking for West Africa.

This study yielded promising results but the tools are not fully operational as yet.

- In East Asia only one important small scale deforestation analysis was carried out so far. It concerned the island of Borneo, where drought and fires damaged 3.5 million hectares in East Kalimantan and more than 1 million hectares in Sabah in 1983 (Malingreau et al., 1989). The use of AVHRR archives in the GAC format (Global Area Coverage, with a 4 km resolution) and the analysis of the Global Vegetation Index added to the study of thermal anomalies (using AVHRR channel 3) in the form of saturated pixels resulted in a first evaluation of the disaster.

14.3.2 Studies at medium or national scales

Studies at scales ranging between 1:250,000 to 1:1,000,000 can be undertaken with Landsat MSS. At such scales the adopted classification system has to be broad (usually based on physiognomic or ecologic criteria). The delay between two complete Landsat MSS coverages and inventories, at the scale of large nations, such as India, Indonesia and Brazil, is usually about 10 years.

Unless the tree coverage is 30% or more, it is almost impossible to analyze it through satellite data. When the tree cover is less than 30%, the spectral response of each pixel is an integration of trees and grasslands (or trees and barren soils). This spectral response does not permit the monitoring of fuel wood extraction in dry or arid areas. At medium or national scales the discrimination between forest and non-forest is easy and technically simple. Most figures provided in this paper have been partly obtained in individual countries using LANDSAT or SPOT data.

14.3.3 Studies at local scales

Local studies and large scale inventories generally have scales between 1:50,000 and 1:100,000. The potential uses and limitations of High Resolution Satellite data (SPOT-HRV, LANDSAT-TM) are quite well known. These tools play an essential role in many fields, such as agricultural land use investigations and forest management. None of these applications, however, can be extended to global monitoring activities, at the scale of a continent.

14.4 CONCLUSIONS

The interpretation of AVHRR data is riddled with major ambiguities and impossibilities. It does not permit the delineation of open forests (65% of the global area of open forest occur in Africa) and shrublands (70% in Africa).

A comparison of several estimates involving satellite technology and conventional means, leads to the following conclusions:

Remaining total of dense humid tropical forests: 8,000,000 km²
Remaining almost untouched humid tropical forests: 5,000,000 km²
Annual deforestation rate in the tropics :   200,000 km²

In less than 20 years most of the remaining lowland evergreen and semi-evergreen forests and corresponding soils will have been either degraded or transformed by man. An exception may be the Amazon basin.

REFERENCES

Achard, F. (1989) Dynamique de la végétation tropicale par télédétection spatiale. Ph. D. thesis, University Paul Sabatier, Toulouse (France).

Blasco, F. (1988) The international vegetation map (Toulouse, France). In A.W. Kuchler and I.S. Zonneveld (Eds.) Vegetation Mapping, p. 443-460, 591-622, Kluwer Academic Publishers, Dordrecht.

FAO (1981) Tropical forest resources assessment project (GEMS), tropical Africa, Tropical Asia, Tropical America, 4 vols. (Lanly Ed.), FAO/UNEP, Rome.

Guppy, N. (1984) Tropical deforestation: a global view. Foreign Affairs, 62:928-965.

Houghton, R.A., R.D. Boone, J.M. Melillo (1985) Net flux of carbon dioxide from tropical forests in 1980. Letters to nature 316:617-620.

Laumonier, Y. (1988): International map of the vegetation - "Sumatra" at 1:1,000,000 scale, Biotrop, Bogor (Indonesia) and ICIV, Toulouse (France).

Malingreau, J.P., C.J. Tucker and N. Laporte (1989): AVHRR for monitoring global tropical deforestation. International Journal of Remote Sensing, 10:855-867.

Myers, N. (1986) Tropical forests: patterns of depletion. In G.T. Prance (Ed.): tropical rain forests and the world atmosphere. American Association for the Advancement of Science, Washington D.C.

UNESCO (1981) Les zones tropicales humides: un monde en changement - Programme MAB: l'écologie en action, avec aperçu de 36 affiches - Paris.

Woodwell, G.M., J.E. Hobbie, R.A. Houghton (1983) Global deforestation: contribution to atmospheric carbon dioxide. Science 222:1081-1086.

CHAPTER 15

Global Data Bases for Evaluating Trace Gas Sources and Sinks

E. MATTHEWS

Centel - Sigma Data Services
NASA - Goddard Institute for Space Studies
2880 Broadway, New York, N.Y.10025, USA

ABSTRACT

The design and development of global digital data bases and their integrated use in studies of terrestrial sources and sinks of trace gases are discussed. Generic design characteristics that expand the utility of these data bases are outlined including: hierarchical classification systems that provide for encoding and accessing data at various levels of detail, and fine spatial and topical resolution which accommodate adaptations to a broad base of applications. The discussion follows the sequence of (i) developing primary data bases on major surface characteristics, (ii) selectively integrating combinations of data bases to produce secondary data sets of source categories and (iii) combining information on source categories with fluxes to produce tertiary data sets of emissions. These procedures are demonstrated with the example of estimating global methane emission from natural wetlands.

15.1 INTRODUCTION

Characteristics of the earth's atmosphere and surface exhibit inherent heterogeneities that vary differently over space and time. Although the concept of scale is commonly associated with spatial resolution, topical and temporal resolution are closely related to spatial scale and can influence the spatial resolution necessary in data formulated for a specific problem. While the issue of 'appropriate' resolution of data bases for global studies is tied to the scope and character of individual problems, there are some fundamental design characteristics of digital data bases that provide for their adaptation to a variety of problems.

In many cases, global studies of trace gas emissions require that point measurements, limited in space and time, be extrapolated over large areas and over seasons. How representative is a single measurement or series of measurements? How can we interpolate or extrapolate information from point measurements to the globe, or from measurements at a single time to the year? What other measurements can

Soils and the Greenhouse Effect. Edited by A.F. Bouwman
© 1990 John Wiley & Sons Ltd.

be drawn upon to evaluate and constrain global estimates of trace gas emissions?

Global scale studies of trace gas sources and sinks present special data requirements, the most obvious being global coverage. The most extensive and well designed series of terrestrial field measurements probably cannot provide needed information about pertinent characteristics over significant portions of the earth. At present, data available from satellites generally are not sufficiently precise or explicit measures of trace gas fluxes. Instruments aboard aircraft or installed at ground stations provide measurements of atmospheric trace gas concentrations that reflect the combined influences of local sources and sinks, and transported constituents.

One mechanism for integrating and globalizing data from measurements localized in space or time is through the development of global data bases that define fundamental features affecting the production, evolution or flux of trace gases. These data bases are compiled in digital form, commonly at fixed resolution. When combined with each other or complemented with supplementary information, they can be employed to evaluate distributions and characteristics of terrestrial sources of trace gases.

The primary focus of this paper is on the design and development of several types of global digital data bases and their integrated use in studies of terrestrial sources of trace gases.

15.2 DATA BASES AND METHODOLOGY

We initiated a broad-based effort at the Goddard Institute for Space Studies (GISS) to compile global data bases useful for deriving the sources and sinks of trace gases. Our global analyses of distributions, magnitudes and seasonality of terrestrial emissions of trace gases draws heavily on an evolving series of data bases tailored for individual applications, in conjunction with measurements comprising representative components of major sources and sinks. The considerable investment of effort required to compile global, high resolution data bases highlights the importance of designing highly flexible data sets that can be successively adapted for a variety of applications (Matthews, 1983). Generic design characteristics that amplify the usefulness of these data bases are hierarchical classification systems, and fine spatial and topical resolution.

The GISS data bases commonly are global in scope and compiled at 1° resolution, using hierarchical classification schemes where appropriate. This combination of features allows aggregation to coarser spatial and topical resolutions. The choice of 1° resolution is a compromise among several considerations such as quality of available source information and scale of natural heterogeneities in the features of interest. Appropriate hierarchical classification schemes sometimes are available (e.g., the UNESCO [1973] system for vegetation and the

FAO [FAO, 1971–1981] system for soils) or are devised for individual data sets (e.g., the land use classification of Matthews [1983]). Some of the data bases used in studies of biogenic sources of methane are listed in Table 15.1. These data bases can be separated into three major groups or levels. Examples of Level 1 data bases are vegetation (Matthews, 1983), soils (Zobler, 1986), land use (Matthews, 1983), fractional inundation (Matthews and Fung, 1987), topography (Gates and Nelson, 1975), political units (Lerner et al., 1988), temperature and precipitation (Shea, 1986) and NDVI (Normalized Difference Vegetation Index from NOAA-AVHRR). They define topically detailed, large scale characteristics of the earth's surface and are the basic starting points for the investigation of trace gas emissions.

The vegetation data were recorded using codes of the hierarchical UNESCO (1973) vegetation-classification system. This five-tiered system is built on a series of classification criteria including life form, density, seasonality, climate, plant architecture, altitude and environmental setting. An abbreviated outline of the classification system is given in Table 15.2 with selected subdivisions shown for several types of forests. Vegetation types can be distinguished, in order of increasing detail, as formation class, formation subclass, formation group, formation and subformation. The first and simplest level (formation class) has five groups while the fifth and most detailed stratification (subformation) distinguishes 225 types. Globally, 178 vegetation types are identified in the Matthews (1983) data base. The classification hierarchy of this system accommodates recording and retrieving data at various levels of detail. For example, the global array of vegetation data was collapsed to a simple nine-component form to prescribe land-surface boundary conditions such as surface albedo and field capacity in a coarse-resolution version of the GISS General Circulation Model (Hansen et al., 1983); the global distribution was used at intermediate topical resolution (32 groups, generally at the formation group level) to delineate variations in net primary productivity and seasonal carbon fluxes (Fung et al., 1987). The selective version tailored for the study of methane emission from natural wetlands is discussed in the following section.

The soils data, compiled from the globally uniform FAO (1971–1981) map series at 1:5M scale, encompasses 106 soil units, grouped into 26 major soil units on the basis of soil formation principles; additional information on texture, phase and slope is included where available (Zobler, 1986).

The land use data of Matthews (1983) were classified using a multi-level system designed specifically for the compilation. Classification criteria of this three-tiered system emphasize variations in the intensity and permanence of modifications caused by agricultural activities and allow for inclusion of crop combinations. Table 15.3 shows an abbreviated outline of the land-use classification indicating the 12

Table 15.1   Global digital data bases for the study of global methane cycles. This is a partial list of data bases completed or available at GISS. Level 1 (primary) data bases are 1° latitude x 1° longitude resolution unless otherwise noted. Higher level data bases are generally compiled at 1° resolution and interpolated to coarser resolutions

---

Level 1. Primary Data Bases

VEGETATION[1]	UNESCO hierarchical classification system; varied sources; 178 vegetation types; source: Matthews (1983)
SOILS[1]	FAO soil classification; FAO soil maps; 106 soil units with slope, phase and texture information; source: Zobler (1986)
LAND USE[1]	hierarchical classification system; varied sources; 119 types including crop and livestock combinations; source: Matthews (1983)
COUNTRY[1]	187 countries, with subdivisions for Australia, Brazil, Canada, China, India, U.S.A. and U.S.S.R.; source: Lerner et al. (1988)
TOPOGRAPHY	topographical heights and land/water distribution source: Gates and Nelson (1975)
FRACTIONAL INUNDATION[1]	from Operational Navigation Charts; percent inundation in 1° cells; source: Matthews and Fung (1987)
NDVI	weekly composites from April 1983 to present; (~0.2° or ~20km resolution); source: NOAA/NESDIS
PRECIPITATION, TEMPERATURE	30-year mean monthly and annual values; (2° x 2.5° resolution); source: Shea (1986)

Level 2. Data Bases of Source Categories

	Level 1 Data Used	Supplemental Information
WETLANDS	Vegetation, Soils, Inundation	
INUNDATION PERIOD	Temperature, Precipitation	Climate/flux thresholds
ANIMAL POPULATIONS	Country, Land Use	FAO[2] and other statistics on animal populations
HUMAN POPULATION	Country, Land Use	Country statistics on urban and rural populations
IRRIGATED RICE	Land Use, Country	FAO[2] rice areas, statistics on multiple cropping

Level 3. Data Bases on Trace Gas Sources

CH_4	Level 1+2 Data Used	Supplemental Information
Wetlands	Wetlands, Inundation Period	Flux measurements
Animals	Animal Populations	Emission rates for animal types
Irrigated Rice (monthly)	Irrigated Rice	Flux measurements, crop calendars for each country

---

[1]   primary data bases compiled at GISS
[2]   Food and Agriculture Organization of the United Nations (FAO) data available annually by country

Table 15.2 Abbreviated outline of the UNESCO vegetation classification system showing the five-level hierarchical structure and associated codes. Types marked with an asterisk are examples of wetlands

1. CLOSED FOREST
    A. Evergreen
        1. tropical ombrophilous (rainforest)
            a. lowland
            b. submontane
            c. montane
            d. subalpine
            e. cloud
        * f. alluvial
                * (1) on frequently flooded riverbanks (Igapo)
                * (2) on occasionally flooded dry terraces
                * (3) seasonally waterlogged (Varzea)
        * g. swamp (non-riverine)
        * h. bog (organic surface deposits)
        2. tropical/subtropical seasonal
        3. tropical/subtropical semi-deciduous
        4. subtropical ombrophilous
        5. mangrove
        6. temperate/subpolar ombrophilous
        7. temperate seasonal broadleaved, summer rain
        8. winter-rain sclerophyllous
        9. tropical/subtropical needleleaved
        10. temperate/subpolar needleleaved
            a. evergreen giant forest
            b. evergreen with rounded crowns
                (1) with evergreen sclerophyllous understorey
                (2) without evergreen sclerophyllous understorey
            c. with conical crowns
            d. with cylindrical crowns (boreal)
        * e. on waterlogged soils
    B. Deciduous
        1. tropical/subtropical drought-deciduous
        2 cold-deciduous with evergreens
        3. cold-deciduous without evergreens
    C. Xeromorphic
        1. sclerophyllous
        2. thorn
        3. succulent
2. WOODLAND
3. SHRUBLAND
4. DWARF SCRUB AND RELATED COMMUNITIES
5. HERBACEOUS VEGETATION

Table 15.3   Abbreviated outline of the hierarchical land-use classification system developed for the digital land-use data base of Matthews (1983)

100 NOMADIC HUNTING OR HERDING
200 LIVESTOCK GRAZING
300 EXTENSIVE MIXED SUBSISTENCE FARMING
    310 Shifting cultivation of tropical crops
    320 Maize/beans dominant
    330 Grain dominant
    340 Tropical crops dominant
    350 Rice dominant
    351 with livestock
    352 with cotton
    353 with wheat
    354 with sugercane and mixed tropical crops
    355 with mixed crops and lumbering
    356 with maize
    357 with tea and jute
400 INTENSIVE SUBSISTENCE FARMING
    410 Irrigated rice dominant
    411 with mixed tropical crops
    412 with mixed tropical crops and livestock
    420 Wheat dominant
    430 Maize/beans dominant
500 PLANTATIONS
600 MEDITERRANEAN
700 LARGE-SCALE COMMERCIAL MIXED FARMING
800 SMALL-SCALE COMMERCIAL MIXED FARMING
    810 Cotton dominant
    820 Tobacco dominant
    830 Fruits and vegetables dominant
    840 Grain dominant
    850 Cropland and livestock dominant
    860 Wheat dominant, with livestock
    861 30-60% cropland with grazing
    862 with barley or oats
    863 with maize and barley or oats
    864 with maize and sugar beets
    865 with barley or oats or rye
    866 with maize and fruits or vines
    870 Maize dominant, with livestock
    880 Cotton dominant, with livestock
    890 Forestry with cropland or livestock
900 COMMERCIAL DAIRYING
950 FORESTRY/LUMBERING
998 URBAN
999 NO USE NOTED

major land-use systems, with selected subdivisions shown for extensive (300) and intensive (400) subsistence farming and small-scale commercial mixed farming (800). Globally, the twelve major divisions expand to 119 types elaborated with information on crop and livestock combinations.

Most Level 1 data bases are nominal data compiled in digital form from maps. Uncertainties arise from many areas including errors in source materials used in the compilations or in the translations of source legends to the classification systems used to record data. Uncertainties are difficult to evaluate and vary among the data sets; for example, vegetation data may be more reliable than soils data simply because vegetation is easier to observe, survey and map. However, vegetation is subject to human-induced modifications which may or may not be reflected in source information. Uncertainties also vary regionally. The soils data were compiled from a global series of FAO maps (FAO, 1971-1981) which are accompanied by maps indicating three levels of reliability classes; class I is most reliable and class III least reliable. Dudal (1978, cited in Gardiner, [1982]) presented areal percentages of regions in each reliability class indicating that 76% of Europe's area is class I and 0% is class III while only 7% of the African continent is considered well surveyed (class I) and 55% is covered by scanty field observations indicated by class III.

Uncertainties can also be introduced in the process of interpolation and/or gridding. The temperature and precipitation data sets of Shea (1986) were produced by gridding long-term average values at 2° × 2.5° resolution from an extensive set of station data. In this case, the level of uncertainty introduced in the interpolation of point data to surrounding regions is probably greater for the precipitation data than for the temperatures since temperature exhibits more coherent and predictable patterns over large areas than does precipitation (Hansen and Lebedeff, 1987).

Level 2 data bases are derived from Level 1 data bases in conjunction with supplemental information (Table 15.1). These data bases provide distributions of source categories identified by properties important for a particular trace gas; some of them also include a temporal component. The Level 2 series has a mix of data sets in nominal, interval, and ordinal form. The data base of natural wetlands was obtained from a selective integration of data sets on vegetation, soils and fractional inundation; the data base on inundation period is a series of twelve monthly data sets of inundation locations determined from the combination of temperature and precipitation data with climate thresholds for methane fluxes; and the series of data bases on domestic animal populations was derived by distributing FAO and other country statistics of individual animal populations over areas selected according to the land use and country data bases.

Level 3 data bases represent global distributions of trace gas emissions for individual sources. These data bases are derived from the integration of Level 1 and Level 2 data sets with emission characteristics. For example, the spatial and temporal distribution of methane emission from natural wetlands was obtained from the wetland data base using information on ecosystem fluxes and inundation periods; the distribution of methane emissions from domestic animals was obtained from application of data on methane emission per animal to the data bases of animal populations.

Uncertainties are associated with each data set and with each integration procedure. Brief mention of potential uncertainties in the data sets was made earlier. The largest uncertainty probably lies in the emission characteristics and in the relative contribution of individual sources because (1) there are large variations in the magnitude of emissions over small spatial scales in apparently similar environments, (2) most environments such as natural wetlands are not covered by measurements carried out throughout the growing season, and (3) there are no flux measurements at all for many environments and/or conditions. For example, published measurements of methane emissions from flooded rice fields have been restricted to a few field studies in temperate zone environments but over 90% the world's rice is grown in tropical and subtropical Asia. At present, these flux rates are applied to global rice harvest areas, sometimes with a simple temperature dependence (e.g., Holzapfel-Pschorn and Seiler, 1986). (The present volume contains the first published measurements of methane fluxes from rice fields in Japan [Yagi and Minami, this volume]). For natural wetlands, measured methane fluxes vary by orders of magnitude among ecosystems and seasons (e.g., Harriss et al., 1985; Barber et al., 1988; Whalen and Reeburgh, 1988). Although these fluxes have been correlated with temperature, moisture, nutrient input and organic matter accumulation, water and peat content, and vegetation characteristics by various authors, no general quantitative relationships have been found or are expected to be found that apply to all wetland environments.

At present, the topical extent of information available for Level 1 and Level 2 data bases is greater than that available for Level 3 data sets on trace gas emissions. For instance, we have more complete information about environmental and ecological characteristics of the global distribution of wetlands than we have available measurements of methane fluxes in representative ecosystems or information on quantified relationships between fluxes and the factors that modulate them in various ecosystems. This abundance of data on primary characteristics results in topical aggregation of the data sets on the simple levels before they are combined with supplementary information to produce the Level 3 data bases on emissions. As increased information on emission characteristics and the factors that control them become available from field studies, more of the complexity of the primary data bases may be employed.

15.3 METHANE EMISSION FROM NATURAL WETLANDS

A global data base of wetlands at 1° resolution was developed from the selected integration of three independent global digital data sources: (1) vegetation, (2) soil properties, and (3) fractional inundation in 1° cells (Matthews and Fung, 1987). The integration yielded a global distribution of wetland sites identified with in situ ecological and environmental characteristics.

The development of the wetlands data base involved using data bases with different strengths and weaknesses to compile the global distribution of wetland characteristics. Distinct wetland ecosystems were first identified in the vegetation data base of Matthews (1983) using descriptions from the UNESCO classification system; some examples of these wetlands are listed in Table 15.2. Twenty-eight wetland types, occupying about 700 1° cells globally, were identified by the vegetation data set. In Africa, these UNESCO wetlands (Figure 15.1A) identify major wetland features such as the internal Niger Delta, the Chad, Sud and Okavango Swamps, and the forested swamps of the Congo Basin. The distribution of ponded soils was derived from the multi-parameter soils data base of Zobler (1986) digitized from the global series of FAO maps at 1:5M scale (FAO, 1971-1981). On a global scale, close to 1300 1° cells are occupied by ponded phase soils; about 20% of these locations are coincident with the UNESCO wetland vegetation sites. The African distribution of ponded soils is shown in Figure 15.1B. The major wetlands of the vegetation map (Figure 15.1A) are identified. In addition, many smaller locations are targeted, such as the coastal swamps of Mozambique, South Africa and the Guineas, and swamps associated with Lake Chilwa, Lake Zwai, Lake Tana and the Nile Headwaters. Since spatial dominance was the criterion for recording vegetation and soils information for both of these data sets, a large number of small scattered wetland locations were under-represented. To capture these sites and to introduce intra-cell fractional inundation information into the procedure for calculating areas, we compiled a new 1° data base of fractional inundation derived from a global series of ~ 250 1:1M scale Operational Navigation Charts (ONC). The areal percentage of inundation in 1° cells was derived from the ONC's which indicate inundation with overprint symbols. The global wetland distribution derived from this source totalled ~ 2500 1° cells. About 40% of them corroborated wetland sites targeted by the vegetation and/or soils data bases but the remaining 60%, composed primarily of small wetlands, were uniquely identified with the ONC's. Figure 15.1C shows the distribution of ONC inundation sites in Africa, indicating fractional inundation values for the locations. Major wetland areas, coincident with those in Figure 15.1A and 15.1B, are generally associated with high inundation values, gradating into surrounding low-inundation sites identified by neither the vegetation nor the soils data bases. All locations targeted as wetlands by any of these three independent data sets were

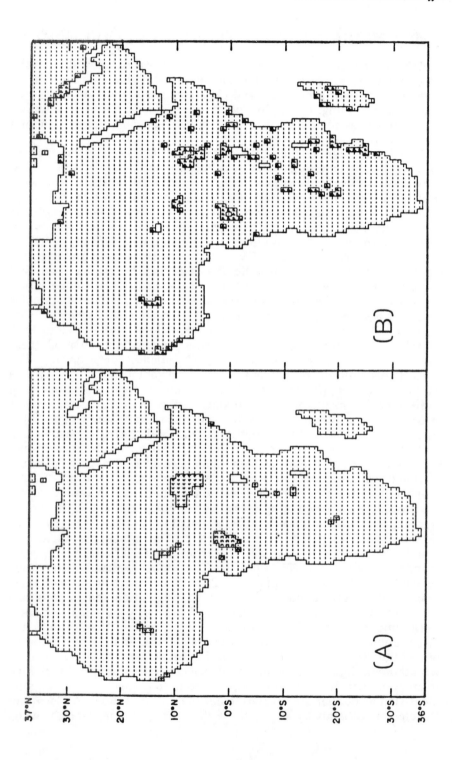

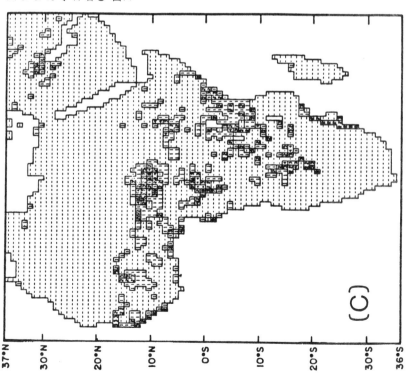

Figure 15.1 Distribution of African wetland sites derived from three independent data sources: (A) UNESCO wetland vegetation types from the vegetation data base of Matthews (1983). Symbols: 1, forested bog; 2, non-forested bog; 3, forested swamp; 4, non-forested swamp; 5, alluvial formations. (B) ponded soils from the soils data base of Zobler (1986) compiled from FAO maps. (C) inundation sites derived from Operational Navigation Charts (Matthews and Fung, 1987). Symbols indicate areal proportion of 1° cells that is inundated. Symbols: 1, 1-10%, 2, 11-20%, 3, 21-30%, ..., 9, 81-90%, *, 91-100%.

incorporated into the final wetlands data set. Integration of these data bases resulted in a geographic distribution totalling 3233 1° wetland locations associated with information on vegetation, soils, and inundation extent. These sites encompassed 130 vegetation types occupying over 100 soils with fractional inundation ranging from 1 to 100%.

As suggested above, the data compiled on the distribution and environmental characteristics of wetland ecosystems is more extensive than the array of ecosystems in which methane fluxes have been measured. We therefore 'expanded' the available measurements in several ways. First, we grouped the large number of wetland sites into five on the basis of vegetation structure and environmental characteristics that affect methane emission. These simple divisions were between organic-rich and organic-poor environments (bogs and swamps/alluvial formations, respectively) and vegetative cover (forested and non-forested) which partially indicates the amount of fresh material available for decomposition by methanogens. The groups were (1) forested bogs, (2) non-forested bogs, (3) forested swamps, (4) non-forested swamps, and (5) alluvial formations. This simplification in topical resolution reduced the data to a form approximating the topical resolution of the field measurements of methane fluxes in wetland ecosystems.

The global wetland area derived from this study is ~5.3 × $10^{12}m^2$. About one-half of the total lies between 50°-70°N, of which >95% is forested and non-forested peat-rich bogs. This high-latitude region is characterized by a relatively short thaw season resulting in highly seasonal emissions of methane. Approximately 35% of the global wetland area is broadly distributed in the latitude zone extending from 20°N to 30°S. This region is co-dominated by forested and non-forested swamps, with a minor contribution from alluvial formations. Methane emission in these low latitudes is governed by inundation periods that may last for 6 months or more of the year.

Methane emission was calculated using 'characteristic' methane fluxes for the major wetland groups along with simple assumptions about the duration of the methane production season. Fluxes ranged from 0.03 g CH_4 $m^{-2}d^{-1}$ for alluvial formations to 0.2 g $m^{-2}d^{-1}$ for forested and non-forested bogs; methane production seasons were estimated to be 180 days in the tropics and declined to 100 days poleward of 60°. This analysis gave an initial estimate of 110 Tg for the global annual emission of methane from natural wetlands (Matthews and Fung, 1987).

The data bases were later expanded for alternative and progressive investigations (Fung et al., 1989). Long-term monthly precipitation and temperature data, gridded to ~2° resolution from station data by Shea (1986) were used to estimate the duration of the inundation season: in high latitudes where temperature is understood to limit methane release, the inundation period of CH_4 production was assumed to be a function of freeze-thaw dynamics; in lower latitudes where moisture is presumed to be the limitation to methane flux, precipitation excess over potential evaporation was the determinant of the methane production season. A further investigation to modify the simple on/off mechanism for methane production in high latitude bogs involved a dependence of the magnitude and seasonality of methane flux on local temperature, giving a gradual increase and decline of the flux during the seasonal thaw season (Fung et al., 1989).

15.4 CONCLUDING REMARKS

Since terrestrial fluxes of trace gases, such as methane from wetlands or nitrous oxide from soils, are strongly influenced by local environment and ecology, series of compatible data bases such as those described here provide a framework for coherent and systematic estimates of trace gas emissions and for the study of the role of various sources in the global distribution of atmospheric trace gases. While the data bases can be used to estimate sources and sinks directly, they can also be used to provide information on potential major sources and therefore to define measurement strategies.

Presently, available information on some features of major sources of trace gases exceeds information on emission characteristics and the factors that control them. It is unlikely that the full complexity of some of these data bases will ever be used to estimate global emissions of some trace gases. Ecologically reasonable estimates of typical fluxes and their seasonality may be chosen to yield a range of plausible estimates of the global annual emission of various trace gases. However, refinement of the estimate of global emissions, particularly for biogenic sources, will be achieved primarily through a coherent program of long-term atmospheric, isotopic and field measurements covering representative sources and environments.

Additional data bases have been developed for investigating possible source distributions of other components of the methane cycle (Fung et al., 1989) and cycles of other trace gases such as nitrous oxide. Some of the primary and secondary data sets discussed here have been adapted for these investigations. The new data bases include information on methane emission associated with (1) municipal solid waste, (2) natural gas production, transmission and consumption, (3) production and consumption of coal, (4) tropical biomass burning and (5) termites. Work in progress includes data bases on global distributions of commercial nitrogenous fertilizer consumption and soil fertility for estimating emission of nitrous oxide.

These data bases can be used to evaluate the relative contribution of various sources of the methane budget in terms of magnitude, location and season, as well as to indicate major potential contributors within individual sources. Evaluation and synthesis of various scenarios of methane emissions from all of these individual sources, using a 3-D atmospheric chemical tracer model, has been carried out by Fung et al.

(1989). Constraints on the role of individual sources are strengthened by comparing geographic and seasonal aspects of model-simulated concentrations of atmospheric methane with measurements of atmospheric methane concentrations (Steele et al., 1987). Additional constraints on sources are based on measurements of stable isotopes and carbon-14 content of methane sources (Cicerone and Oremland, 1988; Wahlen et al., 1989; Fung et al., 1989).

ACKNOWLEDGMENTS

The author would like to thank I. Fung for helpful discussions and E. Devine for technical assistance.

REFERENCES

Barber, T.R., R.A. Burke Jr. and W.M. Sackett (1988) Diffusive flux of methane from warm wetlands. Global Biogeochemical Cycles 2:411-425.

Cicerone, R.J. and R.S. Oremland (1988) Biogeochemical aspects of atmospheric methane. Global Biogeochemical Cycles 2:299-327.

Dudal, R. (1978) Land resources for agricultural development. Paper presented at 11th International Society of Soil Science Congress, Edmonton, Alberta.

FAO (1971-1981) Soil Map of the World (1:5M scale). Volumes 1-10 and accompanying texts. UNESCO, Paris.

Fung, I., C.J. Tucker and K.C. Prentice (1987) Application of advanced very high resolution radiometer vegetation index to study atmosphere-biosphere exchange of CO_2. Journal of Geophysical Research 92(D3):2999-3015.

Fung, I. et al. (1989) Atmospheric methane: 1. Global budget derived from a 3-D model simulation. In preparation.

Gardiner, M.J. (1982) Use of regional and global soils data for climate modelling. In: P. S. Eagleson (Ed.), Land Surface Processes in General Circulation Models, p. 361-393. Cambridge University Press, Cambridge.

Gates, W.L. and A.B. Nelson (1975) A new tabulation of the Scripps topography on a 1° global grid. Part I. Terrain heights. Rep. R-1276-1-ARPA, Rand Corporation, Santa Monica, California, 132 pp.

Hansen, J.E., G. Russell, D. Rind, P. Stone, A. Lacis, S. Lebedeff, R. Ruedy and L. Travis (1983): Efficient three-dimensional global models for climate studies: Models I and II. Monthly Weather Review 111:609-662.

Harriss, R.C., E. Gorham, D.I. Sebacher, K.B. Bartlett and P.A. Flebbe (1985) Methane flux from northern peatlands. Nature 315:652-653.

Holzapfel-Pschorn, A. and W. Seiler (1986) Methane emission during a cultivation period from an Italian rice paddy. Journal of Geophysical Research 91:11803-11814.

Lerner, J., E. Matthews and I. Fung (1988) Methane emission from animals: a global high-resolution data base. Global Biogeochemical Cycles 2:139-156.

Matthews, E. (1983) Global vegetation and land use: New High-resolution data bases for climate studies. Journal of Climate and Applied Meteorology 22:474-487.

Matthews, E. and I. Fung (1987) Methane emission from natural wetlands: global distribution, area, and environmental characteristics of sources. Global Biogeochemical Cycles 1:61-86.

Shea, D.J. (1986) Climatological Atlas: 1950-1979, Surface Air Temperature, Precipitation, Sea-Level Pressure, and Sea-Surface Temperature (45°S-90°N). NCAR Technical

Note NCAR/TN-269 + STR. Atmospheric Analysis and Prediction Division, National Center for Atmospheric Research, Boulder, Colorado.

Steele, L.P., P.J. Fraser, R.A. Rasmussen, M.A.K. Khalil, T.J. Conway, A.J. Crawford, R.H. Gammon, K.A. Masarie and K.W., Thoning (1987) The global distribution of methane in the troposphere. Journal of Atmospheric Chemistry 5:125-171.

UNESCO (1973) International Classification and Mapping of Vegetation. Unesco, Paris.

Wahlen, M., N. Tanaka, R. Henry, B. Deck, J. Zeglen, J.S. Vogel, J. Southon, A. Shamesh, R. Fairbanks and W. Broecker (1989) Carbon-14 in methane sources and in atmospheric methane: the contribution from fossil carbon. Science 245:286-290.

Whalen, S.C. and W.S. Reeburgh (1988) A methane flux time series for tundra environments. Global Biogeochemical Cycles 2:399-409.

Zobler, L. (1986) A world soil file for global climate modelling. NASA Technical Memorandum 87802.

Part VI

Partitioning of Solar Energy

CHAPTER 16

The Effect of Land Use Change on Net Radiation and its Partitioning into Heat Fluxes

H.-J. BOLLE

Institut für Meteorologie, Freie Universität Berlin
Dietrich-Schäfer-Weg 6-10, 1000 Berlin 41, F.R. Germany

ABSTRACT

The interaction of the land-surfaces with climate is established by the exchange of heat, water, momentum and trace gases. In this paper the importance of the exchange of heat and the consequences of a change of radiative, thermal and hydrological properties of the land-surfaces is discussed. For Global Change studies reliable area averaged quantities are required rather than detailed information about local characteristics. These quantities are identified by considering the various interaction processes. A short overview is finally given of the research activities initiated within the framework of the World Climate Research Programme and the International Geosphere-Biosphere Programme to validate the parameterizations and informations about changes of land-surface characteristics that are based upon experimental data and observations from space.

16.1 INTRODUCTION

Climate interacts with the continental surfaces in two ways. The first is the impact that a change of global climate will have on the land-surfaces. Meteorological processes affect the growth of the diverse species of vegetation and their productivity. They are also responsible for hydrological soil properties, the tendency for erosion and in extreme cases desertification. Changes of these processes may also influence the use of the land by mankind. The second way is the effect that a change of the land-use may have on climate. Man is continuously altering the surface of the Earth by converting the primary natural vegetation into agricultural land which may be converted into a "second hand" natural vegetation on an as well changed soil if man looses interest in its use.

The interaction of the land-surfaces with the climate system is maintained by radiative exchange and the transfer of substances, in addition to energy and momentum between the surface and the atmosphere. With respect to the substances water and radiatively active gases like carbon dioxide, methane and nitrous dioxide are of primary

Soils and the Greenhouse Effect. Edited by A.F. Bouwman
© 1990 John Wiley & Sons Ltd.

interest. Interrelations exists between these substances, such as the link between water vapour and carbon dioxide via photosynthesis. In the context of this paper we shall, as far as the exchange of substances is concerned, deal exclusively with water. The flux of water vapour is at the same time also a flux of latent heat. The other important heat fluxes are those of sensible heat into the ground and into the atmosphere. Other energy fluxes such as that of the photosynthetic active radiation (PAR) into the plants or of kinetic energy into the ground are negligible with respect to the energy budget. However, the flux of momentum that accompanies the exchange of kinetic energy between the surface and the atmosphere is of importance. It is a key factor for the structure of the Planetary Boundary Layer (PBL) and therefore for the transport of heat away from the surface.

Which land-surface characteristics determine the fluxes of water vapour, heat and momentum?

The surface albedo α_s determines the amount of solar energy available for conversion into heat. It is the ratio of outgoing to incoming shortwave (300 - 2,500 nm) radiative flux. The net longwave radiative flux from the surface reduces the amount of radiative energy available for conversion into heat. Thermal properties of the soil are responsible for the heat flux into and the intermediate storage of heat in the soil. Soil moisture, the humidity structure of the lowest layer of the atmosphere and the presence of productive plants govern the partitioning of the available radiative energy into latent and sensible heat. This partitioning is important because the heating effects of the sensible and the latent heat are different. While the sensible heat is primarily heating the PBL, the sensible heat is only released in condensation processes that generally start at the top of the PBL and continue in the free troposphere. The final factor which is essential for the turbulent transport of heat into the planetary boundary layer is surface roughness.

The precise understanding and mathematical formulation of these processes at the scale of the climate models is a prerequisite to understanding climate and its modification by the action of man.

16.2 LAND-SURFACE PROPERTIES OF IMPORTANCE FOR THE NET RADIATION FLUX AT THE SURFACE

The albedo depends on the composition and moisture content of the top surface layer of the soil, the structure of the surface and its vegetation, the solar zenith angle and the ratio between direct and diffuse solar radiation. Its spectral distribution is a function of the wavelength dependent refractive index of the material. The most prominent feature of vegetated surfaces is the strong chlorophyll absorption at wavelengths shorter than 700 nm with a secondary minimum absorption in the green part of the spectrum. At longer wavelengths water absorption bands can eventually be present at 1.55, 1.96 and 3 μm as well as absorption features caused by minerals (Figure 16.1). In most cases the albedo, and its spectral distribution, is a mixture from soil and vegetation signatures. If a canopy is closed, but not optically thick, the reflectance of the soil will still have an effect on the radiation returning to space. There might even be an indirect influence in so far as the appearance of the same species of vegetation can be different depending on the quality of the underlying soil.

The variations in optical properties of soils is large. It ranges from dark lateritic soils e.g. in Africa with albedo values around 12% (Figure 16.2) and rich agricultural soils to bright desert sands with albedo values

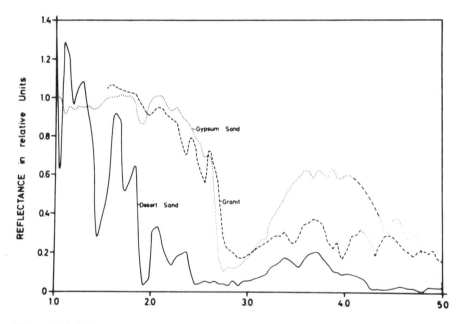

Figure 16.1 Reflectance spectra in the middle infrared of three different soil materials.

around 40% and more than 50% at wavelengths of 1500 nm. The albedo at short wavelengths is characteristic for the colour of the soil. A very white soil was found in Spain with nearly uniform reflectance values over a wide spectral range. The more vegetation covering the surface the deeper will be the absorption around 650 nm, until a minimum spectral albedo values are reached of less than 5% at 690 nm for crops in their period of most intensive growth (Figure 16.3). The total, or "meteorological" albedo, can be measured directly or alternatively computed from the spectral albedo by first multiplying each spectral value with the spectral irradiance, and then integrating over the wavelengths 300 – 2,500 nm finally dividing by the integral value of the irradiance.

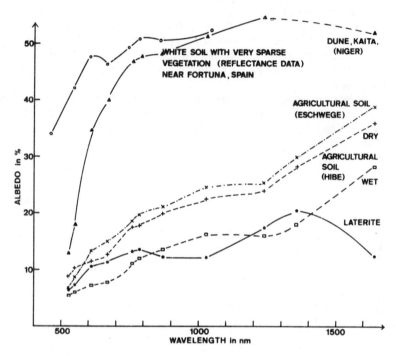

Figure 16.2   Spectral albedo of different soils.

The net radiation at the surface, M_{R^*}, is given by

$$M_R^* = M_{SW}^-(1 - \alpha_s) + M_{LW}^- - M_{LW}^+ \qquad (16.1)$$

where M is the flux area density (W m^{-2}), SW stands for shortwave, LW for longwave (thermal IR), - for downwelling and + for upwelling flux.

The downwelling longwave flux is a complicated function of the vertical distribution of temperature, water vapour, aerosols and clouds. The upwelling longwave radiation depends on the surface or "skin" temperature in the following way:

$$M_{LW}^+ = \varepsilon_s \sigma T^4 + (1 - \varepsilon_s)M_{LW}^- \qquad (16.2)$$

σ is the Stefan Boltzmann constant and ε_s the emissivity of the surface. This emissivity is close to one for all wet surfaces and for structured dark appearing surfaces that act as traps for longwave radiation. ε_s is less than one for most dry soils with low organic matter and high mineral content. The reason is the excitation of vibrational bands in the crystals of minerals causing so-called "Reststrahlenbanden" with high reflectance in well defined spectral regions just near 10 μm where the maximum of the emission occurs for normal surface temperatures (Figure 16.4). If the emissivity is less than one, the surface emits less

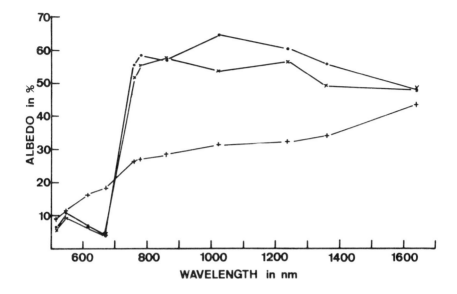

Figure 16.3 Spectral albedo of agricultural vegetation. The measurement is from the Hildesheimer Börde, Federal Republic of Germany. Dots: green fertilization, crosses: cabbage, + - signs: nearly bare soil with very little plants. Date: 7.9.1988, 16:10 MESZ.

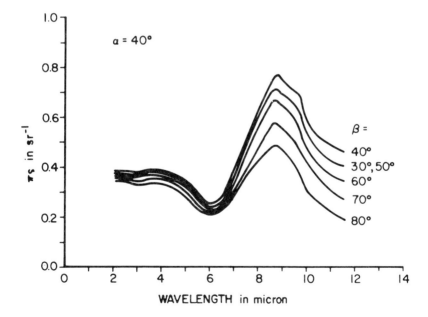

Figure 16.4 Reflectance (multiplied by π) of white quartz sand from the Negev for an incidence angle of 40° and different reflectance angles. The figure shows the strong "Reststrahlen" reflection of quartz.

than the maximal possible flux but reflects also a fraction of the downwelling sky radiation, which partly compensates for the reduced emission. Consequently the net longwave radiative loss becomes smaller than expected for the thermodynamic temperature of the surface during daytime. This effect is less pronounced when at night the surface cools down substantially below the air temperature and the reflected fraction becomes relatively higher.

Complete sets of radiative fluxes are shown in Figure 16.5. One example is for Niger, Africa, the other one from the German test area Hildesheimer Börde. Under the dry African conditions with high albedo and high irradiance the net shortwave flux is of the same magnitude as in Germany but the net outgoing longwave flux is larger than in mid latitudes. This is caused by the higher surface temperatures determined by the various soil properties and notwithstanding the lower emittance. The result is a much smaller total net flux in Niger as compared with the higher and wetter latitudes.

The temperature to which the surface rises under the impact of the net radiation flux depends on the magnitude of the heat fluxes that dissipate from the surface. These in turn depend on the gradients that build up between on the one hand the surface, and the atmosphere, the surface and adjacent soil layers on the other hand.

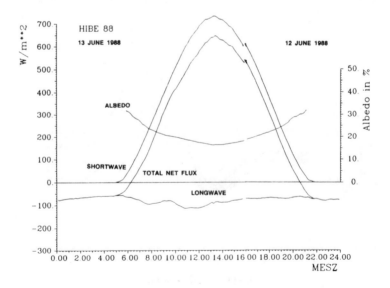

Figure 16.5a Radiation fluxes at the surface: Radiation fluxes measured over agricultural land in the Hildesheimer Börde, Federal Republic of Germany. Composite of the cloudless hours of two days.

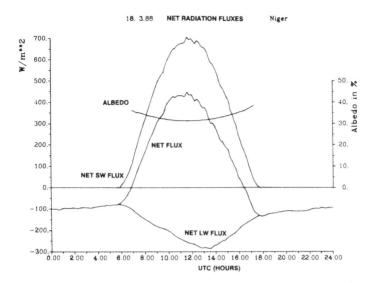

Figure 16.5b   Radiation fluxes at the surface: Radiation fluxes measured over dry grassland in Ibecetene, Niger. Sky covered with very thin cirrus.

16.3 THE HEAT FLUX INTO THE SOIL

The temperature at the surface can be measured by satellites for large areas. Its diurnal variation can be expressed by a Fourier expansion:

$$T(0;t) = T_0 + \Sigma \, \Delta_n T \, \exp(\sin \omega[t - t_0]) \qquad (16.3)$$

In a first approximation the forcing function can be regarded as sinusoidal with an amplitude $\Delta_0 T$:

$$T(0;t;1) = T_0 + \Delta_0 T \, \sin(\omega[t - t_0]) \qquad (16.4)$$

The heat flux into the soil is conductive. It depends on the value of its thermal conductivity λ (which may include other mechanisms such as water movement),

$$M_G = \lambda dT/dz. \qquad (16.5)$$

The divergence of this heat flux or, in case of a horizontally homogenous soil, its vertical gradient, equals the rate at which the soil is heated. This can be expressed for constant thermal conductivity by

$$dT/dt = (1/C_s) \, dM/dz = D_T(d^2T/dz^2) \qquad (16.6)$$

and depends on the thermal diffusivity $D_T = \lambda/C_s$. $C_s = c_s\rho_s =$ heat capacity. C_s is a function of the volume fractions of minerals (F_m), organic material (F_o) and water content (F_w): $C_s \approx (1.94 \, F_m + 2.5 \, F_o + 4.19 \, F_w)10^6$

J m^{-3} K^{-1} (Brutsaert, 1984). The density varies between 1.3 kg m^{-3} for organic material and 2.6 kg m^{-3} for minerals. Average values for C$_s$ are 2.0 - 2.5×10^6 J m^{-3} K^{-1}.

The solution of equation (16.6) results in an exponential decline of the temperature amplitude with depth:

$$T(z) = T(0) \exp(-\delta z), \qquad (16.7)$$

$$\delta = (\omega \, c_s \, \rho_s / 2\lambda)^{1/2} = (\omega / 2D_T)^{1/2} = I/(\lambda 2^{1/2})$$

and a heat flux

$$M_G(z;t) = \Delta_0 T \exp(-\delta z) \, I \sin(\omega[t-t_0] + \pi/4 - \delta z) \qquad (16.8)$$

[I = $(\lambda c_s \, \rho_s \, \omega)^{1/2}$ = "thermal inertia"], and a phase shift of -δz. The reciprocal value of δ is the depth at which the temperature wave is reduced to 1/e of its value at the surface.

The soil characteristics δ and D_T respectively C$_s$ and λ can be obtained from the decline of the temperature and flux amplitudes with depth, the phase shifts, and the ratio between flux and temperature amplitude. Another possibility is the physical analysis of soil samples in the laboratory. Both methods have their advantages and drawbacks. While in the field temperature and particularly flux measurements become problematic in inhomogeneous soils, the laboratory measurements are often destructive for the soil sample.

The energy stored during the day in the soil is diverted from the heat flux that would otherwise have heated the atmosphere and caused convection. This energy is released to the atmosphere during the night, when it counteracts the cooling of the atmosphere, the reduction of the PBL height and the production of dew.

16.4 RESISTANCES THAT CONTROL THE HEAT FLUXES INTO THE ATMOSPHERE

A flux into the atmosphere can only develop if a gradient (or a difference in the respective potential) is built up of the quantity to be transported. In the case of sensible heat this is a gradient of the potential temperature and in the case of the latent heat flux a gradient of the specific moisture. Except for a very thin layer around each obstacle the transport from one atmospheric level to the other is of turbulent nature and can be caused by free or forced convection. Free convection develops in still air if density gradients are present while wind enforces the turbulent exchange. In the case of evaporation and latent heat the necessary gradient of the specific moisture will only build up above the surface if water and the energy for its evaporation are available. If water is present at the surface a maximum possible amount determined by the net radiation flux will be evaporated (potential evaporation). If present in deeper layers of the soil it may be evaporated by the intruding heat wave and the vapour travels through the pores and other micro-pathways to the surface. From still lower layers water may be extracted by deep rooting plants and then released through the cuticula and, more effectively, through the leaf stomata. At the location of the water one can always expect saturated air with a strictly temperature regulated water vapour pressure $e_s(T)$. Between this saturated zone and the free boundary layer the water vapour has to pass through different "resistances" such as the stomata resistance r_{st}, the resistance r_i of the interfacial sublayer (or canopy-air resistance) where conduction still plays a role and the resistance r_a from the top of the canopy to a reference layer in the surface sublayer of the PBL. In parallel to this way through the plants there is (except for very dry conditions) always the direct way from the soil to the atmosphere with a sequence of different resistances. For climate application it is important to find a formulation for the area averaged fluxes that is compatible with the grid sizes of climate models. In practice water vapour routes are often combined to a "canopy" resistance r_c and an aerodynamic resistance r_a or the stomata resistance r_{st} and a resistance r_{av} from the leaf to the surface sublayer. The latent heat flux into the atmosphere can then be represented by

$$M_{LH} = \rho L(q_s^{\cdot} - q_r)/(r_{st} + r_{av}) \qquad (16.9)$$

(L is the latent heat of vaporization) and the sensible heat flux by

$$M_{SH} = \rho c_p(\Theta_0 - \Theta_r)(1/r_a) \qquad (16.10)$$

where in the most simple form

$$(1/r_a) = k^2 u(z)/\ln\{(z - d)/z_0\} \qquad (16.11)$$

with k = von Karman constant (≈ 0.4), u(z) = wind velocity at a reference height z, d = displacement height of zero wind level, z_0 = roughness length.

The latent and the sensible heat fluxes can furthermore be related to each other by the Bowen Ratio which is, under the assumption that $K_{SH} = K_{LH}$ given by:

$$\text{ß} = M_{SH}/M_{LH} = (c_p/L)[(\Theta_0 - \Theta_r)/(q_0 - q_r)][(r_{st} + r_{av})/r_a] \qquad (16.12)$$

16.5 INTERACTION WITH CHANGING CLIMATE

Climatic parameters influencing the processes at the surface are cloudiness - with its effect on the radiation regime - and precipitation - that affects the hydrological regime, the Bowen ratio as well as the albedo. The absolute value of the temperature is often of secondary importance for the processes at the land surfaces because only temperature gradients determine the magnitude of the fluxes. The absolute temperature may, however, affect the vegetation and this would indirectly have impact on the partitioning of the solar energy into the heat fluxes. The stomata resistances are also temperature dependent.

The reduction of the solar input caused by an increase in surface albedo and as a consequence the reduction of the sensible heat flux from a dry surface causes a relative cooling inside the PBL where the sensible heat is deposited, and as a consequence sinking aloft and moisture divergence at the surface which suppress moist convection and reduces local rainfall (Sud and Molod, 1988).

Water often invades the continents from the sea with humid air forced upward over the continents due to heating or by mountains providing precipitation in coastal zones. As it enters central parts of the continents the air would dry out if not new water vapour is provided from the surface with its vegetation, which is partly intercepted rainwater but partly also stems from underground aquifer systems. If this vegetation is removed or changed the supply of water to the atmosphere is also reduced and in lee of this zone the precipitation will change. This may lead to a change in the local climate, a drying out of such areas or in a final stage to desertification. If at the same time the precipitation falls at high intensities, the sudden impact of rain on an area of reduced vegetation leads to erosion. This can manifest itself in a lack of minerals or even structural changes of the land surfaces. During long dry periods the land is in addition less protected against wind erosion.

From a moist bare surface water is evaporated before a significant rise of the surface temperature takes place. Wet surfaces generally have lower albedo as in their dry state (Figure 16.2). Consequently more shortwave radiative energy is absorbed and converted into heat. Because most of this energy is latent heat, the surface temperature is not rising as high as it would under dry conditions. Therefore also the outgoing longwave flux is reduced and even more energy must leave the surface as sensible heat than under dry conditions. This albedo effect may even be enhanced if vegetation with its low shortwave reflectance is present. However, the opposite can also be true, especially for dark soils, because of the high near infrared reflectance of green and productive plants. If wet soil is covered by (dense) vegetation, this vegetation also increases the area for evaporation and transpiration, thus enhancing the latent heat flux. At the same time the canopy temperature rises above the air temperature resulting from the biological activity, which increases also the sensible heat flux on the account of the soil heat flux. Moist soil also has a higher heat capacity than dry soil material with the consequence that during the summer more heat will be accumulated. This compensates for the reduced incoming radiation in autumn. In a wetter area the seasonal amplitude is therefore less pronounced due to the storage of heat in the early summer and its release in fall.

In addition one has to take into account the already indicated feedbacks of changing climate on the action of man or, starting with the action of man, its influence on the retardation or enhancement of the environmental change and the consequences for climate. Reduced precipitation may initially increase the demand for irrigation water, which for some time provides for additional evaporation but draws on the water reserves and finally causes a drop of the ground water level.

In dry climates certain plants with Crassulacean Acid Metabolism (CAM plants) favourably grow. These CAM plants inhale CO_2 during the night and close their stomata during daytime to reduce transpiration. The CO_2 stored during the night is fixed during the day when photosynthetically active radiation (PAR) is available. If plants, which cycle the whole photosynthesis during daytime (C_3 and C_4 plants), are replaced by CAM plants, the evaporation would be reduced.

The change of vegetation at the continents does also have an effect on the roughness of the surface. The roughness enters the equations of the heat fluxes into the atmosphere by the z_0 value (equation 16.4). This value determines the height of the boundary layer and the transfer of momentum between the surface and the atmosphere. This momentum transfer is important for the tropospheric wind system and thus for the general circulation. A decrease of the bulk aerodynamic transport resistance of the PBL by a smoother surface would reduce all heat fluxes. In order to maintain the rate of heat transport that is required for an equal amount of net radiative flux it would tend to increase the temperature and moisture gradients above the surface. At the same time the wind stress is reduced and windspeed increased. This produces the moisture convergence in the PBL into a surface low which, in turn, reduces moist convection and rainfall.

A final aspect is the effect of a large scale albedo change on the global climate. This can be estimated by a very simple radiative climate model that assumes radiative equilibrium at the top of the atmosphere: the outgoing thermal radiation equals the solar radiation absorbed in the Earth's system determined by the planetary albedo α_p [eq. (16.1) in Figure 16.6]. If one introduces for the longwave fluxes equivalent emission temperatures and emissivities (" stands for tropospheric, ' for middle atmospheric quantities) it is possible to express via eq. (16.2)-(16.4) in Figure 16.6 the surface temperature by the effective planetary emission temperature [which is defined by the absorbed solar radiation, eq. (16.1)] and inter alia by the surface albedo. If eq. (16.6) in Figure 16.6 is differentiated with respect to the fraction $A_e/4\pi R^2$ of the surface albedo that represents cloudfree land, it results

$$dT/d\alpha_{s,1} = - \tau^*/(1 - \alpha_p)(T_{eff}/T)^4(1 - \alpha_{s,1})(A_e/4\pi R^2)(T/4) \qquad (16.19)$$

Inserting the values for the present climate in eq. (16.19) yields approximately

$$dT_o = - 0.8 \, d\alpha_{s,1}/\alpha_{s,1} \qquad (16.20)$$

With an average land-surface albedo of 0.25 the global temperature effect of an albedo change of +1% would be only - 0.032 K. Only a 15% albedo increase would have an effect on the temperature of the same order of magnitude with opposite sign as the 70 ppm-v increase of CO_2 within the last 100 years.

ENERGY FLUXES IN A SIMPLE RADIATION-TRANSFER MODEL

TOP OF THE ATMOSPHERE:

(1) $\quad T_{eff} = \sqrt[4]{\frac{1}{4\sigma} S_0 (1-\rho_p)} = \sigma \, T_0^4 \, \tau_L \cdot \sigma T''^4 \epsilon'' \tau_L' \cdot \sigma T'^4 \epsilon'$

AT THE TROPOPAUSE:

(2) $\quad -\frac{1}{4} S_0 \tau_K' \, (1-\rho_t) - \epsilon' \sigma T'^4 \cdot \sigma T_0^4 \tau_L'' \cdot \sigma \epsilon'' T''^4 = 0$

AT THE EARTH-SURFACE:

(3) $\quad -\frac{1}{4} S_0 \tau_K (1-\rho_0) - \epsilon' \sigma T'^4 \, \tau_L'' - \epsilon'' \sigma T''^4 \cdot \sigma T_0^4 \cdot \Phi_H = 0$

ROUGH APPROXIMATION OF HEAT FLUXES:

(4) $\quad \Phi_H = \lambda \frac{S_0}{4} \, \tau_K \, (1-\rho_0)$

FROM (2) AND (3) FOLLOWS WITH (1) AND (4):

(5) $\quad T_0 = T_{eff} \sqrt[4]{\left[\tau_K \frac{1-\rho_0}{1-\rho_p} (1-\lambda) \cdot \tau_K' \frac{1-\rho_t}{1-\rho_p} \right] \frac{1}{1+\tau_L''} \cdot \epsilon' \frac{T'^4}{T_{eff}^4}}$

NEGLECTION OF STRATOSPHERE/MESOSPHERE

$(\tau' = 1, \ \epsilon' = 0)$ **RESULTS IN:**

(6) $\quad T_0 = T_{eff} \sqrt[4]{\left[\tau_K \frac{1-\rho_0}{1-\rho_p} (1-\lambda) \cdot 1 \right] \frac{1}{1+\tau_L''}}$

$$\left(\tau = \tau' \tau'' \quad K = \text{shortwave} \quad L = \text{longwave} \right)$$

Figure 16.6 Basic equations of a simple radiative climate model.

16.6 CONCLUSION

The coupling between the soil, the vegetation and the atmosphere by exchange of water, heat and momentum under the influence of the solar radiation, precipitation and wind is well understood at the small scale in uniform terrain. It is much more difficult to obtain large area averages of the quantities that determine these interactions and information about their changes as needed for global change studies. The individual micro-processes have for this purpose to be averaged and parameterized for areas in the order of $10^4 - 10^5$ km². This seems at present to be feasible only if continuous observations from space can be used to obtain information about parameters that are directly or indirectly related to these quantities and the fluxes. Though changes of land-surface characteristics such as albedo will probably not have effects on global climate that are comparable to the current increase of the concentrations of greenhouse gases, their regional impact e.g. on the water cycle can be considerable.

The following quantities averaged over 10^4 to 10^5 km² and representative for major ecosystems are in this respect of primary importance.

- Surface albedo and its seasonal variability with vegetation and soil moisture.
- Surface emissivity in the longwave infrared part of the spectrum.
- Thermal properties of the soil and their variability with soil moisture:
  - thermal conductivity
  - volumetric heat capacity (or specific heat and density separately) or the combinations of these quantities that appear in eq. (16.6) to (16.8) as thermal diffusivity, thermal inertia or the depth δ at which the temperature amplitude is reduced to $1/e$ of the surface value.
- Resistances for the transport of heat into the atmosphere for at least the major ecosystems in dependence of the wind velocity at a reference height and in the case of the area averaged effective stomatal resistances on irradiance, temperature and humidity.
- Surface roughness, which has a large effect on the structure of the PBL.

Of secondary but long term influence on the exchange of energy between the surface and the atmosphere are changes of the soil quality and mineral content as a consequence of e. g. changes of fertilization and irrigation practices. There may on a longer term affect the vegetation cover or lead to erosion and desertification.

It is clear that the determination of surface characteristics needed for climate studies requires a global observation system in which observations from space play an important role. A set of experiments will be necessary to improve the understanding of the interaction between the surface and climate at scales of 100×100 km², and to validate models and parameterizations that describe these interactions, as well as algorithms needed for the interpretation of satellite data. To improve the use of satellite data for this purpose - which is completely different from the thematic interpretation for e. g. agricultural purposes - the International Satellite Land-surface Climatology Project was established under the sponsorship of UNEP, COSPAR, IAMAP and different national bodies. Field experiments to validate models and algorithms are ongoing in a number of countries and are now being planned in the boreal forests of Canada, the tropical forests of Brazil, and along a transect in Australia. The Commission of the European Communities consider within its EPOCH programme an international project called European Climatic Hydrological Interaction between Vegetation, Atmosphere and Land-surfaces (ECHIVAL) with a first large experiment in Spain.

ACKNOWLEDGEMENT

Studies reported here are sponsored by the Commission of the European Communities (Contracts EV4C-0012-D and ST 2 J 0222-1-D) and the Bundesministerium für Forschung und Technologie (Contract LOF 32/87). In the experiments participated a group of colleagues of the Institut für Meteorologie der Freien Universität Berlin under the leadership of Dipl. Met. M. Eckardt. The drawings are made by Mrs. S. Fleischer and part of the photographic work by Mr. W. Müller.

REFERENCES

Brutsaert, W. (1984) Evaporation into the Atmosphere, D. Reidel Publ. Comp., Dordrecht.
Sud, Y.C. and A. Molod (1988) A GCM Simulation Study on the influence of Saharan Evapotranspiration and Surface-Albedo Anomalies on July Circulation and Rainfall. Monthly Weather Review 116:2388-2400.

CHAPTER 17

Quantification of Regional Dry and Wet Canopy Evaporation

A. HENDERSON-SELLERS and A.J. PITMAN

School of Earth Sciences, Macquarie University
North Ryde, New South Wales 2109, Australia

ABSTRACT

A review is undertaken of the current status of the land-surface parameterization schemes available for incorporation into global climate models (GCMs). Particular attention is focused upon one type of vegetation, the tropical moist forest, and the determination of the amount of rainfall intercepted by the canopy which is re-evaporated. Intercomparisons show that other poorly simulated components of the GCM impact the land-surface climate to at least as large an extent as the choice of land-surface parameterization scheme. In particular, it has been shown that incoming solar radiation is very much too high during rain events in one particular GCM and the spatial and temporal distribution of rainfall is rather poorly simulated. In stand-alone sensitivity tests, it has also been shown that arbitrarily selected precipitation regimes can lead to highly misleading evaluations of interception loss.

Thus, the first generation of land-surface models to incorporate a canopy seem to warrant more detailed testing, but the prescribed conditions used to evaluate the models must first be subjected to rigorous examination.

17.1 INTRODUCTION

The energy budget of the continental surface, and hence the land-surface climate, is controlled by the net surface radiation and the partitioning of available energy between latent and sensible heat fluxes. In order to try to predict changes due to, for example, deforestation, desertification or greenhouse warming we must first be able to simulate this partitioning and hence we must develop methods of evaluating regional-scale evaporation. Global climate models (GCMs), in particular, require adequate means of representing regional-scale evaporation.

At present there is considerable effort focused upon improving the land-surface parameterization schemes which are designed to offer a quantified evaluation of regional-scale evaporation. It is now acknowledged that transpiration is controlled by the soil moisture deficit on monthly to seasonal timescales and by the atmospheric vapour pressure deficit on diurnal timescales (e.g. Avissar et al., 1985).

Soils and the Greenhouse Effect. Edited by A.F. Bouwman
© 1990 John Wiley & Sons Ltd.

The current status of models of transpiration from vegetation was adequately reviewed by Jarvis and McNaughton (1986) and their treatment has since been elaborated upon by Paw U and Gao (1988) and Martin (1989). The deceptively simple problem of predicting re-evaporation of rainfall from a fully or partially wetted canopy has received somewhat less attention from climate modellers.

In this paper we review the current status of these land-surface schemes concentrating particularly on one type of vegetation and status: evaporation from the wetted canopy in moist tropical forest.

17.2 REPRESENTING THE LAND SURFACE IN GCMS

In early GCMs, the land surface was simply a reflector of solar radiation and an emitter of infrared radiation. The first hydrological parameterization scheme for GCMs, that of Manabe (1969), has been termed the 'bucket' model. In this highly simplified scheme, the 'soil' had a 'field capacity' of 15 cm. The bucket filled with water when precipitation exceeded evaporation, and after the bucket became full the overflow was runoff. Evapotranspiration occurred in these models at its potential rate when the soil was at, or close to, saturation. When soil moisture dropped below some critical value, the evapotranspiration was proportional to potential evapotranspiration, the proportionality factor being set equal to the ratio of the current soil moisture to the critical soil moisture.

Hunt (1985) reviewed single- and multiple-layer soil parameterization schemes for climate modelling, considering in particular their applicability to drought-prone semi-arid regions. He compared the schemes of Manabe (1969) with that of Hansen et al. (1983) for mid-latitudes and with that of Deardorff (1978), assuming bare soil and a diurnal mean solar radiation. He found the Deardorff model to have a faster initial response to drying conditions followed by a slow release of deeper water, from which he inferred it to be superior to the other two models.

The empirical basis for the bucket parameterization is diurnally averaged data. It is therefore most useful in GCMs that use diurnally averaged solar heating, but it may be inappropriate for climate models that include a diurnal cycle of solar radiation at the surface. Once the diurnal cycle is included, it is necessary to include realistic descriptions of the sensible and latent heat fluxes. Recognition of this requirement prompted Dickinson (1984) to develop soil-water and canopy energy-balance parameterizations. The treatment of soil water is inferred from a much more detailed soil model and includes the diffusion limitation of evaporation that often occurs around midday.

The treatment of the canopy energy and moisture balance includes: (i) interception of precipitation by vegetation and subsequent evaporative loss and leaf drip; (ii) moisture uptake by plant roots, distributed between the upper and full soil columns; and (iii) stomatal resistance to transpiration. This land-surface scheme, referred to as the 'biosphere-atmosphere transfer scheme' (BATS), in common with the 'simple biospheric model' (SiB) of Sellers et al. (1986) and 'bare essentials of surface transfer' (BEST) of Pitman (1988), can represent a wide range of vegetation-soil coupled systems by selection of the appropriate land-cover and soil-description class (e.g. Dickinson et al., 1986; Wilson et al., 1987).

BATS, SiB and BEST generalize the aerodynamic transfer formulations usually used in GCMs to allow for the presence of multiple surfaces, i.e. canopy and soil, that are considered separately in the energy balance calculation. Aerodynamic drag coefficients are calculated for the canopy according to mixed-layer theory, as a function of the reference height at which atmospheric variables are available, roughness length and stability. For describing the transfer of heat and moisture from the multiple surfaces, additional resistances are calculated on the basis of the structure and biophysiology of the canopies.

BEST is similar conceptually to Deardorff's (1978) canopy model and Dickinson et al.'s (1986) soil model. However, it incorporates a more advanced soil hydrology parameterization and an improved calculation of canopy temperature. Considerable care has been taken to include each element of the land-surface parameterization to an appropriate level of complexity, while some of the elements of the earlier land-surface schemes which could not be supported by observational data for all ecotypes were removed (see Pitman, 1988).

These three schemes, BATS, SiB and BEST, include an explicit canopy parameterization and have been incorporated into one of more GCMs and used, in this coupled mode, for sustained (i.e. longer than a few weeks) climatic integrations. Other, somewhat similar, land-surface schemes (e.g. Abramopoulos et al., 1988) have yet to be examined in anything but a stand-alone mode.

Dickinson and Henderson-Sellers (1988) undertook a detailed set of intercomparisons of a two-layer, bucket scheme and the BATS parameterization. The BATS and the two-layer, bucket models were run to steady state, assuming that showers occurred every third day, beginning at 00, 06, 12 and 18 hours, local time, and lasting for 0.5 h with instantaneous rainfall rates of 0.003 mm s-1, corresponding to 5.4 mm per shower and ~220 mm month^{-1}, that is, relatively dry conditions for a tropical forest. The most notable difference was that the BATS model gives significantly larger sensible heat fluxes and smaller latent heat fluxes. The peak afternoon dry day differences in both fluxes are about 150 W m^{-2}, and the mean difference in latent heat flux averaged over wet and dry days is about 22 W m^{-2} or about 0.4 mm d^{-1} of evapotranspiration. The fluxes from BATS appear more realistic than those from the two-layer, bucket model, in comparison with those observed by Shuttleworth et al. (1984a). The relatively low dry day evapotranspiration of the BATS model that results from its canopy resistance to water flux is compensated, in part, by increased interception loss during the rainy days. The mean difference in evapotranspiration is compensated by less runoff in the two-layer, bucket model.

During the rainy day, both models give similar flux patterns except that during rain the latent fluxes of the BATS model, as a result of interception loss, exceed somewhat the net radiative heating, whereas the two-layer, bucket model exhibits a reduced latent flux in response to the assumed reduction in solar heating. This large evaporative flux in the BATS model is compensated by negative fluxes of sensible heat. Shuttleworth et al. (1985) show that total evaporation from a fully-wetted canopy in Amazonia often exceeds 'potential evaporation', the additional energy being obtained from air advected into the wetted area from adjacent dry canopy forest.

17.3 GCM SIMULATIONS OF TROPICAL MOIST FOREST

Traditionally, GCMs have been validated in terms of temperatures and winds away from the surface, radiative fluxes at the top of the atmosphere and broad annual average patterns of precipitation. Surface conditions, except for sea level pressure patterns, have largely been ignored in validation of model outputs for two reasons: (a) lack of data and (b) concern for subgrid scale biases in the data. However, validation against observed surface conditions and processes is crucial to studying the climatic role of soils and vegetation. Observations from the Reserva Florestal Ducke, 25 km from Manaus, Amazonas, Brazil taken between September 1983 and August 1985 are described by Shuttleworth (1988), Shuttleworth et al. (1984a; 1984b) and Lloyd et al. (1988).

Dickinson (1988) undertook a comparison between two versions of the NCAR Community Climate Model (CCM)'s representation of the Amazonian forest and these observations. He found that the transpiration term in the two model simulations differ more from each other than from the observations of Shuttleworth (1988). These differences seem to result from the large difference in simulated rainfall over the Amazon region in the two CCM simulations. The more recent version of the CCM has a greatly lowered transpiration suggesting that the vegetation is water stressed for much of the year. Such stress could be the result of the low rainfall and excess radiation but also may in part result from an inadequate soil model, both in texture as modified by biological processes, and in the depth to which forest roots can access soil moisture.

It was also clear from this comparison with observations that there are several features of the BATS canopy model that may significantly exaggerate the amount of rainfall intercepted. An important term is the water holding capacity of the foliage; BATS has a prescribed capacity of 0.2 mm per unit leaf area. The modelled tropical forest has a leaf area index of 6 and assumes 0.9 of a grid square covered by vegetation giving 1.08 mm total water capacity compared to the 0.74 mm reported by Shuttleworth (1988) for the Amazon site. A potentially more serious source of model error is the neglect of spatial variability of intensity and timing of rainfall events which effectively greatly reduces the water holding capacity of the foliage over a model grid square (Lloyd et al., 1988).

The NCAR CCM formulation for convective rainfall assumes a maximum of 30% cloud cover to allow for the spatial variability of convective processes insofar as they affect clouds, but this cloud prescription has the undesirable side effect of providing at least 70% clear sky (sunny) conditions during tropical rainfall and hence there is an exaggerated supply of radiative energy to drive interception. Dickinson's (1988) comparison of model and observed interception indeed showed a large model overestimate of interception, with the largest values of the earlier version of the CCM being nearly twice that observed, and those of the later version being up to 50% too large. This model bias is not out of line with the model bias towards excess solar radiation, in general, and excess radiation during rain events, in particular. Thus, the effect of the error in effective water holding capacity of foliage, although probably present, is not clearly distinguishable in this comparison.

Dickinson (1988) also compared the simulated canopy transpiration with observed values. He found that the earlier version of the CCM produced transpirations which exceeded those observed over most months and by up to 50 mm during the peak wet season but which were in good agreement with Shuttleworth's (1988) data during the dry season. By contrast, the later version of the CCM gives good simulation of transpiration during the wet season but values too low by up to 50 mm during the dry season.

17.4 DEPENDENCE OF EVAPORATION ON THE RAINFALL REGIME

The durations of rain storms and the time of day at which they occur are important factors in a description of the land-surface climate. However detailed information about storm frequency and intensity, especially in tropical forests, is rarely available. Lloyd et al. (1988) describe a 2-year rainfall climatology for the Ducke Reserva and, based on these data, Lloyd (1989) has evaluated the importance of prescribed rainfall patterns on computed values of interception loss.

In the absence of better information, Dickinson and Henderson-Sellers (1988) were obliged to prescribe precipitation for their stand-alone comparison by assuming that four showers of 5.4 mm and lasting 0.5 h occurred every third day, beginning at 00, 06, 12 and 18 hours local time. Analysis of the rainfall data from the Amazon for the two years 1983-1985 by Lloyd et al. (1988) showed that the rainfall rate for hours with rainfall greater than 0.5 mm was 5.15 mm h^{-1}, implying an average of 9.27 mm for the average 1.8 hour daily storm. These results led Lloyd (1989) to propose a simple but reasonably realistic pattern of the rainfall climate which is to assume one storm of 2 hours duration occurring between 1300 and 1500 hours, on three out of every five days, at a rate of 5.15 mm h^{-1}.

Lloyd (1989) evaluates the interception loss using an analytical model (Gash, 1979 and Gash et al., 1980) which is a storm based simplification of the earlier Rutter model (Rutter et al., 1971; 1975). Interception loss is calculated on the basis of storm rainfall pattern, average rates of rainfall and evaporation during rainfall, and a knowledge of the forest canopy structure parameters. This model (run over a period of a year for each of the two rainfall regimes and using forest structure parameters obtained by Lloyd et al., 1988) produces very different estimates of the rainfall interception loss. The prescribed rainfall scheme of Dickinson and Henderson-Sellers (1988) produced 482 mm of interception loss from a modelled rainfall input of 2628 mm (i.e. 18.3% of rainfall). The Lloyd rainfall on the other hand produced 285 mm interception loss from an input of 2377 mm of rainfall (i.e. 12.0% of rainfall). The measured interception loss by Lloyd et al. (1988) was 8.9 ± 3.6% of rainfall. These results led Lloyd (1989) to conclude that great care must be exercised when performing tests of land-surface scheme predictions of evaporation.

The BATS and BEST models have recently been examined to determine their sensitivity to rainfall intensity. By simulating the tropical forest ecotype in a stand-alone mode, it was possible to investigate how the models' simulation of re-evaporation of intercepted precipitation performed. Five rainfall intensities were simulated from which the gross precipitation (mm y^{-1}) and total interception loss (mm y^{-1}) were derived (Figure 17.1). In general, BEST simulates a slightly lower interception loss (1-2% lower) at high precipitation intensities (>3.0x10^{-3} mm s^{-1}). However, the difference between the two models increases as the precipitation intensity decreases, such that there is a difference of almost 4% at lower rainfall intensities (<1.5x10^{-3} mm s^{-1}).

BEST and BATS incorporate fundamentally similar models of precipitation interception. However, one of the areas in which the two models differ is in their formulation and calculation of canopy temperature. The differences seen in the interception loss might well be explained by slight differences in canopy temperature, affecting the re-evaporation of precipitation.

Although the model of interception incorporated by a given land-surface scheme is important, other factors are probable at least as important. In particular, Figure 17.1 shows that when forcing stand-alone models, the prescribed intensity of rainfall is extremely significant. The interception loss ranges between 4% and 20% of gross precipitation with intensities varying between 1-4x10^{-3} mm s^{-1}. Considerable care must therefore be taken in prescribing model precipitation. This is not difficult, providing some observational data exist.

Figure 17.1 also shows an observed estimate of interception loss (from Lloyd et al., 1988). BEST's simulation of interception loss is closer to the observed than BATS. However, it must be recognized that these observed data are from one site, in one tropical forest (see Shuttleworth et al., 1984a; 1984b). BEST and BATS are specifically designed to represent interception loss at the spatial scales appropriate to the GCM of around 300 km by 500 km or 1.5×10^5 km^2. The observations and modelled results are therefore spatially incongruent (cf. Jarvis and McNaughton, 1986) and thus the data cannot necessarily be used to evaluate the relative performance of two such similar surface schemes. The observed data can, however, be used as a guide, which suggests that both schemes produce reasonable estimates of interception loss, although qualitatively BATS seems to overestimate this flux a little.

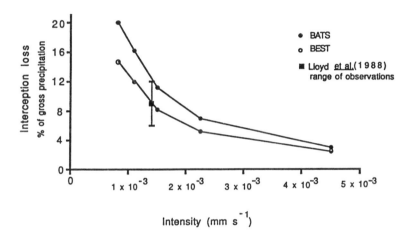

Figure 17.1 The relationship between precipitation intensity (mm s-1) and interception loss (% of gross precipitation) for tropical moist forest from two land surface models: BATS and BEST. An observed value of 8.9% (from Lloyd et al., 1988) is shown, with an error bar of ±3%.

The implications to regional evaporation of using canopy models in GCMs are considerable. It is not possible to control the precipitation intensities in a GCM. Hence, unrealistic simulations of intensity will lead to unrealistic simulations of interception, re-evaporation and, eventually, the whole regional hydrological cycle. Note that in Figure 17.1, the amounts of precipitation in the BATS and BEST models are the same for all precipitation intensities. Thus it follows that a GCM may simulate a realistic annual amount of precipitation and yet lead to a totally unrealistic simulation of the hydrological cycle, because the precipitation intensity was poorly represented.

17.5 DISCUSSION

The present generation of GCMs incorporate a range of land-surface parameterization schemes. It is clear that simple bucket schemes can no longer be considered a reasonable representation of the land-surface. Inclusion of any canopy parameterization requires additional computations for both the transpiration and re-evaporation of intercepted water. The recently developed land-surface schemes such as BATS, BEST and SiB do include these terms, albeit in a somewhat stylized form.

Increasing complexity of the land-surface parameterization demands comparison with observations in order to test the formulations. Preliminary attempts to examine the performance of the BEST and BATS schemes, for just one vegetation type, have been reviewed here. These intercomparisons have produced a rather interesting result: other components of the GCM are at least as important to the land-surface climate as the land-surface parameterization itself. In particular it has been shown that incoming solar radiation is very much too high during rain events in the CCM GCM.

It has also been shown that arbitrarily selected precipitation regimes can lead to highly misleading evaluations of interception loss.

The initial testing of these land-surface schemes has also shown that, when incorporated into a GCM, the simulations produced are likely to be highly sensitive to variables rarely analyzed (e.g. precipitation intensity). It is no longer tenable to compare GCMs against only a few atmospheric variables. In order to support the sophistication offered by the new generation of land-surface models and in order to use the surface-based predictions generated by coupled land-surface-GCM models, it is necessary to investigate new methods of testing their simulations.

Overall, the first generation of land-surface models to incorporate a canopy seem to warrant more detailed testing. It is imperative, however, that the prescribed conditions used to evaluate the models be subjected to rigorous examination. High resolution data sets can be coupled to land-surface models to provide valuable information concerning the models' success at simulating specific ecotypes. However, any temptation to "tune" surface schemes to represent a specific observation site must be avoided since the spatial scales of the model and observations are, as yet, incompatible.

REFERENCES

Abramopoulos, F., C. Rosenzweig and B. Choudhury (1988) Improved ground hydrology calculations for global climate models (GCMs): soil water movement and evapotranspiration. Journal of Climate 1:921-941.

Avissar, R., P. Avissar, Y. Mahrer and B.A. Bravdo (1985) A model to simulate response of plant stomata to environmental conditions. Agricultural and Forest Meteorology 34:21-29.

Deardorff, J. (1978) Efficient prediction of ground temperature and moisture with inclusion of a layer of vegetation. Journal of Geophysical Research 83:1889-1903.
Dickinson, R.E. (1984) Modeling evapotranspiration for three-dimensional global climate models, 58-72 in Climate Processes and Climate Sensitivity. Geophysical Monographs 29, (eds. J.E. Hansen and T. Takahashi), American Geophysical Union, Washington DC.
Dickinson, R.E. (1988) Implications of tropical deforestation for climate -a comparison of model and observational descriptions of surface energy and hydrological balance. Proceedings of the Royal Society of London (submitted).
Dickinson, R.E. and A. Henderson-Sellers (1988) Modelling tropical deforestation: a study of GCM land-surface parameterizations. Quarterly Journal of the Royal Meteorological Society 114:439-462.
Dickinson, R.E., A. Henderson-Sellers, P.J. Kennedy and M.F. Wilson, (1986) Biosphere-atmosphere transfer scheme (BATS) for the NCAR Community Climate Model, National Center for Atmospheric Research, Boulder, CO. Technical Note/TN-275+STR.
Gash, J.H.C. (1979) An analytical model of rainfall interception in forest. Quarterly Journal of the Royal Meteorological Society 105:43-55.
Gash, J.H.C., I.R. Wright and C.R. Lloyd (1980) Comparative estimates of interception loss from three coniferous forests in Great Britain. Journal of Hydrology 48:89-105.
Hansen, J., G. Russell, D. Rind, P. Stone, A. Lacis, S. Lebedeff, R. Ruedy and L. Travis (1983) Efficient three-dimensional global models for climate studies: Models I and II. Monthly Weather Review 111:609-622.
Hunt, B.G. (1985) A model study of some aspects of soil hydrology relevant to climatic modelling. Quarterly Journal of the Royal Meteorological Society 111:1071-1085.
Jarvis, P.G. and K.G. McNaughton (1986) Stomatal control of transpiration: scaling up from leaf to region. Advances in Ecological Research 15:1-49.
Lloyd, C.R. (1989) Amazonian rainfall distribution and forest rainfall interception, submitted to Quarterly Jounral of the Royal Meteorological Society.
Lloyd, C.R., J.H.C. Gash, W.J. Shuttleworth and A. de O. Marques (1988) The measurement and modelling of rainfall interception by Amazonian rainforest. Agricultural and Forest Meteorology 43:277-294.
Manabe, S. (1969) Climate and ocean circulation. I: The atmospheric circulation and hydrology of the earth's surface. Monthly Weather Review 97:739-774.
Martin, P. (1989) The significance of radiative coupling between vegetation and the atmosphere. Agricultural and Forest Meteorology (in press).
Paw U, K.T. and W. Gao (1988) Applications of solutions to non-linear energy budget equations. Agricultural and Forest Meteorology 43:121-145.
Pitman, A.J. (1988) The development and implementation of a new land surface scheme for use in GCMs, unpublished Ph.D. Thesis, Liverpool University, 481pp.
Rutter, A.J., K.A. Kershaw, P.C. Robins and A.C. Morton (1971) A predictive model of rainfall interception in forests, I. Derivation of the model from observations in a stand of Corsican pine. Agricultural and Forest Meteorology 9:367-384.
Rutter, A.J., A.C. Morton and P.C. Robins (1975) A predictive model of rainfall interception in forests, II. Generalization of the model and comparison with observations in some coniferous and hardwood stands. Journal of Applied Ecology 12:367-380.
Sellers, P.J., Y. Mintz, Y.C. Sud and A. Dalcher (1986) A simple biosphere model (SiB) for use within general circuation models. Journal of the Atmospheric Sciences 43:505-531.
Shuttleworth, W.J. (1988) Evaporation from Amazonian rain forest. Proceedings of the Royal Society of London, Series B 233:321-346.
Shuttleworth, W.J., J.H.C. Gash, C.R. Lloyd, C.J. Moore, J. Roberts, A.de O. Marques Filho, G. Fisch, V. de Paula Silva Filho, M.N.G. Ribeiro, L.C.B. Molion, L.D. de

Abreu S'a, J.C.A. Nobre, O.M.R. Cabral, S.R. Patel and J.C. de Moraes (1984a) Eddy correlation measurements of energy partition for Amazonian forest. Quarterly Journal of the Royal Meteorological Society 110:1143-1162.

Shuttleworth, W.J., J.H.C. Gash, C.R. Lloyd, C.J. Moore, J. Roberts, A.de O. Marques Filho, G. Fisch, V. de Paula Silva Filho, M.N.G. Ribeiro, L.C.B. Molion, L.D. de Abreu S'a, J.C.A. Nobre, O.M.R. Cabral, S.R. Patel and J.C. de Moraes (1984b) Observation of radiation exchanges above and below Amazonian forest. Quarterly Journal of the Royal Meteorological Society 110:1163-1169.

Shuttleworth, W.J., J.H.C. Gash, C.R. Lloyd, C.J. Moore, J. Roberts, A.de O. Marques Filho, G. Fisch, V. de Paula Silva Filho, M.N.G. Ribeiro, L.C.B. Molion, L.D. de Abreu S'a, J.C.A. Nobre, O.M.R. Cabral, S.R. Patel and J.C. de Moraes (1985) Daily variations of temperature and humidity within and above Amazonian forest. Weather 40:102-108.

Wilson, M.F., A. Henderson-Sellers, R.E. Dickinson and P.J. Kennedy (1987) Sensitivity of the biosphere-atmosphere transfer scheme (BATS) to the inclusion of variable soil characteristics. Journal of Climate and Applied Meteorology 26:341-462.

Part VII

Concluding Address

CHAPTER 18

Concluding Remarks

F. BRETHERTON

Space Science and Engineering Center
University of Wisconsin
1225 West Dayton Street, Madison WI 53706, USA

PREAMBLE

For the next 27 minutes I am going to give you a highly biased, idiosyncratic, commentary on what's really been happening this week. Wait, I don't think I said that right! It should have been - I plan to give you a carefully balanced, objective, summary of the conclusions of this meeting. Of course, you have just heard the carefully considered formal conclusions of the expert working groups, and it would be quite inappropriate for me to comment further on those, except perhaps to point out where they are being pig-headedly stupid or clearly haven't the least idea of what they are talking about. The chairman has just outlined my qualifications for this delicate diplomatic task. So that you understand where I am coming from, I should perhaps add that, thanks to my excellent British education, I took my one and only course in chemistry and biology when I was aged 13, and have been having my first introduction to soil science the last few days. However, as you see, I have a loud voice and a determined manner and it's my job to tell it to you as it is.

THE HIDDEN AGENDA

What have we been doing all this week?

Before discussing that I must repeat the comment that has already been made so many times, on the superb job done by the conference organizers, not just on the logistics but also before the meeting in the selection of the invitees and in the preparation of background papers. From my experience this conference ranks among the best in that regard. I think the group of people assembled here has indeed provided a remarkably stimulating set of perspectives. They are struggling to come to grips with some fundamental issues, and, with admittedly a few exceptions, even I could understand what they were talking about.

Soils and the Greenhouse Effect. Edited by A.F. Bouwman
© 1990 John Wiley & Sons Ltd.

The most important conclusions of this meeting are not written down, and were mostly not even discussed, in the public sessions at least. Actually they were arrived at over the lunch and coffee tables, where individuals, many of whom had never even met before, started to talk to each other seriously. They began to understand why, what just seemed so obvious to one of them didn't quite make sense from the perspective of the other, and they began to reconcile those differences on a one-on-one basis. This is an absolutely crucial first step in a dialogue which is not concluded at this moment, but is rather only just beginning. This dialogue is between, on the one hand, soil and land use scientists and, on the other hand, the community that for the last 20 years has been articulating "what scientists know and don't know" about the issues we have come to know as the greenhouse warming or, more generally, as global change. This particular dialogue has never happened before in a serious way, to my knowledge at least, and I can already glimpse some of the irreversible changes taking place in our global environment, that is in the environment of global scientific thinking, due to the accumulation, little by little, of emissions from this meeting.

Up to this time, in the United States and in the World Climate Research Program at least, and possibly more generally in the IGBP, thinking about the land surface has been dominated by meteorological modellers, some of whom are doubling as plant physiologists. In addition, there has been a group in the IGBP of ivory tower ecosystem biologists. What I sense happening at this meeting is the injection into all that of a new group, of a group which is rooted in communities that have traditionally solved practical problems, in the soils, in agriculture and in forestry, and are experienced in actually managing a segment of our environment. Albeit this management is not on a global scale and albeit the research workers themselves are often not taking the practical decisions, but nevertheless the tradition and the approaches there come from practical problem solving.

In illustration of this consider the paper by Sanchez, to be found elsewhere in this volume. Forgive the ad hominem references. They are intended to be specific, not personal. Sanchez has presented a proposal, based on extensive on site research, not for complete elimination but for a serious reduction in the destruction by slash and burn cultivators of primary tropical forest. This proposal, I think it is fair to say, may just have a chance of proving realistic in the economic and social environments for which it is intended, and could help stabilize a serious situation pending long term solutions. Now there are heaps of pitfalls in such initiatives, but it is nevertheless very important that proposals like this are beginning to surface in the global change program. They are critical in bridging the gap, so apparent in recent years, between the process of understanding the physical, chemical and biological science of global change and the process of devising practical, real world solutions to the issues the science raises.

SCIENCE

Let me touch now on the science itself. Here again I am going to be very selective.

The first point I would like to make is about modelling. The principal, indeed the only, way that scientists of different disciplines can communicate effectively in a thoroughly multi-disciplinary program like this is through the use of quantitative modelling. It is simply impossible for a meteorologist (say) to duplicate, in a reasonable time, the judgments of a soil scientist based on 20 years of tramping through the undergrowth and looking at soil horizons. This experience can only be shared by boiling it down to formulae that relate to physical or chemical processes and to the numbers to go into that formulae, in other words by incorporating it in a model of what is going on. Yet if we take a model such as Van Breemen's for soil carbon storage which has been validated at a few individual locations and attempt to extrapolate to a global scale, this can only be done by drawing upon the experience of soil scientists from many different nations. We must find survey data, such as a soils map or database, from which to estimate the key model parameters area by area and then assume, often with little justification, that the same processes should be valid elsewhere on the globe.

Now Baumgardner has proposed a global soil and terrain database, SOTER, and described the prototype data set and operations manual. Thus a critical, near term, agenda item is for van Breemen to sit down with Baumgardner - unfortunately he has already left so Sombroek will have to do instead - and thrash out:

- First, is there soil data for the needed parameters or items from which they could be inferred? In other words, is the relevant information actually in the various national soil databases?
- Second, if the information is there how credible is it, quantitatively, for this purpose, and
- Third, if there is credible information is it being preserved in the process of correlating the different national classification schemes and aggregating their contents into a single global database?

In other words, we need to take the SOTER prototype and test it - test it against a variety of real modelling problems before the excruciatingly laborious operational phase begins.

A second science topic concerns evapotranspiration. Comments were made in the discussion of the Working Group 1 report that it is important to obtain reliable data in relation to estimating this quantity over the land areas. I would like to reinforce those comments. The strongest greenhouse gas is not CO_2, but H_2O, both in its gaseous form and, potentially even more important, as cloud. It is not only stronger but also, in atmospheric terms, much less well understood. The only reason we don't refer to water as a greenhouse gas is because, assuming that the evaporation from land and water is internally determined, the concentration in the atmosphere is not subject to direct human influence, being controlled by other atmospheric processes. However, if the atmosphere warms due to increasing CO_2 concentrations it will hold more water vapour, which reinforces the CO_2 heating in a positive feedback. This effect is included in the usual estimates for global temperature rise, and indeed it is the modelling of water and cloud that is the source of much of the uncertainty in those estimates. Note that, to the extent that human alteration of the land surface modifies the evaporation, it is an additional, direct, cause of climate change.

Global climate modellers are acutely conscious that the most important manifestations of climate change are on the regional scale, with precipitation at least as important as temperature, and globally connected changes in wind regime probably contributing as much locally as global warming. However, their present models are very bad at predicting even present day regional climate, which is why you never hear much about it. The primary reason appears to be lack of computing resources to integrate the models for many years at sufficiently high spatial resolution. This obstacle is disappearing as computers become more powerful. However, it is widely believed that immediately behind is another problem, the inadequate treatment of water, particularly the processes governing rainfall and evaporation from the land surface. Reliable data to feed into existing models of these processes is critical. Meteorologically driven, internationally coordinated field experiments to validate local versions of those models have been completed or will take place in the near future, but the bottleneck is the global data bases to infer basic properties such as soil albedo and porosity, as well as other measures of vegetation cover.

Thus Bolle, who has been leading these efforts, also needs to sit down with Baumgardner/Sombroek and thrash out the same questions on SOTER, but for a quite different set of model requirements. Of course, I am not imagining that it is all going to be straightforward, but we do need to establish as quickly as possible whether key information is or is not likely to be available on a global basis. Above all, in view of the immense effort required to compile these global databases, we need to be sure we are doing this compilation as right as we possibly know how. Incidentally, does anyone have any ideas about how to infer rooting depths in semi-arid regions, particularly near and away from water courses?

POLICY

The last point I want to make relates to policy. I would like to reinforce the conclusions of Working Group 4, though I will put a slightly different gloss on them.

Scientists participating in discussions of potential policy actions face a real dilemma. On the one hand their professional effectiveness depends upon maintaining as objective a view as possible of what is, and what is not, known about the physical, chemical and biological aspects of the problem. On the other hand they must participate as informed individuals in a much wider debate that deeply touches human values and the economic and political systems in which we live. In reality, when communicating with non-scientists there is no such thing as scientific objectivity. Even the selection of what is perceived to be important is tinged by individual biases and social assumptions. Recommendations for action, in particular, are highly dependent on the background of the group making them. In my experience much conflict, apparently about technical matters, derives in large measure from different values and social priorities, rather than from the ostensible substance. I believe it is more efficient to recognize this openly, and to attempt to state these underlying broad assumptions as clearly as possible as a precursor to any recommendations, even if the latter are apparently only technical in nature.

In this spirit I must admit that during the past few days I have been waging a quiet campaign, well maybe not exactly quiet, against the organizers of this meeting, and indirectly against the Intergovernmental Panel on Climate Change (IPCC). This campaign was based on my experience from many years ago that people who ask stupid questions get stupid answers. This conference is organized around the title Soils and the Emission of Greenhouse Gases. It is, indeed, important to answer the precise questions posed by the organizers, but I believe that, if you get trapped into phrasing the whole research agenda and the way you address the problem in terms of these questions, you will not be doing yourselves justice in terms of the real issue. Because the real issue is not the greenhouse effect. At least that's my view.

The greenhouse effect is, I believe, just a symptom of the fact that there are now so many people in the world and the number seems to be going on up endlessly, coupled with the fact that to varying degrees the individuals in that world are using, or think they need to use, larger and larger quantities of energy and non-renewable resources. We as a world community are now in the business, whether we like it or not, of managing the global environment as a whole, and we have very little idea of how to do it. We cannot separate actions relating to greenhouse gases from those addressing acid rain, or land degradation, or deforestation. They are really just aspects of one issue, and what's more, it starts on the global scale, extends down through the regional scale and to the very local scale. There are interconnections between all those aspects, and we had better know what they are, otherwise we are in for some very unpleasant surprises. Thus we had better get out of this habit of compartmentalized thinking, in which you have one group of experts dealing with this problem, another dealing with that, and they don't really communicate. Global change is an issue on which there are no experts, that is who are really knowledgeable across the whole range of scientific, economic and political considerations that have to be blended together. A technical solution to a problem that is too narrowly conceived must be expected to backfire. Whereas progress can only come incrementally, each step as it becomes feasible, an overall strategy must always be clearly in mind.

This broadening of the issue should not be a surprise. It has been my experience, going back close to 20 years in what you might generically call the climate problem, that every time a new community of experts is seriously engaged in the debate, the establishment view of the problem changes in significant ways. I can go through a list of meetings year by year when the scales dropped from my eyes and I realized: "we got the problem wrong". For example I remember vividly the occasion, in Annapolis, Maryland in 1979, at which, for the first time meteorologists and just one representative from the agricultural research community got together and he remarked: "what's all this talk about carbon dioxide changing our climate and ruining our agriculture? There are million of hectares out there under cultivation in an artificial environment, and most of them are enriched in CO_2. Why? Because they grow better! It helps! What's all this gloom and doom?" Of course, the issue is a bit more complicated than that, but suddenly, for the meteorologists concerned it was a brand new perspective. Or how, also in 1979, a committee of the U.S. National Research Council suddenly realized the heat storage in the ocean is a critical factor in the time development of the warming associated with a given concentration of atmospheric CO_2. Every one of these new areas broadened our perspective and opened up major new research areas, of which many are not yet settled.

At this point I would much rather stop. However, I realize that logic demands at least a brief discussion of what, in addition to improving our understanding of how the atmosphere, oceans, ice, land surface and biota all interact, the overall strategy might be. This has to be a very personal statement, in which I claim no expert knowledge and indeed am conscious of much naive and woolly thinking. I am also aware of its extreme sensitivity. My starting point is the seemingly exponential rise in world population. I know of just three ways in which it might cease. The first is mass starvation, the second is global war. These two I regard as unacceptable. The third is to raise the standard of living of those societies most affected by increasing population to the point at which they voluntarily bring it under control. This has happened in a number of countries in the past. There is no guarantee that it would happen in the future. However, it is the only strategy I know that gets to the heart of the issue. If adopted, its consequences are frightening in their implications. It demands a massive and immediate transfer of technology and resources to developing nations to ensure their real economic advance, and acceptance for the time being of a substantial increase in their per capita energy and resource use, though that increase would start of course from a relatively low base. Measures by developed countries to limit greenhouse emissions and use of non-renewable resources would have to be part of the package. They must also recognize that, given the alternatives, it is in their selfish national interest to expedite, not drag their feet on, third world development, because every year that passes increases the magnitude of the global environmental problems that will eventually have to be faced, and paid for largely by them. Of course, this agenda is idealistic in the extreme, and there are at this time no plausible mechanisms by which it might be implemented. There are some who find the global environment a convenient rallying cry, yet assume the experts will come up with painless technical fixes. I know of no such painless fixes. Maybe this proposal will provoke them to a dialogue.

CONCLUSIONS

I have tried to make three points:

Understanding of global change issues and perceptions of the relative importance of different facts are changing. This change is partly as a result of new research, but even more rapidly as new communities of experts learn to communicate with each other.

Soil and land use scientists have much to contribute. However, the case in terms of greenhouse gas emissions is but a small part of the total.

The greenhouse warming is but a symptom. The real issue is managing the global environment in relation to world population and resource use. We have much to learn about the linkages between the natural science and human dimensions necessary to achieve such management.

The seeds of something important have been germinating at this meeting. I look forward to see how they develop.

Part VIII

Considerations for Studying Global Change Effects Relative to Terrestrial Ecosystems

D.S. OJIMA and Th. ROSSWALL

IGBP Secretariat, The Royal Swedish Academy of Sciences, Box 50005, 10405 Stockholm, Sweden.

ABSTRACT

Global Change is a complex set of interactive responses which include the climate system, the chemical atmosphere system, the ecosystems, and the human system. The objective of the International Geosphere-Biosphere Programme: A Study of Global Change (IGBP) is to describe and understand the interactive physical, chemical, and biological processes that regulate the total Earth system, the unique environment that it provides for life, the changes that are occurring in this system, and the manner in which they are influenced by human activities.

The primary goal of the IGBP is to develop a predictive understanding of the Earth system, especially in relation to changes that affect the biosphere. In this context, the IGBP has defined a number of research priorities, within which core research projects are being defined. The research priorities addresses how the global atmospheric chemistry is regulated and interacts with biological processes; how vegetation interacts with the hydrological cycle; and how terrestrial ecosystems interact with global changes.

The research programme will be, by necessity, an integrated, interdisciplinary programme that will take advantage of recent advances in remote sensing and data handling. Process studies in the field and laboratory will play a major role in developing a basic understanding of key processes related to global change. Recent development of numerical modelling and conceptual development of linkages between various facets of the global system have increased the possibility of developing models to be used in predicting changes in the Earth system.

INTRODUCTION

The Earth science community is faced with an enormous challenge of developing an understanding of the "cause-and-effects" relationships of global change. Marked alterations in the Earth's environment are being predicted in patterns of climate, atmospheric chemistry, natural biosphere, and the human system during the coming century (World Resources Institute, 1986 and 1988; Malone and Roederer, 1985; NRC, 1986; IGBP, 1986, 1988 and 1989), and these presage even greater change as the role of human interference in natural processes increases. Global change encompasses the totality of these alterations (Table 19.1.1). A major concern over global change is centred on the dramatic increases of "greenhouse" gases. This paper will outline the major

Soils and the Greenhouse Effect. Edited by A.F. Bouwman

research areas of the International Geosphere-Biosphere Programme of ICSU on the way terrestrial ecosystems respond to and interact with global changes and make specific note of those interactions related to greenhouse gas fluxes.

Natural changes to terrestrial ecosystems have been dramatic since the last glacial maxima resulting from global climatic shifts. As the ice-sheets retreated, ecosystems were reshaped to what we see today. Human activities recently have played a large part in accelerating changes to the composition and distribution of plant communities through both purposeful and inadvertent actions. Land use changes associated with agriculture, forestry, and other land management practices have resulted in totally new ecosystems composed of specialized plant and animal communities and have displaced many of the natural ecosystems throughout the world (Ehrlich et al., 1977; Pimentel and Pimentel, 1979; Malone and Roederer, 1985; Clark et al., 1986). Inadvertent human activities have resulted in introduction of "exotic" species of plants and animals, rerouting of hydrological flows, and industrial contamination of land, air, and water (Lovelock, 1979; SMIC, 1971; Kellog, 1988).

Table 19.1.1 Global Change is a composite phenomenon which involves changes not only in the physical climate system (e.g., global temperature and precipitation) but entails alterations in the atmospheric chemistry, the ecosystems, and the human system of the Earth, as well. These subsystems are intertwined in a complex web of interactions which taken together is Global Change

"GLOBAL CHANGE"

PHYSICAL-CLIMATE SYSTEM

Precipitation
Temperature
Radiant Energy
Humidity
Lightning

ATMOSPHERIC-CHEMICAL SYSTEM

H_2O, CO_2, N_2O, CH_4, O_3, etc.

ECOSYSTEMS

Production systems
Structure
Process rates

HUMAN SYSTEM

Land-use Systems
Urban-Industrial Systems

The atmospheric composition reflects, in part, the characteristics of the current ecosystems and the changes resulting from these land use alterations in the recent past. Other human activities are also contributing to global change. Industrial emissions during the last century have rapidly altered atmospheric composition as evidenced most dramatically by the amount of acid deposition falling in the northern hemisphere and by increasing concentrations of CO_2 and CFC in the atmosphere. Land use management has greatly altered not only plant communities, but also, soil characteristics, topography, nutrient inputs and outputs, and water regimes. Growing populations have exerted greater pressure on the Earth's ecosystems (both marine and terrestrial). These activities are viewed by many to be overriding natural controls of the global environment, and accelerating the rate of global change to an unprecedented level (SMIC, 1971; Bolin and Cook, 1983; Bolin et al., 1986; Clark et al., 1986; Kellog, 1988; Pearman, 1988).

   The International Geosphere-Biosphere Programme (IGBP) was initiated by the International Council of Scientific Unions (ICSU) to develop an understanding of the effects of global change and to develop the interdisciplinary research programme that is necessary to meet this goal. In principle, the IGBP should be viewed as one project: the study of the basic biological, chemical and physical processes and their interactions that regulate the Earth system as well as the linkages between the different parts of the system (Figure 19.1.1). The interactive geosphere-biosphere in the global Earth system is extremely complex. For practical reasons it is necessary to structure the programme in a manner that will ensure the development of well-defined and manageable core projects. Within the IGBP, there are two planning committees specifically concerned with the development of core projects which relate to greenhouse gas fluxes from the land surface and the interactions with the atmosphere. The areas of research they cover are:

- Terrestrial Biosphere-Atmospheric Chemistry Interactions
- Effects of Global Change on Terrestrial Ecosystems

TERRESTRIAL BIOSPHERE-ATMOSPHERIC CHEMISTRY INTERACTIONS

The linkage between the biosphere and the composition of the global atmosphere is now well recognized (Lovelock, 1979; Bolin and Cook, 1981; Clark, 1985; Malone and Roederer, 1985; Dickinson, 1986; Goudriaan, 1986; NRC, 1986; Strain, 1986; Pearman, 1988; IGBP, 1986, 1988 and 1989). The atmospheric content of oxygen, carbon and nitrogen are related to biological processes as well as physical-chemical properties of the atmosphere. Examples include CO_2 fluxes from metabolic processes, production of methane by methanogenic bacteria, and exchange of N compounds by bacteria in the soil. Plants are also active as sources and sinks of various trace gases. On the other hand, atmospheric deposition can also have a severe impact on the biosphere

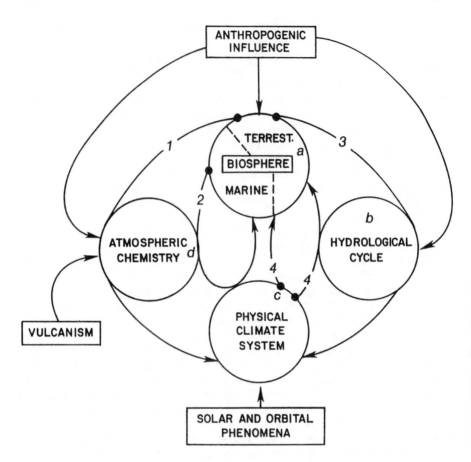

Figure 19.1.1 The interactions among the various components of the Earth system is the focus of the research programme of the IGBP. The numbers among the interconnecting lines correspond to the planning groups of the IGBP (see IGBP, 1988, Report No. 4 for further explanations).

as pollutants are accumulated on plant surfaces, in soils, and in water reserves.

During the past two decades our understanding has developed in regard to the basic chemical reactions that transform the many compounds that are brought into the atmosphere by emissions from the biosphere and by agricultural and industrial activities. However, quantitative knowledge of these important atmospheric chemical processes is still incomplete.

Human activities, both industrial and agricultural, are now altering fluxes of a number of trace gas species. For instance, worldwide concentrations of methane (CH_4) have been increasing by about 1% annually over the past decades. Because methane oxidation leads to carbon monoxide (CO), and because CO and CH_4 are the main reactant gases of hydroxyl (OH) in the background troposphere, atmospheric concentrations of OH may be on the decline. In fact, as both CH_4 and CO are removed mainly by reactions with OH from the atmosphere, there exists the potential for instability in the global atmospheric photochemical system (Crutzen, 1988); more CH_4 leading to less OH, less OH leading to more CH_4 and CO, and so on. As a consequence, there is concern that the oxidative power of the atmosphere may gradually decline, leading to the accumulation not only of CH_4 and CO, but also of many other important atmospheric trace gases. The recorded increase in the global background concentrations of CH_4 is primarily due to increased emissions at the Earth's surface by several human-influenced activities (e.g., rice agriculture, cattle-holding, land-filling) (Cicerone and Oremland, 1988), but may well be significantly enhanced by the loss of OH radicals.

These effects hold for the background, relatively pristine, troposphere. In atmospheric environments that are influenced by industrial fossil fuel and agricultural biomass burning activities, major emissions of nitric oxide lead to the opposite effect, i.e. enhancements in the concentrations of OH, and especially tropospheric ozone (O_3). The resulting increases in background tropospheric ozone concentrations may well have had a negative influence on the productivity of some temperate forest systems. Furthermore, in tropical ecosystems the emissions of NO by biomass burning have been shown to lead to greatly enhanced tropospheric ozone concentrations.

It is estimated that the enhanced combined climate warming caused by greenhouse gases other than CO_2, especially CH_4, O_3, N_2O, $CFCl_3$, and CF_2Cl_2, will match that of CO_2 in coming decades (Ramanathan, 1988). Only the increase in the atmospheric burdens of $CFCl_3$ and CF_2Cl_2 can be traced entirely to industrial chemical synthesis and release to the atmosphere. However, in the case of CO_2, CH_4, N_2O and NO, both industrial, agricultural, and other biospheric processes play important roles in establishing emissions to the atmosphere.

The atmospheric concentrations of each of these gases are increasing at significant annual rates. The industrial input of NO to the atmosphere has risen particularly rapidly over the past three to four decades, such that the global industrial NO source is now comparable to that produced by the combination of all natural sources, such as lightning discharges and soil emissions. In order to be able to estimate future increases in trace gases, and possibly lower their rates of increase, it is important to greatly improve our knowledge of their biogenic, agricultural and industrial sources and sinks. Although current attention is mostly focused on CO_2 and CH_4, alterations in the atmospheric and biospheric nitrogen cycle are particularly large. This demands strong research efforts to acquire a greatly improved understanding of the processes regulating the global biogeochemical nitrogen cycle. The production of N_2O and NO is of particular interest. On the whole, it is clear that there is still a basic lack of understanding regarding the biospheric processes that regulate the emissions of these and other chemically important trace gases to the atmosphere. One of the main objectives of the IGBP will be to acquire wider improved knowledge on the response of terrestrial ecosystems to anticipated atmospheric and climatic changes.

GLOBAL CHANGE EFFECTS ON AND INTERACTIONS WITH TERRESTRIAL ECOSYSTEMS

Changes in the global environment such as climate and nutrient deposition are known to affect ecosystem processes such as primary production and decomposition, and these, in turn, feed directly back into the climate and atmospheric systems. Processes which contribute to this feedback include plant production, decomposition, nutrient losses and gains (via gaseous fluxes, fluvial processes, aeolian movement, and leaching), evapotranspiration, run-off and run-on. Changes in the rates of these ecosystem processes directly influence the rates of change in the global environment. These processes can be characterized as having a high sensitivity to changes in the environment and have relatively fast response time (days to years).

In addition, there are feedback mechanisms between natural ecosystems and global climate which are less well understood, but may be no less important in their influence of global change in climate and subsequent changes in ecosystem properties. These feedback mechanisms result from changes in ecosystem development which alter their physiognomic structure for example, shifting from forest to savannah-woodlands or grasslands (Eagleson, 1986; Salati, 1986). Interference with the structure of vegetation, altering the roughness or reflectivity of the land surface, can modify regional and global scale climate through changes in albedo, humidity, and ground-level wind patterns (Dickinson, 1986; Sellers, 1986). Eco-physiology of systems may shift, from perennial C4 plants to annual C3 plants. Other subtle changes in ecosystem properties caused by climate change such as litter quality and soil properties may also affect greenhouse gas fluxes (Pastor and Post, 1988; Schimel et al., 1989).

Ecosystem responses to environmental factors such as water availability, temperature, and nutrient availability, have a considerable spatial variability. The exertion of limitations to production and decomposition by one or more factors will depend on climate-ecosystem interrelationships (Strain, 1986;Pastor and Post, 1988; Schimel et al., 1989). Soil properties are important features of the ecosystem which set limits on production and decomposition (Jenny, 1980;Parton et al., 1987). Water and nutrient availability, redox, soil structure, pH, and other soil factors play an important role in determining organic matter stabilization, soil flora and fauna distribution, and periods of active production or consumption of biogenic trace gases.

Land use patterns are also important factors in determining ecosystem dynamics throughout the world. Land management practices such as fire, grazing and cultivation affect ecosystem composition, cycling of nutrients and of organic matters, and other factors which influence rates of net trace gas flux. Global change can affect land use patterns, and in turn, land use changes may be an important factor in affecting global change by modifying land surface characteristics and net greenhouse gas emissions.

The manner in which global change will affect terrestrial ecosystems is a complex set of interacting factors of the global changes themselves, the response of the ecosystems to change, and the feedback responses back to the global environment. The development of core research projects which link ecosystem processes and the changes in the atmospheric chemistry is currently underway in the IGBP.

RESEARCH THRUSTS

The overall objectives of the projects are:

- To develop a predictive understanding of the role and occurrence of atmospheric oxidants and the chemical impact on the climate.
- To critically assess our understanding of the controls and interactions in the plant-microorganism-soil-water system which regulate trace gas exchange with the atmosphere.
- To develop a predictive understanding of the effects of global change phenomenon on terrestrial ecosystems.

RESEARCH PRIORITIES

The research programme will investigate controls on fluxes relative to production and consumption processes of three classes of biogenic trace gases. These are nitrogen species of N_2O, N_2, and NO, CH_4, and non-methane hydrocarbons (NMHC). The global distribution of these important gases in the atmosphere will also be documented.

The interdisciplinary nature of the project makes it imperative that a collaborative effort be undertaken which include active participation of atmospheric chemists, meteorologists, ecologists, pedologists, microbiologists, plant physiologists, geochemists, engineers and others. The development of the research project will incorporate analytical methods to determine the relevant gas fluxes and atmospheric concentrations; processes studies to study physiological, biochemical, and photochemical exchange processes affecting fluxes and concentrations of trace gases; field studies in key regions identified as source or sink regions for the key trace gas species; integrated field (micrometeorological techniques) and remote sensing campaign studies; and model development for extrapolation and predicting regional and global characteristics of the atmospheric composition and biogenic contribution.

Isotopic studies provide an additional method for determining global budgets of trace gases. Accurate estimates of the isotopic composition of sources to the atmosphere are essential for developing isotope-based models. This is currently being studied for methane. Integration of isotopic measurements with coordinated measurement campaigns will strengthen the linkage between local studies and global estimates.

The diverse nature of trace gases studied within these core projects allows for a geographic or regional division of the core activities in the following areas. These include:

High latitude wetland ecosystems;
Rice cultivation systems;
Land use systems in the tropics; and
Mid-latitude forests ecosystems.

In these regions, specific data and monitoring needs are required for development of models and for extrapolation studies. Information on land surface characteristics are very important, especially topography, vegetation type, climatic variables, and soil characteristics. In addition, climatic variables such as surface temperature, rainfall amounts and distribution, and atmospheric chemical composition are required. Data collection will be both land-based and derived via satellite observations.

In summary, given projected changes in greenhouse gas concentrations in the next century (Table 19.1.2), there is an urgent need to understand the processes responsible for these increases. The IGBP is developing an integrated, interdisciplinary programme which will attempt to develop a predictive understanding of greenhouse gas fluxes. The interplay of land-use change, soil properties, and ecosystem dynamics must be studied in an integrated fashion and in cooperation among the scientists within these and associated disciplines.

Table 19.1.2 Greenhouse gas effects for 2100 (taken from Dickinson, 1987)

Species	Current Level	Expected Increase	Likely Cause
CO_2	345 ppm	720 ppm	Fossil fuel, soil, and biomass burning
CH_4	1.7 ppm	3.0 ppm	Change in source or OH levels
N_2O	0.3 ppm	0.5 ppm	Fossil fuel, soils
CFC's	0.1 - 0.6 ppb	1 - 4 ppb	Human use

REFERENCES

Bolin, B., B.R. Döös, J. Jäger and R.A. Warrick (1986) *The Greenhouse Effect, Climate Changes, and Ecosystems. A Synthesis of Present Knowledge.* Wiley and Sons, Chichester, London, 541 p.

Bolin, B. and R.B. Cook (eds.) (1983) *The Major Biogeochemical Cycles and Their Interactions.* John Wiley and Sons, New York, p. 532.

Cicerone, R.J. and R.S. Oremland (1988) Biogeochemical aspects of atmospheric methane. Global Biogeochemical Cycles 2:299-328.

Clark, W.C., J. Richards and E. Flint (1986) Human transformations of the earth's vegetation cover: Past and future impacts of agricultural development and climatic change. In: Rosenzweig, C. and R. Dickinson (eds.): *Climatic-Vegetation Interactions.* Proceedings of a workshop held at NASA/Goddard Space Flight Center, Greenbelt, Maryland, Jan 27-29, 1986 p. 54-59.

Clark, W.C. (1985) Scales of climate impacts. *Climatic Change* 7:5-27.

Crutzen, P.J. (1988) Variability in atmospheric-chemical systems. In: Rosswall, T., R.G. Woodmansee and P.G. Risser (eds.): *Scales and Global Change: Spatial and temporal variability in biospheric and geospheric processes. SCOPE 35.* Wiley and Sons, Chichester, UK, p. 81-108.

Dickinson, R. (1986) Global climate and its connections to the biosphere. In: Rosenzweig, C. and R. Dickinson (eds.): *Climatic-Vegetation Interactions.* Proceedings of a workshop held at NASA/Goddard Space Flight Center, Greenbelt, Maryland, Jan 27-29, 1986, p. 5-8.

Eagleson, P.S. (1986) Stability of tree/grass vegetation systems. In: Rosenzweig, C. and R. Dickinson (eds.): *Climatic-Vegetation Interactions.* Proceedings of a workshop held at NASA/Goddard Space Flight Center, Greenbelt, Maryland, Jan 27-29, 1986 p. 106-109.

Ehrlich, P.R., A.H. Ehrlich and J.P. Holdren (1977) *Ecoscience: Population, resources, environment.* W.H. Freeman and Co., San Francisco, USA. P. 1051.

Goudriaan, J. (1986) Simulation of ecosystem response to rising CO_2, with special attention to interfacing with the atmosphere. In: Rosenzweig, C. and R. Dickinson (eds.): *Climatic-Vegetation Interactions.* Proceedings of a workshop held at NASA/Goddard Space Flight Center, Greenbelt, Maryland, Jan 27-29, 1986 p. 49-53.

IGBP (1986) The International Geosphere-Biosphere Programme: A Study of Global Change. IGBP Report N˚1. Final Report of the Ad Hoc Planning Group, ICSU 21st General Assembly, Berne, Switzerland 14-19 September, 1986.

IGBP (1988) The International Geosphere-Biosphere Programme. A Study of Global Change (IGBP). IGBP Report N˚4. A Plan for Action. A Report Prepared by the Special Committee for the IGBP for Discussion at the First Meeting of the Scientific Advisory Council for the IGBP, Stockholm, Sweden 24-28 October, 1988.

IGBP (1989) Effects of Atmospheric and Climate Change on Terrestrial Ecosystems. IGBP Report N˚5. Report of a Workshop Organized by the IGBP Coordinating Panel on Effects of Climate Change on Terrestrial Ecosystems at CSIRO, Division of Wildlife and Ecology, Canberra, Australia 29 February - 2 March, 1988. Compiled by B.H. Walker and R.D. Graetz.

Jenny, H. (1980) *The Soil Resource: Origin and Behavior.* Springer-Verlag, New York, p. 377.

Kellog, W.W. (1988) Human impact on climate: The evolution of an awareness. In: Glantz, M.H. (ed.): *Societal Responses to Regional Climatic Change: Forecasting by Analogy.* Westview Press, Boulder, CO, USA. pp.9-40.

Lovelock, J.E. (1979) *GAIA: A New Look at Life on Earth.* Oxford University Press, 157 p.

Malone, T.F. and J.G. Roederer (eds.) (1985) *Global Change: Proceedings of a Symposium Sponsored by ICSU during its 20th General Assembly in Ottawa, Canada.* ICSU Press Symposium Series No. 5. Cambridge University Press.

National Research Council (1986) *Global Change in the Geosphere-Biosphere: Initial Priorities for IGBP.* Commission on Physical Sciences, Mathematics, and Resources. National Academy Press, Washington, DC.

Pastor J. and W.M. Post (1988) Response of northern forests to CO_2-induced climate change. *Nature* 334:55-58.

Parton, W.J., D.S. Schimel, C.V. Cole and D.S. Ojima (1987) Analysis of factors controlling soil organic matter levels in Great Plains Grasslands. *Soil Science Society of America Journal* 51:1173-1179.

Pearman, G.I. (1988) Greenhouse gases: Evidence for atmospheric changes and anthropogenic causes. In: *Greenhouse: Planning for Climate Change.* Pearman, G.I (ed.). CSIRO Australia, p. 3-21.

Pimentel, D. and M. Pimentel (1979) *Food, Energy and Society.* In: *Resource and Environmental Sciences Series.* Wiley and Sons, New York, 165 p.

Ramanathan, V. (1988) The greenhouse theory of climate change: A test by an Inadvertent global experiment. *Science* 240:293-299.

Salati, E. (1986) Amazon: Forest and hydrological cycle. In: Rosenzweig, C. and R. Dickinson (eds.). *Climatic-Vegetation Interactions.* Proceedings of a workshop held at NASA/Goddard Space Flight Center, Greenbelt, Maryland, Jan 27-29, 1986 p. 110-112.

Schimel, D.S., W.J. Parton, T.G.F. Kittel, D.S. Ojima and C.V. Cole (1989) Grassland biogeochemistry: Links to atmospheric processes. *Climatic Change, (In press).*

Sellers, P.J. (1986) The simple biosphere model (SiB). In: Rosenzweig, C. and R. Dickinson (eds.): *Climatic-Vegetation Interactions.* Proceedings of a workshop held at NASA/Goddard Space Flight Center, Greenbelt, Maryland, Jan 27-29, 1986 p. 87-90.

SMIC (1971) Inadvertent climate modification. A *Report of the Study of Man's Impact on Climate (SMIC).* Sponsored by the Massachusetts Institute of Technology. Hosted by the Royal Swedish Academy of Sciences and the Royal Swedish Academy of Engineering Sciences. The MIT Press, Cambridge, Massachusetts, 308 p.

Strain, B.R. (1986) The biosphere and links to climate. In: Rosenzweig, C. and R. Dickinson (eds.): *Climatic-Vegetation Interactions.* Proceedings of a workshop held at NASA/Goddard Space Flight Center, Greenbelt, Maryland, Jan 27-29, 1986 p. 9-12.

World Resources Institute and the International Institute for Environment and Development (1986) World Resources 1986: An assessment of the resource base that supports the global economy. World Resources Institute and the International Institute for Environment and Development. Basic Books, Inc., New York, USA. P. 353.

World Resources Institute and the International Institute for Environment and Development (1989) World Resources 1988-1989: An assessment of the resource base that supports the global economy. World Resources Institute and the International Institute for Environment and Development. Basic Books, Inc., New York, USA. P. 372.

EXTENDED ABSTRACT 19.2

Deforestation Reduction Initiative: an Imperative for World Sustainability in the Twenty-first Century[1]

P.A. SANCHEZ

Tropical Soils Research Program
North Carolina State University, Box 7619, Raleigh, N.C. 27695, USA

ABSTRACT

Major adverse effects of global warming are predicted for the United States and other mid-latitude countries. Within that, 15 to 25% of global warming results from clearing of tropical rainforests. Third world population growth forces landless rural populations to migrate and over exploit tropical rainforests, a problem exacerbated by government colonization policies in such countries as Brazil, Peru, and Indonesia. The resulting agriculture is unsustainable and leads to further deforestation and migration to urban centres.

Research has shown that these trends can be reversed. An integrated approach consisting of development and application of sustainable management technologies for tropical soils and appropriate government policies will eliminate the pressure for further deforestation. Some management technologies are available and other evolving which allow continuous production. For every hectare put under sustainable management five to ten hectares of forest are saved each year.

Twelve of the 64 countries with tropical rainforests contribute 75 percent of carbon resulting from clearing. Ten others account for much of the balance. A worldwide, coordinated *Deforestation Reduction Initiative* involving research, education, and policy formulation within and among source countries is proposed and will be discussed.

THE PROBLEM

Major adverse effects of global warming are expected to take place in the United States and other mid-latitude countries within the next 30 to 60 years. Increased concentration of greenhouse gases in the atmosphere (carbon dioxide, nitrous oxides and methane) will raise mean annual temperatures in Continental U.S. by 4 to 10°F, flood some coastal areas due to a 4 foot rise in sea level and increase the incidence of droughts. Little effect is expected in the tropics, but the centre of gravity of

---

[1] Presented at the Bureau of Science and Technology, U.S. Agency for International Development, Washington, D.C. July 22, 1988.

Soils and the Greenhouse Effect. Edited by A.F. Bouwman

agricultural production may shift northwards towards Canada and the Soviet Union.

Recent EPA estimates indicate that 15 to 25% of the global warming is due to the clearing of tropical rainforests, which is occurring now at an annual rate of 7 million hectares. Current magnitude of these emissions (1 billion tons of carbon per year) is equivalent to current emissions from industrial combustion in the United States. Research has shown that something can be done to decrease the rate of tropical deforestation, and thus attenuate the expected adverse effects.

Deforestation is driven by a complex set of demographic, biological, social and economic forces described in Figure 19.2.1. Third World population growth continues at a high rate, while most of the fertile and accessible lands are intensively utilized. Government policies often exacerbate land scarcity by allowing gross inequities in land tenure. These factors result in an increasing landless rural population which

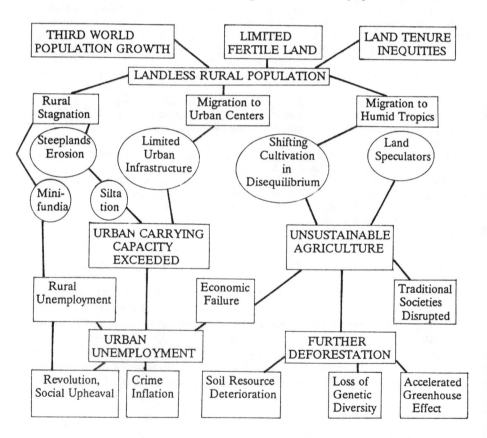

Figure 19.2.1 Cause-effect relationships related to tropical deforestation in developing countries.

essentially has three choices: stagnate where they are, migrate to the cities, or migrate to the rainforests that constitute the frontier of many Third World countries. Although urban migrations are spontaneous, national policies in key countries include the occupation of their tropical rainforests, notably Brazil, Peru and Indonesia via colonization programs.

Densely populated rural environments such as the Andean valleys, Northeast Brazil, and Java suffer from an ever-decreasing farm size and the overuse of steepland areas. This results in widespread soil erosion, siltation of reservoirs and other adverse off-site effects to urban centres. Migration to the cities in search of a better life results in bitter disappointments, and coupled with limited urban infrastructure, produces unmanageable cities with populations far exceeding their carrying capacity and infrastructure.

Migration to the humid tropics seldom results in a bountiful cornucopia (with few notable exceptions). An equilibrium between the rainforest and shifting cultivation by traditional humid tropical societies is broken by the colonists, and in some countries by land speculators as well. The result is shifting cultivation in disequilibrium which quickly turns into various forms of unsustainable agriculture. Traditional societies are disrupted, economic failures abound and migration to urban centres follows.

The two end points are urban unemployment and further deforestation. The consequence of the former is abject urban poverty which leads to widespread crime, poor health and in many cases social upheaval. Deforestation depletes the ecosystems' limited nutrient capital, decimates plant and animal genetic diversity and accelerates global warming due to carbon dioxide and nitrous oxide emissions.

Can these trends be alleviated and eventually reversed? Our answer, based on long-term research is an emphatic *yes*. The key is an integrated approach consisting of 1) development and application and sustainable management technologies for tropical soils and 2) appropriate government policies that will provide incentives to discourage the need for further deforestation.

People do not cut tropical rainforests because they like to; they clear land out of sheer necessity to grow more food. Deforestation, therefore, can be reduced by the widespread adoption of sustainable management practices that permit the use of cleared land on an indefinite basis. Sustainable management options for acid soils of the humid tropics have been developed at Yurimaguas, Peru and elsewhere to fit different landscape positions, soils and levels of socioeconomic infrastructure development (Figure 19.2.2).

Most of the options are based on low-input systems which serve as transition technology for other options such as agroforestry systems, legume-based pastures, and continuous crop rotations. Other options developed in humid tropical Asia and Africa, include paddy rice, perennial crop production and forest plantations. In spite of the continuing need for additional research, there is no question that the technological basis for sustainable management options for acid soils of the humid tropics is available now.

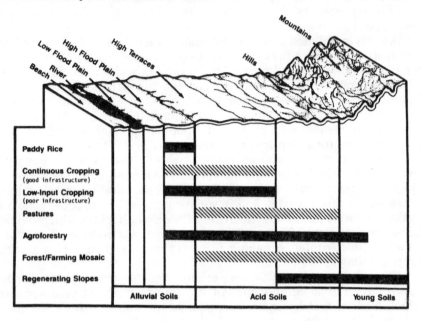

Figure 19.2.2 Some soil management options for humid tropical landscapes dominated by Oxisols and Ultisols (3).

For every hectare put into these sustainable soil management technologies by farmers, five to ten hectares per year of tropical rainforests will be saved from the shifting cultivator's ax, because of their higher productivity. Estimates at Yurimaguas for the various management options are as follows:

1 hectare in sustainable management options	Equals: hectares saved from deforestation annually
Flooded rice	11.0
Low-input cropping (transitional)	4.6
High-input cropping	8.8
Legume-based pastures	10.5
Agroforestry systems	not determined

Such technologies are equally applicable to secondary forest fallows, where clearing does not contribute significantly to global warming because of the small tree biomass. Nevertheless the use of secondary forest fallows is of very high priority, because in many areas is an excellent alternative to primary forest clearing. Many of the degraded or unproductive pastures or croplands resulting from extremely poor management practices can also be reclaimed using some, but not all of these available technologies.

Such technologies, however, are useless without effective government policies that encourage, support and regulate them. Likewise, well conceived policies will fail without sustainable technologies. Therefore the hope lies on a joint policy-technology approach: The Deforestation Reduction Initiative.

The present situation is analogous to when the world technical assistance community launched the Green Revolution in the late 1960's. At that time sustainable technologies for high-yielding rice and wheat production were sufficiently developed to be used. Key government officials were convinced of their importance by leading scientists, and instituted the necessary policies to make massive farmer adoption possible in India, Pakistan, Philippines and other countries. The Green Revolution became a worldwide success during the next twenty years and the goal of arresting worldwide famine was definitely achieved.

The Deforestation Reduction Initiative is a similar challenge but with the major difference that it affects directly the livelihood of both Third World humid tropical countries and the developed countries as well. It is also likely to be more difficult to implement since the technology and policy issues are more complex than implementing high-input crop production technologies on the better lands of the tropics.

THE GOAL

A worldwide, coordinated Deforestation Reduction Initiative is proposed to simultaneously accomplish the following goals during the next 25 years:

1. Increase and stabilize crop, pasture and tree production in areas presently under shifting cultivation through the massive adoption of sustainable management technologies supported by appropriate government policies.
2. Decrease deforestation of primary tropical rainforest from the present annual rate of 7 million hectares to 1 million hectares.
3. Preserve the genetic diversity of the majority of the remaining areas of tropical rainforests by decreasing the demand for further land clearing.

4. Conserve the natural resource base of humid tropical regions through a reasonable, stable balance between sustainable agriculture and preserved rainforests.

APPROACH

Priority countries. There are 64 countries in the world with tropical rainforests and estimates of net carbon emissions due to land clearing in 1980 have been made by the Woods Hole Research Centre. The breakdown by region, including emissions for primary forests, secondary forests and woodlands is:

Region	Forest Cover billion has.	Net carbon emissions in 1980 million tons	%
Tropical America	1.2	665	40
Tropical Asia	0.4	621	37
Tropical Africa	1.3	373	23

Source: Houghton et al., 1987

Twelve out of these 64 countries account for three quarters of the net carbon emissions from clearing primary forests:

Country	Net carbon emissions in 1980 from primary forests (million tons)
Brazil	207
Colombia	85
Indonesia	70
Malaysia	50
Côte d'Ivoire	47
Mexico	33
Thailand	33
Peru	31
Nigeria	29
Ecuador	28
Zaire	26
Philippines	21

Ten other countries account for much of the remaining deforestation: Burma,Cameroon, Venezuela, Madagascar, India, Nicaragua, Laos, Vietnam, Kampuchea and Bolivia.

Just like with the Green Revolution, dialogue may be initiated with top science policy-makers in selected countries, to discuss the possibility of a cooperative program. In countries where the possibilities of collaboration are good, a dialogue between senior government officials should follow, with agreements signed and commitments for a 25 year program.

Within each participating country, "hot spots" of deforestation can be identified where technology transfer and government policies would be focused. Examples of these hot spots are Rondonia, Acre and Southern Bahia in Brazil; Caquetá in Colombia; transmigration areas in Sumatra and Kalimantan in Indonesia; the Huallaga Valley, Pucallpa, and Pichis Palcazu in Peru, Chapare in Bolivia, Coca-Lago Agrio in Ecuador, and many others.

PROGRAM COMPONENTS

- An international workshop to launch the Tropical Deforestation Initiative.
- Selection of participating countries and "hot spots" or target areas. Protocols signed.
- Closer monitoring of deforestation rates in participating countries and target areas.
- Dynamic, well funded sustainable technology validation programs in each target area, including training of local research and extension specialists.
- Assistance to participating countries in developing appropriate policies to support the Initiative.
- Support continuing research on developing stable management options with reinforcement of one major research centre in tropical America, Asia and Africa. Emphasis will be on sustainability in the long run.
- Support on-site research to quantify the nature of gas emissions from various sustainable management options, including clearing and common mismanagement options. Major questions about nitrous oxides emissions exist. For some reason air above the Amazon has one of the highest nitrous oxides concentrations in the world. Understanding the processes that regulate gaseous emissions at the ground level needs to be researched and linked with atmospheric measurements.
- Support research to determine how the adoption of successful management technologies affect forest and soil preservation, conservation of genetic diversity and the hydrologic balance.

REFERENCES

EPA (1988) Meeting summary. Stabilization Report. Workshop on Agriculture and Climate Change. February 29-March 1, 1988. (draft). Office of Policy Analysis, EPA, Washington.

Houghton, R.A., R.D. Boone, J.R. Fruci, J.E. Hobbie, J.M. Melillo, C.A. Palm, B.J. Petersen, G.R. Shaver and G.M. Woodwell (1987) The flux of carbon from terrestrial ecosystems to the atmosphere in 1980 due to changes in land use: geographic distribution of the global flux. Tellus 39B:122-139.

McElroy, M.B. and S.C. Wofsy (1986) Tropical forests: interactions with the Atmosphere. In: G.T. Prance (ed): Tropical Rain Forests and the World Atmosphere, pp. 33-60. Westview Press, Boulder, Colorado.

Mooney, H.A., P.M. Vitousek and P.A. Matson (1987) Exchange of materials between terrestrial ecosystems and the atmosphere. Science 238:926-932.

Sanchez, P.A. and J.R. Benites (1987) Low-input cropping for acid soils of the humid tropics. Science 238:1521-1527.

Sanchez, P.A. and E.R. Stoner and E. Pushparajah (eds) (1987) Management of Acid Tropical Soils for Sustainable Agriculture. IBSRAM Proceedings No.2, IBSRAM, Bangkok, 299 pp.

Deep Weathering at Extra-tropical Latitudes: a Response to Increased Atmospheric CO_2

M. BIRD[1], B. FYFE[1], A. CHIVAS[2] and F. LONGSTAFFE[1]

[1]*Department of Geology, University of Western Ontario, London, Ontario N6A 5B7, Canada*
[2]*Research School of Earth Sciences, Australian National University, G.P.O. Box 4, Canberra 2601, Australia*

ABSTRACT

The $\delta^{18}O$ composition of clay minerals from deeply weathered regolith profiles formed at high-latitudes indicate that many profiles formed under cool to cold conditions, rather than in a 'tropical' climate as previously thought. An increased atmospheric CO_2 concentration, coupled with warmer than modern temperatures, abundant rainfall and increased biological productivity, provides a plausible mechanism for deep-weathering at extra-tropical latitudes.

Deeply weathered regolith formed at extra-tropical (>30° north or south) latitudes has been described from many parts of the world, from Palaeozoic to Tertiary times (Table 19.3.1; Dury, 1971; Fitzpatrick, 1963; Goldich, 1938; Hall, 1985; McAllister and McGreal, 1983; Rosenqvist, 1975; Sapojnikov, 1979). These regolith profiles commonly contain abundant kaolinite and in the case of laterites and bauxites they also contain abundant secondary Fe and Al minerals such as hematite, goethite and gibbsite. Such profiles are common over much of the Australian continent (e.g. Butt and Smith, 1980) and may well have been considerably more extensive in the northern Hemisphere, were it not for the scouring of much of the landscape by the icesheets of the Quaternary. The apparent contradiction associated with the abundant evidence for high-latitude lateritization/kaolinization in Australia has previously been recognised (Schmidt and Ollier, 1988). However, the usual interpretation of such phenomena has been that the presence of deeply weathered regolith profiles indicates that tropical or sub-tropical climates prevailed at high latitudes, rather than that the presence of such profiles at high latitudes indicates that tropical conditions are not neccessary for the formation of the profiles (Ollier, 1984; Biscaye, 1965; Stein and Robert 1985; Singer, 1984; Parra et al., 1986).

The oxygen-isotope composition ($\delta^{18}O$) of a clay mineral formed in a weathering environment is dependent primarily upon the isotopic composition of the meteoric water from which the mineral formed (Lawrence and Taylor, 1971), which in turn is primarily a function of

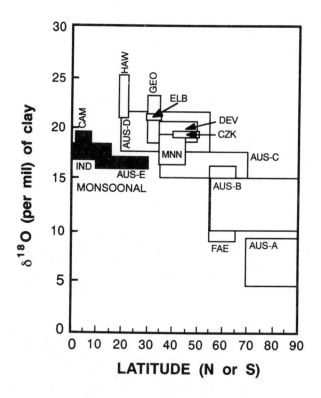

Figure 19.3.1   Oxygen-isotope composition of clays from deeply weathered regolith profiles as a function of palaeolatitude of formation of the profiles. See Table 19.3.1 for explanation of abbreviations and description of localities.

air temperature (Dansgaard, 1964; Yurtsever and Gat, 1981). Therefore, in a weathering environment the $\delta^{18}O$ of a mineral such as kaolinite can be related in general terms to the temperature at which that mineral formed, such that the lower the $\delta^{18}O$ of the mineral the lower the temperature of formation and vice versa. Figure 19.3.1 indicates that the $\delta^{18}O$ values of the minerals decrease dramatically with increasing palaeolatitude of formation of the profiles, suggesting decreasing temperatures of formation. Clays from India, Cameroon and northern Australia have anomalously low $\delta^{18}O$ values in comparison with Hawaiian clays which have high $\delta^{18}O$ values consistent with warm temperatures of formation. This behaviour results from the monsoonal nature of the climate in the former regions, where rainfall is anomalously depleted in ^{18}O as a result of strong heating and convective uplift of rain-bearing airmasses (Bird, 1988b; Bird and Chivas, 1988a).

Table 19.3.1 Age, palaeolatitude of formation and $\delta^{18}O$ composition of clays from deeply weathered regolith profiles

Location (prefix)	Weathering age	Palaeo-lat.°	$\delta^{18}O$	Description
Australia (AUS-A)	E. Permian	70-90	+4.4 to +9.0	laterite and kaolinitic clay rocks in the Sydney Basin, eastern Australia Bird and Chivas (1988a), Loughnan (1975)
Australia (AUS-B)	pre L. Meso-zoic	55-90	+10.0 to +15.0	relict deep weathering in Australia Bird and Chivas (1988b, in press)
Australia (AUS-C)	L. Mesozoic-E. Tertiary	35-70	+15.0 to +17.5	deep weathering widespread in Australia Bird and Chivas (1988b, in press)
Australia (AUS-D)	post mid-Tertiary	15-55	+17.5 to 21.4	deep weathering widespread in Australia Bird and Chivas (1988b, in press)
Australia (AUS-E)	Tertiary to Recent	10-30	+16.0 to 17.0	deep weathering in monsoonal north Australia; Bird and Chivas (1988b, in press)
Minnesota (MNN)	early L.Creta-ceous	35-45	+16.3 to +19.3	deeply weathered Archaean basement and Upper Cretaceous sediments; Parham (1970), Goldich (1938), Dury and Knox (1971)
Faeroe Is. (FAE)	E. Tertiary	55-65	+8.8 to +16.4	intrabasaltic kaolinitic and smectitic deep weathered profiles wiedespread in Faeroe Islands; Nilsen and Kerr(1978), Parra et al. (1986, 1987)
Czeckoslo-vakia(CZK)	mid Tertiary	40-50	+19.0 to +19.3*	deeply weahtered granite, Savin and Epstein (1970)
Devon, U.K.(DEV)	L. Mesozoic-E.Tertiary	35-50	+18.5 to +20.5*	deeply weathered granite, Sheppard (1977)
Cameroon (CAM)	Tertiary - Recent	2-7	+17.0 to +19.5*	thick lateritic profile, Clauer et al. (1989)
Georgia (GEO)	L. Mesozoic-E. Tertiary	30-35	+18.5 to +23.1*	thick sedimentary kaolinite deposits from deeply weathered source; Hassanipak and Eslinger (1985)
Georgia (ELB)	L.Cretaceous-E.Tertiary	30-35	+20.9 to 21.3*	deeply weathered Elberton granite; Lawrence and Taylor (1971)
Hawaii (HAW)	Quaternary	19-22	+21.0 to +25.0*	deeply weathered basalts; Lawrence and Taylor (1971)
India (IND)	E.Tertiary-mid-Tertiary	0-15	+16.8 to +18.3	deeply weathered basement and sedimentary kaolin; Soman and Slukin (1985); Bird (1988)

* isotopic data taken from the published study referenced;
° palaeolatitudes taken from Condie (1982)

Early Permian to mid-Tertiary samples from Australia and early Tertiary samples from the Faeroe Islands all have $\delta^{18}O$ values which are too low to have resulted from formation in a monsoonal climatic regime, and therefore the observed low values are indicative of formation at low temperatures. For the southern Australian samples, calculations suggest temperatures of the order of 5°C cooler than present (present average sea-level temperature ~16 to18°C) during regolith formation in the late Mesozoic - early Tertiary and 10 to 17°C cooler during pre-late Mesozoic periods of weathering (see Bird, 1988a; Bird and Chivas, in review, for details). The pre-late Mesozoic temperatures are consistent with recent palaeoclimatic information for the period (Kemper, 1987; Frakes and Francis, 1988; Gregory et al., 1989). Kaolinites formed by weathering in the early Permian have such low $\delta^{18}O$ values that the influence of glacial meltwaters, accompanying the deglaciation of the continent at that time, are implicated in the formation of the profiles (Bird and Chivas, 1988b). The increase in $\delta^{18}O$ of clays forming in Australian regolith profiles in response to the drift of the Australian continent to lower (warmer) latitudes can be used to provide an estimate of the age of weathering for profiles where there is no independent age control (Bird and Chivas, 1988a, in press).

The oxygen-isotope data presented above suggests that at least some, and probably most, deeply weathered regolith profiles which formed at extra-tropical latitudes formed in cool to cold and presumably moist climatic conditions, even though the mineral assemblages are typical of tropical weathering. Paton and Williams (1972) have previously suggested that tropical temperatures are not neccessary for the formation of deeply weathered regolith profiles, but rather that a suitable parent lithology, high rainfall and efficient leaching are the only requirements for their formation.

As it is apparent that such profiles are not currently forming at extra-tropical latitudes, there must be another important factor controlling the formation of deeply weathered regolith, which has changed since the major periods of deep weathering which occurred in the Mesozoic and Tertiary. One factor that may be important in the controlling the formation of deeply weathered regolith, particularly at high latitudes, is atmospheric carbon-dioxide concentration, via CO_2-HCO_3 equilibria in infiltrating groundwater, and subsequent reaction of bicarbonate with rock forming minerals.

It is probable that the pCO_2 of at least the Cretaceous and Tertiary atmosphere was significantly higher than that of the modern atmosphere (Berner et al., 1983; Walker et al., 1981; Barron and Washington, 1985; Volk, 1987), and this is likely to have led to increased weathering rates in the past. Because of biological activity, soil pCO_2 is usually conside rably higher than atmospheric pCO_2, and therefore the increase in weathering rate would primarily result from increased productivity, leading to higher root and microbial CO_2 production from respiration. The possibility that weathering on a global scale is partially modulated by feedback between atmospheric CO_2 content, biological productivity and weathering rate has previously been suggested (Lovelock and Whitfield, 1982; Volk, 1987).

In the context of high latitude weathering, higher atmospheric pCO_2 in the past would have resulted in much higher HCO_3^- concentrations in weathering solutions, and this coupled with wetter and slightly warmer than modern high latitude climates may have been sufficient to produce deeply weathered regolith at relatively low temperatures given an extended period of time. By analogy with the modern situation, warmer summer periods at high latitudes in the past may have been accompanied by a dramatic increase in productivity and therefore in soil pCO_2. If winter temperatures were low enough to allow significant snow accumulation, then snowmelt (coincident with the rapid increase in soil pCO_2 at the onset of summer warming) would provide an abundant water supply for infiltration of the weathering bedrock and removal of solutes generated by weathering. This process, in effect, mimics a low-latitude 'wet' and 'dry' season. In addition, the onset of warming following a period of glaciation is accompanied by increased atmospheric CO_2 concentrations at a time when abundant glacial meltwaters are available to both stimulate productivity and flush through weathering bedrock. It is also worth noting that CO_2 solubility in water iincreases with decreasing temperature (Butler, 1982). One response of weathering in such a scenario may be that it may be more dependent on lithology, topography and vegetation type than comparable processes at higher temperatures. Weathering would also be more dependent upon factors such as nutrient availability, which will ultimately determine productivity limits and thereby soil pCO_2 (e.g. Pastor and Post, 1988).

The conclusion that deep weathering at extra-tropical latitudes has occurred at comparatively low temperatures in the geologic past, probably in response to elevated atmospheric CO_2 levels, highlights the importance of weathering as a sink for atmospheric CO_2 and the potential of the large area of comparatively unweathered land currently at high latitudes to act as a significant CO_2 sink in the future. It also underscores the importance of biological activity in modulating weathering rate.

REFERENCES

Barron, E.J. and W.M. Washington (1985) Warm Cretaceous climates: High Atmospheric CO_2 as a plausible mechanism. In: Sundquist, E.T. and W.S. Broeker (Eds.) The carbon cycle and atmospheric CO_2: natural variations Archaean to Recent. American Geophysical Monograph. 32:546-553

Berner, R.A., A.C. Lasaga and R.M. Garrels (1983) The carbonate-silicate geochemical cycle and its effect on atmospheric carbon dioxide over the past 100 million years. American Journal of Science 283:641-683

Bird, M.I. (1988a) An Isotopic Study of the Australian Regolith. Ph.D. dissertation, Australian National University, Canberra (unpubl.)

Bird, M.I. (1988b) Isotopically depleted rainfall and El Niño. Nature 331:489-490

Bird, M.I. and A.R. Chivas (1988a) Oxygen-isotope dating of the Australian regolith. Nature 331:513-516

Bird, M.I. and A.R. Chivas (1988b) Stable-isotope evidence for low-temperature weathering and post-formational hydrogen-isotope exchange in Permian kaolinites. Chemical Geology (Isotope Geoscience Section) 72:249-265

Bird, M.I. and A.R. Chivas (in press) Stable-isotope geochronology of the Australian regolith. Geochimica et Cosmochimica Acta.

Bird, M.I. and A.R. Chivas (in review) Application of stable-isotope geochronology to the Australian regolith. Australian Journal of Earth Sciences

Biscaye, P.E. (1965) Mineralogy and sedimentation of Recent marine deep sea clay in the Atlantic Ocean and adjacent seas and oceans. Geological Society of America Bulletin 76:802-832.

Butler, J.N. (1982) Carbon dioxide equilibria and their applications. Addison-Wesley, Reading, Massachusetts, 259pp.

Butt, C.R.M. and R.E. Smith (1980) Conceptual models in geochemical exploration: Australia. Journal of Geochemical Exploration, 12:89-365

Clauer, N., J.-P. Muller and J.R. O'Neil (1989) Oxygen-isotope signature of successive generations of kaolinite in a laterite. Geochemical implication. Abstract, 9th International Clay Conference, Strasbourg.

Condie, K.C. (1982) Plate tectonics and crustal evolution. Pergamon, N.Y., 310pp.

Dansgaard, W. (1964) Stable isotopes in precipitation. Tellus 16:436-468

Dury, G.H. (1971) Relict deep weathering and duricrusting in relation to the palaeoenvironments of middle latitudes. Geographical Journal 137:511-521

Dury, G.H. and J.C. Knox (1971) Duricrusts and deep-weathering profiles in southwestern Winsconsin. Science 174:291-292

Fitzpatrick, E.A. (1963) Deeply weathered rock in Scotland, its occurrence, age and contribution to the soils. Journal of Soil Science 14:33-43

Frakes, L.A. and J.E. Francis (1988) A guide to Phanerozoic cold polar climates from high-latitude ice-rafting in the Cretaceous. Nature 333:547-549

Goldich, S.S. (1938) A study in rock weathering. Journal of Geology 46:17-58

Gregory, R.T., C.B. Douthitt, I.R. Duddy, P.V. Rich and T.H. Rich (1989) Oxygen isotopic composition of carbonate concretions from the Lower Cretaceous of Victoria, Australia: implications for the evolution of meteoric waters on the Australian continent in a palaeopolar environment. Earth and Planetry Science Letters 92:27-42

Hall, A.M. (1985) Cenozoic weathering covers in Buchan, Scotland, and their significance. Nature 315:392-395

Hassanipak, A.A. and E.V. Eslinger (1985) Mineralogy, crystallinity, 18O/16O and D/H ratios of Georgia kaolins. Clays and Clay Minerals 33:99-106

Kemper, E. (1987) Das klima der Kreide-Zeit. Geologisches Jahrbuch, Reihe A, 96: 399pp.

Lawrence, J.R. and H.P. Taylor Jr. (1971) Deuterium and oxygen-18 correlation: Clay minerals and hydroxides in Quaternary soils compared to meteoric waters. Geochimica et Cosmochimica Acta 35:993-1003

Loughnan, F.C. (1975) Laterites and flint clays in the early Permian of the Sydney Basin, Australia, and their palaeoclimatic implications. Journal of Sedimentary Petrology 45:591-598

Lovelock, J.E. and M. Whitfield (1982) Lifespan of the biosphere. Nature 296:561-563

McAllister, J.J. and W.S. McGreal (1983) An investigation of deep-weathering products from a fossil laterite horizon in central Antrim, N. Ireland. in: Melfi, A.J. and A. Carvalho (eds.) Lateritization processes - Proceedings of the 2nd International Seminar on Lateritization processes, Sao Paulo. 345-358

Nilsen, T.H. and D.R. Kerr (1978) palaeoclimatic and palaeogeographic implications of a lower Tertiary laterite (latosol) on the Iceland-Faeroe Ridge, North Atlantic region. Geological Magazine 115:153-182

Ollier, C.D. (1984) Weathering. 2nd edition, Elsevier, New York, 270pp.

Parham, W.E. (1970) Clay mineralogy and geology of Minnesota's kaolin clays. Minnesota. Geological Survey Special Publication 10: 142pp.

Parra, M., C. Puechmaille, J.C. Dumon, P. Delmont and A. Ferragne (1986) Geochemistry of Tertiary alterite clay phases on the Iceland - Faeroe Ridge (Northern Atlantic), Leg 38, Site 336. Chemical Geology 54:165-176

Parra, M., P. Delmont, J.C. Dumon, A. Ferragne and J.C. Pons (1987) Mineralogy and origin of Tertiary interbasaltic clays from the Faeroe Islands, northeastern Atlantic. Clay Mineralogy 22:63-82

Pastor, J. and W.M. Post (1988) Response of northern forests to CO_2-induced climate change. Nature 334:55-58

Paton, T.R. and M.A.J. Williams (1972) The concept of laterite. Annals Association American Geographers 62:42-56

Rosenqvist, I.Th. (1975) Origin and mineralogy of glacial and interglacial clays of southern Norway. Clays and Clay Mineralogy 23:153-159

Savin, S.M. and S. Epstein (1970) Oxygen and hydrogen isotope geochemistry of clay minerals. Geochimica et Cosmochimica Acta 34:25-42

Sapojnikov, D.G. (1979) Lateritic formations of the U.S.S.R. Proceedings of an International Seminar on Lateritization Processes, Trivandrum, India. 185-189

Schmidt, P.W. and C.D. Ollier (1988) Palaeomagnetic dating of late Cretaceous to early Tertiary weathering in New England, N.S.W., Australia. Earth Science Reviews 25:363-371

Sheppard, S.M.F. (1977) The Cornubian Batholith, southwest England: D/H and 18O/16O studies of kaolinite and other alteration minerals. Journal of the Geological Society of London 133:573-591

Singer, A. (1984) The palaeoclimatic interpretation of clay minerals in sediments - A review. Earth Science. Reviews 21:251-293

Soman, K. and A.D. Slukin (1985) Lateritization cycles and their relation to the formation and quality of kaolin deposits in south Kerala, India. Proceedings of an International Seminar on Lateritization Processes, Tokyo: 319-332

Stein, R. and C. Robert (1985) Siliciclastic sediments at sites 588, 590 and 591: Neogene and Paleogene evolution of the southwest Pacific and Australian climate. Initial Reports of the Deep Sea Drilling Programme 90 U.S. Govt. Printer, Washington, 1437-1455

Walker, J.C.G., P.B. Hays and J.F. Kasting (1981) A negative feedback mechanism for the long-term stabilization of the Earth's surface temperature. Journal of Geophysical Research 86:9776-9782

Volk, T. (1987) Feedbacks between weathering and atmospheric CO_2 over the last 100 million years. American Journal of Science 287:763-779

Yurtsever, Y. and J.R. Gat (1981) Atmospheric waters. In: Gat, J.R. and R. Gonfiantini, (eds). Stable isotope hydrology: Deuterium and oxygen-18 in the water cycle. International Atomic Energy Agency, Vienna, 103-142

Considerations for Modelling Carbon Interactions between Soil and Atmosphere

H.L. BOHN

Department of Soil & Water Sciences
University of Arizona, Tucson, AZ 85721 USA

ABSTRACT

As an approximation over periods of a decade or so, the soil is at a steady state with respect to total carbon and atmospheric CO_2. Longer scale models should consider changes in both soil organic carbon (SOC) and inorganic carbon (SIC). The soil also stabilizes the trace gas composition of the atmosphere. This role is more obvious in arid regions where soil-gas reactions are less shielded by plants. The soil's large surface area, active microbial population, moisture content, and pH buffering are well suited to buffering C, N, and S in the atmosphere. Although soils emit large amounts of CO_2 and small quantities of other C, N, and S gases, the soil's primary role is as a C, N, and S sink.

Soil organic carbon is the largest, ca. 15×10^{14} kg C, carbon reservoir at the earth surface. This reservoir responds positively to increased photosynthesis, increased moisture, and perhaps directly to increased atmospheric CO_2 by decreasing the SOC decomposition rate. The SOC responds negatively to increased temperature. Aerobic soils are actively removing CO, hydrocarbons, and other volatile organic compounds from the atmosphere. The removal rate of trace gases is limited by a physical boundary problem rather than by chemical reaction rates.

Soil inorganic carbon may be as large a sink, ca. 10^{15} kg, as SOC but responds only slowly although positively to increased CO_2 and temperature, and negatively to increased moisture. Irrigation can both add and remove CO_2 from the atmosphere.

The amount of soil organic nitrogen, ca. 15×10^{13} kg N, is smaller than atmospheric N_2, but larger than the amounts of N in biomass and the surface oceans. Soil is an active sink for atmospheric NH_3, N_2O (alkaline soils), NO, and NO_2-N_2O_4. The size of the organic N reservoir in soil should respond similarly to soil organic C.

The amount of sulfur in soils, ca. 2×10^{12} kg S, is the largest S reservoir at the earth's surface and soil is an active sink for SO_2. The ratio of organic to inorganic S in soils is smaller than the organic N/inorganic N ratio. Organic S responds to climate changes like organic C and N in soils and biomass.

Carbon budgets are generally only moderately accurate, particularly in accounting for the half of fossil fuel C released which rapidly disappears from the atmosphere. Probably the biggest problem is the lack of understanding of the soil's role in C turnover. Many carbon models have almost ignored the soil. This paper attempts to identify some reactions which need quantification if C budgets are to be more accurate.

Soils and the Greenhouse Effect. Edited by A.F. Bouwman
© 1990 John Wiley & Sons Ltd.

Soil carbon is by far the largest active C reservoir in the carbon cycle (Bohn, 1986; Holser, 1989) and these estimates may have underestimated soil organic carbon (SOC) in tropical soils (Sanchez et al., 1982). Oxidation, or decay, of SOC is the largest source of CO_2 input to the atmosphere; the latest estimates for long team release are $6\text{-}8 \times 10^{13}$ kg (Houghton, 1985). The soil is also a large sink for the trace carbon, nitrogen, and sulfur gases in the atmosphere. Modelers heretofore have considered SOC and its CO_2 production only as steady states with respect to net primary photosynthesis (NPP) and organic decay. Better models will have to include non-steady state behavior. The Carboniferous Era is an obvious example in geologic time of long-term period of nonsteady state behavior -- SOC accumulation and reduced oxidation rates relative to NPP. The continuing accumulation of peat in northern latitudes since the last glaciation is a current example of the soil's natural non-steady state behavior in the carbon cycle. Half of the 4×10^{12} kg C y^{-1} of the annual fossil fuel consumption disappears from the atmosphere. This represents about 0.2% change of SOC.

Modelers may have ignored SOC because of its common misnomer, dead biomass, and therefore assumed that SOC is directly dependent on the amount of live biomass or on the magnitude of NPP. While this assumption simplifies their models, it neglects:

A. Increased oxidation of SOC to CO_2 by:

- Cultivation of new lands which generally reduces SOC by 1/3 and converts it to CO_2 ($1\text{-}7 \times 10^{11}$ kg y^{-1} (Bohn, 1978)).
- Increased temperature which accelerates SOC decay rates.
- Conversion of tropical forests to grasslands probably decreases SOC, but forest harvest and regeneration is a CO_2 sink if mature, slow-growing trees are removed and their C is stored in buildings.
- Decreased rainfall will convert SOC to CO_2 by increasing soil temperature and aerobiosis as well as be reducing NPP input to the soil.

Conversely, soil can remove CO_2 from the atmosphere by:

B. Accumulating SOC beyond receiving increased NPP, via:

- Increased rainfall which lowers oil temperatures and decreases 0_2 diffusion rates, thereby reducing oxidation rates.
- Minimum- or no-tillage agriculture which should increase SOC.
- Peat accumulation in northern latitudes (offset slightly by harvesting peat for fuel and loss of peatlands in warmer regions by cultivation). The annual growth of peatlands since the last glaciation is a loss of $1\text{-}3 \times 10^{11}$ kg C y^{-1} from the atmosphere in the peatlands themselves and probably an equal amount of C fixation in the much larger areas of adjoining mineral soils. The gradual filling-in of northern lakes and the growth of "blanket bogs" indicate that this process of C accumulation is continuing (Bohn, 1978).

- Increasing atmospheric CO_2 can repress SOC degradation rates. Since most SOC degradation occurs on the soil surface, the effect of CO_2 diffusion through the soil should be minimal and the negative feedback inhibition of increasing atmospheric CO_2 could be directly proportional to the CO_2 increase, or about 7×10^{12} kg C y^{-1} accumulating as undecomposed SOC in soils. This inhibition might be particularly important because it is largely independent of climate changes and soil nutrient availability.

In addition, soils in arid regions engage in:

C. Inorganic reactions involving CO_2:

- Natural precipitation and dissolution of CO_2 in arid soils.
- Release of CO_2 by irrigation with alkaline and CO_2-rich waters which precipitate $CaCO_3$ in soils and release CO_2.
- Removal of CO_2 by $CaCO_3$ dissolution and HCO_3^- accumulation as a result of irrigation.
- Release of CO_2 by upwelling of CO_2-rich waters in humid regions.

Schlesinger (1982) estimated the amount of soil inorganic carbon (SIC) to be about 10^{15} kg, composed of calcite ($CaCO_3$) and dissolved HCO_3^-, in the world's arid soils. The amount of $CaCO_3$ in semiarid grasslands has not been estimated but is probably of the same order of magnitude. Dating SOC by isotopic methods is difficult (Amundson, et al, 1989) but the natural flux of C from the atmosphere to SIC in arid soils is probably appreciably less than the SOC-atmosphere flux; Schlesinger (1985) estimated it to be $1-2 \times 10^{10}$ kg C y^{-1}. Although the amount of SIC responds to climate change, the change of deserts to semiarid grassland and vice versa during the last glacial epoch probably entailed more downward and upward movement of the zone of $CaCO_3$ accumulation rather than great losses or gains of $CaCO_3$ in the profile. The gains and losses would have been greater in the soils that cycled from semiarid to subhumid to semiarid climate at that time.

The SIC is as difficult to measure as SOC. The accumulation of $CaCO_3$ in soils is climate-dependent and seems to reach a maximum at perhaps 400 mm rainfall. With less rainfall, the soil weathering rate, and hence the Ca^{2+} supply rate, diminish and therefore so does the rate of $CaCO_3$ precipitation. At higher rainfall, the amount of Ca^{2+} leached from the soil profile begins to increase so $CaCO_3$ precipitation also decreases.

The distinctions between in-place formation, water transport and marine or littoral precipitation, and wind transport of $CaCO_3$ are inexact. Large amounts of $CaCO_3$ are, for example, in the soils and parent material surrounding the Arabian/Persian Gulf. The $CaCO_3$ also accumulates at the margins of intermittent lakes in arid regions. Both these SIC accumulations result from Ca^{2+} input from elsewhere rather than from minerals weathering in place. Wind erosion also transports considerable $CaCO_3$ in arid regions.

Regarding proposition C2, CO_2 release by $CaCO_3$ precipitation in irrigated soils, about 2×10^{14} m^2 of land is irrigated annually. Assuming 1 m of irrigation water (1 m^{3}m^{-2}), 500 mg dissolved HCO_3^-, CO_2, and loss of half of that carbon as CO_2 due to $CaCO_3$ precipitation and CO_2 evaporation, the irrigation of arid lands contributes about 10^{11} kg C (4 $\times 10^{11}$ kg CO_2) annually to the atmosphere.

The CO_2 fluxes due to propositions C3 and C4 are probably very much smaller than that due to C2. Irrigation waters in arid regions are more often alkaline thus causing more CaCO3 precipitation than dissolution. The amount of CO_2 released to the atmosphere by slow upwelling of subterranean water in soils counters, and perhaps cancels, that of the flux of C3.

REFERENCES

Amundson, R.G., O.A. Chadwick, J.M. Sowers and H.E. Doner (1989) The stable isotope chemistry of pedogenic carbonates at Kyle Canyon, Nevada. Soil Science Society of America Journal 53:201-210.

Bohn, H.L. (1978) On organic soil carbon and CO_2. Tellus 30:472-475. The paper contains a typographical error. The amount of SOC in peat soils should be 200 kg C m^{-2}.

Bohn, H.L. (1982) Estimate of organic carbon in world soils II. Soil Science Society of America Journal 46:1118-1119.

Holser, W.T., M. Schlidlowski, F.T. Mackenzie and J.B. Maynard (1989) Biogeochemical cycles of carbon and sulfur. In: B.C. Gregor et al. (Eds.): Chemical Cycles in the Evolution of the Earth, p 170-174. Wiley, NY.

Houghton, R.A, W.H. Schlesinger, S. Brown and J.F. Richards (1985) Carbon dioxide exchange between the atmosphere and terrestrial ecosystems. In: J.A. Trabalka (Ed.), Atmospheric Carbon Dioxide and the Global Carbon Cycle, p 113-140. U.S. Dept. of Energy, Washington, D.C.

Sanchez, P.A., M.P. Gichuru and L.B. Katz (1982) Organic matter in major soils of the tropical and temperate regions. Twelfth International Congress of Soil Science 1:99-114.

Schlesinger, W.H. (1985) The formation of caliche in soils of the Mojave Desert, California. Geochimica et Cosmochimica Acta 49:57-66.

An Approach to the Regional Evaluation of the Responses of Soils to Global Climate Change

J.J. LEE[1] and D.A. LAMMERS[2]

[1]US Environmental Protection Agency
[2]US Department of Agriculture Forest Service
[1,2]Environmental Research Laboratory, Corvallis, Oregon, USA, 97330

ABSTRACT

The world's soils will respond to global climate change primarily through changes in soil moisture, soil temperature and, on a somewhat longer time-scale, soil organic matter content (e.g. O horizons). These changes can affect the process of global climate change directly by affecting radiation balance (albedo, temperature modulation), hydrologic cycling, and production of greenhouse gases. Changes in soils might have indirect effects on climate by causing changes in vegetation.

How soils change will depend upon local features, such as topography, soil properties, and vegetation. Nevertheless, for use in assessing impacts on global climate, the changes must be quantified on regional and global scales. The US EPA's Direct/Delayed Response Project developed and successfully implemented an approach for integrating local and regional scales in the context of the effects of acidic deposition on lakes and streams. In this paper, this approach is described and examples of how it might be applied to the issue of global climate change are given.

The world's soils will respond to global climate change primarily through changes in soil moisture, soil temperature and, on a somewhat longer time-scale, soil organic matter content (e.g. O horizons). These changes will, in turn, have a dynamic role in influencing further changes in global climate and determining the effects of those changes. Soils can affect the process of global climate change directly by affecting radiation balance (albedo, temperature modulation), hydrologic cycling, and production of greenhouse gases. For example, in areas that have (or will have) sparse vegetation, soil color affects albedo. In the short term, soil color changes in response to moisture. In the longer term, changes in litterfall could affect the color of the soil surface. Decreased vegetation cover would result in increased exposure of the soil surface and could increase erosion, thus exposing subsoil with, possibly, different color. Soils can also act as heat-sinks, with important effects on regional climates and local soil processes. Soil particle-size distribution, permeability, structure, depth, slope, and landscape position affect waterflow across the surface and through the soil and

Soils and the Greenhouse Effect. Edited by A.F. Bouwman
© 1990 John Wiley & Sons Ltd.

thus affect hydrologic cycling. Changes in soil moisture could cause a soil to change from a reducing to a oxidizing regime, or vice versa, on a seasonal or annual basis. This would affect the emissions of the greenhouse gases methane and nitrous oxide. Changes in soil temperature and, possibly, organic matter content, would also affect these emissions. In addition to these direct effects, changes in soil moisture and temperature can directly cause changes in vegetation. Changes in the physical status of soils can indirectly affect vegetation through changes in soil bio-geo-chemical cycling, pathogen populations, etc. The changes in vegetation can, in turn, affect the process of global climate change by affecting vegetation cover and/or composition, evapotranspiration, and greenhouse gas emission. There are undoubtedly many more examples of ways in which changes in soils could affect global climate change either directly or indirectly through changes in vegetation. Thus it is necessary to model these interactive soil processes and to be able to project, on regional and global scales, how they will change and how these changes will affect global climate change.

One approach to regional assessment is to interpret general maps of soils, vegetation, climate, etc. This is very useful for defining regions of interest; i.e. those where changes might, on average, be expected to be greatest or most important. It is not, however, adequate for quantitative assessment of response and feedback to global climate change. For this, it is necessary to interpret the characteristics of systems at a local (i.e. large) scale. This is especially true if the responses are highly non-linear, as would be true if system response is disproportionately sensitive to relatively rare and/or small spatial features.

An example of the need for local scale interpretation is an increase in the extent of poorly drained soils resulting from increased precipitation. The distribution of precipitation within a watershed depends on surface and bedrock topography, and on soil properties. Depressions would accumulate water and might become wetlands, especially if they lack deep, permeable soils. Existing wetlands might be enlarged, or might be converted to lakes or streams. Thus, the net change in the spatial and temporal distribution of soils with reducing regimes will depend upon local features. In general, it will be important to develop and use models of catchment hydrology to predict the extent, duration, and amplitude of soil moisture deficit, flooding, and soil saturation. For some applications, local models of soil temperature will be needed. Quantitative models of, for example, the relationship between greenhouse gas emission and soil moisture and temperature will be needed to provide the link to effects on global climate change.

In practical application, the need for local interpretation conflicts with the goal of regional assessment. With limited resources, better resolution at the local scale can be achieved only by including fewer systems at the regional scale. Thus, it is not possible to optimize for both. It is, however, possible to have adequate resolution for a specific purpose at both scales, by constructing a suitable bridge between local and regional scales.

The regional assessment problem was successfully addressed in the context of acidic deposition by the US Environmental Protection Agency's Direct/Delayed Response Project (DDRP)(Church, 1989; Church et al., 1989; Lee et al, 1989a, 1989b). The central question was: Within the regions of concern, how many lakes and streams will become acidic due to current or altered levels of acidic deposition, and on what time scales? The DDRP considered three regions in the eastern US that, all together, extended about 1500 km, from Maine to Georgia. These regions were defined by using existing small-scale maps (e.g. maps of Major Land Resource Areas, regional/state soil maps) and other existing data (e.g. existing water chemistry data) to separate areas in which lakes and streams would be likely to differ in their responses to acidic deposition. For each region (or stratum within a region), 35 to 50 systems (lakes or streams) were randomly selected from all systems in the target population; this ensured adequate statistical characterization at the regional scale. The watershed of each lake or stream was mapped at a large scale (1:24,000) for soils, vegetation, depth to bedrock, land use, and drainage; this defined the occurance of these features within each watershed and within each region. Soil scientists, in conjunction with modelers and statisticians, used the known properties of the soils identified in each region to group soils into classes according to their probable roles in surface water acidification. Each of these classes was then sampled several times across the region, using a randomized sampling scheme. This yielded regionally applicable estimates of the means and variances of the relevant soils properties. The data on the regionally defined sampling classes were combined with the spatial distribution of soils from the mapping of each watershed and entered into models to predict the response of each watershed. Finally, the statistical structure was used to combine the predicted watershed responses into regional characterizations of response.

Within the DDRP, the regional soil classes provided the bridge between the local and regional scales. In this approach, the individuality of watersheds was ascribed to differences in the amounts of different classes of soils present. Thus, it was assumed that watershed-specific differences in the properties of soils within classes could be ignored within the resolution required at a regional scale. Because individual watershed responses were extrapolated to the regional scale, it was not necessary to have detailed, large scale, maps of soils, vegetation, etc. for entire regions. It was adequate to have these for a sample of the specific population of watersheds under study. This approach also had the advantages of preserving the relationships among watershed characteristics and of allowing investigation of the importance of sub-watershed areas, such as areas adjacent to lakes and streams. This approach can be summarized as follows:

- Identify the potential effect of interest (e.g. nitrous oxide emission);

- Develop hypotheses as to what soil processes are important (e.g. denitrification);
- Identify soil properties and landscape features that control or correlate with these processes (e.g. soil drainage, moisture, and temperature; vegetation);
- Use general (small scale) maps and imagery and other data from remote sensing to define regions with potential for greatest or most important change (e.g. use global/regional soil maps; ecoregion maps);
- Randomly select a statistically adequate number of systems within these regions (e.g. watersheds; square km of forest or agricultural land)
- Develop local (large scale) maps of relevant attributes of each system (e.g. 1:24,000 maps of soils, vegetation, landuse);
- For each region, develop soil/landscape classes that discriminate the important features for the potential effect under study (e.g. soil classes);
- Obtain regional data on the important properties (identified in step 3) of each class;
- Use models, regional data, and maps of individual systems to predict responses of individual systems (e.g. local changes in nitrous oxide emission per unit area);
- Use statistical design to extrapolate to regions; combine regions to extrapolate to globe (e.g. regional/global changes in nitrous oxide emissions associated with particular climate change scenarios).

Given the scale of the evaluation (ultimately, global), it will be necessary to use existing data to the maximum extent possible. Fortunately, many of the relevant soil properties are available from existing soil surveys. Examples include drainage class, hydrologic characteristics, texture, permeability, depth, soil available water capacity, soil temperature, and organic matter content. One potential problem is correlation of soils if multiple classification systems are used within a region of interest. In this case, classes should be defined so that they can use the information provided by all taxonomic systems, but that do not depend on the hierarchial structure of any one system; a similar approach was used by DDRP. Preliminary quantitative evaluations can be made by selecting points on small-scale maps and using existing large-scale soil maps (possibly in conjunction with remote imagery) to predict the responses of individual systems. In regions that lack adequate large-scale soil and vegetation maps, areas around randomly selected points will need to be mapped to provide a regional estimate of the occurance of soils and vegetation.

REFERENCES

Church, M.R. (1989) Predicting the future long-term effect of acidic deposition on surface water chemistry: The Direct/Delayed Response Project. EOS, Transactions, American Geophysical Union (in press).

Church, M.R., K.W. Thornton, P.W. Shaffer, B.P. Rochelle, G.R. Holdren, M.G. Johnson, J.J. Lee, R.S. Turner, D.L. Cassell, D.A. Lammers, W.G. Campbell, C.I. Liff, C.C. Brandt, L.H. Liegel, G.D. Bishop, D.C. Mortenson, S.M. Pierson, D.D. Schmoyer (1989) Future Effects of Long-Term Sulfur Deposition on Surface Water Chemsitry in the Northeast and Southern Blue Ridge Province (Results of the Direct/Delayed Response Project). US Environmental Protection Agency, Environmental Research Laboratory, Corvallis, OR.

Lee, J.J., D.A. Lammers, M.G. Johnson, M.R. Church, D.L. Stevens, D.S. Coffey, R.S. Turner, L.J. Blume, L.H. Liegel and R.H. Holdren (1989a) Watershed surveys to support an assessment of the regional effects of acidic deposition on surface water chemistry. Environmental Management 13:95-108.

Lee, J.J., D.A. Lammers, D.L. Stevens, K.W. Thornton and K.A. Wheeler (1989b) Classifying soils for acidic deposition aquatic effects: A scheme for the northeastern U.S. Soil Science Society of America Journal 53:1153-1163.

EXTENDED ABSTRACT 19.6

Changes in Soil Organic Carbon and Nitrogen as a Result of Cultivation

W.M. POST and L.K. MANN

Environmental Sciences Division, Oak Ridge National Laboratory
P.O. Box 2008, Bldg. 1000, Oak Ridge, Tennessee 37831-6335, USA

ABSTRACT

We assembled and analyzed a data base of soil organic carbon and nitrogen information from over 1100 profiles in order to explore factors related to the changes in storage of soil organic matter resulting from land conversion. The relationship between cultivated and uncultivated organic carbon and nitrogen storage in soils can be described by regression lines with uncultivated storage on the abscissa, and cultivated storage on the ordinate. The slope of the regression lines is less than 1 indicating that the amount of carbon or nitrogen lost is an increasing fraction of the initial amount stored in the soil. Average carbon loss for soils with high initial carbon is 23% for 1-meter depth. Average nitrogen loss for the same depth is 6%. In addition, for soils with very low uncultivated carbon or nitrogen storage, cultivation results in increases in storage. In soils with the same uncultivated carbon contents, profiles with higher C:N ratios lost more carbon than those with low C:N ratios, suggesting that decomposition of organic matter may, in general, be more limited by microbial ability to break carbon bonds than by nitrogen deficiency.

Because of increasing concern about atmospheric levels of CO_2, the role of soil in storing and releasing organic C must be more accurately assessed (Houghton et al., 1987). We need to be able to extrapolate in a systematic way from comparisons of C storage in cultivated and uncultivated soil samples to C losses at the landscape level. Although there is an abundance of literature quantifying losses of C due to cultivation, coverage of major agricultural soils has been uneven (Schlesinger, 1985; Mann, 1986). Our approach was to assemble a large data base and determine whether any general patterns of C and N storage changes as a result of cultivation could be inferred.

We have used statistical analyses to derive general estimates of changes in C and N storage across a broad range of soil types and amounts of organic matter. Data used in these analyses are from two types of sources - published paired plot studies and survey pedon data from the U.S. Department of Agriculture Soil Conservation Service (USDA-SCS) National Soils Analytical Laboratory. A critical part of the analysis was determining the suitability of the SCS data for quantifying changes due to land use. By comparing organic C and N storage in cultivated and uncultivated pedons within the same soil series, we

Soils and the Greenhouse Effect. Edited by A.F. Bouwman
© 1990 John Wiley & Sons Ltd.

hypothesized that we would be able to predict changes in soil organic C storage on a regional basis.

We extracted from the literature 625 "paired plots." Paired plots were carefully selected cultivated and uncultivated sites that were assumed to be similar before cultivation and were usually contiguous or close to one another. The data from the SCS files were not collected for the purpose of cultivated-uncultivated comparisons. Instead, they represent survey samples which we handled in the following manner. For each soil series, profile data were divided into two groups – cultivated and uncultivated. Pedons were considered cultivated if they had an Ap horizon at the surface and if vegetation was either row crop or pasture. The data for each soil series group were averaged to form a cultivated-uncultivated pair. Data from over 800 profiles were used to calculate 120 soil series pairs representing major agricultural soils in the United States. Despite variation introduced by a lack of control of sample locations, by limited sample numbers, and by limited information on past cultural treatments, regression analysis showed the SCS data to be similar enough to literature paired-plot data to be treated as samples from the same population.

Soils high in C and N tended to lose the largest amounts when cultivated. Soils low in C and N showed gains, though these gains were not consistent. That is, when gains occurred, they were not statistically significant. Cultivated C or N storage at 15, 30, or 100 cm can each be represented as a linear functions of uncultivated C and N (Figure 19.6.1). At low initial amounts, cultivation increases both C and N. Above a certain threshold (a different one for each element and depth considered) fractional loss of C or N caused by cultivation increases with the initial content. The estimated maximum loss of C resulting from cultivation is 29% for 0 to 15 cm, 22% for 0 to 30 cm, and 23% for 0 to 100 cm. Mean changes in C storage for soil orders demonstrated the overall pattern of change in C storage, with cultivation showed by pedon pairs. Means for Aridisols and Inceptisols, however, appeared to be slightly different from the overall pattern. For all soil orders the estimated maximum losses for N are 8% for 0 to 15 cm, 4% for 0 to 30 cm, and 6% for 0 to 100 cm. Average N changes are approximately one-fourth the corresponding C changes at all depths.

In the top 15 cm of uncultivated soils C:N ratios range from a mean of 12 in Mollisols to a mean of 22 in Ultisols (Figure 19.6.2). After cultivation, the range in surface-soil C:N ratio narrows, with most soil means ranging between 11 and 14. Some of the reduction from high C:N ratios to lower ones at the surface is caused by mechanical mixing of deeper soil (which usually has a lower C:N ratio) with the surface soil. This, however, can explain only a small portion of the changes observed. From the data we can infer changes in production and decomposition rates of different classes of organic matter.

The highest surface levels of N occur in Mollisols. These soils also have high amounts of C, but C:N ratios are low. In comparison with the usually forested soils (e.g., Alfisols, Ultisols, etc.), these prairie soils support productive vegetation that produces litter with high N contents and low lignin contents (high litter quality). After cultivation, both C and N are lost to the same extent, so C:N ratios are approximately the same. Soils that support native woody vegetation have a different

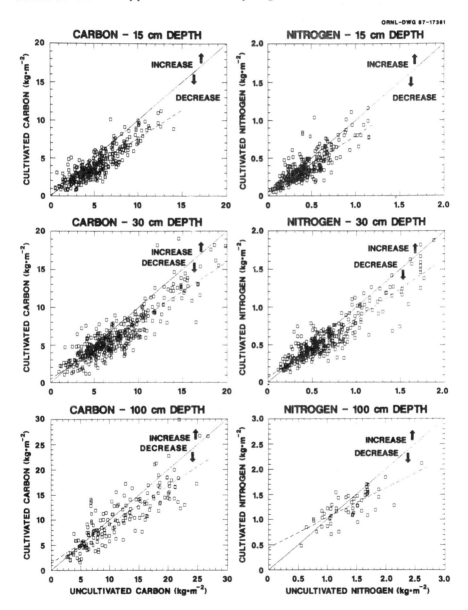

Figure 19.6.1 The relationship betwen carbon or nitrogen storage in the surface to a depth of 15, 30, and 100 cm before and after cultivation. A solid line indicates points of no change; a dashed line indicates the linear regression line.

pattern, which is most clearly seen in Ultisols. Here, C:N ratios are large (average C:N = 20.9) because of the production of low-quality leaf litter and woody material that decompose slowly. After cultivation, the C:N ratio is significantly lower (average C:N = 14.7). During this process, very little N is lost while organic C is converted to CO_2. If the mean uncultivated C:N ratio for a soil series is plotted as the abscissa, and the mean cultivated C:N ratio of a soil series is plotted as the ordinate, the 1:1 line is the set of points where uncultivated C:N = cultivated C:N. Mollisols are clustered near the 1:1 line with C:N = 12 (Figure 19.6.2). Other soil groups show significant reductions in C:N ratio with cultivation and lie largely below the 1:1 line. There is also a tendency for the variance of C:N ratios to decrease. As a result, cultivated soil C:N ratios lie in a narrower range than uncultivated ones.

The interactions between vegetation and climate determines litter quality, vegetation productivity, and decomposition rates (Meentemeyer, 1978; Melillo et al., 1982; Pastor et al., 1984; Parton et al., 1987). These in turn determine C and N storage in these various soil organic matter pools, as well as the interactions between C and N (Post et al., 1982, 1985; Wessman et al., 1988). Excluding Vertisols, for which we do not have a sufficient number of samples for adequate analysis, Mollisols

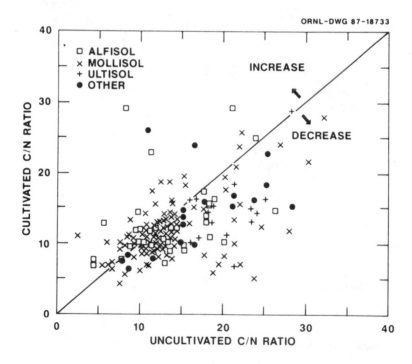

Figure 19.6.2 The relationship between carbon/nitrogen ratios in the surface depth of 15 cm before and after cultivation. A solid line indicates points of no change.

show the greatest accumulation of C and N. Organic matter in these soils is derived from high-quality litter, low in lignin and high in N. Cultivation results in loss of organic matter but not much change in the C:N ratio. Organic matter in forest soils, on the other hand, is derived from lignified woody material that is of low litter quality and contains high C:N ratios. As a result, these soils have a larger proportion of material in a slowly decomposing organic matter fraction (with higher C:N ratios) than in an actively decomposing fraction. Conversion to cultivation results in significant reduction of C:N ratios, which indicates compositional changes in soil organic matter, probably a significant reduction of a slowly decomposing organic matter fraction. This pattern is particularly striking in highly weathered soils such as Ultisols. Less-weathered forest soils such as Alfisols are intermediate between Ultisols and Mollisols. Soils that are initially very low in organic matter because of climatic factors (Aridisols), show gains in C and N upon cultivation.

Initial N content or concentration in relation to C does not contribute much to the variation in the magnitude of C change after conversion to agriculture, except that in soils with very high C:N ratios, more C may be mineralized than in soils with low C:N ratios. This contradicts the hypothesis that N enhances C loss by facilitating decomposition by microorganisms. Rather, it appears that microbial activity is, in general, C-limited. Replacement of forests by crops results in a change in litter: woody material with a high proportion of slowly decomposing organic matter is replaced by herbaceous crops with a lower proportion of slowly decomposing organic matter. The net effect is that the slowly decomposing organic matter fraction in soils where there were large inputs to the slow fraction from native woody vegetation is reduced by conversion to more herbaceous crop material that produces litter with lower lignin:N ratios. This reduces high C:N ratios so that eventually all C:N ratios in cultivated soils converge on the ratio typical of prairie soils (agricultural crops and pastures can be thought of as producing litter similar in quality to prairie vegetation). The final organic matter content is then determined by the microorganism activity that is related to soil-forming factors independent of initial N concentration. Thus, decomposition of recalcitrant organic matter is probably not significantly accelerated, but over a long time (50 to 100 years) organic matter will eventually be reduced and not be replenished by sufficiently large additions of new recalcitrant organic matter to maintain this important pool of organic matter.

ACKNOWLEDGEMENTS

Research sponsored by the U.S. Department of Energy, Carbon Dioxide Research Program, Ecological Research Division, Office of Health and Environmental Research, Budget Activity Number KP 05 00 00 0, and the National Science Foundation's Ecosystem Studies Program under

Interagency Agreement BSR-8417923, under contract DE-AC05-84OR21400 with Martin Marietta Energy Systems, Inc. Publication No. 3412, Environmental Sciences Division, Oak Ridge National Laboratory.

REFERENCES

Houghton, R.A., R.D. Boone, J.R. Fruci, J.E. Hobbie, J.M. Melillo, C.A. Palm, B. Peterson, G. Shaver, G.M. Woodwell, B. Moore, D.L. Skole and N. Myers (1987) The flux of carbon from terrestrial ecosystems to the atmosphere in 1980 due to changes in land use: Geographic distribution of global flux. Tellus 39B:122-39.

Mann, L.K (1986) Changes in soil carbon storage after cultivation. Soil Science 142:279-88.

Meentemeyer, V. (1978) Macroclimate and lignin control of litter decomposition rates. Ecology 59:465-72.

Melillo, J.M., J.D. Aber and J.F. Muratore (1982) Nitrogen and lignin control of hardwood leaf litter decomposition dynamics. Ecology 63:621-26.

Parton, W.J., D.S. Schimel, C.V. Cole and D.S. Ojima (1987) Analysis of factors controlling soil organic matter levels in Great Plains grasslands. Soil Science Society of America Journal 51:1173-79.

Pastor, J., J.D. Aber, C.A. McClaugherty and J.M. Melillo (1984) Aboveground production and N and P cycling along a nitrogen mineralization gradient on Blackhawk Island, Wisconsin. Ecology 65:256-68.

Post, W.M., W.R. Emanuel, P.J. Zinke and A.G. Stangenberger (1982) Soil carbon pools and world life zones. Nature 298:156-59.

Post, W.M., J. Pastor, P.J. Zinke and A.G. Stangenberger (1985) Global patterns of soil nitrogen. Nature 317:613-16.

Schlesinger, W.H. (1985) Changes in soil carbon storage and associated properties with disturbance and recovery. In J. R. Trabalka and D. E. Reichle (eds.), The Changing Carbon Cycle: A Global Analysis. Springer-Verlag, New York.

Wessman, C.A., J.D. Aber, D.L. Peterson and J.M. Melillo (1988) Remote sensing of canopy chemistry and nitrogen cycling in temperate forest ecosystems. Nature 335:154-56.

The Rates of Carbon Cycling in Several Soils from AMS ^{14}C Measurements of Fractionated Soil Organic Matter

S.E. TRUMBORE[1], G. BONANI[2] and W. WÖLFLI[2]

[1]*Lamont-Doherty Geological Observatory of Columbia University, Palisades, NY 10964, USA*
[2]*Insitut für Mittelenergiephysik, ETH-Hönggerberg, 8093 Zürich Switzerland*

ABSTRACT

^{14}C mean residence times (MRT) of fractionated organic matter are reported for three pre-bomb soil profiles. Comparisons of organic matter extracted with acid and base showed that the longest MRTs were associated with the non-acid-hydrolysable fraction.

The MRT of organic matter in a soil layer represents a combination of the rates of several processes, including decay to CO_2 and transport out of the layer. In some instances (notably in the A horizon of the Podzol soil studied here), the MRT is dominated by the rate of transport, rather than the rate of decay. Thus it is important to use the distribution and balance of carbon in the soil profile to assess the meaning of the MRT with respect to influencing atmospheric CO_2.

INTRODUCTION

To better understand the role of soil organic matter in the global carbon cycle, it is necessary to determine the residence time of carbon in soils, and its variation with factors such as climate, vegetation and topography. The ^{14}C age of organic matter extracted from soils is interpreted as a mean residence time (MRT) for carbon. MRT's reported for soil organic matter range from between several hundred and several thousand years, with mean residence time increasing with depth in the soil profile (Scharpenseel et al., 1968). The interpretation of the ^{14}C age as a MRT is complicated in modern soils by the presence of ^{14}C produced during testing of thermonuclear weapons in the atmosphere. Since any modern soil sample will be contaminated to an unknown degree with bomb ^{14}C, it is important to measure MRT in soils which were collected before the peak of ^{14}C production, in 1963-1964.

Chemical and physical fractionation of organic matter separates it into components with widely differing MRTs (Campbell et al., 1967; Goh et al., 1984). As we wish to study the carbon cycle in soils in order to predict its response to a perturbation (such as climate change), it is important to be able to separate carbon consistently and quantitatively into more labile and more refractory components. This approach can

Soils and the Greenhouse Effect. Edited by A.F. Bouwman
© 1990 John Wiley & Sons Ltd.

also yield information on the processes and rates of formation of the complex, heterogeneous suite of compounds which make up soil organic matter.

We report MRT's derived from measurements of ^{14}C in fractionated soil organic matter for three soils. The samples were collected prior to 1960, so that problems of bomb ^{14}C contamination are avoided, and the ^{14}C age can be interpreted correctly as a residence time for carbon in the soil. Because of the limited availability and low carbon contents of the stored soil samples, we used Accelerator Mass Spectrometry (AMS) to measure ^{14}C. The results will be used to illustrate the range of stability of organic compounds in soils, and to show differences in rates and processes of carbon cycling in soil profiles from different climate regimes.

SAMPLES AND FRACTIONATION PROCEDURE

The profiles studied represent three important soils types: Mollisol, Ultisol, and Podzol. Pertinent information is given in Table 19.7.1.

Table 19.7.1 Description of the soil profiles obtained for this study

Soil Type	Mollisol	Ultisol	Podzol
Sample location	Tama County, Iowa USA	Amazon Basin, Brazil	near Leningrad, USSR
Vegetation type	Cultivated	Tropical Forest	Temperate Forest
Year collected	1959	1959	1927
Obtained from	Dr. T. Fenton Iowa State University	Dr. W. Sombroek ISRIC Wageningen	Dr. W. Sombroek ISRIC Wageningen
Information available (ref.)	Iowa soil survey Profile S59Iowa-86-1	Profile 303 Sombroek (1966)	Glinka Memorial Collection (#4) Mokma & Buurman (1982)

All soils were dried, ground, and sieved (<2mm). To remove small fragments of vascular plant matter (not already removed by sieving), 3 to 4g aliquots of soil were repeatedly shaken with $ZnBr_2$ solution (1.6 g cm^{-3}), then centrifuged. Undecayed plant matter, charcoal, and easily soluble organic matter were decanted with the $ZnBr_2$ solution. The >1.6 g cm^{-3} material (referred to here as the dense fraction) was rinsed, dried, re-ground, and used for subsequent fractionation steps.

Several fractionation procedures, including hydrolysis in acid (hot 6N HCl) and base (cold 0.1N NaOH - 0.1N $Na_5P_2O_7$), were tested to determine the most consistent method for separating organic matter into labile and refractory components. Details of these procedures are given in Trumbore (1988). Table 19.7.2 shows a comparison of acid and base hydrolysis for selected samples from the three horizons studied here. In each case, the residue after acid (hot 6N HCl) hydrolysis yield the longest MRT. This confirms previous findings by Campbell et al. (1967) and Trumbore (1988), comparing acid and base hydrolysis of fractionated organic matter from other soils.

AMS ^{14}C measurements were made at the ETH/SIN facility in Zürich, Switzerland (Suter et al., 1984), for the unfractionated soil, the dense fraction (residue after $ZnBr_2$), and the non-acid-hydrolysable residue. The carbon inventory and MRT for the low density and hydrolysable fractions were calculated from the mass balance of ^{14}C and C.

RESULTS

The inventory of carbon (as g C m[-2] per cm depth) and the ^{14}C derived MRT for all three soil profiles are plotted in Figure 19.7.1. Open triangles represent the unfractionated organic matter; open and filled circles represent the dense and acid residue fractions, respectively. The 'error bars' represent the depth range integrated by each sample. Table 19.7.3 gives the total carbon inventory, average age, and calculated fluxes of carbon for the unfractionated soil, and hydrolysable and non-hydrolysable components, for the A and B horizons of the three soils.

Details of the profiles shown in Figure 19.7.1 reflect differences in the processes of storage and vertical transport of organic carbon. The amount of carbon stored in the soil decreases with depth (and MRT increases) for the Mollisol and Ultisol, which lack sharply defined horizon boundaries. By contrast, a maximum in carbon (and a local minimum in MRT) is observed in the B horizon (22-37cm) in the Podzol profile. This maximum consists of mostly soluble and hydrolysable organic matter, since the non-hydrolysable carbon remains low in abundance through this layer.

The ratio of hydrolysable to non-hydrolysable carbon ranges from a maximum of 7.7 in the Podzol B horizon to 0.5 in the Ultisol (see Table 19.7.3), demonstrating a large range in the composition of the organic matter contained in the three soils. Differences in the MRT of the soil organic matter fractions are not so striking. Although the residue after acid hydrolysis remains low in abundance at all depths (with the exception of the Ultisol A1 horizon), the MRT of this fraction increases with depth. It is probable that the acid hydrolysis procedure is not completely efficient at separating refractory and labile organic matter, and that the gradients seen in the soil profiles result from varying degrees of 'contamination' with a younger component.

Table 19.7.2 Comparison of acid and base hydrolysis in fractionating soil organic matter by MRT. For each fraction, the gravimetric %C, the per cent of the total carbon contained in the soil layer (%C$_T$), and the MRT are given. The error in %C determinations is 10% (up to 50% for values less than 0.2%). MRT is the ^{14}C age, errors are ±70 years or less. Values with asterisks were calculated from mass balance of other measurements

Soil	Horizon	Depth (cm)	Unfractionated			Dense (ZnBr$_2$ Residue)			Base Extract Residue			Acid Extract Residue		
			%C	%C$_T$	MRT	%C	%C$_T$	MRT	%C	%C$_T$	MRT	%C	%C$_T$	MRT
Mollisol	A1	0-16.5	2.5	100	770	2.4	80	873	1.1	28	850	0.7	18	1240
	A3	28-42	1.5	100	2190	1.4	92	2580	0.6	32	2760	0.5	30	3515
Ultisol	A1	0-22	3.5	100	860*	3.0	85	940	-	-	-	2.4	50	2550
Podzol	E1	0-12	1.5	100	1550	1.2	80	1550	0.7	46	2020	0.5	33	2250
	B1	22-30	2.6	100	2000	1.3	50	2380	0.2	10	2570	0.2	10	2800

Table 19.7.3 Carbon inventory and averaged MRT for A and B horizons of the three soils investigated. Fluxes are calculated as the product of inventory times the reciprocal of the MRT for each layer. The ratios of hydrolyzable to non-hydrolyzable properties are also given. "Unfractionated" refers to the dense fraction in the Mollisol and Ultisol, where differences between total and dense fractions are small. The 60-95cm portion of the B horizon of the Ultisol (*) was assumed to have the same characteristics as the 95-150cm portion. ND = not determined

	Mollisol			Ultisol			Podzol		
	Inventory (kgC m⁻²)	Annual Flux (gC m⁻²y⁻¹)	MRT (y)	Inventory (kgC m⁻²)	Annual Flux (gC m⁻²y⁻¹)	MRT (y)	Inventory (kgC m⁻²)	Annual Flux (gC m⁻²y⁻¹)	MRT (y)
A Horizon	(0-42cm)			(0-60cm)			(0-18cm)		
Unfractionated	13.8	10.1	1380	16.0	10.1	1580	3.8	2.0	1940
Hydrolysable	10.7	8.9	1200	5.4	4.5	1200	2.0	1.4	1430
Non-hydrolysable	3.1	1.2	2580	10.6	5.6	1890	1.8	0.6	2960
Hyd/Non-hyd.	3.4	7.4	0.5	0.5	0.8	0.6	1.9	2.3	0.5
B Horizon	(42-64cm)			(60-150cm)*			(19-37cm)		
Unfractionated	3.3	1.0	3440	7.6	ND	ND	6.1	2.6	2270
Hydrolysable	2.6	0.9	2980	3.6	ND	ND	5.4	2.4	2220
Non-hydrolysable	0.7	0.1	5420	4.0	0.7	6100	0.7	0.2	2800
Hyd/Non-hyd.	3.7	9.0	0.5	0.9	ND		7.7	12.0	0.8

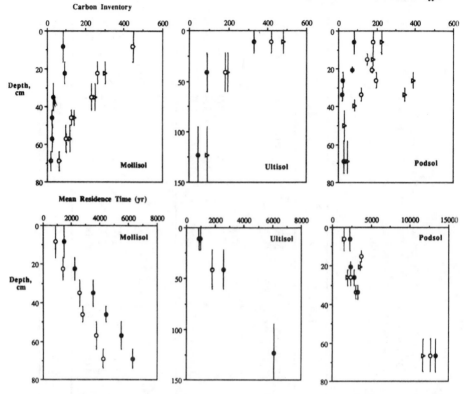

Figure 19.7.1 Summary of the measurements of carbon and ^{14}C MRT for fractionated organic matter in the three soil profiles. The top three figures show carbon inventory (in units of gC m^{-2} per cm of depth) as a function of depth in the soil for the Mollisol (left), Ultisol (centre) and Podzol (right). MRT (in years) as a function of depth is shown as the bottom three figures. The different fractions are identified as follows:

    open triangles = unfractionated soil
    open circles   = dense fraction (residue after ZnBr$_2$ separation)
    filled circles = non-hydrolysable carbon (residue after acid hydrolysis of the dense
                               fraction).

The "error" bars indicate the depth range integrated by the sample.

DISCUSSION

The MRT of carbon in fractionated soil organic matter reflects the combined rates of several processes occurring in the soil. These processes, illustrated schematically in Figure 19.7.2, include decay to CO_2, transport out of the profile (e.g. as dissolved organic matter), and transformation to other fractions. At steady state, when the carbon is neither accumulating in nor being lost from the profile, inputs equal losses, and the reciprocal of the mean residence time equals the sum of the rates of loss by decay, transport and transformation. The relative importance of these processes can be assessed from considering the distribution of carbon in the soil profiles.

A comparison of the fluxes of carbon from the A and B horizons from the three soils (Table 19.7.3) can show the degree to which transport may be important as a means of removing carbon from the A horizon. If we assume that the input of fresh vascular plant material does not extend into the B horizon, then the flux of carbon transported from the A horizon must equal the net flux out of the B horizon (assuming steady state). Since the carbon flux from the Podzol A horizon approximately equals the required input to B, the MRT of carbon in the A horizon is primarily a reflection of the time required to transport carbon downward. Exported carbon is not directly interacting with atmospheric CO_2, so that interpretation of the MRT in this soil as a time scale for organic matter decay would be misleading.

The Mollisol and Podzol soil profiles have both developed since the end of the last glaciation. Organic matter in these soils with MRT's longer than about 5000 years (mostly found deeper in the soils), may not

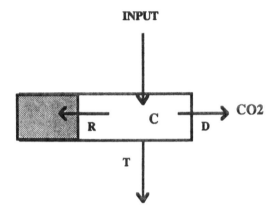

$$\frac{dC}{dt} = P - (T + R + D)\,C = 0 \text{ (steady state)}$$

and

$$\frac{1}{MRT} = T + R + D$$

if $T*C_n$ is the sole input to underlying layer,

$$T*C_n = \frac{C_{n+1}}{MRT_{n+1}}$$

Figure 19.7.2 Factors affecting the balance of carbon in an organic matter fraction in a layer of soil, where: C = the pool of carbon in the fraction, and D, T and R represent the rates of decay (to CO_2), transport from the layer, and transformation to another fraction, respectively. At steady state, the rate of input of carbon equals the rate of loss, so that 1/MRT equals D+T+R.

yet have achieved steady state. In these fractions, the MRT may reflect the average ^{14}C content in carbon which has accumulated through time. For example, the ^{14}C age of the hydrolysis residue of organic matter from the C horizon of the Podzol soil is 13,360 years. As very little carbon has been added to this layer, carbon transported from the A horizon may be passively accumulating in the B horizon, making this soil an efficient sink of carbon.

CONCLUSIONS

The study of ^{14}C in fractionated mineral soil organic matter provides valuable information regarding the variability of carbon cycling in soils. This preliminary study of profiles from three important soil types illustrates the need to understand the processes of decay, transport and transformation of soil organic matter. Careful study of soil profiles can help quantify the relative importance of these loss processes in determining the MRT.

The majority of carbon stored as organic matter in the soils measured in this study has MRT longer than ocean circulation time scales. An estimate of the potential role of soil carbon on a global scale, however, is not possible without a more extensive data set. Measurements of ^{14}C in pre-bomb soils can be used to study carbon cycling in the long-term (several hundreds to thousands of years). Careful comparison of pre and post bomb soil profiles will give information on carbon cycling on shorter time scales (O'Brien and Stout, 1978; Trumbore, 1988).

REFERENCES

Campbell, C.A., E.A. Paul, D.A. Rennie and K.J. McCallum (1967) Applicability of the carbon-dating method of analysis to soil humus studies. Soil Science 104: 217-224.

Goh, K.M., T.A. Rafter, J.D., Stout and T.W. Walker (1976) The significance of fractionation dating in dating the age and turnover of soil organic matter. New Zealand Journal of Soil Science 27:89-100.

Mokma, D.L. and P. Buurman (1982) Podzols and Podzolization in temperate regions. International Soil Museum Monograph 1, Wageningen, The Netherlands.

O'Brien, B.J. and J.D. Stout (1978) Movement and turnover of soil organic matter as indicated by carbon isotopic measurements. Soil Biology and Biochemistry 10:309-317.

Scharpenseel, H.W., M.A. Tamers and F. Pietig (1968) Altersbestimmung von Boden durch die Radiokohlenstoffdatierungsmethode. Zeitschrift für Pflanzenernährung und Bodenkunde 119:34-52.

Sombroek, W.G., 1966. Amazon soils. Centre for Agricultural Publication and Documentation, Wageningen, The Netherlands.

Suter, M. and others (1984) Precision measurements of 14C in AMS - some results and prospects. Nuclear Instruments and Methods B5:117-122.

Trumbore, S.E. (1988) Carbon cycling and gas exchange in soils. PhD thesis, Columbia University.

Positive and Negative Feedback Loops within the Vegetation/Soil System in Response to a CO_2 Greenhouse Warming

G.H. KOHLMAIER, A. JANECEK, and J. KINDERMANN

Institut für Physikalische und Theoretische Chemie
Johann Wolfgang Goethe-Universität, Niederurseler Hang
D-6000 Frankfurt 50, F.R.G.

ABSTRACT

This paper focuses on the significance of the land biota and their role as a source or sink of atmospheric CO_2 in a changing world. Three areas of concern are addressed:
1) The response of the vegetation and soils to temperature increases as expected within the CO_2 greenhouse effect tending to release carbon dioxide to the atmosphere, thus amplifying the occurring greenhouse effect in a positive feedback loop.
2) A regulating and stabilizing effect of intact land biota on the climate. There are indications that the present atmospheric levels of carbon dioxide, about 25% above the preindustrial value, already have led to a "CO_2 fertilization effect" of significant magnitude for the atmospheric carbon contents. Such negative feedback control can lead to additional storage of atmospheric CO_2 in the form of wood and organic matter in the underlying soils. Destruction of biota may alter or stop this stabilizing effect.
3) Cybernetic feedback and gain factors, with the corresponding temperature increases for the described processes are calculated using different future scenarios, comparing steady state conditions to the dynamic approach.

MODEL DESCRIPTION

Both net primary production of plants and the organic carbon components of the soil are known to respond to temperature changes, often characterized by their specific Q_{10} values. The Q_{10} value describes the rate increase of a given process for a temperature increase of 10 K. We can assume that for most cases the Q_{10} values for net primary production are lower (1.0 to 1.5) than those for heterotrophic respiration and soils (1.3 to 4.0), with the consequence that carbon is lost to the atmosphere when the temperature rises (e.g. Kohlmaier 1981, Schleser 1981), amplifying the CO_2 greenhouse effect in a positive feedback loop.
    For a complete analysis we need to consider at least two negative feedbacks: the direct effect of CO_2 on the net primary production ("fertilization effect") and the enhancement of plant growth by increased temperatures within a limited range, as demonstrated by e.g., Idso et al.

Soils and the Greenhouse Effect. Edited by A.F. Bouwman
© 1990 John Wiley & Sons Ltd.

(1989). In a preliminary model study we suggest to express this combined effect by the following relation ($Q = Q_{10} - 1$):

$$NPP = NPP^0(1 + \tilde{Q}(AB)/10 \times \Delta T) \times (1 + \beta \ln[CO_2/CO_2^0]) \quad (19.8.1)$$

In Figure 19.8.1 we show the dynamic equations (19.8.2), (19.8.3) and (19.8.4) which describe the carbon release or uptake of the biota-soil system as induced by temperature changes, assuming a two layer structure for the soils. It is important to emphasize here that we need to integrate over a variety of biomes with different Q_{10} responses, each of which still has some considerable uncertainties.

A: Atmosphere   H1: litter and upper soil $\tilde{Q}$: Q_{10}-value -1

B: Biota   H2: lower soil ΔT: Temperature change

β: CO_2 fertilization factor κ : fraction destinated to H2 $\kappa = 0.2$

$$\frac{dB}{dt} = NPP^0(1 + \frac{\tilde{Q}(AB)}{10} \Delta T(t))(1 + \beta \ln \frac{CO_2}{CO_2^0}) - \frac{NPP^0}{B^0} B \quad (2)$$

$$\frac{dH1}{dt} = \frac{NPP^0}{B^0} B - (1-\kappa)(1 + \frac{\tilde{Q}(H1A)}{10} \Delta T(t)) \frac{NPP^0}{H1^0} H1$$
$$- \kappa (1 + \frac{\tilde{Q}(H1H2)}{10} \Delta T(t)) \frac{NPP^0}{H1^0} H1 \quad (3)$$

$$\frac{dH2}{dt} = \kappa (1 + \frac{\tilde{Q}(H1H2)}{10} \Delta T(t)) \frac{NPP^0}{H1^0} H1$$
$$- \kappa (1 + \frac{\tilde{Q}(H2A)}{10} \Delta T(t)) \frac{NPP^0}{H2^0} H2 \quad (4)$$

$A^0 = 600$ Gt C; $B^0 = 700$ Gt C; $H1^0 = 55$-330 Gt C; $H2^0 = 1320$-$H1^0$ Gt C; $F_{ab}^0 = 55$ Gt C yr^{-1}

Figure 19.8.1 Systems equations.

STEADY STATE CONSIDERATIONS

Eq. (19.8.1) can be simplified for small changes to read

$$NPP = NPP^0(1 + \tilde{Q}(AB)/10 \times \Delta T + \beta \ln[CO_2/CO_2^0]) \quad (19.8.5)$$

As for steady state conditions the temperature change ΔT with respect to a CO_2 increase can be expressed approximately by a logarithmic expression we write:

$$\Delta T^{st} = C \times \ln (1 + \Delta CO_2^{st}/CO_2^0) \qquad (19.8.6)$$

where C refers to the climate sensitivity, which e.g. can be derived from 3-D general circulation models (GCMs). Depending on the model assumptions, temperature increases between 1 and 5 K for a CO_2 doubling have been suggested. The uncertainty is caused by very difficult estimation of the various climatic feedback factors, which still exclude the coupling of the atmosphere and the biota soil system. Since both the fertilization effect and the temperature increase show a logarithmic dependence on CO_2, the fertilization and the temperature term can be expressed by just one effective parameter, β_{eff} or Q_{eff}, respectively:

$$NPP = NPP^0(1 + \beta_{eff} \ln[CO_2/CO_2^0]) \qquad (19.8.7)$$

with $\qquad \beta_{eff} = \beta + C \times \tilde{Q}(AB)/10 \qquad (19.8.7a)$

or equivalently

$$NPP = NPP^0(1 + \tilde{Q}_{eff}(AB)/10 \times \Delta T) \qquad (19.8.8)$$

with $\qquad \tilde{Q}_{eff}(AB)/10 = \tilde{Q}(AB)/10 + \beta/C \qquad (19.8.8a)$

In order to obtain a steady state solution, the model was additionally simplified, unifying H1 and H2 in a single soil carbon compartment H. Within the above made assumptions the differential equations of the biota-soil system are given by

$$dB/dt = NPP^0 (1 + \tilde{Q}_{eff}(AB)/10 \times \Delta T(t)) - (NPP_0/B^0)B \qquad (19.8.9)$$

$$dH/dt = (NPP^0/B^0)B - (1 + [\tilde{Q}(HA)/10] \Delta T(t)) \times (NPP^0/H^0)H \quad (19.8.10)$$

The equilibrium response of the biota and the soils is thus obtained by the stationary state approximation (dB/dt and dH/dt equal to zero):

$$B^{st} = B^0(1 + \tilde{Q}_{eff}(AB)/10 \times \Delta T) \qquad (19.8.11)$$

$$H^{st} = H^0 (1 + \tilde{Q}_{eff}(AB)/10 \times \Delta T) / (1 + \tilde{Q}(HA)/10 \times \Delta T) \quad (19.8.12)$$

The CO_2 release from the biota-soil system is calculated by the following equation:

$$\Delta CO_2^{st}{}_{bio} = (H^0 - H^{st} + B^0 - B^{st}) \times a_f \qquad (19.8.13)$$

where a_f is the airborne fraction of CO_2 released, here taken to be constant and equal to 0.6.

In Table 19.8.1 we summarize the results for fertilization and for a CO_2 release from the soils within the equilibrium considerations for $\Delta T^{st}(2 \times CO_2)$ of 1, 3 and 5 K, according to different climate sensitivities to CO_2, as given by Tricot and Berger (1987). The CO_2 release from the biota-soil system is considerably reduced if we assume a CO_2 fertilization effect or any other change in environmental parameters which increases production including the warming in the boreal and tundra regions; e.g. for a 3 K temperature increase and a Q(HA) value of 1 for the soils the fraction of carbon dioxide released is reduced from

15% ($\tilde{Q}_{eff}(AB)=0$) to −5% ($\tilde{Q}_{eff}(AB)=0.8$), implying that strong fertilization can inverse the flux.

Table 19.8.1 Sensitivity of the Equilibrium Model: Response of the biota/soil system to a change in temperature and a change in net primary production, characterized by the specific $\tilde{Q}$ values ($\tilde{Q} = Q_{10}$-1). $\tilde{Q}(HA)$ describes the temperature dependence of the soils, while $\tilde{Q}_{eff}(AB)$ comprises the effects on net primary production (temperature, CO_2 fertilization effect)

ΔT_{eq} =	1 K			3 K			5 K		
$\tilde{Q}$ (HA) =	1	1.5	2	1	1.5	2	1	1.5	2
a) fraction released $\tilde{Q}_{eff}(AB)=0.0$ (NPP = const.)	0.06	0.09	0.11	0.15	0.20	0.25	0.22	0.28	0.33
b) fraction released $\tilde{Q}_{eff}(AB)=0.5$ (NPP increases with ΔT)	0.01	0.04	0.06	0.02	0.08	0.13	0.02	0.10	0.16
c) fraction released $\tilde{Q}_{eff}(AB)=0.8$ (NPP increases with ΔT)	-0.02	0.01	0.04	-0.05	0.01	0.06	-0.10	-0.01	0.06
d) fraction released $\tilde{Q}_{eff}(AB)=-0.5$ (NPP increases with ΔT)	0.11	0.13	0.15	0.25	0.32	0.36	0.41	0.46	0.50
case a) Gt C released $H^\circ = 330$ Gt C $B^\circ = 700$ Gt C	30	43	55	76	102	124	110	141	165
Case a) Gt C released $H^\circ = 1320$ Gt C $B^\circ = 700$ Gt C	120	172	220	305	410	495	440	566	660

DYNAMIC RESPONSE

In the equilibrium considerations presented above, we calculated the CO_2 release or uptake of the biota/soil system for a given external temperature variation. In a complete analysis of the effect, we need, however, to take into consideration the feedback between the additional CO_2 released (absorbed) and the corresponding additional temperature increase (decrease).

GCMs correlate a surface temperature increase with a prescribed external atmospheric CO_2 concentration. As shown above in eq. (19.8.6), steady state or equilibrium models lead to a logarithmic dependence of the temperature increase with increasing levels of CO_2. In a first order approximation this logarithmic dependence can be maintained in the dynamic calculations if a time delay τ of approximately 15 years (Tricot and Berger, 1987) is introduced:

$$\Delta T(t) = C.\ln(1 + \alpha.\Delta CO_2(t-\tau)/CO_2^0) \qquad (19.8.14)$$

where

$$\Delta CO_2(t) = \Delta CO_{2\,ext}(t) + \Delta CO_{2\,bio}(t, \Delta T, \Delta CO_2(t)) \qquad (19.8.15)$$

is the total atmospheric CO_2 due to the prescribed external carbon input (fossil fuel plus biogenic release from deforestation) and the carbon released or absorbed by the soil-biota system. $\alpha > 1$ transforms the CO_2 increase into an equivalent CO_2 increase, which in this study is chosen $\alpha = 1.3$, referring to a scenario, where the additional greenhouse gases refer to methane predominantly, while the chlorofluorocarbons are phased out by the end of this century. ΔCO_2 ext was chosen to be representative of an intermediate CO_2 release, with an expected CO_2 doubling occurring in 2070, while the equivalent CO_2 concentration doubles in 2050. The final external ceiling $\Delta CO_2(\infty)$ has been chosen to be 500 ppm excess CO_2, or 780 ppm total within the suggested logistic function:

$$\Delta CO_{2\,ext}(t) = \Delta CO_2(\infty)/(1+(\Delta CO_2(\infty)/\Delta CO_2(0)-1).\exp(a.(t-t_0)) \qquad (19.8.16)$$

with $a = 0.025$ and $\Delta CO_2(0) = 2.7$ ppm. In the feedback analysis ΔCO_2 bio of eq. (19.8.15) was obtained by numerical integration using eq. (19.8.2-4) to determine the differential CO_2 release and eq. (19.8.14) to obtain the corresponding feedback temperature.

FEEDBACK FACTORS FOR EQUILIBRIUM AND DYNAMIC RESPONSE

In the equilibrium case of chapter 2 we have determined the CO_2 release for a prescribed reference temperature change, assumed here, as shown in Table 19.8.2, to be 1, 3 and 5 K for an equivalent CO_2 doubling. Using the cybernetic feedback idea the additional CO_2 will lead to a total equilibrium temperature change, which includes all the CO_2 released under the new equilibrium temperature. The total CO_2 in the atmosphere is computed in an iterative procedure, in which the excess CO_2 and the temperature change ΔT converge to their corresponding final values, noting that

$$f_{eq}(\Delta CO_2{}^*{}_{ext}) = \Delta T(\Delta CO_2{}^*{}_{ext} + \Delta CO_2{}^*{}_{bio})/\Delta T(\Delta CO_2{}^*{}_{ext}) \qquad (19.8.17)$$

with $\Delta CO_2{}^*{}_{ext}/CO_2^0 = 1/\alpha$ = equivalent CO_2 doubling and $\Delta CO_2{}^*{}_{bio}$ corresponding to an equivalent CO_2 doubling.

We distinguish within each climate sensitivity three possible responses of the biota-soil system, namely one case, in which the vegetation is not affected by additional CO_2 ($\beta=0$), while for the other two cases β has been chosen to be equal to 0.15 and 0.5. The temperature sensitivity of the soils has been chosen for all three cases equal, using a Q value of 1.5 both for the top soil and deeper soils as well as for the transfer between them.

Table 19.8.2 Comparison of dynamic and equilibrium feedback factors for an equivalent CO_2 doubling

$\Delta T_{ext}^{(eq)}$ [K]	$\Delta T_{ext}^{(dyn)}$ [K]	β	$\cong$ $\tilde{Q}_{eff}$(AB)	f_{eq}	f_{dyn}	$\Delta T_{feedb}^{(eq)}$ [K]	$\Delta T_{feedb}^{(dyn)}$ [K]
1	0.87	0	0	1.18	1.08	1.18	0.94
		0.15	1.05	0.97	0.98	0.97	0.85
		0.50	3.47	0.66	0.80	0.66	0.70
3	2.60	0	0	1.41	1.25	4.23	3.25
		0.15	0.35	1.23	1.13	3.69	2.94
		0.50	1.15	0.88	0.91	2.64	2.37
5	4.33	0	0	1.54	1.40	7.70	6.06
		0.15	0.21	1.39	1.28	6.95	5.54
		0.50	0.70	1.05	1.03	5.25	4.46

$\tilde{Q}$(HA) = $\tilde{Q}$(H1H2) = $\tilde{Q}$(H1A) = $\tilde{Q}$(H2A) = 1.5

It is important to note that the fertilization effect or an equivalent stimulating effect through increasing surface temperature plays an important role in the total carbon budget between atmosphere and biota. This is true not only because the living standing biomass is increased under CO_2 fertilization, but also because through increased litter production the carbon loss through enhanced decomposition is partly compensated by the input of decaying organic material.

Considering dynamic feedback factors we note that due to the time delays in the biota-soil system, a smaller quantity of carbon dioxide is released and therefore also a smaller increase in temperature is to be expected. The dynamic feedback factor can be immediately obtained from the calculations described in the preceding chapter by defining

$$f_{dyn}(t^*) = \Delta T(\Delta CO_{2\,ext}(t^*) + \Delta CO_{2\,bio}(t^*))/\Delta T(\Delta CO_{2\,ext}(t^*)) \quad (19.8.18)$$

where t^* corresponds to the time of equivalent CO_2 doubling.

We recognize from Table 19.8.2 that a strong fertilization can compensate (f ≈ 1) or overcompensate (f < 1) the positive feedback (f > 1) between temperature increase and CO_2 release from soils. We note further that the dynamic feedback factors are closer to unity than the corresponding equilibrium values due to the time delays in the biota-soil system. It is very difficult to know at the moment, what will really happen in the future greenhouse climate; if we take an intermediate climatic sensitivity of 3 K and a modest combined β factor of 0.15, we expect a dynamic feedback of f = 1.13, implying that the temperature increase will be higher by 13% than the temperature increase without the atmosphere-biota coupling for an equivalent CO_2 doubling.

REGIONAL RESPONSE OF THE BIOTA/ SOIL SYSTEM TO A GREENHOUSE WARMING

As GCMs predict a greater increase in temperature for the higher latitudes and as a substantial fraction of soil carbon is stored in the higher latitudes it is to be expected that these areas will make the greatest contribution to described feedback effect. In Table 19.8.3 we show a short summary for the soil carbon masses of the higher latitude

Table 19.8.3
a. Soil carbon inventory of the higher latitudes (modified after Lashof, 1988)

Region	Soil carbon mass (Gt)
Tundra	200
Taiga	178
Northern Taiga	92
Conifer	23*
Temperate broad-leaved forest	59
	552 ≈ 42.4%**
Total of all biomes	1302

b. Bog areas in the Northern Hemisphere (Matthews and Fung, 1987)

Bog areas between:	Million km^2
90° - 66.5°N	106×10^3
66.5° - 50.8°N	2082×10^3
50.8° - 0°N	692×10^3
Total bog area	2880×10^3

* from total of 71; ** of total biomes

biomes and conclude, using the data of Lashof (1988), that more than 40% of the soil carbon is stored in these areas. In addition we need to consider the higher latitude bog areas which have been estimated by Matthews and Fung (1987) to sum up to 2.88 million km^2 for the Northern Hemisphere, out of which 2.19 mill. km_2 are located between 50.8° and 90°N.

From Table 19.8.2 we recognize that both the equilibrium and the dynamic feedback factor f increases with rising temperature when both Q_{10} values and the soil carbon mass are assumed to remain unchanged, which implies that the higher latitudes make the greatest contribution to the biotic feedback. We are presently in preparation of calculating the differential release or uptake of CO_2 from the different biomes to get a more detailed picture of the feedbacks involved. We can already state now that the response of the higher latitudes is very important for the total effect. Similarly we are trying to estimate the increased release of methane from the bog lands especially important since the CH_4 has a 32-fold higher climatic sensitivity than CO_2. We finally would like to emphasize that with the increased length of the growing season in the higher latitudes and with no expected water shortage, primary productivity should increase also which implies a higher litter production and a less pronounced loss of soil carbon.

REFERENCES

Idso, S.B., B.A. Kimball and M.G. Anderson (1989) Greenhouse warming could magnify positive effects of CO_2 enrichment on plant growth. CDIAC Communications. ORNL, Oak Ridge, TN,USA.

Kohlmaier, G.K., H. Bröhl, E.O. Sire, G. Kratz, U. Fischbach and Jiang Yunsheng (1981) The response of the biota-soil-system to a global temperature change. Commission of the European Communities, Climatologie Programme Meeting.

Lashof, D.A. (1988) The dynamic greenhouse: Feedback processes that may influence future concentrations of atmospheric trace gases. U.S. Environmental Protection Agency, Washington, D.C. 20460

Matthews, E. and I. Fung, (1987) Methane emission from natural wetlands: global distribution, area, and environmental charactristics of sources. Global Biogeochemical Cycles 1:61-86.

Schleser, G.H. (1981) The response of CO_2 evolution from soils to global temperature changes. Zeitschrift Naturforschung 22:338-345.

Tricot, C. and A. Berger (1987) Modelling the equilibrium and transient responses of global temperature to past and future trace gas concentrations. Climate Dynamics 2:39-61.

Effects of Elevated Atmospheric CO_2-Levels on the Carbon Economy of a Soil Planted with Wheat

L.J.A. LEKKERKERK[1], S.C. VAN DE GEIJN[2] and J.A. VAN VEEN[1]

[1]Research Institute Ital, P.O. Box 48,
6700 AA Wageningen, The Netherlands
[2]Centre for Agrobiological Research, P.O. Box 14,
6700 AA Wageningen, The Netherlands

ABSTRACT

Changes in the amount of soil organic matter are determined by input of plant material on one hand and by microbial decomposition on the other hand. The predicted increase in atmospheric CO_2 from 350 to 700 ppm CO_2 for about the middle of the next century will lead to an increase in plant biomass. The aim of this study was to investigate the consequences of an increased concentration of CO_2 in the atmosphere for the dynamics of soil organic matter. Wheat plants were grown in a $^{14}CO_2$-labelled atmosphere containing either 350 or 700 ppm CO_2. After 22, 35 and 49 days of growth in the $^{14}CO_2$-climate chambers soil columns with plants were harvested from both chambers and dry weight of the plants and the amount of ^{14}C and total-C in shoots, roots, respired CO_2 and soil were determined.

The dry weight of the plants was higher at the elevated CO_2-level. The input of ^{14}C-labelled root-derived material was increased at 700 ppm CO_2 compared with 350 ppm CO_2. Because of the preference of microorganisms for the easily decomposable root-derived material the turnover rate of the more resistant native soil organic matter was reduced at the highest CO_2-level. More non-soluble ^{14}C-residue was present in the soil at 700 ppm CO_2.

INTRODUCTION

Soil organic matter is an essential factor for the functioning of terrestrial ecosystems. Its continuous turnover controls the cycling of nutrients and the supply of mineral nutrients to plants. Soil organic carbon is mainly derived from carbon which has been photosynthetically fixed by plants. The plant carbon comes into the soil as litter from above ground plant material or, after translocation of assimilates from the shoot to the root, as root derived material (soluble root exudates, mucilage and dead parts of roots). The proportion of carbon being released from the roots into the soil as respired CO_2 and organic compounds varies between 10 and 40% of total net carbon assimilation for arable crops (Van Veen et al., submitted). For a growing crop this

Soils and the Greenhouse Effect. Edited by A.F. Bouwman
© 1990 John Wiley & Sons Ltd.

equals a carbon input into the soil of 900 to 3000 kg C per ha per year (Van Veen et al., 1989).

The quantity and quality of the root derived material depend on several parameters, such as plant species, development stage of the plant and environmental factors, e.g. temperature, nutrient status and soil moisture (Merckx et al., 1987).

The soil microbial population is considered to be the driving force for the turnover of soil organic matter. A large proportion of the root derived material is easily decomposable for microorganisms, which utilize the carbon compounds as the main source for biosynthesis and energy production. The utilization of this root-derived material by soil microbes is controlled by various soil related factors like the nutrient status (Merckx et al., 1986; 1987) and soil texture (Merckx et al., 1985).

With regard to the effects of elevated CO_2-concentrations in the atmosphere most research has been directed towards the effects on above ground plant processes, such as photosynthesis, transpiration and biomass accumulation. According to Cure (1986) a doubling of the atmospheric CO_2-concentration from 350 to 700 ppm will cause a higher assimilation rate, resulting in an average increase in yield of 41% for C-3 crops. C-4 plants respond in a different way because of differences in carbon metabolism. The extent of plant response to CO_2-enrichment also depends on environmental factors. Goudriaan and de Ruiter (1983) showed that the absolute increase in photosynthetic rate was always highest under non limiting conditions of nutrients and light. However, the demand for nutrients by crops will be greater and may induce an increased competition for nutrients between plants and soil microorganisms at elevated CO_2-levels.

The effects of elevated atmospheric CO_2 on below ground processes, which in view of the relation between primary production, carbon input into the soil, microbial decomposition and soil organic matter are of prime importance for the soil ecosystem have received much less attention to date. It is of great importance to know whether the changes in carbon inputs into the soil, both with respect to quantity and quality give rise to changes in the rate of turnover or sizes of the soil carbon pools.

In the present paper some preliminary results of a short term experiment with CO_2 and wheat plants will be discussed, with emphasis on soil/root-respiration and turnover of soil organic matter.

AIM AND HYPOTHESIS

The aim of this study was to investigate the influence of an elevated concentration of CO_2 in the atmosphere on the quantity and turnover of soil organic matter as a consequence of changes in the production of below ground plant biomass and root-derived material.

It is reasonable to assume that because of a higher plant production the input of organic matter into the soil both as litter and root derived material will increase. Whether this higher input of organic matter will lead to an equal increase in the content of soil organic matter largely depends on the activity of microbes.

One possibility is that the microbial activity will be stimulated, because more easily decomposable carbon is available. In this case the extra carbon input causes a priming effect, resulting in an enhanced soil organic matter turnover.

The amount of organic material in the soil will then increase to a smaller extent than expected according to the extra carbon input. Simultaneously the nutrient availability will increase.

Another possibility is that the microbial activity will not be affected by the extra carbon input caused by other limitations such as nutrient availability and water. Provided plants may successfully compete for limiting nutrients and water with microbes through extension of the root surface and, possibly by the help of symbiotic and associated microbes in the rhizosphere, the plant-carbon input will be increased, but its turnover may be reduced. Consequently the decomposition of recalcitrant, structural plant debris and native soil organic matter will be reduced, resulting in a relative increase in the amount of organic material in the soil.

MATERIALS AND METHODS

Experimental Design

Wheat plants were grown at 350 and 700 ppm CO_2 in 2 identical growth chambers in a $^{14}CO_2$-containing atmosphere with a specific activity of 1.5-3.0 kBq mg^{-1} C. Each chamber contained 14 soil columns with wheat plants. The climate conditions for both chambers were equal and controlled by computer. At the start of the experiment the day length was 12 hours. After 2 weeks the light period was extended to 14 hours; after 5 weeks to 16 hours. Light intensity was between 20000 and 25000 lux. The relative humidity was about 80%. The shoot- and root-compartment were separated by a base plate, which made it possible to adjust the temperature of the 2 compartments independently. The temperature of the shoot-compartment was 18°C during day time and 14°C during night time. For the root-compartment the temperature was 16°C and 14°C respectively. Water was given to the plants daily by connected irrigation tubes.

During growth soil/root-respiration was measured 3 times a week for each column. Plants (n=3) were harvested at 3 different dates, 22 (T1), 35 (T2) and 49 (T3) days after starting the experiment. At harvest plant dry weight, the release of carbon components into the soil and the amount of ^{14}C and C-total within shoots, roots, soil/root-respiration and soil were measured.

Soil

Before starting the experiment the soil, a loamy sand was sieved (<2 mm). The water content was set to 14% w/w and the soil was mixed with nutrients. Per kg moist soil a fertilizer mixture was added containing 30 mg N (NH_4NO_3), 70 mg P and 149 mg K ($K_2HPO_4.3H_2O$ and KH_2PO_4), 1.2 mg Cu ($CuSO_4$), 0.3 mg B (H_3BO_3), 2.0 mg Mo ((NH_4)$_6Mo_7O_{24}.4H_2O$), 1.6 mg Mn ($MnSO_4.H_2O$), 0.4 mg Zn ($ZnSO_4.7H_2O$) and 0.9 mg Fe (Fe-EDTA). An extra amount of nitrogen (60 mg N per kg soil) was added to the planted pots in the irrigation water after both 4 and 6 weeks of growth in the climate chambers. Plant growth was supposed not to be limited by nutrients or water in this experiment. The final bulk density of the soil after filling the columns was 1.2 g cm^{-3}.

Growth of the Plants

After germinating seeds of springwheat (Triticum aestivum L. cv. Ralle) in glass tubes for 10 days, the seedlings were transferred to columns, each containing 2.7 kg of loamy sand.

Ten days later the columns were sealed at the shoot-root interface, using a silicone rubber (Dow Corning, Q3-3481). This was necessary to avoid CO_2-exchange directly with the atmosphere, so as to be able to measure soil/root-respiration and to make a balance of the ^{14}C-partitioning. The columns were placed in the ESPAS (Experimental Soil Plant Atmospheric System) growth chambers and after closing the systems $^{14}CO_2$ was added to the atmosphere and the experiment was started.

Measurements

- Soil/root-respiration. The columns were flushed with CO_2-free air twice a day. The CO_2 of the soil atmosphere was collected in a 0.5 N NaOH- solution. Three times a week the carbon (total-C and ^{14}C) in the NaOH-solution was determined by titration with 0.5 N HCl, after precipitation with $BaCl_2$. $^{14}CO_2$-measurements were done by scintillation counting.
- At harvest time shoots were clipped at soil level and roots were carefully removed from the soil by handpicking. To remove adhering soil particles the roots were washed with water. The plant material was dried at 80°C and the dry weights were determined.
- The dried plant material was ground and the ^{14}C-content of shoots, roots and dried soil were determined by a wet combustion procedure according to Dalal (1979).
- Easily decomposable ^{14}C-compounds released by the roots were determined by shaking 25g of fresh root-free soil with 50 ml 0.5 M K_2SO_4 (pH 6.8) for 1 hour. The soil suspensions were centrifuged (10000g, 10 min.) and ^{14}C was measured by liquid scintillation counting.

RESULTS AND DISCUSSION

Plant Biomass

The dry weight of both shoots and roots of the plants grown at 700 ppm CO_2 was higher at all harvest times compared with the plants grown at 350 ppm CO_2. These differences were statistically significant (P<0.05) for the shoot at T_2 and T_3 and for the root at T_2. After 49 days of growth in the climate chambers with elevated CO_2-concentration the shoot/root-ratio of the plants was significantly higher than the shoot/root-ratio of the plants grown at 350 ppm CO_2 (Table 19.9.1).

Table 19.9.1 Dry weight (g) and shoot:root ratio (S/R) of wheat plants grown at 350 and 700 ppm CO_2

	T_1(22d)		T_2(35d)		T_3(49d)	
	350	700	350	700	350	700
SHOOT	1.54	1.94	5.03	7.69*	14.02	20.96**
ROOT	0.57	0.80	2.42	3.50*	4.15	4.87
TOTAL	2.11	2.74	7.45	11.19**	18.17	25.83**
S/R	2.72	2.50	2.10	2.21	3.29	4.33**

* : P<0.05; ** : P<0.001

Distribution of ^{14}C

When comparing the distribution pattern of ^{14}C over the whole plant no significant differences were found between the two CO_2-treatments in the amount of ^{14}C in shoot, root, soil/root-respiration and soil as a percentage of ^{14}C fixed by the plants. This suggests that no changes in the distribution pattern of carbon within the plant occur due to elevated CO_2-levels. Thus an increased photosynthetic fixation of carbon by plants at higher CO_2-levels resulted in a proportional increase of carbon transported to the different plant- and soil- compartments. As a consequence the carbon input into the soil was also higher at elevated CO_2 concentration.

However,when comparing the distribution of ^{14}C in root, soil/root-respiration and soil as a percentage of the amount of ^{14}C being translocated to the roots, relatively more $^{14}CO_2$ was respired and a lower percentage of ^{14}C was retained in the roots at 700 ppm CO_2. These differences were significant at T_2 and T_3 (P<0.05).

Soil/Root-Respiration

The soil/root-respiration expressed as accumulated amount of $^{14}CO_2$ respired per column was higher when plants were grown at 700 ppm

CO_2. From day 15 onwards this difference was significant. The amount of total-CO_2 respired, including respiration of native soil organic matter differed also between the two CO_2-treatments. The accumulated amount of respired CO_2 was significantly higher at 700 ppm CO_2 after 22 days of growth in the ESPAS (Table 19.9.2). At T_3 the total-CO_2 respiration at elevated CO_2 was 57% higher than at 350 ppm CO_2, whereas the $^{14}CO_2$ soil/root-respiration of the 700 ppm CO2-treatment 74% higher was than of the 350 ppm CO2-treatment.

Table 19.9.2 Accumulated $^{14}CO_2$ soil/root-respiration (kBq/column) and accumulated total-CO_2 soil/root-respiration (mg C/column) of planted soils after 22, 35 and 49 days of exposure to 350 and 700 ppm CO_2

	T_1(22d)		T_2(35d)		T_3(49d)	
	350	700	350	700	350	700
$^{14}CO_2$-resp.	270	500**	1470	2580**	2050	3570**
Tot-CO_2-resp.	261	399*	843	1282*	1090	1707**

* : $P<0.05$; ** : $P<0.001$

Soil Organic Matter Dynamics

Part of the CO_2 collected from the soil columns during the experiment was generated by decomposition of the native soil organic matter. On the basis the specific activity of the plant material and the specific activity of the respired CO_2 the contribution of mineralization of native organic matter could be calculated. Until T_1, unlabelled soil organic matter was decomposed at the same rate at both CO_2-levels, and contributed 25-40% of the total soil/root-respiration.

At T_3 the decomposition of native soil organic matter accounted for 27% and 10% of the total amount of respired CO_2 for the 350 and 700 ppm CO_2 treatments respectively. Extraction of the soil with K_2SO_4 showed an increased input of easily decomposable ^{14}C-labelled carbon compounds into the soil when plants were exposed to an elevated CO_2-concentration. When combining these results one might hypothesize that microorganisms preferred the easily decomposable root-derived material more than the more resistant native soil organic matter.

When comparing the carbon coming from the roots and remaining in the soil as ^{14}C-labelled material with the output of carbon by microbial decomposition of native soil organic matter ($^{12}CO_2$-respiration) a net balance of the organic carbon content of the soil can be made. At T_3 there was a net decrease of the soil organic matter when plants were exposed to 350 ppm CO_2 (-108 mg C per column) and a net increase in soil organic matter when plants were grown at 700 ppm CO_2 (+84 mg C per column). Moreover, the root biomass at 700 ppm was larger than or equal to the biomass at 350 ppm CO_2. These roots will also be incorporated in the soil organic matter pool after harvest. However, as only the first part of the growing season before flowering is considered here, extrapolation of these short term effects of elevated CO_2 on soil organic matter dynamics to long term carbon accumulation in arable soils would be erroneous. It should be noted that part of the initial decomposition of soil organic matter in both cases was probably caused by disturbance of the soil when mixing the soil with nutrients and filling the columns, which is known to stimulate microbially mediated processes in the soil. However, the tendency shown should be further investigated, as changes in soil organic matter could play a quantitatively important role in the global carbon budget.

CONCLUSIONS

The input of easily decomposable root-derived material in the soil was higher when plants were exposed to 700 ppm CO_2, compared with exposure to 350 ppm CO_2. Because microorganisms seemed to prefer this material as their energy source, the turnover of the more resistent native soil organic matter was reduced at 700 ppm CO_2. Also the non-soluble ^{14}C-residue in the soil was increased at the highest CO_2-level.

LITERATURE

Cure, J.D. (1986) Crop responses to carbon dioxide doubling: a literature survey. Agricultural and Forest Meteorology 38:127-145.

Dalal, R.C. (1979) Simple procedure for the determination of total carbon and its radioactivity in soils and plant materials. Analyst 104:151-154.

Goudriaan, J. and H.E. de Ruiter (1983) Plant growth in response to CO_2 enrichment, at two levels of nitrogen and phosphorus supply. 1. Dry matter, leaf area and development. Netherlands Journal of Agricultural Science 31:157-169.

Merckx, R., A. Dijkstra, A. Den Hartog and J.A. Van Veen (1987) Production of root derived material and associated microbial growth in soil at different nutrient levels. Biology and Fertility of Soils 5:126-132.

Merckx, R., J.H. Van Ginkel, J. Sinnaeve and A. Cremers (1986) Plant-induced changes in the rhizosphere of maize and wheat. I. Production and turnover of root-derived material in the rhizosphere of maize and wheat. Plant Soil 96:85-93.

Merckx, R., A. Den Hartog and J.A. Van Veen (1985) Turnover of root derived material and related microbial biomass formation in soils of different texture. Soil Biology and Biochemistry 4:565-569.

Van Veen, J.A., E. Liljeroth, L.J.A. Lekkerkerk and S.C. Van de Geijn: Carbon translocation in plants and turnover in the rhizosphere: ecosystem controls and consequences for microbial activity and soil organic matter accumulation at elevated atmospheric CO_2-levels. (submitted to Oikos)

Van Veen, J.A., R. Merckx and S.C. Van de Geijn (1989) Plant-and soil related controls of the flow of carbon from roots through the soil microbial biomass. Plant and Soil 115:179-188.

Food or Forest? Can the Tropical Forests Survive Along with Continuing Growth of Population and Economy?

R.J. SWART and J. ROTMANS

National Institute for Public Health and Environmental Protection
P.O.Box 1, 3720 BA Bilthoven, the Netherlands

ABSTRACT

The rapid conversion of tropical forests is an important contributor to the emissions of greenhouse gases, second only to the combustion of fossil fuels. Soil emissions play an important role during and after the conversion next to the direct contribution to trace gas emissions by burning of vegetation. The rate of deforestation is strongly tied to population growth and economic development. The Integrated Model for the Assessment of the Greenhouse Effect (IMAGE) has been developed as a policy tool to evaluate the climate impact of different global developments. Within the carbon cycle module of this model deforestation has been simulated as a function of food production, livestock development, commercial wood production and other factors. The authors find that survival of tropical forests is unlikely unless a dramatic productivity increase in agriculture and forestry or limitation of population growth can be achieved within the coming decades. Land expansion should take place according to land suitability and fuelwood from plantations could be used efficiently as a replacement of fossil fuels.

INTRODUCTION

Growth of the human population and its restless pursuance of economic growth are changing the face of the earth. The most pregnant symptom of the associated global environmental problems is the anticipated climate change. Next to the emissions of greenhouse gases by the combustion of fossil fuels the conversion of tropical forests is probably the most important contributor to the global greenhouse effect. But the 10 - 20 million hectares of closed tropical rainforests that are annually destroyed do also increase erosion, modify local climate, lead to the distinction of countless species and threaten the indigenous population. As covered extensively elsewhere in this conference deforestation influences the greenhouse effect in a multitude of ways. Emissions of CO_2 increase by decomposition of organic material in the soils. Increasing numbers of termites may be instrumental in increasing methane emissions, while also N_2O-emissions are likely to rise after forest conversion. Furthermore deforestation influences the world's

Soils and the Greenhouse Effect. Edited by A.F. Bouwman
© 1990 John Wiley & Sons Ltd.

albedo and the hydrological cycle, both of which are important players in the climate game.

At the Dutch National Institute for Public Health and Environmental Protection (RIVM) the Integrated Model for the Assessment of the Greenhouse Effect (IMAGE) was developed to evaluate global policies with regard to climate change and to increase the awareness amongst different groups of the seriousness of the problem.

The model includes causes, processes and impacts (see figure 19.10.1, Rotmans et al., 1989). A crucial component of this model is the carbon cycle, which has been modelled according to Goudriaan and Ketner (1984). Deforestation is integrated into this module by way of a land transfer matrix, in which the elements represent shifts from one ecosystem (e.g. tropical forest) to another (e.g. agricultural land). Since

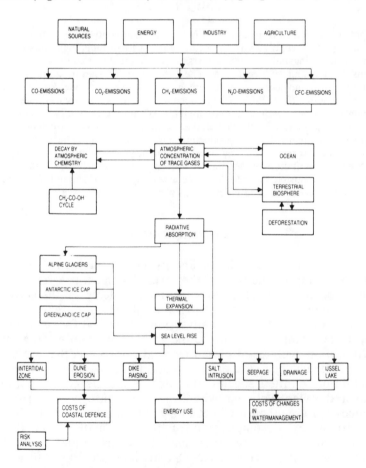

Figure 19.10.1 Structure of the Integrated Model for the Assessment of the Greenhouse Effect (IMAGE)

deforestation is presently the type of land use change that is the most relevant for climate change, a deforestation module was developed to describe the relationship between human activities and resulting forest destruction, making the matrix elements dynamic. The work on this module is intended to serve as a feasibility study for the future development of a global land use change module, including also other potentially important processes like desertification, forest dieback and the drainage of wetlands and agricultural soils. In combination with regionalized data on soils, climate and other environmental factors such a module enables the development of emission scenarios for global man induced emissions by land use changes. A more detailed description of the present work can be found in Swart and Rotmans (1989).

SCOPE

In the module only the three major tropical forest areas are distinguished separately. Although we realize that for a more reliable study on deforestation the module should have been disaggregated to a national or district scale, the available information and the purpose of this exercise prevented us from doing so. Therefore we do not intend to present projections of forest conversion for the coming years, but do merely assess the long term fate of the forests as a function of a number of developments in the three regions.

In the Goudriaan carbon cycle model only six types of ecosystems are distinguished: tropical forests, temperate forests, grasslands, agricultural land, semi-desert/tundra and human area. For the purpose of this work the tropical forests were subdivided into closed and open forests, the latter to include logged forests, forest plantations and forest fallow.

The following processes are distinguished to simulate future deforestation:

- expansion of permanent agriculture into forests
- expansion of shifting or pioneer cultivation
- expansion of pasture land
- logging
- fuelwood gathering
- implementation of industrial or mining projects
- development of forest plantations
- degradation of arable land
- degradation of pasture land
- conversion of non-forest land for agriculture

The way these elements are structured in the module is presented in figure 19.10.2. FAO statistical information or recent estimates of past and present ecosystem areas and deforestation causes were used for validation.

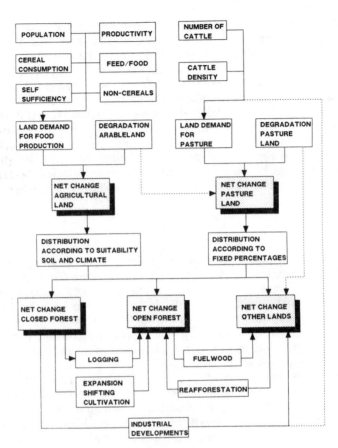

Figure 19.10.2 Structure of the IMAGE-deforestation module

PROCESSES

The most important cause of deforestation is the expansion of agricultural land, both as permanent agriculture and shifting cultivation. Expansion of permanent agricultural land is assumed to be dependent on population, per capita cereal demand, average productivity per hectare, present arable land, fraction of land used for non-cereals, fraction of cereals used for feed, a self-sufficiency ratio to allow for food imports or aid and erosion losses. This land demand has to be taken from virgin ecosystems. For the results in this paper this has been done in a rational way, namely according to soil suitability. Soil suitabilities have been estimated by Bouwman (1989) for this project. Irrigation is not included as a separate parameter but is assumed to be a factor that enables colonization of virgin lands or productivity increases on present agricultural land. Erosion is assumed to cause a shift from arable land to pasture and finally to semi-desert.

We assumed that expansion of traditional shifting cultivation and logging would lead not only to a conversion from closed to open forests, but would also open the forest up to landless farmers: in any year additional demand for permanent agricultural land is first satisfied by taking areas that were converted from closed forests by shifting cultivation or logging the previous year. So far only the increasing number of shifting cultivators has been taken into account, not decreasing fallow times. The impact of logging on tropical forests in the module is related to the expected growth rate of hardwood production and the harvesting efficiency.

Expansion for livestock development is a major agent of deforestation in Latin America. It has been modelled as a function of expanding numbers and cattle density.

Fuelwood gathering is considered in open forest areas only, assuming that in the closed forests regrowth exceeds local demand. In the module deforestation because of fuelwood consumption is taken proportional to the growth of the rural population. It is very difficult to make estimates of the impact of industrial, mining or infrastructural projects. Because of the side effects this impact can be enormous, although the area directly affected by the projects may be small.

Finally reforestation is included in the module by scenario dependent logistic functions.

RESULTS INCLUDING EFFECTS ON CARBON CYCLE AND CLIMATE

Calculations were made for four scenarios, varying from an optimistic case in which forest destruction is reversed to a pessimistic laissez-faire case in which the closed tropical forests disappear in the second half of the next century (figure 19.10.3 to 19.10.5). According to our analyses agriculture (both permanent and shifting) and demand for hardwood are the most important forces that will continue to threat the forests unless major changes will take place (see figure 19.10.6 and 19.10.7, negative values imply area increases). In terms of the greenhouse effect the impact of the different scenarios on CO_2 concentrations and temperature rise has been calculated. The difference between the highest and the lowest deforestation scenarios both combined with a moderate energy scenario (emissions from fossil fuel combustion increase from 5.4 Gt C y^{-1} in 1985 to 12.0 Gt C y^{-1} in 2100, which is an average increase of 0.7% per year) gave a difference of about 60 ppm in 2100 or almost 10% (figure 19.10.8). This is the net effect of the biospheric contribution, including not only deforestation, but also CO_2-fertilization.

In the beginning of the next century the relative difference is somewhat higher because of the impact of the assumed high plantation rates in that period.

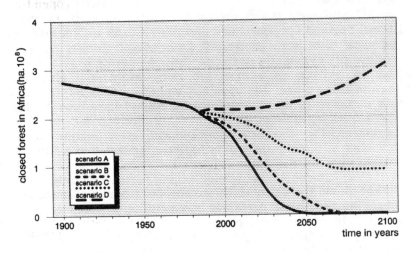

Figure 19.10.3 Simulated conversion of closed tropical forests in Africa.

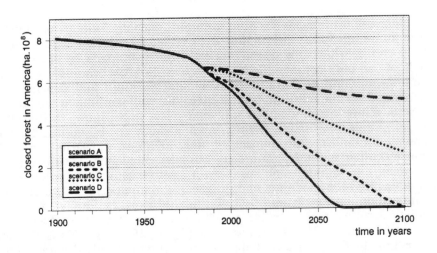

Figure 19.10.4 Simulated conversion of closed tropical forests in Latin America.

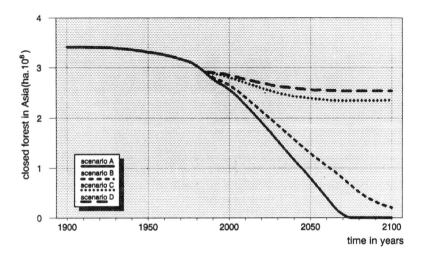

Figure 19.10.5 Simulated conversion of closed tropical forests in Southeast Asia.

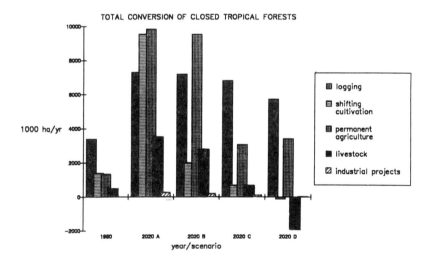

Figure 19.10.6 Example of simulated forests conversion rates by source.

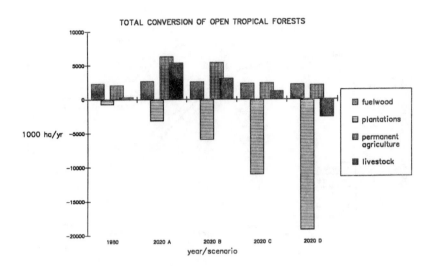

Figure 19.10.7 Example of simulated forests conversion rates by source.

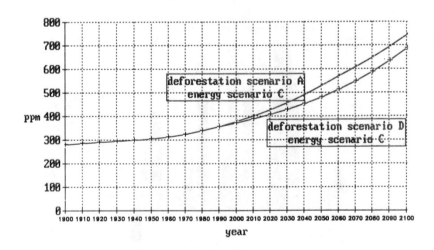

Figure 19.10.8 CO_2 concentration.

CONCLUSIONS AND RECOMMENDATIONS

To facilitate the development of scenarios for climate change by land use modifications a model linking different human activities to different land use would be needed. The deforestation module of the Integrated Model for the Assessment of the Greenhouse Effect shows that such a model is feasible. Preliminary calculations with the module suggest that it will be extremely difficult to halt deforestation without seriously curtailing the development prospects of the developing countries. Rapid and prolonged intensification of agricultural practices will be necessary for both small and large scale farming, taking due account of the economic difficulties involved and the associated environmental problems. Rapidly increasing demand for fuelwood and wood products can only be prevented from further attacking the world's forests by effective plantation programmes. At a global scale in terms of the anticipated climate change the net impact of deforestation seems temporary and small relative to fossil fuel contributions. We calculated a maximum of 10% difference in CO_2-concentrations in 2100 between the highest and lowest deforestation scenario.

REFERENCES

Bouwman, A.F. (1989) Land Evaluation for Rainfed Farming, ISRIC Working Paper 89/1, Wageningen.

Goudriaan, J. and P. Ketner (1984) A Simulation Study for the Global Carbon Cycle, including Man's Impact on the Biosphere, Climatic Change 6, pp. 167-192.

Rotmans, J., H. de Boois and R.J. Swart (1989) The Integrated Model for the Assessment of the Greenhouse Effect, to be published in Climatic Change.

Swart, R.J. and J. Rotmans (1989) The Integrated Model for the Assessment of the Greenhouse Effect: a Module for Tropical Deforestation, RIVM report nr. 758471007 (in preparation), Bilthoven.

A Global Inventory of Wetland Distribution and Seasonality, Net Primary Productivity, and Estimated Methane Emissions

I. ASELMANN and P.J. CRUTZEN

Max-Planck-Institute for Chemistry,
Division of Atmospheric Chemistry,
P.O.Box 3060, 6500 Mainz, FRG

ABSTRACT

A global data set on the geographic distribution and seasonality of freshwater wetlands and rice paddies has been compiled, comprising information at a spatial resolution of 2.5° latitude by 5° longitude. The data set is compiled from wetland monographs, surveys and maps. Information on seasonality has further been inferred from climatological records and hydrological data of major catchment areas. Global coverage totals 5.7×10^6 km^2 for natural wetlands and 1.3×10^6 km^2 for rice paddies. Natural wetlands have been grouped into six categories: bog, fen, swamp, marsh, floodplain, and shallow lake with estimated global areas of 1.87, 1.48, 1.13, 0.27, 0.82, 0.11 $\times 10^6$ km^2 respectively. Published net primary productivity data (NPP) were used to derive latitude dependent mean NPP values for each of the six categories. Estimated global NPP in natural wetlands is between 4000-8000 Tg (1 Tg $= 10^{12}$ grams) of dry matter. NPP of rice paddies calculated from yield statistics is 1400 Tg (dry matter). Measured CH$_4$ fluxes from wetlands can be extrapolated to estimate a worldwide emission of 60-140 Tg per year from natural wetlands and 40-160 Tg for paddies. Major source regions are subtropical Asia due to intense rice cultivation, the temperate - boreal region between 50 and 60°N, and tropical swamps and floodplains between 0 - 10°S. Emissions are highly seasonal.

INTRODUCTION

Methane is one of the major greenhouse gases and strongly influences the photochemistry of the atmosphere (Levy, 1971; Crutzen, 1973; Ramanathan et al., 1987; Brühl and Crutzen, 1988). Its outstanding role in the photochemistry and its persistent increase in the atmosphere has given rise to extensive research into the global biogeochemical aspects of methane, which have recently been summarized by Cicerone and Oremland (1988). Wetlands have been shown to be a major source of atmospheric methane but estimates of their likely source strength still yield rather large ranges (Koyama, 1963; Ehhalt, 1974; Seiler, 1984; Sebacher et al., 1986; Bartlett et al., 1988). Numerous physical, chemical and biological factors will influence methanogenesis in wetlands (Conrad, 1989). The multitude of factors and the fact that in many cases

Soils and the Greenhouse Effect. Edited by A.F. Bouwman
© 1990 John Wiley & Sons Ltd.

methane is consumed by CH_4 oxidizing organisms before reaching the atmosphere has, as yet, baffled a synthesizing description of methane emissions suitable for modelling. Scientific interest in wetlands in the past concentrated mainly on qualitative descriptions and classification, but global quantification were based on rather rough estimates (e.g. Lieth, 1975; Whittaker and Likens, 1975; Ajtay et al., 1979). A recent study showed that wetlands in the world should cover some 5 - 6 × 10^6 km² (Matthews and Fung, 1987), considerably more than previously assumed. Here we present a global data set on wetlands that confirm the estimate by Matthews and Fung, albeit with regional differences.

WETLAND DISTRIBUTION, PRODUCTIVITY, AND METHANE EMISSIONS

Sources of information on the distribution of wetlands are described in Aselmann and Crutzen (1989). Estimated wetland areas (in × 10^6 km²) for different regions are: Canada: 1.268; USA (contiguous states): 0.228; South America: 1.524; USSR: 1.512; Europe: 0.154; Africa: 0.355; South East Asia: 0.241; Australia and New Zealand: 0.015; Other Regions: 0.394.

The data set is spatially resolved on a 2.5° latitude by 5° longitude grid and distinguishes six broad wetland categories by the following common wetland terminology (Zoltai and Pollett, 1983): bogs, fens, swamps, marshes, floodplains, and shallow lakes. Bogs denote raised peatlands in which organic matter (mostly sphagnum moss) has accumulated over a long time. Their peat is strongly acid and mineral input is entirely by precipitation or dry deposition. Fens are also peatlands but may be less acid since their hydrology and topography allows for soil nutrient influence. Swamps denote forested freshwater wetlands on waterlogged or seasonally inundated soils, with little or no peat accumulation. Marshes are herbaceous mires dominated by grasses, sedges or reeds and may be either permanently or seasonally waterlogged. Floodplains primarily comprise seasonally flooded savannas but were also used for non-peaty wetlands which do not match the definitions for swamps or marshes. Shallow lakes are open water bodies, a few meters deep, which are likely to emit methane through the water column.

Times of inundation in cases of seasonal wetlands were taken from the information in the source literature, from hydrological data, i.e. high water periods of the major rivers (Degens, 1982 ff) or deduced from meteorological records (Müller, 1983). Areas with insufficient information were classified as wetlands with unknown seasonality. The distribution of natural wetlands can be depicted from Figure 19.11.1. Major wetland regions are: (1) the boreal and temperate zone between 50 -70°N, in particular West Siberia and East-Central Canada which are dominated by fens and bogs; and (2) the tropics between 0 - 10°S with large contributions from the Amazon lowland swamps. Northern peatlands are highly seasonal due to winter freezing, whereas tropical swamps in many cases undergo seasonal flooding. The data set is supplemented by a compilation of rice paddies, resolved on the same grid size. Areas and time of growing were taken from Darmstädter et al. (1987) and from Huke (1980), and were distributed according to the World Atlas of Agriculture. The total rice area is 1.3×10^6 km^2, of which almost 90 percent is cultivated in Asia. The cultivation period for one crop varies between 3 and 7 months. Double or permanently cropped areas have been considered according to the length of cultivation period.

The wetland data base has been used to estimate the dry matter (dm) net primary productivity (NPP). Mean NPP values for broad climate regions (arctic, boreal, temperate, tropical) for each of the wetland categories were drawn from literature and multiplied with the respective areas. Minimum NPP rates of 100 - 300 g m^{-2}a^{-1} were derived for arctic peatlands and maximum values of 1500 - 4000 gm^{-2}a^{-1} were assigned to tropical marshes. On this basis global NPP of wetlands is estimated to be between 4000 - 8000 Tg a^{-1} (Table 19.11.1). NPP of rice fields is 1400 Tg y^{-1}. This is calculated by multiplying rice yields with a factor of 2.86 (Aselmann and Lieth, 1983). Methane emissions have been studied within the carbon cycling in natural wetlands (Clymo and Reddaway, 1971; Svensson, 1983). Average ratios of CH$_4$/NPP (on a carbon to carbon basis) are 2 - 7 percent. This ratio, when applied to the NPP estimates of natural wetlands, would yield a range of 50 - 360 Tg CH$_4$. This range seems too large to be consistent with the overall methane budget.

Alternatively, data on methane emissions from wetland sites have been compiled to derive a geometric mean emission rate for each wetland category. Differences between the geometric means are apparent, but ranges overlap significantly (Table 19.11.1). Extrapolating the emission rates on the basis of the data set yields a global emission of about 80 (40 - 160) Tg CH$_4$ for natural wetlands and 90 (60 - 140) Tg for rice paddies (Table 19.11.1). Figure 19.11.2 shows the largest source regions between 20 -30°N due to rice cultivation (ca. 40 Tg), between 0 - 15°S (ca. 35 Tg, primarily from tropical swamps), and at northern latitudes dominated by temperate/boreal bogs and fens (ca. 23 Tg). Emissions from paddies peak in August (Figure 19.11.3); those from natural wetlands are highest between June and September (Figure 19.11.4).

Figure 19.11.1a,b Distribution of natural wetlands. Numbers denote the percentages of the 2.5˚ latitude by 5˚ longitude grids that are covered by wetlands. To convert to areas these percentages should be multiplied by the total area of a grid given in the right hand column in km^2. Left hand column and top and bottom row denote the centre coordinates of the grids. (see pages 444-445).

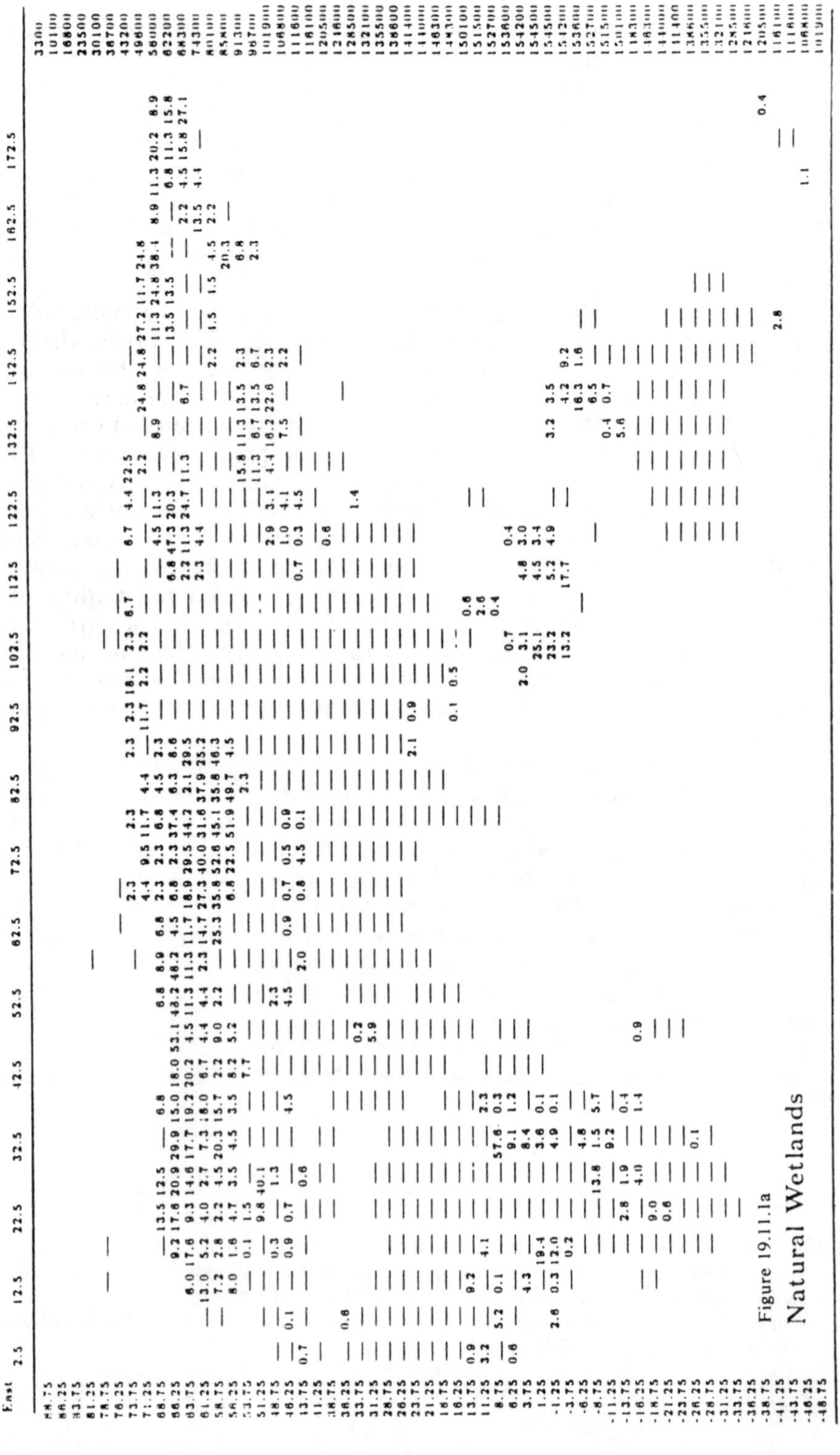

Figure 19.11.1a

Natural Wetlands

Figure 19.11.b
Natural Wetlands

Soils and the Greenhouse Effect

Table 19.11.1 Global Wetland Characteristics

Wetland Categories	Net Primary Productivity[a]	Emission Rate $(mgCH_4 \, m^{-2} \, d^{-1})$	Area $(10^{12}m^2)$	Mean Prod. Period[b] (days)	Methane Emission $(Tg \, y^{-1})$
Bogs	620-1400	15 (1-50)	1.87	178	5 (0.4-18)
Fens	430-970	80 (28-216)	1.48	169	20 (7-52)
Swamps	1600-3220	84 (57-112)	1.13	274	26 (18-35)
Marshes	290-740	253 (137-399)	0.27	249	17 (12-30)
Flood-plains	1170-1990	100	0.82	122	10
Lakes	50-100	43 (17-89)	0.12	365	2 (1-4)
Natural Wetlands	4160-8420		5.69		80 (40-160)
Rice Fields	1350	300-100	1.31	130	92 (60-140)

Figures in parenthesis denote ranges
[a]: in dry matter; [b]: mean CH_4 productive period comprised of both permanent and seasonal wetland areas in the respective wetland categories. The CH_4 productive period is determined either by monthly mean temperatures above 0°C or inundation. Swamps with unknown seasonality have been treated as permanent.

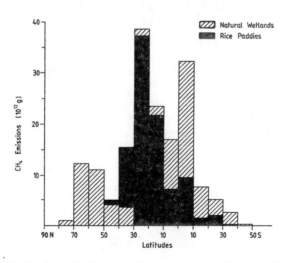

Figure 19.11.2 Latitudinal distribution of the combined CH_4 emissions from natural wetlands and rice paddies.

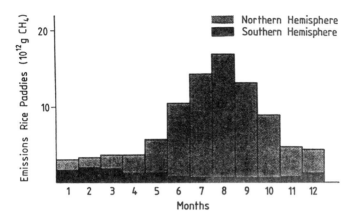

Figure 19.11.3 Monthly distribution of methane emissions from rice paddies computed from linearly temperature dependent CH_4 fluxes in the range from 300 to 1000 mg $m^{-2}d^{-1}$ for temperatures from 20°C to 30°C and constant emission of 300 mg $m^{-2}d^{-1}$ for temperatures below 20°C.

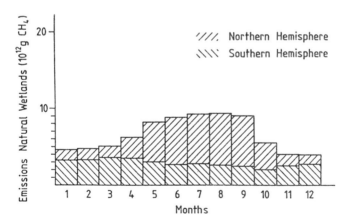

Figure 19.11.4 Monthly distribution of methane emissions from natural welands using the mean CH_4 flux rates from Table 19.11.1.

CONCLUSIONS

The present data show that natural wetlands cover 5.7×10^6 km^2. Almost 40 percent of these wetlands are tropical, which gives tropical wetlands a much higher rank as potential methane sources than previously assumed. Methane estimates are uncertain because single site fluxes remain, until know, unpredictably variable. Methane/NPP ratios in particular seem inappropriate to derive reasonable emission estimates from wetlands. Improved understanding of methane dynamics in different wetland environments may require a more detailed wetland classification in order to apply new flux models to come to global extrapolation.

ACKNOWLEDGMENTS

This work was supported by the Ministry for Research and Technology of the Federal Republic of Germany through grants BMFT/KF 20128.

REFERENCES

Ajtay G.L., P. Ketner and P. Duvigneaud (1979) Terrestrial primary productivity and phytomass. In: B. Bolin, E.T. Degens, St. Kempe and P. Ketner (eds). The Global Carbon Cycle (Scope 13). pp 129-181. John Wiley and Sons, Chichester-New-York-Brisbane-Toronto.

Aselmann I. and H. Lieth (1983) The implementation of agricultural productivity into existing models of primary productivity. In: E.T. Degens, St. Kempe and H. Soliman (eds). Transport of Carbon and Minerals in Major World Rivers -Part 2. Mitt. Geol.-Paläont. Inst. 55. pp 276-291. University of Hamburg; SCOPE/UNEP Special Vol.

Aselmann I. and P.J. Crutzen (1989) Freshwater wetlands: Global distribution of natural wetlands and rice paddies, their net primary productivity, seasonality and possible methane emissions. Journal of Atmospherical Chemistry (in press)

Bartlett K.B., P.M. Crill, D.I. Sebacher, R.C. Harriss, J.O. Wilson and J.M. Melack (1988) Methane flux from the Central Amazonian floodplain. Journal of Geophysical Research 93:1571-1582.

Brühl Ch. and P.J. Crutzen (1988) Scenarios of possible changes in atmospheric temperatures and ozone concentrations due to man's activities, estimated with a one-dimensional coupled photochemical climate model. Climate Dynamics 2:173-203.

Cicerone R.J. and R.S. Oremland (1988) Biogeochemical aspects of atmospheric methane. Global Biogeochemical Cycles, 2:299-328.

Clymo R.S. and E.J.F. Reddaway (1971) Productivity of Sphagnum (bog-moss) and peat accumulation. Hydrobiologia 12:181-192.

Conrad R. (1989) Control of methane production in terrestrial ecosystems. In: M.O. Andreae und D.S. Schimel (eds.). Exchange of Trace Gases Between Terrestrial Ecosystems and the Atmosphere. Dahlem Konferenzen. Chichester, Wiley and Sons (in press).

Crutzen P.J. (1973) A discussion of the chemistry of some minor constituents in the stratosphere and troposphere. Pure Applied Geophysics 106-108:1385-1399.

Darmstädter J., L.W. Ayres, R.U. Ayres, W.C. Clark, R.P. Crosson, P.J. Crutzen, T.E. Greadel, R. McGill, J.F. Richards and J.A. Torr (1987) Impacts of World

Development on Selected Characteristics of the Atmosphere: An Integrated Approach. Oak Ridge National Laboratory, 2 volumes, ORNL/Sub/86-22033/1/V2, Oak Ridge, Tennessee 37831, USA.

Degens E.T. (1982 ff) Transport of Carbon and Minerals in Major World Rivers - Part 1,2,3. Mitt. Geol.-Paläont. Inst. 52. 766 pp. University of Hamburg; SCOPE/UNEP Special Vol.

Ehhalt D.H. (1974) The atmospheric cycle of methane. Tellus 26: 58-70.

Huke R.E. (1980) South Asia; Rice Area Planted by Culture Type. (map 1:1.45m; prepared at the International Rice Research Institute (IRRI), Manila).

Koyama T. (1963) Gaseous metabolism in lake sediments and paddy soils and the production of atmospheric methane and hydrogen. Journal of Geophysical Research 68:3971-3973.

Levy H.II. (1971) Normal atmosphere: Large radical and formaldehyde concentrations predicted. Science 173:141-143.

Lieth H. (1975) Primary production of the major vegetation units in the world. In: H. Lieth and R.H. Whittaker (eds). Primary Productivity of the Biosphere (Ecological Studies 14), pp 203-215. Springer, New York-Heidelberg-Berlin.

Matthews E. and I. Fung (1987) Methane emission from natural wetlands: Global distribution, area and environmental characteristics of sources. Global Biogeochemical Cycles 1:61-86.

Müller M.J. (1983) Handbuch ausgewählter Klimastationen der Erde. 346 pp. Forschungstelle Bodenerosion der Universität Trier, Mertesdorf/Ruwertal. Vol.5.

Ramanathan V., L. Callis, R. Cess, J. Hansen, I. Isaksen, W. Kuhn, A. Lacis, F. Luther, J. Mahlman, R. Reck and M. Schlesinger (1987) Climate-chemical interactions and effects of changing atmospheric trace gases. Rev.Geophys. 25:1441-1482.

Sebacher D.I., R.C. Harriss, K.B. Bartlett, S.M. Sebacher and S.S. Grice (1986) Atmospheric methane sources: Alaskan tundra bogs, an Alpine fen and a subarctic boreal marsh. Tellus 38:1-10.

Seiler W. (1984) Contribution of biological processes to the global budget of CH_4 in the atmosphere. In: M.J. Klug and C.A. Reddy (eds). Current Perspectives in Microbial Ecology. American Society of Microbiology, pp 468-477. Washington D.C.

Svensson B.H. (1983) Carbon fluxes from acid peat of a subarctic mire with emphasis on methane. (Dissertation) Institutionen fór Mikrobiologi. 301 pp. Swedish Univ. Agricultural Science Rapport 20, Uppsala.

Whittaker R.H. and G.E. Likens (1975) The biosphere and man. In: H. Lieth and R.H. Whittaker (eds). Primary Productivity of the Biosphere Ecological Studies 14, pp 305-328. Springer, New York-Heidelberg-Berlin.

Zoltai S.C. and F.C. Pollett (1983) Wetlands in Canada: Their classification, distribution and use. In: A.J.P.Gore (ed). Ecosystems of the World (4B). Mires: Swamp, Bog, Fen, and Moor. Vol.2, pp 245-268. Elsevier, Amsterdam.

Progress Toward Predicting the Potential for Increased Emissions of CH_4 from Wetlands as a Consequence of Global Warming

M.K. BURKE, R.A. HOUGHTON, and G.M. WOODWELL

The Woods Hole Research Center
Box 296, Woods Hole, MA 02543, USA

ABSTRACT

Natural wetlands are major contributors to atmospheric methane. Because methanogenesis in wetlands is temperature sensitive, the warming of the earth may have increased the emissions of methane from wetlands and thereby contributed to the observed rise in atmospheric concentrations of methane. We estimated the emissions of methane likely to have resulted from climatic warming during the last century, and found that this additional source probably has been only a minor contribution. However, future warming may result in much greater emissions from wetlands. Our ability to predict future methane emissions from wetlands will be improved as climate models and the understanding of soil biological and hydrologic processes improve.

The concentration of methane in the atmosphere is currently increasing at an annual rate of 1% (40 to 80 Tg CH_4 y^{-1}) (Blake and Rowland, 1988), and the rate of increase has been accelerating during the past 300 years. The average surface temperature of the earth has also increased, by about 0.5°C in the past 100 years, probably because of the increased concentrations of radiatively active gases. This pattern is similar to the observed parallel rise and fall of global temperatures and methane concentrations during glacial-interglacial cycles (Raynaud et al., 1988). The relationship of temperature and methane concentrations suggests a positive feedback (Houghton and Woodwell, 1989): the radiative properties of methane contributing to the warming, and the warming enhancing the production of methane (methanogenesis). The relatively short atmospheric residence time (8 to 10 years) for methane means that atmospheric concentrations of methane are influenced by short-term fluxes of methane to and from the atmosphere, rather than by accumulations over long periods.

Natural wetlands, always a major source of methane, may have increased their emissions. No data exist on past emission rates from natural wetlands, but if the relationships between methane emissions and temperature documented in field and laboratory studies (e.g. Crill et al., 1988) apply to methane emissions from wetlands in general, the

Soils and the Greenhouse Effect. Edited by A.F. Bouwman
© 1990 John Wiley & Sons Ltd.

climatic warming during the past century may have caused an increase in the emissions of methane. It is the aim of this analysis to estimate the potential for this positive feedback and to identify topics that need further study to provide more accurate estimates.

We estimated changes in the emissions of methane that could have resulted from warming during the last century by using present estimates of the global rates of methane emissions from wetlands (Matthews and Fung, 1987), temperature anomalies during the last century (Hansen and Lebedeff, 1987), and the relationship between temperature and methane production. The relationship was defined using a $Q_{10}=5$, which is approximately the average value reported in the literature for methane flux from wetland soils. In addition, we estimated the potential for methane emissions from natural wetlands during the next century if temperatures rise 3°C as has been predicted for a doubling of atmospheric CO_2 (Schlesinger and Mitchell, 1985).

The results of these calculations (Figure 19.12.1) show annual methane emissions from wetlands to have increased by about 28 Tg (or 34%) between 1880 and 1980. Over the last century the estimated additional release of methane was 1600 Tg (19% above background). This additional source of methane from wetlands is only 13% of the total additional source from all sources (Table 19.12.1). By the year 2080, however, the additional emissions of methane from wetlands could be 280 Tg y^{-1}, greater than the sum of all current additional sources (Table 19.12.1).

Our present ability to predict how methane emissions from natural wetlands will change with climatic warming is limited. We do not have an adequate understanding of the many factors that can influence methane emissions from these systems. There is a need to advance our understanding of the topics discussed below before we can accurately predict the influence of global warming on methane emissions from wetlands.

Table 19.12.1 Methane emissions from identified sources[1] and the amount of change during the last century

SOURCE	1880's Tg y^{-1}	1980's Tg y^{-1}		Additional Source Tg y^{-1}	% of Additional Source
Wetlands	83	111	(11-300)	28	13.2
Animals	21	75	(75-100)	54	25.5
Fossil Fuel	4	50	(30-100)	46	21.7
Rice production	5	59	(30-350)	54	25.5
Biomass burning	65	65	(20-110)	0	0
Landfills	0	30	(30-70)	30	14.1
Termites	28	28		0	0
	206	418		212	100.0

[1] Values from literature

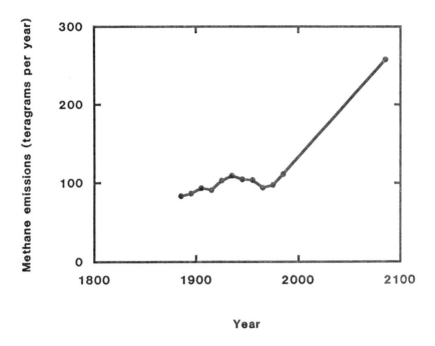

Figure 19.12.1 Estimates of global annual methane emissions from natural wetlands during the last century, averaged for each decade, and for the decade ending in 2080. Values for 2080 were calculated using estimates of a 3°C warming, as projected for an effective doubling of CO_2 concentrations above preindustrial concentrations.

Change in the Area of Wetlands

Both the flooding and draining of land for agricultural purposes has occurred during the last century, and the area and distribution of wetlands will likely change as the climate warms. In fact, predictions of increased precipitation in high latitudes and temperature enhanced evapotranspiration suggest that the area of wetlands may increase in high latitudes, and decrease or become more ephemeral in lower latitudes. Therefore, the importance of wetlands in the global methane budget may change for reasons other than the warming itself.

Thawing of Permafrost

It is unclear whether methane production will increase or decrease as permafrost melts, and the uncertainty is largely a function of hydrology: will areas of melted permafrost be wetter or drier? In addition, it is still uncertain whether a deeper active layer will result in greater methane emissions.

Relationships between Temperature and Methane Emissions

Temperature influences methanogens differently both between populations and across ranges of temperature, and temperature may also enhance the activity of methane oxidizing bacteria, thus reducing the emissions of methane to the atmosphere. The understandinf of the influence of temperature on different biological processes in soil must be improved to predict the changes in methane emissions associated with increasing temperatures more accurately.

Changes in the Length of the Season of Methanogenesis

The length of the season during which methanogenesis can occur is a function of both temperature and hydrology. Better predictions of future regional changes in both temperature and moisture regimes, of the resulting evapotranspiration rates and effects on hydrology are needed to predict regional changes in the length of the season for methanogenesis.

Quality and Quantity of Organic Substrate

The availability of organic substrate influences the production of methane, and may therefore limit the degree to which systems can respond to increases in temperature. The factors that presently limit gaseous carbon production in natural wetlands must be determined before we can estimate the potential for increasing emissions as a result of climate change.

Based solely on present estimates of the relationship between temperature and methane production, climatic warming during the last century seems to have been responsible for only a small additional release of methane from wetlands to the atmosphere. However, future emissions from natural systems may be much higher. Our ability to predict changes in emissions from wetlands will improve as climate models improve and as biological processes in soil processes become better understood.

REFERENCES

Blake, D.R. and F.S. Rowland (1988) Continuing worldwide increase in tropospheric methane 1978 to 1987. Science 239:1129-1131.
Crill, P.M., K.B. Bartlett, R.C. Harriss, E. Gorham, E.S. Berry, I.I. Sebacher, L. Madzar and W. Sanner (1988) Methane flux from Minnesota peatlands. Global Biogeochemistry Cycles 2:371-384.
Hansen, J. and S. Lebedeff (1987) Global trends of measured surface air temperature. Journal of Geophysical Research 92:345-372.
Houghton, R.A. and G.M. Woodwell (1989) Global climatic change. Scientific American 260:36-44.

Matthews, E. and I. Fung (1987) Methane emission from natural wetlands: global distribution, area, and environmental characteristics of sources. Global Biogeochemistry Cycles 1:61-86.

Raynaud, D., J. Chappellaz, J.M. Barnola, U.S. Korotdevich and C. Lorius (1988) Climatic and CH_4 cycle implications of glacial-interglacial CH_4 change in the Vostok ice core. Nature 333:655-657.

Schlesinger, M.E. and J.B.F. Mitchell (1985) Model projections of the equilibrium climatic response to increased carbon dioxide. In: M.C. MacCracken and F.M. Luther (Eds.), Projecting the climatic effects of increasing carbon dioxide, U.S. Department of Energy, Washington D.C. DOE/ER-0237.

Organic Matter Dynamics, Soil Properties, and Cultural Practices in Rice Lands and their Relationship to Methane Production

H.U. NEUE[1], P. BECKER-HEIDMANN[2], H.W. SCHARPENSEEL[2]

[1]*International Rice Research Institute, P.O. Box 933, Manila, Philippines*
[2]*Institute of Soil Science, Univ.of Hamburg, Allende-Platz 2, Hamburg 13, F.R.G.*

ABSTRACT

The global rice area harvested in 1988 has been discriminated with regard to the major rice ecologies which are irrigated, rainfed, deepwater and upland. Net primary production of rice paddies is estimated to be 1,400 g m^{-2}y^{-1} of which 400 g m^{-2}y^{-1} is returned to the soil. Reliable data on additional inputs of organic substrates are not available but the global trend is decreasing. Soil related controlling factors of methane formation have been identified and various soil properties are amalgamated into three crucial parameters (temperature, texture and mineralogy, Eh/pH buffer systems). Considering these soil characteristics and water regimes, only 80 million ha of harvested wetland rice lands are tentatively classified to be potential sources of methane formation. The estimated global annual emission of methane from wetland rice fields is 25-60 Tg.

INTRODUCTION

In recent years methane fluxes have received increased attention, because methane contributes to the greenhouse warming of the troposphere. Compared to carbon dioxide its equivalent warming effect (mol basis) is believed to be 32 times higher (Dickinson and Cicerone, 1986). The methane load of the atmosphere has been increasing with accelerated speed to 1.7 ppmv during the past decades (Khalil and Rasmussen, 1987). The current rate of increase is about 1.1 % per year. Estimates of global sinks amount to a total of 375-475 Tg (Tg = terragram; 1 Tg = 10^{12} g) of methane annually (Bingemer and Crutzen, 1987). Anaerobic decay of organic matter in wetland rice fields appears to be an important source of methane. However, global extrapolation of the few emission data presented to date (Cicerone et al., 1983; Seiler et al., 1984; Holzapfel-Pschorn and Seiler, 1986; Holzapfel-Pschorn et al., 1986) are highly uncertain and very tentative. Emission characteristics of various rice ecologies have not been established and the effect of environmental factors and agricultural practices on methane emission is not well documented. In this paper, the methane emission of different rice ecologies considering ecophysiological and soil biochemical principles are evaluated.

DISTRIBUTION OF RICE LANDS AND
NET PRIMARY PRODUCTIVITY

The global rice area totals 136 million ha with a harvested area of 143 million ha of which almost 90% is cultivated in Asia (Table 19.13.1). Rice is grown under a wider variety of climatic, soil, and hydrological conditions than any other crop. A comprehensive classification of rice environments, proposed by Neue (1989a) illustrates the high diversity of each of the four major rice ecologies: irrigated, rainfed, deepwater and upland rice lands. The cultivation cycle of rice coincides with the rainy season when water supply is sufficient (> 1000 mm/season). Two short duration (< 130 days) rice crops per year are grown where temperatures are high enough and irrigation water is available.

Water regimes vary from epiaquic to aquic and to peraquic. Water losses through percolation are the most variable in wetland rice soils. In East Asia, percolation rates between 9 to 20 mm d^{-1} are considered essential for high fertility. The need for drainage and high percolation rates is associated with accumulation of toxic substances (organic acids, high partial pressure of CO_2, H_2S) and nutrient imbalances related to Fe, Mn, P, Zn, S, and Na. It is estimated that at least 10 million ha of rice

Table 19.13.1 Distribution of ricelands (million ha) by rice ecologies

Region	Irrigated			Rainfed		Deep water	Upland	Total area harvested	Yield (t ha^{-1})	Estimated rice production (million ton)
	Season			Fertile	Low fertility					
	Wet/ early	Summer	Dry/ late							
East Asia[a]	9.6	14.6	9.8	2.8	-	-	-	36.8	5.4	200.0
Southeast Asia[b]	9.0	-	4.9	4.8	8.9	3.75	4.65	36.0	2.9	102.5
South Asia[c]	15.5	-	3.9	8.5	11.5	7.3	6.70	53.4	2.0	105.5
Near East[d]	1.25	-	-	-	-	-	-	0.75	3.3	2.45
South/Central America, Caribbean,USA	2.5	-	0.5	-	0.4		5.65	9.2	2.9	26.5
Africa	0.9	-	0.75	1.2	-		2.70	5.45	1.8	9.9
USSR	0.66	-	-	-	-	-	-	0.65	4.1	2.7
Europe	0.42	-	-	-	-	-	-	0.42	5.4	2.3
Oceania	0.12	-	-	-	-	-	-	0.12	6.6	0.79
Australia	0.11	-	-	-	-	-	-	0.11	7.1	0.76
World	73.76			38.95		11.45	19.7	142.9	3.3	477.0

[a] China, Taiwan, China, Korea DPR, Korea RP, Japan;
[b] Burma, Cambodia, Indonesia, Laos, Malaysia, Philippines, Thailand, Vietnam;
[c] Bangladesh, Bhutan, India, Nepal, Pakistan, Sri Lanka;
[d] Afghanistan, Iran, Iraq.
Source: FAO, 1988; Misra et al., 1986; Huke, 1982; N.N., 1987.

lands in East Asia have high percolation rates or are periodically drained during the growing season.

A survey of paddy soils in tropical Asia revealed that 20% had a pH < 5 and 48.2% had either coarse texture, high content of kaolinitic clays, or low base status. In 49% of the soils total N content was < 0.1%, which is equivalent to < 1% organic carbon (Kawaguchi and Kyuma, 1977; Kyuma, 1985). More than 50% of rainfed rice is grown on low fertility soils and/or is prone to periods of drought.

The net primary production of wetland rice soils may be deduced from yield statistics (Table 19.13.1) and estimates of aquatic biomass and weed growth during fallow periods. In 1988 worldwide wetland rice production was 477 million tons of which upland rice contributed about 28 million tons. Based on a shoot:grain ratio of 3:2 (Ponnamperuma, 1984) and a root:shoot ratio of 0.17 (Yoshida, 1981; Watanabe and Roger, 1985) the total dry matter production of wetland rice amounts to 1,151 million tons. Adding 74 million tons (600 kg ha^{-1} per season) dry matter of aquatic biomass (Roger and Watanabe 1984; Watanabe and Roger, 1985) and 200 million tons of weed dry matter (2 t ha^{-1} during fallow periods) yields a total dry matter production of 1,425 million tons or 1,400 g m^{-}y^{-1}. It is assumed that on an average 15% of the straw and all roots, aquatic biomass and weeds amounting to 392 million tons dry matter or (157 million tons of carbon) are returned to the soil. If a maximum of 30% of the returned carbon is transformed to methane as found by Martin et al. (1983) and Neue (1985) in studies with ^{14}C labelled straw in soils prone to methane formation 63 Tg y^{-1} of methane would be produced in wetland ricelands globally.

No exact data on organic amendments of rice soils per country is available but the global trend is decreasing rather than increasing. In Japan the addition of organic substrates decreased from 6 t in 1965 to 2.7 t in 1980 (Kanazawa, 1984). To sustain or increase soil fertility and rice yields moderate organic amendments seem to be essential.

ORGANIC MATTER DECOMPOSITION

Organic matter decomposition in submerged rice soils is closely related with methane formation. Waterlogging of soils is generally associated with decay retardation and accumulation of organic matter. Neue and Scharpenseel (1987) observed in field experiments that decomposition of ^{14}C labelled straw in the tropics was as rapid in flooded as in aerobic soils. Based on field studies, Mohr and van Baren (1954) suggested that where the mean annual temperature exceeds 32°C, mineralization of organic N in flooded soils proceeds as rapidly as in well aerated soils.

Degradation and mineralization of organic substrates in flooded rice soils are promoted (Figure 19.13.1) (Neue and Scharpenseel [1987], Neue [1989b], and Neue and Snitwongse (1989):

- thorough puddling of the soil before planting of rice;
- soil temperature of the puddled layer of 30-35°C;

- neutral to alkaline soil pH of the flooded soil;
- high soil:water ratio and associated low bulk density (0.2-0.8g cm^{-3}) of the puddled layer;
- shallow floodwater;
- high and well balanced nutrient supply;
- permanent supply of energy rich photosynthetic aquatic and benthic biomass;
- high diversity of micro- and macroorganisms;
- supply of O_2 into the reduced layer by rice root, excretion and diurnal supersaturation of the floodwater with O_2.

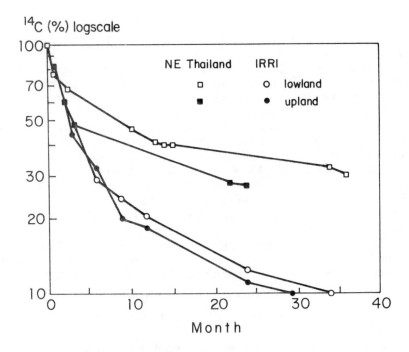

Figure 19.13.1   Decomposition pattern of rice straw in soils with kaolinitic clays (Thailand) and mixed clays (IRRI).

Except for rice paddies with high percolation rates or aeration periods, rapid decay of organic substrates and fast turnover of remaining metabolites are always associated with high methane production. Measurements of the natural distribution of carbon isotopes in wetland rice soils confirmed the above relationships. As anaerobic bacteria produce methane with a very low $\delta^{13}C$ value down to -90‰ (Rosenfeld and Silverman, 1959), causing in turn ^{13}C enriched residues, the $\delta^{13}C$ values of the organic matter of wetland rice soils can be a measure of their methane producing capacity. Methane formation through CO_2 reduction or decarboxylation of acetate enrich ^{13}C in the remaining metabolites (Meinschein et al., 1974). By ^{13}C analysis using the thin-layer sampling method of complete soil monoliths cut into 2 cm slices (Becker-Heidmann and Scharpenseel, 1986) it is possible to distinguish between organic matter of different decomposition stages within a profile. Volatile decomposition products, normally $\delta^{13}C$ enriched with respect to the source as well as to the gaseous products (especially methane), percolate down to and the deeper part of the profile. Rice soils that have rapid anaerobic turnover of organic matter reveal high enrichments of [13]C. In a Tropaquept we found an enrichment of 7‰ in relation to the source, which is a comparatively high value (Figure 19.13.2).

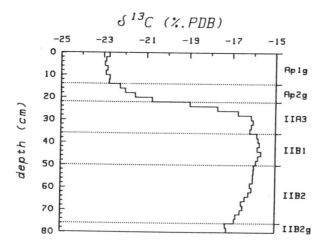

Figure 19.13.2 Distribution of $\delta^{13}C$ in a methane prone Tropaquept at Los Banos, Philippines (adapted from Becker-Heidmann & Scharpenseel, 1989).

CRUCIAL SOIL CHARACTERISTICS FOR METHANE PRODUCTION

Apart from carbon supply and hydrologic conditions, there are 4 major groups of controlling factos of methane formation: temperature, texture and mineralogy; Eh/pH buffer, and salinity. The inhibitory effect of salinity and sulfate reduction on methane production has been discussed elsewhere.

Soil temperature

The soil temperature of the puddled layer in submerged soils fluctuates diurnally in response to the meteorological regime acting upon the atmosphere-floodwater and floodwater-soil interface. Soil and floodwater properties interact with the meteorological regime by temporal changes in reflectivity, heat capacity, thermal conductivity, incoming water temperature and water flow. In tropical lowlands the mean annual temperature of the puddled soil equals or exceeds the mean annual maximum air temperature in most months of the year if the flow of the shallow floodwater is minimal and the soil:water ratio is high. In puddled soils with a high content of 2:1 lattice clays and low bulk density the annual mean maximum temperatures was found to be 33°C at a mean maximum air temperature of 33.2°C (Neue, 1989b). Hackmann (1979) reported that floodwater temperature are above minimum air temperature, but below maximum air temperature if daily amplitudes of air temperature are high, while water temperatures are above maximum air temperature if daily fluctuations are low.

At low latitudes methane production may be more affected by soil parameters because annual fluctuations of the air temperature are small. At higher latitudes, for example East Asia, lower air and soil temperatures should restrict methane production in the early period of rice cultivation. With increased temperature (10-35°C) organic acids and H_2 persist for a shorter time and more methane is produced (Yamane and Sato, 1967).

Emission rates were also observed to increase exponentially with rising temperature (Holzapfel-Pschorn and Seiler, 1986). Contrarily, Davies (1988) claims that rising temperature may increase oxidation of methane.

Texture and mineralogy

Soils with high content of expanding lattice clays and adequate organic matter contents are likely sources of methane. Flooding and puddling disrupts soil aggregates and disperses soil colloids, thereby increasing the active surface area and the actual pH/Eh buffer capacity. The resulting wide soil:water ratio and low bulk density is maintained as long as the soil is kept flooded (Neue, 1989b).

Soils that are more sandy or silty in texture or have high contents of kaolinitic clays are difficult to puddle. The bulk density of these soils often increases upon puddling resulting in reduced speed of Eh and pH changes, increased concentrations and residence time of organic acids and retarded organic matter decomposition. To overcome these adverse growth conditions, the aeration can be increased through higher percolation rates, drainage or by maintaining the soil moisture content near the saturation point. Under such conditions redox potentials hardly become negative and methane formation should be suppressed considerably.

pH/Eh buffer capacity

The magnitude and speed of reduction and pH changes in flooded rice soils is determined by the amount of easily degradable organic substrates, their rate of decomposition, and the amount and kinds of reducible nitrates, manganese and iron oxides, sulfate and organic compounds. The dominant poise in soils containing free $Fe(III)$ oxyhydroxides is the ferric-ferrous reaction. Soils low in active iron with high organic matter attain Eh values of -200 to -300 mV within 2 weeks after submergence (Ponnamperuma, 1972). In soils with high contents of iron and organic matter the Eh falls to -50 mV and may then slowly decline over a period of month to -200 mV. Redox potentials (corrected to pH 7) of -150 to -190 mV are needed for the formation of methane. The addition of amorphous iron oxide in anaerobically incubated soils depressed the formation of volatile fatty acids and methane (Asami and Takai, 1970).

Takai (1961) showed for a range of Japanese rice soils that the ratio of CO_2 to CH_4 is depending on the ratio of oxidizing capacity (electron accepting) to reducing capacity (Table 19.13.2). The amount of O_2, NO_3, $Mn(IV)$ and $Fe(III)$ corresponds to the oxidizing power. Formation of NH_4 was used as an index of the reducing power.

Table 19.13.2 CO_2 and CH_4 production as affected by oxidizing and reducing power

Soil fertility	Organic matter metabolism	Relative	
		Oxidation/ reduction capacity	$CO_2:CH_4$
High	Facultative	100	100
	Anaerobic	54	30
		57	37
		24	19
		23	17
		27	14
	Strictly	11	11
Low	Anaerobic	09	07

Adapted from Takai (1961), quoted by Watanabe (1984).

The pH values at steady state of flooded alkaline, calcareous, and acid soils are highly sensitive to the partial pressure of CO_2. After addition of organic substrates the partial pressure of CO_2 may reach peaks up to almost 100 kPa in flooded soils while typical values at steady state range from 5-20 kPa. One to three tons of CO_2 are produced in the puddled layer of 1 ha during the first weeks of flooding (Ponnamperuma, 1972). The decrease of CO_2 in favor of CH_4 after the initial phase is probably caused by assimilation of CO_2 and precipitation of carbonates, rather than reduction of CO_2 to CH_4.

ESTIMATES OF GLOBAL METHANE EMISSION FROM FLOODED RICE FIELDS

It is suggested that rice soils with dry fallow periods between rice growing seasons are likely to show high methane production if one or more of the following conditions are met:

- EC > 4 mS cm^{-1} while flooded;
- acidic or allic reaction;
- ferritic, gibbsitic, ferruginous or oxidic mineralogy;
- > 40% of kaolinitic or halloysitic clays;
- < 18% clay in the fine earth fraction if the water regime is epiaquic;
- occasional drying up of the soil during the cultivation period.

These soils comprise Oxisols, most of the Ultisols, and some of the Aridisols, Entisols and Inceptisols.

Rice soils that are prone to methane production mostly belong to the orders of Entisols, Inceptisols, Alfisols, Vertisols, and Mollisols.

Harrison and Aiyer (1913) did pioneer work on the gases liberated in rice paddies. They showed that the gas escaping from the reduced layer was methane, with minor amounts hydrogen and carbon dioxide. They also established that no methane escaped directly from flooded soils because methane oxidation takes place at the soil surface or floodwater.

Since the transpiration stream is the major pathway for methane (Holzapfel-Pschorn et al., 1985) its emission should be low in deepwater rice, where most of the plants are submerged. In floating rice (> 1 m floodwater depth) roots in the soil and lower parts of the stem are mostly physiological dead, so that the pathway for methane via the rice plant is blocked. Since dry seeding is the main method of deepwater rice establishment, soil reduction and soil organic matter decomposition proceeds more slowly, thus limiting methane production.

Global estimate of methane emission are highly uncertain and tentative. The characterization and distribution of environmental conditions and cultural practices that limit or favour methane production and emission are not well established at present. Merely gathering more data on methane fluxes in the field may not improve estimates. We need more knowledge on the principles that explain methane fluxes, a better data base of crucial parameters and simulation models to predict global emission rates.

Nevertheless, based on above data and principles only 80 million ha of harvested wetland rice lands are estimated to be potential sources of methane. Assuming an average methane emission rate of 200-500 mg $CH_4 m_{-2}$ during an average growing period of 130 days would give a global emission of 21-52 Tg per year. In 30 million ha methane production and emission may occur only for short periods while in the remaining 14 million ha of wetland rice emission of methane may be negligible. The global annual emission of methane from wetland rice field are estimated to be in the order of 25-60 Tg only.

REFERENCES

Asami, T. and Y. Takai (1970) Relation between reduction of free iron oxides and formation of gases in paddy soil (in Japanese). Journal of Science Soil Manure 41:48-55.

Becker-Heidmann, P. and H.W. Scharpenseel (1986) Thin layer $\delta^{13}C$ and $D^{14}C$ monitoring of "Lessivé" soil profiles. In: Stuiver, M. and Kra, R.S.: Proceedings 12th International ^{14}C Conference, Radiocarbon 28:383-390.

Becker-Heidmann, P. and H.W. Scharpenseel (1989) Carbon isotope dynamics in some tropical soils. In: Long, A. and Kra, R.S.: Proceedings 13th International ^{14}C Conference, Radiocarbon 31:3 (in press).

Bingemer, H.G. and P.J. Crutzen (1987) The production of methane from solid wastes. Journal of Geophysical Research 92:181-2,187.

Cicerone, R.G., G.D. Shetter and C.C. Delwiche (1983) Seasonal variation of methane flux from a Californian rice paddy. Journal of Geophysical Research 88:022-024.

Davies, M.B. (1988) Ecological systems and dynamics. *in:* Toward an understanding of global change. Initial priorities for U.S. contributions to the International Geosphere-Biosphere Programme. National Academy Press, Washington D.C., p. 92.

Dickinson, R.E., R.J. Cicerone (1986) Future global warming from atmospheric trace gases. Nature 319:109-114.

FAO (Food and Agricultural Organization) (1988) Quarterly Bulletin of Statistics, Vol. 1 #4.

Hackman, Ch.W. (1979) Rice field ecology in Northeastern Thailand. The effect of wet and dry season on a cultivated aquatic ecosystem. Monographiae Biologicae Vol. 34, J. Illies (Ed.), 22p. W. Junk Publisher.

Harrison, W.H. and P.A.S. Aiyer (1913) The gases of swamp rice soil. I. Their composition and relationship to the crop. Memoires Department of Agriculture India. Chemistry Series 5:65-104.

Holzapfel-Pschorn, A. and W. Seiler (1986) Methane emission during a cultivation period from an Italian rice paddy. Journal of Geophysical Research 91:11,803-11,814.

Holzapfel-Pschorn, A., R. Conrad and W. Seiler (1985) Production, oxidation and emission of methane in rice paddies. FEMS Microbiology Ecology 31:343-351.

Holzapfel-Pschorn, A., R. Conrad and W. Seiler (1986) Effects of vegetation on the emission of methane from submerged paddy soil. Plant and Soil 92:223-233.

Huke, R.E. (1982) Rice area by type of culture: South, Southeast and East Asia. International Rice Research Institute, P.O. Box 933, Manila, Philippines.

Kanazawa, N. (1984) Trends and economic factors affecting organic manures in Japan. In: Organic Matter and Rice, p. 557-568. International Rice Research Institute, P.O. Box 933, Manila, Philippines.

Kawaguchi, K. and K. Kyuma (1977) Paddy soils in tropical Asia, their material, nature, and fertility. 258 p, University Press of Hawaii, Honolulu.

Khalil, M.A.K. and R.A. Rasmussen (1987) Atmospheric methane: Trends over the last 10,000 years. Atmospheric Environment 21:2,445-2,452.

Kyuma, K. (1985) Fundamental characteristics of wetland soils. In: Wetland Soils: Characterization, Classification and Utilization, p 191-206. International Rice Research Institute, P.O. Box 933, Manila, Philippines.

Martin, U., H.U. Neue, H.W. Scharpenseel and P. Becker (1983) Anaerobe Zersetzung von Reisstroh in einem gefluteten Reisboden auf den Philippinen. Mitteilungen Deutsche Bodenkundliche Gesellschaft 38:245-250.

Meinschein, W.G., G.G.L. Rinaldi, J.M. Hayes and D.A. Schoeller (1974) Intramolecular isotopic order in biologically produced acetic acid. Biomedical Mass Spectrometry 1:172-174.

Misra, B., S.K. Mukhodpadhyay and J.C. Flinn (1986) Production constraints of rainfed rice in Eastern India.

Mohr, E.C.J. and P.A. Van Baren (1954) Tropical Soils. Interscience Ltd., London.

N.N. (1987) China Agricultural Yearbook. Agricultural Publishing House, Liangma Bridge, Chaoyang District, Beijing, China.

Neue, H.U. (1984) Gaseous products of the decomposition of organic matter in submerged soils. In: Organic Matter and Rice, p 311-328. International Rice Research Institute, P.O. Box 933, Manila, Philippines.

Neue, H.U. (1985) Organic matter dynamics in wetland soils. In: Wetland Soils: Characterization, Classification and Utilization, p. 110-122. International Rice Research Institute, P.O. Box 933, Manila, Philippines.

Neue, H.U. (1989a) Rice growing soils: constraints, utilization and research needs. In: Classification and management of rice growing soils. FFFTC Book Series, Food and Fertilizer Technology Center for the ASPAC Region, Taiwan, R.O.C. (in press).

Neue, H.U. (1989b) Holistic view of chemistry of flooded soils. In: Proceedings of the First International Symposium on Paddy Soil Fertility, 33 p. International Board for Soil Research and Management (IBSRAM), P.O. Box 9-109, Bangkok, Thailand. (in press).

Neue, H.U. and H.W. Scharpenseel (1987) Decomposition pattern of 14C-labeled rice straw in aerobic and submerged rice soils of the Philippines. The Science of the Total Environment 62:431-434.

Neue, H.U. and P. Snitwongse (1989) Organic matter and nutrient kinetics. In: Proceedings of the International Rice Research Conference 1988, 22 p. International Rice Research Institute, P.O. Box 933, Manila, Philippines. (in press).

Ponnamperuma, F.N. (1972) The chemistry of submerged soils. Advances in Agronomy 24:29-96.

Ponnamperuma, F.N. (1984) Straw as a source of nutrients for wetland rice. In: Organic Matter and Rice, p 117-136. International Rice Research Institute, P.O. Box 933, Manila, Philippines.

Roger, P.A. and I. Watanabe (1984) Algae and aquatic weeds as source of organic matter and plant nutrients for wetland rice. In: Organic Matter and Rice, p. 147-168. International Rice Research Institute, P.O. Box 933, Manila, Philippines.

Rosenfeld, W.D. and S.R. Silverman (1959) Carbon isotopic fractionation in bacterial production of methane. Science 130:1658-1659.

Seiler, W., A. Holzapfel-Pschorn, R. Conrad and D. Scharffe (1984) Methane emission from rice paddies. Journal of Atmospheric Chemistry 1:241-268.

Takai, Y. (1961) Reduction and microbial metabolism in paddy soils (3) (in Japanese) Nogyo Gijitsu (Agricultural Technology) 19:122-126.

Watanabe, I. (1984) Anaerobic decomposition of organic matter in flooded rice soils. In: Organic Matter and Rice, p 237-258. International Rice Research Institute, P.O. Box 933, Manila, Philippines.

Watanabe, I. and P.A. Roger (1985) Ecology of flooded rice fields. In: Wetland Soils: Characterization, Classification and Utilization, p. 229-246. International Rice Research Institute, P.O. Box 933, Manila, Philippines.

Yamane, I. and K. Sato (1967) Effect of temperature on the decomposition of organic substances in flooded soil. Soil Science and Plant Nutrition 13:94-100.

Yoshida, S. (1981) Fundamentals of rice crop science. 269 p, International Rice Research Institute, P.O. Box 933, Manila, Philippines.

EXTENDED ABSTRACT 19.14

Effects of Organic Matter Applications on Methane Emission from Japanese Paddy Fields

K. YAGI AND K. MINAMI

National Institute of Agro-Environmental Sciences
Tsukuba 305, Japan

ABSTRACT

The CH_4 fluxes were measured using the chamber method in rice paddy fields at four different sites (Ryugasaki, Kawachi, Mito and Tsukuba) located in Ibaraki prefecture, Japan. Successive application of organic matter, such as rice straw and its compost, have been performed at three of these sites. Soil types are different in the individual paddy fields including alluvial soil, peaty soil and soils on volcanic ash (Andosol).

Emission rates of CH_4 differed markedly with the soil types. The highest rate was observed in the peaty soil. The second highest was the alluvial soil. The emission rates in the Andosol were low. The difference in annual emission rates between the peaty soil and the light-coloured Andosol was approximately a factor of 40. Application of rice straw to the paddy fields significantly increased the CH_4 emission rates at all the sites. Application of the compost did not remarkably enhance the CH_4 emission. We found a positive correlation between the annual emission rate of CH_4 and the contents of readily mineralizable carbon (RMC) in precultivated soils.

INTRODUCTION

The concentration of atmospheric methane (CH_4) is increasing by about 1% per year. CH_4 is one of the greenhouse gases, which have strong infrared absorption bands and trap part of the thermal radiation from the earth's surface, and the increasing concentration of CH_4 has significant effects on the global heat balance. In addition, CH_4 plays an important role in the photochemical reactions of the troposphere and the stratosphere, and change of its concentration strongly influences atmospheric chemical reactions.

Atmospheric CH_4 is produced both by natural and anthropogenic processes (Cicerone and Oremland, 1988). Of the wide variety of sources of atmospheric CH4, paddy fields are considered as one of the most important. The harvested area of paddy rice is increasing steadily. Field measurements of CH_4 emission from paddy fields have been carried out in California (Cicerone and Shetter, 1981; Cicerone et al, 1983) and in southern Europe (Seiler et al., 1984; Holzapfel-Pschorn and Seiler, 1986; Holzapfel-Pschorn et al., 1986). However, there is no field data covering

Asian countries in which 90% of the world paddy area is located. Therefore the estimated value of the global CH_4 emission rate from paddy fields is still highly uncertain. Furthermore, to make an accurate estimation, more information about factors controlling CH_4 production and emission from paddy fields is needed.

In this study, the results are presented of field measurements of annual CH_4 emission and the effects of soil properties, application of organic matter, and cultivation practices on CH_4 emission from Japanese paddy fields.

EXPERIMENTAL METHODS

Study Sites and Soils

Field studies were carried out in 1988 in rice paddies at four different sites (Ryugasaki, Kawachi, Mito and Tsukuba) located in Ibaraki prefecture, Japan. The different soil types used in this study included alluvial soils, peaty soils and Andosols (volcanic ash soil). Successive application of organic matter, such as rice straw and its compost, have been performed at three of these sites.

Soil samples were collected from individual plots before flooding and some properties of the soils are listed in Table 19.14.1. Total carbon and total ·nitrogen content were determined by means of a dry combustion method using a CN analyzer. For measuring contents of the readily mineralizable carbon (RMC), the soil samples in fresh and moist condition were incubated under N_2 atmosphere at 25°C. The RMC was defined as the total amount of CO_2-C and CH_4-C produced after 28 days incubation.

Table 19.14.1 Summary of the paddy fields studied and their soil properties. The soil samples were collected before flooding

Site	Ryugasaki[1]	Kawachi[2]	Mito[1]	Tsukuba[1]
Soil	alluvial	peaty	humic Andosol	light-coloured Andosol
Texture[3]	SCL	CL	L	L
Percolation rate (mm d^{-1})	10	10	30	30
pH	5.9	5.6	6.5	5.9
Total C (%)	1.4	3.4	6.0	2.2
Total N (%)	0.13	0.26	0.40	0.21
C/N	10	13	15	10
RMC[4] (μgC/g/28 days)	78.9	232.9	38.3	19.4

[1]   the data of the mineral fertilizer plot
[2]   the data of the rice straw plot
[3]   L = loam; C = clay; SCL = sandy clay loam; Cl = clay loam
[4]   readily mineralizable carbon

Measurements of CH₄ Fluxes

The chamber method as proposed by Minami and Yagi (1988) was used for the measurements of the CH_4 flux. The chambers were constructed from metal cylinders with effective internal height of 16.7, 66.0 and 102.0 cm, covering a cross-sectional area of 638 cm^2. The CH_4 fluxes were determined by measuring the temporal increase of the CH_4 concentration of the air within the chamber. The air within the chamber was sampled by an air sampling pump and collected in a tedlar bag. All the measurements were performed between ten and noon, except in the case of measuring the diurnal variation of fluxes.

The CH_4 concentration of the samples collected in the tedlar bags were determined by using a gas chromatograph equipped with a gas sampler and a flame ionization detector.

RESULTS AND DISCUSSION

Seasonal variation of the CH_4 fluxes from the paddy field in Ryugasaki is shown in Figure 19.14.1, along with soil Eh. The measurements were carried out at four plots with different treatments of organic matter and nitrogen fertilizer. The CH_4 emission showed an increase as the soil Eh decreased nearly a month after flooding. A remarkable variation of the CH_4 fluxes was observed thereafter. Although the magnitude of the fluxes was different among the plots, three peaks of the flux were found in all plots. The first peaks appeared in the late tillering stage of the rice plants at the end of June. The second peak occurred during the middle of July before supplement application of mineral fertilizer, the flux rate reached a peak higher than 30 mg $m^{-2}h^{-1}$ for the rice straw plot. The third peaks occurred during the middle of August and these were characterized by the relatively high flux rates in all the plots. The flux rates of the rice straw, the compost, the mineral and the non-nitrogen plots reached 32.6, 17.3, 13.3 and 15.3 mg $m^{-2}h^{-1}$, respectively, in this period.

During the midsummer drainage and after the supplement application of mineral fertilizer, two flux minima were observed. Furthermore, when the fields were drained in the end of August, the fluxes rapidly dropped. These reductions in the CH_4 fluxes in response to cultivation practices, such as drainage and the supplement application, suggest that production and emission of CH_4 from paddy soils are significantly influenced by the changes of soil conditions.

The CH_4 flux rate from the rice straw plot was much greater than those from the other plots throughout the flooding period. On the other hand, the application of the compost prepared from rice straw did not remarkably enhance the CH_4 flux as compared with the mineral plot. The difference in the rate and the variation of the CH_4 fluxes between the mineral and the non-nitrogen plot were also small.

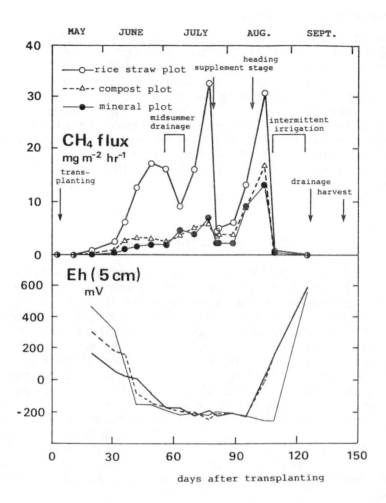

Figure 19.14.1 Seasonal variation of the CH_4 fluxes and soil Eh measured in Ryugasaki paddy field. Open circles with bold line are data from the rice straw plot, open triangles with broken line are from the compost plot and solid circles with fine line are from the mineral fertilizer plot (from Yagi and Minami, 1989).

Insignificant diurnal variations of the CH_4 fluxes were observed during July 26-27 in *Ryugasaki*. Eleven measurements were carried out in each plot of the rice straw, the compost and the mineral treatment. The average flux rates were 4.9±0.5, 4.2±0.2 and 2.3±0.2 mg m^{-2}h^{-1}, respectively.

In *Kawachi*, the flux measurements were carried out in a plot located in a peaty wetland. Rice straw was applied to the paddy field after the previous harvest. The trend of the seasonal variation of the CH_4 flux in

Kawachi was very similar to that observed in Ryugasaki. The flux rates were much greater than those from the other sites studied. The highest CH_4 flux rate was found on July 8, up to 67.2 mg m^{-2} hr^{-1}, which is approximately twice as high as the maximum rate at the rice straw plot in Ryugasaki.

The paddy field studied in *Mito* is situated on a diluvial upland. The soil is a humic Andosol with a rapid percolation rate of not less than 30 mm per day. Five plots with different treatments were used for the flux measurements. The trends of the seasonal variation of the fluxes and the effects of organic matter applications almost corresponded to those observed in Ryugasaki. In addition, the CH_4 flux from the paddy field increased with increasing application of rice straw. The flux rates in Mito were low compared to those in Ryugasaki and Kawachi. The highest flux rate was found during the middle of July, up to 10.2 mg $m^{-2}h^{-1}$ for the rice straw plot.

Flux measurements were also carried out in *Tsukuba* in the mineral fertilizer and the rice straw plots with light-coloured Andosols. The fluxes remained low throughout the cultivation period, ranging between <0.1 and 0.7 mg m^{-2} hr^{-1}.

From the integration of the observed CH_4 flux rates, the total amount of CH_4 emission from the individual plots were calculated. Since these paddies are never flooded after harvest until the next growing season, the integrated value can be considered as the annual emission. The results are listed in Table 19.14.2, along with the average and the maximum value of the observed flux rates.

Table 19.14.2 Methane flux and annual emission rates from the individual plots on the paddy fields

Site	Plot	Flux		Annual Emission rate (g $m^{-2}y^{-1}$)
		Average (mg $m^{-2}h^{-1}$)	Maximum	
Ryugasaki	non-nitrogen	2.8	15.3	8.0
	mineral	2.9	13.3	8.2
	compost (12t ha^{-1})	3.8	17.3	10.5
	rice straw (6t ha^{-1})	9.6	32.6	27.0
Kawachi	rice straw (6t ha^{-1})	16.3	67.2	44.8
Mito	non-nitrogen	1.4	5.1	4.1
	mineral	1.2	5.7	3.6
	compost (12t ha^{-1})	1.9	5.7	5.9
	rice straw (6t ha^{-1})	3.2	10.2	9.8
	rice straw (9t ha^{-1})	4.1	8.9	12.6
Tsukuba	mineral	0.2	0.5	0.6
	rice straw (6t ha^{-1})	0.4	0.7	1.1

Emission rates of CH_4 differed markedly between soil types. The highest rate was measured in the paddy field with the peaty soil. Since the management of irrigation water and other cultivation practices in Kawachi were practically similar to those in Ryugasaki, the difference in the CH_4 emission rate at the two sites is probably caused by the difference in soil organic matter contents. High rates of CH_4 production and emission also have been reported in studies of peaty wetlands (William and Crawford, 1984; Harris et al., 1985). The emission rates in the Andosols were low. The annual emission rates flux for the peaty soil were 40 times higher than for the light-coloured Andosol. Low emission from the paddies in Andosols may be caused, in part at least, by the rapid percolation rates of water in these paddies. Actually, the soil Eh at 5 cm depth in these paddy fields did not reach -200 mV.

Application of rice straw to the paddy fields significantly increased the CH_4 emission rates at all the sites. Annual emission rates from the plots receiving 6,000 kg ha^{-1} of rice straw in addition to the mineral fertilizer increased approximately 2 to 3 fold as compared with the mineral fertilizer plots. The effects of the compost application and the non-nitrogen fertilization on CH_4 emission were both small. It is well known that addition of various types of readily decomposable organic matter increased CH_4 production in paddy soils from laboratory incubation experiments (Yamane and Sato, 1963; Tsutsuki and Ponnamperuma, 1987). Tsutsuki and Ponnamperuma (1987) also showed that compost addition did not enhance the formation of anaerobic decomposition products. The relation between the availability of soil organic matter and the CH_4 emission is illustrated in Figure 19.14.2,

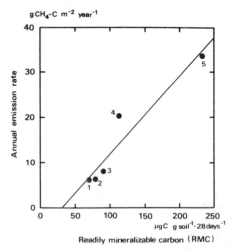

Figure 19.14.2 The relationship between the contents of the readily mineralizable carbon (RMC) and the annual emission rates of CH_4 for Ryugasaki (1-4) and Kawachi (5). The soil samples were collected before flooding. Individual data are: 1. the non-nitrogen plot; 2. the mineral fertilizer plot; 3. the compost plot and 4. and 5. the rice straw plots (from Yagi and Minami, 1989).

which shows the correlation of the annual emission rate of CH_4 with the RMC for Ryugasaki and Kawachi. The percolation rate of water and the cultivation practices are almost identical in these sites. The results summarized in Figure 19.14.2 suggest that the RMC in paddy soil is one of the principal factors affecting CH_4 emission from flooded soils.

Annual emission rates of CH_4 from paddy fields measured in California and southern Europe ranged between 12 and 54 $g.m^{-2}$. Our data in Japanese paddies were slightly lower than previous studies. The difference of these rates may be caused by the differences in the soil properties and cultivation practices.

REFERENCES

Cicerone, R.J. and J.D. Shetter (1981) Sources of atmospheric methane: measurements in rice paddies and a discuss. Journal of Geophysical Research 86:7203-7209.

Cicerone, R.J., J.D. Shetter and C.C. Delwiche (1983) Seasonal variation of methane flux form a California rice paddy. Journal of Geophysical Research 88:11022-11024.

Cicerone, R.J. and R.S. Oremland (1988) Biogeochemical aspects of atmospheric methane. Global Biogeochemical Cycles 2:299-327.

Harris, R.C., E. Gorham, D.I. Sebacher, K.B. Bartlett and A. Flebbe (1985) Methane flux from northern peatlands. Nature 315:652-653.

Holzapfel-Pschorn, A. and W. Seiler (1986) MEthane emission during a cultivation period from an Italian rice paddy. Journal of Geophysical Research 91:11803-11814.

Holzapfel-Pschorn, A., R. Conrad and W. Seiler (1986) Effect of vegetation on the emission of methane from submerged paddy soil. Plant and Soil 92:223-233.

Minami, K. and K. Yagi (1988) Method for measuring methane flux from rice paddies. Japanese Journal of Soil Science and Plant Nutrition 59:458-463 (in Japanese).

Seiler, W., A. Holzapfel-Pschorn, R. Conrad and D. Scharffe (1984) Methane emission from rice paddies. Journal of Atmospheric Chemistry 1:241-268.

Tsutsuki, K. and F.N. Ponnamperuma (1987) Behaviour of anaerobic decomposition products in submerged soils. - Effects of organic material amendment, soil properties, and temperature. Soil Science and Plant Nutrition 33:13-33.

Yagi, K. and K. Minami (1989) Effects of mineral fertilizer and organic matter applications on the emission of methane from some Japanese paddy fields. (Submitted to Soil Science and Plant Nutrition).

Yamane, I. and K. Sato (1963) Decomposition of plant constituents and gas formation in flooded soil. Soil Science and Plant Nutrition 9:28-31.

Williams, R.T. and R.L. Crawford (1984) Methane production in Minnesota peatlands. Applied Environmental Microbiology 47:1266-1271.

Factors Influencing the Fraction of the Gaseous Products of Soil Denitrification Evolved to the Atmosphere as Nitrous Oxide

J.R.M. ARAH AND K.A. SMITH

The Edinburgh School of Agriculture
West Mains Road, Edinburgh EH9 3JG, UK

ABSTRACT

Nitrous oxide (N_2O) is produced in the soil by microbial denitrification, which also produces dinitrogen (N_2). Many macroscopic field properties influence the fraction of the gaseous products of denitrification evolved as N_2O - the "N_2O fraction"; there is no universal relationship between this fraction and total denitrification flux. Low pH, high nitrate concentration, low moisture content, and low availability of oxidisable organic matter all tend to increase N_2O fractions, but the relative importance of these parameters is difficult to assess. Denitrification in agricultural soils occurs predominantly in anaerobic microsites in an otherwise well-aerated medium; the process may be modelled by numerical solution of the partial differential equations governing simultaneous diffusion and reaction in a spherical environment. More work is required to characterise the relevant microsite parameters in typical agricultural soils.

INTRODUCTION

Nitrous oxide (N_2O) is an important "greenhouse gas" produced by natural processes in soil. The most important of these processes are nitrification (microbial oxidation of ammonium to nitrite and nitrate) and denitrification (reduction of nitrate, via nitrite and N_2O, to dinitrogen, N_2). Various factors affect the relative significance of nitrification and denitrification as sources of N_2O, but it has been suggested on the basis of measurements of the ^{15}N content of atmospheric N_2O that denitrification is marginally the larger global source (Yoshida and Matsuo, 1983). Denitrification occurs predominantly through the activity of anaerobic microorganisms (Firestone, 1982). N_2O produced by denitrification may either escape to the atmosphere or be further reduced to N_2. The proportion of the two gases evolved as N_2O - the "N_2O fraction" - varies greatly with environmental conditions; in order to predict N_2O fluxes it is necessary to understand denitrification.

Soils and the Greenhouse Effect. Edited by A.F. Bouwman
© 1990 John Wiley & Sons Ltd.

MEASUREMENT OF DENITRIFICATION FLUXES
AND NITROUS OXIDE FRACTIONS

N_2O and N_2 fluxes may be measured using ^{15}N-labelled fertilizer (Siegel et al., 1982), or by inhibiting the reduction of N_2O to N_2 with acetylene and determining the consequent enhanced N_2O flux (Ryden et al., 1979). [15]N techniques assume a rapid equilibration between the labelled fertilizer and any soil-derived nitrate (Boast et al., 1988). The acetylene technique depends on the attainment throughout the denitrifying soil of inhibitory concentrations of acetylene, and the absence of any undesirable secondary effects of the inhibitor (Ryden et al., 1979). If either assumption is violated the probable effect is an underestimate of denitrification and an overestimate of the N_2O fraction. Spatial and temporal variability bedevil the quantification of denitrification (Folorunso and Rolston, 1984), whatever the measurement technique.

N_2O and total denitrification fluxes measured using a simple closed chamber and acetylene inhibition on a Winton series clay loam under grass, subject to three different compaction treatments (causing differences in air-filled porosity), were employed to calculate the N_2O fractions plotted in Figure 19.15.1. Measurements were made on duplicate 0.05 m^2 microplots between 19/4/89 and 6/7/89; all treatments received 150 kg N as ammonium nitrate on 31/3/89 and again on 9/6/89. N_2O accounted for between 5% and 100% of the gaseous products of denitrification, with little correlation between this fraction

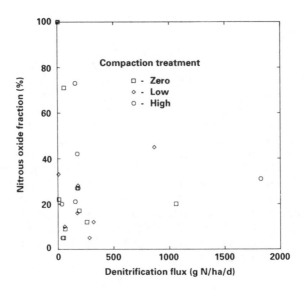

Figure 19.15.1 Measured N_2O fractions against total denitrification flux (acetylene inhibition technique).

and the total flux. These measurements are not especially representative or interesting, but they provide a typical picture of an extremely variable process with few obviously dominant controlling parameters.

FACTORS AFFECTING NITROUS OXIDE FRACTIONS

High soil water contents and poor soil structure (both of which result in low air-filled porosity) favour denitrification but result in low N_2O fractions (Focht, 1978; Smith, 1989). The primary cause is reduced diffusivity of oxygen (resulting in more extensive anaerobic regions) and of N_2O (retarding N_2O escape and promoting its reduction to N_2). Concentrations of water-soluble and mineralisable carbon in the soil are highly correlated with denitrification activity (Burford and Bremner, 1975), and there is evidence that increasing carbon availability decreases the N_2O fraction (Firestone, 1982). Acid conditions (pH < 5) inhibit denitrification and result in an increase in the N_2O fraction (Blackmer and Bremner, 1978; Christensen and Tiedje, 1988). High concentrations of nitrate also increase the N_2O fraction (Blackmer and Bremner, 1978).

MODELLING NITROUS OXIDE FRACTIONS

McConnaughey and Bouldin (1985) developed a model in which nitrate is reduced (via nitrite) to N_2O, which either escapes to the atmosphere or is reduced in turn to N_2; nitrate reduction is inhibited by oxygen, N_2O reduction by oxygen and nitrate; enzyme-mediated reactions obey Michaelis-Menten kinetics. This scheme ignores nitric oxide (NO) as an intermediate, and assumes no additional inhibitory effects (e.g. of nitrite on oxygen reduction). The medium in which the process occurs is physically homogeneous; spatial variation is ignored.

The spatial variability of denitrification has been attributed to "hot-spots" associated with high local concentrations of oxidisable organic matter (Parkin, 1987), and to the presence in the soil of macroscopic aggregates in which microbial activity produces anaerobic and hence potentially denitrifying microsites (Smith, 1980; Arah and Smith, 1989). In either case we can model denitrification as the simultaneous diffusion and reduction of oxygen, nitrate, and N_2O in a spherically symmetrical environment. In the hot-spot model the denitrifying system is a zone of indeterminate radius centred on the hot-spot; in the aggregate model it is the aggregate itself.

The denitrification process within a spherical microsite may be described by a partial differential equation the relevant controlling parameters of which are: the Michaelis constant for the reaction, the microsite radius and diffusion constant, the microsite reaction potential (the rate at which the reaction would proceed in the absence of any diffusive limitation or inhibition), the external substrate concentration, and any inhibitory effects of other substances (this last is particularly relevant to the inhibition of denitrification by oxygen, and of N_2O reduction by nitrate). The diffusion-reaction equation can be solved by numerical methods and the effects of changes in the various controlling parameters examined.

Figure 19.15.2 shows the effect on microsite N_2O fractions of varying one or other of these controlling parameters while all else remains constant. The "standard" microsite in this Figure is an unsaturated aggregate 1 cm in radius, with Michaelis constants, reaction potentials and external concentrations (described by Arah, 1989) fairly typical of those encountered in soil. This standard microsite does not in fact denitrify. The major importance of microsite radius is apparent from the Figure, as is that of the reaction potential. Here, gaseous diffusion constants and external oxygen concentrations are less significant; their effects on a different standard may be greater. Results plotted are for the "pseudo-equilibrium" case, in which the internal state of the microsite remains constant. This may be expected whenever the microsite has suffered no major perturbations for 24 hours or so (Arah, 1989).

A real soil contains many potentially denitrifying microsites, characterised by different values of the controlling parameters identified above. Such a soil may be represented by an array of microsites, with

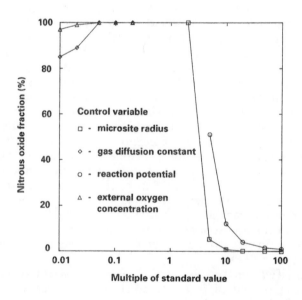

Figure 19.15.2 Calculated N_2O fraction of spherical microsites - sensitivity analysis. Standard values of control parameters are: microsite radius 1 cm; gaseous diffusion constant 5×10^{-9} m^2s^{-1}; reaction potentials 10^{-4} mol $m^{-3}s^{-1}$ (for oxygen and N_2O) and 2×10^{-5} mol $m^{-3}s^{-1}$ (for nitrate); and external oxygen concentration 1 mol m^{-3}. Details in Arah (1989).

each of the controlling parameters obeying an independent probability distribution, which in a real soil is likely to be approximately lognormal. Properties of the assembly of microsites may be calculated by integrating over these independent probability distributions.

DISCUSSION

How do the microsite parameters of the model correspond to those macroscopic properties which are known to affect the N_2O fraction? Michaelis constants and inhibition functions may be expected to remain relatively constant from soil to soil, and external substrate concentrations are directly measurable; microsite radius is determined by soil structure, diffusion constants by the pore-size distribution and the moisture content; reaction potentials may be related to the local concentration of oxidisable organic matter; and the effects of soil acidity may be interpreted either as a general inhibition of all microbial processes or as a more specific inhibition of nitrous oxide reduction. Thus all the generally recognised factors influencing the N_2O fraction may be incorporated within the microsite model.

In order to be able to employ a model based on microsites it is necessary to know the probability distributions of the various microsite parameters. Much work remains to be done to characterise these probability distributions in real soils.

We have discussed the modelling of denitrification in spherically symmetric environments, which are perhaps the best approximation to the dominant sites of denitrification in unsaturated, cultivated soils. Other symmetries may predominate in other situations; where roots are the main source both of oxygen and of organic substrate (in a paddy soil, for example) a cylindrical model may be more appropriate. The mathematical methods described here for spherical symmetry are quite capable of extension to the cylindrical case. The methods may satisfactorily indicate qualitative trends; what remains to be tested is the extent to which they yield useful quantitative predictions.

REFERENCES

Arah, J.R.M. (1989) Diffusion/reaction models of denitrification in soil microsites. In: J. Sorensen and N.P. Revsbech (Eds.), Denitrification in Soil and Sediment. Plenum Press, London (in press).

Arah, J.R.M. and K.A. Smith (1989) Steady-state denitrification in aggregated soils: a mathematical model. Journal of Soil Science 40:139-149.

Blackmer, A.M. and J.M. Bremner (1978) Inhibitory effect of nitrate on reduction of nitrous oxide to dinitrogen gas by soil microorganisms. Soil Biology and Biochemistry 10:187-191.

Boast, C.W., R.L. Mulvaney and P. Baveye (1988) Evaluation of nitrogen-15 tracer techniques for direct measurement of denitrification in soil: I. Theory. Soil Science Society of America Journal 52:1317-1322.

Burford, J.R. and J.M. Bremner (1975) Relationships between the denitrification capacities of soils and total, water soluble and readily decomposable soil organic matter. Soil Biology and Biochemistry 7:389-394.

Christensen, S. and J.M. Tiedje (1988) Denitrification in the field, analysis of spatial and temporal variability. In: D.S. Jenkinson and K.A. Smith (Eds.), Nitrogen Efficiency in Agricultural Soils, p 295-301. Elsevier, London.

Firestone, M.K. (1982) Biological denitrification. In: F.J. Stevenson (Ed.), Nitrogen in Agricultural Soils, p 289-326. American Society of Agronomy, Madison, Wisconsin.

Focht, D.D. (1978) Methods for analysis of denitrification in soils. In: D.R. Nielsen and J.G. MacDonald (Eds.), Nitrogen in the Environment, vol 2, p 433-490. Academic Press, New York.

Folorunso, O.A. and D.E. Rolston (1984) Spatial variability of field-measured denitrification gas fluxes. Soil Science Society of America Journal 48:1214-1219.

McConnaughey, P.K. and D.R. Bouldin (1985) Transient microsite models of denitrification: I. Model development. Soil Science Society of America Journal 49:886-891.

Parkin, T.B. (1987) Soil microsites as a source of denitrification variability. Soil Science Society of America Journal 51:1194-1199.

Ryden, J.C., L.J. Lund, J. Letey and D.D. Focht (1979) Direct measurement of denitrification loss from soils: II. Development and appication of field methods. Soil Science Society of America Journal 43:110-118.

Siegel, R.S., R.D. Hauck and L.T. Kurtz (1982) Determination of $^{30}N_2$ and application to measurement of N_2 evolution during denitrification. Soil Science Society of America Journal 46:68-74.

Smith, K.A. (1980) A model of the extent of anaerobic zones in aggregated soils, and its potential application to estimates of denitrification. Journal of Soil Science 31:263-277.

Smith, K.A. (1989) Anaerobic zones and denitrification in soil: modelling and measurement. In: J. Sorensen and N.P. Revsbech (Eds.), Denitrification in Soil and Sediment. Plenum Press, London (in press).

Yoshida, N. and S. Matsuo (1983) Nitrogen isotope ratio of atmospheric N_2O as a key to the global cycle of N_2O. Geochemical Journal 17:231-239.

Seasonal Nitrous Oxide Emissions from an Intensively-Managed, Humid, Subtropical Grass Pasture

E.A. BRAMS[1], G.L. HUTCHINSON[2], W.P. ANTHONY[1], and G.P. LIVINGSTON[3]

[1]CARC, Prairie View A & M University, Prairie View, TX;
[2]USDA-ARS, Fort Collins, CO;
[3]NASA, Ames Research Center, Moffett Field, CA.

ABSTRACT

Nitrous oxide emissions from a bermuda grass (Cynodon dactylon) pasture on sandy loam in a humid, subtropical region of southern Texas were monitored for one year using a closed soil cover technique. Emissions from plots under minimum cultural management (spring harvest followed by a single annual maintenance application of ammonium sulphate) were significantly lower than from plots under intensive cultural management (harvest and fertilization repeated on a 9 week cycle throughout the growing season). Highest fluxes occurred during the warmer seasons immediately after the first rainfall following application of N fertilizer. Seasonal changes in soil temperature and soil water content indicated that higher values for both parameters were a necessary, but not sufficient, condition to enhance the flux. Measured fluxes more closely resembled the low fluxes characteristic of temperate region grasslands than high fluxes characteristic of the tropics.

INTRODUCTION

Nitrous oxide (N_2O) is a long-lived trace atmospheric constituent, the concentration of which has risen steadily (0.2% per year) over the past three or four decades (Weiss, 1982). It functions both as a potentially important greenhouse gas in various global climate change scenarios and as a participant in reactions leading to destruction of the stratospheric ozone that shields life forms on the planetary surface from solar ultraviolet radiation. Soil microbial processes and combustion processes have long been recognized as the major sources of N_2O in the atmosphere, but because of the complexity of those microbial processes and the paucity of combustion emissions data, there remains great uncertainty regarding specific causes of the observed increase in the gas's atmospheric concentration (McElroy and Wofsy, 1986).

Tropical forests (Keller et al., 1986) and tropical grasslands (Luizão et al., 1989) have been shown to be among the more important biogenic sources of N_2O, and it has been postulated that subtropical ecosystems

Soils and the Greenhouse Effect. Edited by A.F. Bouwman

may also represent a globally-significant seasonal source (McElroy and Wofsy, 1986). The geographic extent of subtropical ecosystems is large, and most have been subject to intense anthropogenic disturbance. Specific objectives of our research were (i) to quantitatively determine the seasonal and annual rates of N_2O emission from minimally- and intensively-managed grass pastures in a humid, subtropical climate, and (ii) to measure the response of soil N_2O emissions to changes in soil N pool sizes and transformation rates, to various cultural practices such as harvest and fertilization, and to selected environmental parameters such as precipitation and seasonal and diel changes in soil temperature. Seasonal differences in the diel variability of soil N_2O emissions and the important controlling influence of soil N transformation rates will be discussed in separate publications.

METHODS

Two management schemes were imposed on 10 m × 30 m plots in an established 30ha bermuda grass (Cynodon dactylon) pasture on well-drained, very uniform Kenney sandy loam (a member of the loamy, siliceous, thermic, Grossarenic Paleudalfs) in southern Texas. Selected soil and climatic data are given in Table 19.16.1. The two treatments - minimum cultural management (spring harvest followed by a single annual maintenance application of N fertilizer) and intensive cultural management (harvest and fertilization repeated on a 9 week cycle throughout the growing season) - were replicated four times in a completely randomized block design. The source of fertilizer N was ammonium sulphate; the amount applied on each fertilization date (which always followed harvest by 5 days) was that recommended for maximum protein production based on soil tests performed by Texas A & M University (Table 19.16.2).

To begin each measurement of N_2O flux, a vented, cylindrical soil cover (30 cm diam. × 30 cm tall) was mounted atop a permanently-installed ring (30 cm diam. × 7.5 cm tall) driven 5 cm into the soil, and the two were sealed together by overlapping the seam between them with an external large rubber band. The soil covers (about 25 L total volume) were constructed from rigid polyvinyl chloride pipe using design criteria suggested by Hutchinson and Mosier (1981), then insulated with polyurethane foam and covered with reflective aluminized polyester film to minimize internal heating by solar radiation. Accumulation of N_2O beneath each cover after 10 and 20 minutes was determined by drawing 30 mL air samples through the covers' sampling ports using 60 mL polypropylene syringes fitted with nylon stopcocks. The N_2O concentration at the time of each cover's installation was assumed the same as that of an ambient air sample collected at the field site. All samples were transported to the laboratory for analysis within 12 hours by gas chromatography using electron capture detection (Mosier and Mack, 1980).

Table 19.16.1 Selected soil and climatic data for the experimental site

Parameter	Value	
Soil (0-16 cm)		
pH (1:1 water)	6.1	
CEC	0.0048	cmol g-1
sand	770	g kg^{-1}
silt	130	g kg^{-1}
clay (predominantly kaolinite)	90	g kg^{-1}
organic matter	17	g kg^{-1}
water retention @ 0.00 MPa	250	g kg^{-1}
@ 0.03 MPa	70	g kg^{-1}
@ 0.20 MPa	40	g kg^{-1}
Climate		
rainfall (avg. annual)	102	cm
air temperature (avg. annual max.)	26.7	°C
(avg. annual min.)	13.0	°C
growing season	304	days

Table 19.16.2 Seasonal N fertilizer application and N$_2$O evolution from bermuda grass plots under minimum cultural management (fertilizer applied in spring only) or intensive cultural management

Season	Interval	Fertilizer Applied	N$_2$O Emissions Minimum	Intensive
	days	------------------------ kg N/ha ------------------------		
late summer	63	117	0.30	0.56
fall	63	82	0.07	0.10
winter	105	---	0.08	0.13
spring	63	112	0.24	0.23
early summer	63	52	0.20	0.35

Soil N$_2$O emission rates were measured at four locations in each plot; the scheduled frequency of flux measurements was highest immediately following harvest and fertilization when highest N$_2$O emission rates were expected. Soil moisture and temperature sensors were buried at 2 cm and 10 cm depths in each plot and their outputs recorded each sampling day. Rainfall, air temperature, and wind speed and direction were also monitored.

RESULTS AND DISCUSSION

Nitrous oxide emissions during the four 9 week harvest/fertilization cycles that spanned the growing season, as well as during the longer winter period when there was little grass growth, are plotted in Figures 19.16.1a-19.16.1e. Rainfall measurements are included to emphasize the importance of individual precipitation events in determining the N_2O flux. In general, highest fluxes occurred during the warmer seasons immediately after the first rainfall following application of N fertilizer. The most striking example occurred during the late summer period following fertilization on August 9, 1988 (Figure 19.16.1a). Soil temperature at 2 cm was about 34 °C, and 6.4 mm rain fell both on the day of fertilization and again 3 days later. The N_2O flux rose rapidly to 42 g N ha^{-1} d^{-1} on the third day following fertilization, then fell sharply as the soil dried, but remained significantly higher ($p < 0.05$) than the flux from the minimum cultural treatment for at least 30 days. The absence of a response to rainfall events during the remainder of this period, as well as other periods well-removed from a fertilization date, indicate that the source of N_2O was probably nitrification, rather than denitrification (Hutchinson and Mosier, 1979). The emissions burst from the minimum cultural treatment that coincided with that from the intensive cultural treatment immediately following late summer fertilization is unexplained, but was repeated following fertilization in the early summer period.

Another example of the importance of precipitation was observed in early summer (Figure 19.16.1e) when elevation of the N_2O flux by fertilization was relatively small until the first rainfall occurred 9 days later. In this case it is likely that the peak emission rate (which probably occurred 2-3 days following rainfall) was not measured on scheduled sampling dates 15 and 19 days after harvest (1 and 5 days following precipitation). Inadequate sampling frequency probably also prevented observing the peak "first rainfall" response during the fall period (Figure 19.16.1b) when there was no precipitation for 20 days following fertilization. Although the N_2O flux was not measured for eight days after that rainfall, there was still a significant difference ($p < 0.05$) between N_2O fluxes from minimum and intensive cultural treatments.

Flux data for the entire year are consolidated in Figure 19.16.2 to better show the importance of seasonal changes in soil temperature and soil water content as controllers of soil N_2O emissions. Warmer soil temperature was not a sufficient condition to enhance the emission of N_2O, but was necessary to allow the soil microbial population to respond to other perturbations such as fertilization or rainfall, and especially to a combination of these two. Similarly, higher soil water content appeared to be a necessary, but not sufficient, condition to drive the flux. An additional feature of the N_2O emissions data not directly attributable to changes in soil temperature, soil water content, or fertilizer supply was the tendency for higher emissions from both treatments beginning about 40 days after the spring harvest and lasting

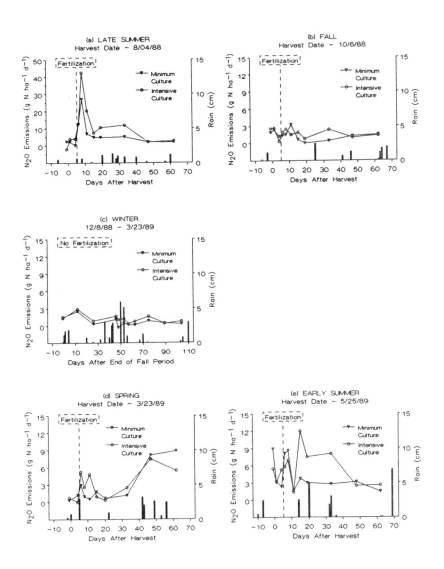

Figure 19.16.1 Nitrous oxide flux and precipitation during five seasonal periods from August, 1988 through July, 1989. Note that the scale of the ordinate is different for (a) than for the other graphs.

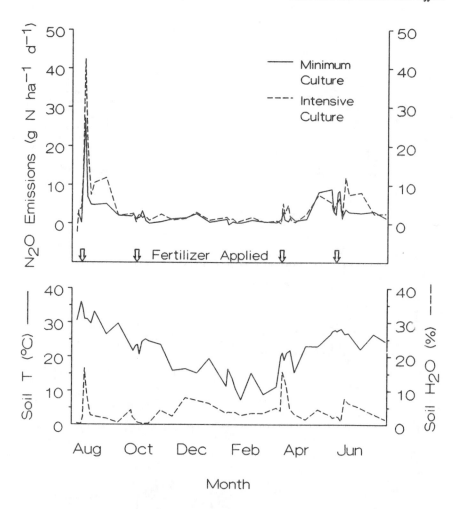

Figure 19.16.2 Nitrous oxide flux as influenced by seasonal changes in soil temperature and soil water content both measured at 2 cm depth.

into early summer (Figures 19.16.1d and 19.16.1e). A similar observation was even more apparent in the spring of 1988 (data not shown). We suspect that the spring flush of microbial growth responsible for decay of detritus accumulated during the winter may have increased the competition for oxygen sufficiently to enhance the potential for denitrification.

The area beneath each emissions curve in Figure 19.16.1 was integrated to provide a seasonal estimate of N_2O flux from this humid, subtropical, grassland ecosystem during the unusually dry year in which it was sampled (Table 19.16.2). Except during the spring period when both treatments were fertilized similarly, the flux from plots under intensive cultural management was significantly higher ($p < 0.05$) than from plots under minimum cultural management. However, N_2O emissions represented less than one percent of the applied N fertilizer in every season. Our measured fluxes more closely resemble the low emission rates characteristic of temperate region grasslands than high emission rates characteristic of the tropics, probably because lower rainfall, rapid drainage, and lower available C at our research site caused less frequent occurrence of conditions supporting denitrification than typically occurs at tropical sites.

ACKNOWLEDGEMENTS

This work was supported by NASA Grant NAG 2-400. The authors sincerely thank Mr. Yung Ping Chang, Mr. Jesse Trevino and Mrs. Adela Walker for invaluable technical and secretarial assistance and gratefully acknowledge the support provided by Dr. Alden H. Reine and Col. Jiles P. Daniels, the Cooperative Agricultural Research Center's Director and Assistant Director.

REFERENCES

Hutchinson, G.L. and A.R. Mosier (1979) Nitrous oxide emissions from an irrigated corn field. Science 205:1125-1127.

Hutchinson, G.L. and A.R. Mosier (1981) Improved soil cover method for field measurement of nitrous oxide fluxes. Soil Science Society of America Journal 45:311-316.

Keller, M., W.A. Kaplan and S.C. Wofsy (1986) Emissions of N_2O, CH_4 and CO_2 from tropical forest soils. Journal of Geophysical Research 91:11791-11802.

Luizão, F., P. Matson, G.P. Livingston, R. Luizão and P. Vitousek (1989) Nitrous oxide flux following tropical land clearing. Global Biogeochemical Cycles (In press).

McElroy, M.B. and S.C. Wofsy (1986) Tropical forests: interactions with the atmosphere. In: G.T. Prance (Ed.) Tropical Rain Forests and the World Atmosphere, p. 33-60, Westview Press, Boulder, CO.

Mosier, A.R. and L. Mack (1980) Gas chromatographic system for precise, rapid analysis of nitrous oxide. Soil Science Society of America Journal 44:1121-1123.

Weiss, R.F. (1982) The temporal and spatial distribution of nitrous oxide. Journal of Geophysical Research 86:7185-7195.

Nitrous Oxide Emissions from the Nitrification of Nitrogen Fertilizers

B.H. BYRNES, C.B. CHRISTIANSON, L.S. HOLT and E.R. AUSTIN

International Fertilizer Development Center
P.O. Box 2040, Muscle Shoals, Alabama 35662, U.S.A.

ABSTRACT

Emissions of N_2O during nitrification were measured in three soils in laboratory incubations employing small closed containers and techniques intended to maximize N_2O efflux. Nitrogen fertilizer was surface applied either as an ammonium hydroxide solution (to simulate anhydrous ammonia) or as urea to each soil. Emissions of N_2O accounted for about 1.5% of the fertilizer on one soil and 0.1% of the fertilizer or less on the other two soils. There was no significant difference in the rate of N_2O emission from the two N sources. One soil acted as an N_2O sink, producing essentially no N_2O emissions from the ammonia source. This experiment indicates that N_2O emissions may be more closely related to soil properties than to the N source that is applied.

INTRODUCTION

The concentration of nitrous oxide (N_2O) in the atmosphere has been increasing because of increased anthropogenic emissions (Bolle et al., 1986) and, by one projection, may double by the year 2100 (U.S. EPA, 1988). Because N_2O has been implicated in the destruction of the stratospheric ozone layer (Crutzen, 1981), research interest has been generated in identifying the sources of N_2O and quantifying its emissions, even though the contribution of N_2O may be quite small relative to that of other anthropogenic gases. Initially, denitrification studies concentrated on reducing N losses in order to improve fertilizer efficiencies. More recently, efforts have been focused on understanding emissions produced from denitrification of nitrate and from nitrification of ammonium sources of fertilizers in order to estimate the contribution of fertilizers to the N_2O emitted to the atmosphere.

The fertilizer-induced emissions (FIE) have been estimated as contributing from 5% to 25% of the total N_2O efflux on a global scale (Bolle et al., 1986; U.S. EPA, 1988); however, these emissions might be altered by technology or by policy decisions that could promote use of one fertilizer over another or use of a particular fertilizer management practice to reduce N_2O production.

Nitrous oxide produced during nitrification has a greater probability of being lost from the soil than does N_2O produced during

Soils and the Greenhouse Effect. Edited by A.F. Bouwman
© 1990 John Wiley & Sons Ltd.

denitrification. Because soil pores are generally open during nitrification and diffusion is allowed to the soil surface, there is less water present to dissolve N_2O. The N_2O emissions from denitrification are normally short-termed, episodic events occurring during the initial development of soil anaerobiosis. Conversely, emissions during nitrification in soils are relatively constant and may actually decrease temporarily following rainfall.

Numerous researchers have measured N_2O efflux from soils during nitrification processes (Bremner and Blackmer, 1981). Losses were found to increase greatly with increasing soil temperature and additions of nitrifiable N sources, including plant and other organic materials. Few studies have attempted to separate the FIE of N_2O due to nitrification from those due to denitrification.

Estimates of the amount of the N_2O efflux during nitrification of fertilizers are variable, but generally such efflux accounts for less than 1% of the fertilizer added (Breitenbeck et al., 1980). In extreme cases, losses of 6% or 7% of anhydrous ammonia as N_2O have been reported (Smith and Chalk, 1980; Bremner et al., 1981). Magalhaes et al. (1984) reported very little loss in a field experiment with anhydrous ammonia. No estimates of N_2O emissions from urea, which is an increasingly important fertilizer (Lastigzon, 1981), have been reported. Though N_2O efflux has been reported for drainage from clear-cut forest areas, there are no such reports for waters draining from agricultural areas (Bowden and Bormann, 1986).

Decreases in N_2O emissions through the use of nitrification inhibitors have been reported (Magalhaes et al., 1984). However, no other effects of nitrogen fertilizer modification (urease inhibitors or insoluble coatings with urea, granule size, etc.) on N_2O loss have been reported.

The following study was conducted to obtain preliminary data on N_2O emissions during nitrification following application of urea or ammonium hydroxide solution (to simulate anhydrous ammonia application) to soil. In conducting these laboratory studies, the amount of N_2O retention in the soil was minimized so that the maximum N_2O emission possible could be estimated. Subsequent studies will be directed toward establishing the effects of various management practices, additives, or coatings on N_2O emissions from urea applications.

METHODOLOGY

Samples (100 g) of three air-dried soils (described in Table 19.17.1) were placed in 500-mL polypropylene wide-mouth jars (11.2 cm in diameter × 5.1 cm high). The lid of each jar was fitted with a Luer-lok® sampling port with a Mininert® valve attached to facilitate removal of 50-cc gas samples with a syringe (Figure 19.17.1).

The soil treatments were (i) control (with no N addition), (ii) prilled urea applied to the soil surface, and (iii) 1 mL of NH_4OH solution (3.3 $\underline{M}$) applied in equal amounts to three locations at the bottom of the soil layer. Nitrogen was applied at a rate equivalent to 50 kg N ha^{-1} on a surface area basis. The soils were 1.3 to 1.5 cm in depth, and moisture

Table 19.17.1 Characteristics of the Soils Used in Study

Soil	Classification[a]	pH (H$_2$O)	CEC[b]	Clay	Silt	Sand	Organic matter
				--------(g/kg)-------			
Crowley	Typic Albaqualf	6.8	1.59	75	750	175	19
Guthrie	Typic Fragiudalf	6.2	1.06	280	610	110	12
Houston	Udic Pellustert	8.0	4.07	630	220	150	17

a. USDA.
b. Cation exchange capacity--cmol$_c$ kg^{-1}.

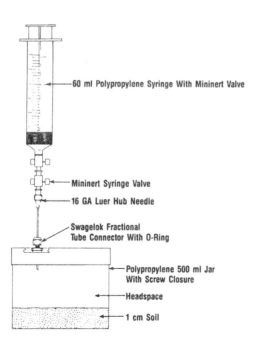

Figure 19.17.1 Soil incubation chamber used in studies.

levels were maintained at 80-90% of field capacity. They were preincubated for 8 weeks at 30°C, and the lids were removed weekly for a short period to allow air exchange.

Gas samples (50 cc) were removed at 2 to 7 day intervals for N_2O analysis. After sampling, the lids of the jars were removed for 10 min, and the atmosphere above the soil atmosphere was allowed to return to ambient conditions. The gas samples were injected into a 50 cc min^{-1} He stream through an ascarite and drierite trap and through a dry ice-acetone trap. The N_2O was condensed in a coil (1.3 m long and 1 mm in diameter) immersed in liquid nitrogen, which served as the sample loop on the injection valve (Figure 19.17.2). After a 5 min condensation period, the injection valve was switched to direct the carrier gas through the coil and then to the gas chromatograph. The liquid nitrogen was removed, and the coil was warmed in water (Delwiche and Rolston, 1976), thereby injecting the contents of the coil into a Tracor Model MF- 150G gas chromatograph equipped with an ultrasonic detector (columns as described by Blackmer and Bremner, 1977). The signal was recorded on a Leeds and Northrup recorder, and peak areas were obtained from a Columbia Scientific Industries Supergrator-2 programmable integrator.

Standard curves were prepared on each sampling date by condensing and then injecting various volumes of a 100 ppm N_2O standard prepared with helium or nitrogen. The N_2O emissions for each period were calculated by subtracting the amount of N_2O in the control from the amount in the fertilized treatments.

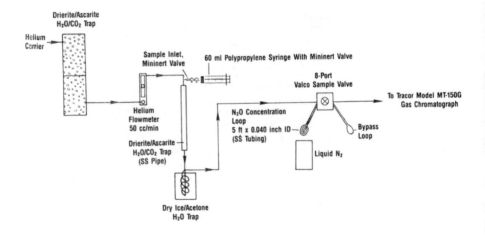

Figure 19.17.2 N_2O concentration and injection system used in studies.

RESULTS AND DISCUSSION

In the Crowley soil, total emission of N_2O amounted to approximately 1.5% for either urea-N or ammonia-N over the 26-day period (Figure 19.17.3). The Guthrie and Houston soils emitted much less N_2O (Figure 19.17.4). Emissions from the two fertilizer sources were similar in both the Crowley and Guthrie soils. In the Houston soil, however, there were no emissions from the ammonia application. Two replications were done, and there was reasonable agreement between replications with the exception of the urea treatment of the Crowley soil. This discrepancy was apparently caused by extreme variability in urea hydrolysis rates between the two replications, though no reason for this variability was observed.

The emissions of N_2O generally increased with increasing nitrification rates for these soils. The complete lack of emission from the ammonium hydroxide treatment in the Houston soil is difficult to explain when the urea in the same soil emitted some N_2O. It is possible that NH_4^+ fixation or direct immobilization occurred, but more likely the N_2O did not evolve because of the placement of the ammonium at the bottom of the soil layer while urea was placed on the surface. At many of the measurement periods, the N_2O concentration of the

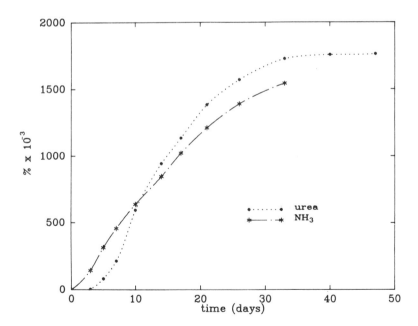

Figure 19.17.3 Rate of N_2O emission from urea and ammonium hydroxide applied to Crowley soil.

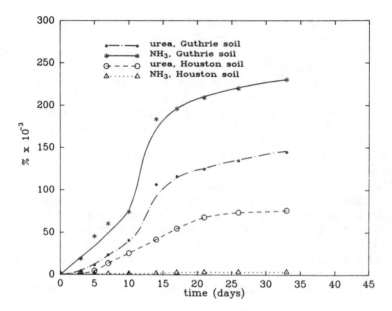

Figure 19.17.4 Rate of N_2O emission from urea and ammonium hydroxide applied to Guthrie and Houston soils. (Note scale differences between Figures 19.17.3 and 19.17.4).

atmosphere above the Houston soil was reduced to less than ambient N_2O concentrations, indicating that this soil was acting as a sink for N_2O production. Anaerobic conditions were not believed to be present.

CONCLUSIONS

The N_2O was produced at similar rates from urea and a simulated anhydrous ammonia application in two of the three soils, and production continued steadily for 26 days. Maximum emission of 1.5% of the fertilizer as N_2O from Crowley soil was less than the amounts reported in some trials (Bremner et al., 1981) but agreed more closely with results of others (Magalhaes et al., 1984). The data indicate that N_2O emissions during nitrification may be more dependent on the soil to which the fertilizer is applied than on the fertilizer type.

REFERENCES

Blackmer, A.M. and J.M. Bremner (1977) Gas chromatographic analysis of soil atmospheres. Soil Science Society of America Journal 41:908- 912.

Bolle, H.J., W. Seiler and B. Bolin (1986) Other greenhouse gases and aerosals, assessing their role for atmospheric radiative transfer. In: B. Bolin, B.R. Döös, J.Jäger and R.A. Warrick (Eds.), The Greenhouse Effect, Climatic Change and Ecosystems, p 157-197. Wiley and Sons, New York.

Bowden, W.B. and F.H. Bormann (1986) Transport and loss of nitrous oxide in soil water after forest clear-cutting. Science 233:867-869.

Breitenbeck, G.A., A.M. Blackmer and J.M. Bremner (1980) Effects of different nitrogen fertilizers on emission of nitrous oxide from soil. Geophysical Research Letters 7:85-88.

Bremner, J.M. and A.M. Blackmer (1981) Terrestrial nitrification as a source of atmospheric nitrous oxide. In: C.C. Delwiche (Ed.), Denitrification, Nitrification, and Atmospheric Nitrous Oxide, p 151-170. Wiley and Sons, New York.

Bremner, J.M., G.M. Breitenbeck and A.M. Blackmer (1981) Effect of anhydrous ammonia fertilization on emission of nitrous oxide from soils. Journal of Environmental Quality 10:77-80.

Crutzen, P.J. (1981) Atmospheric chemical processes of the oxides of nitrogen, including nitrous oxide. In: C.C. Delwiche (Ed.), Denitrification, Nitrification, and Atmospheric Nitrous Oxide, p 17-44. Wiley and Sons, New York.

Delwiche, C.C. and D.E. Rolston (1976) Measurement of small nitrous oxide concentrations by gas chromatography. Soil Science Society of America Journal 40:324-327.

Lastigzon, J. (1981) World fertilizer progress into the 1980's. Tech. Bull. T-22, Int. Fertilizer Development Center, Muscle Shoals, Alabama, U.S.A.

Magalhaes, A.M.T., P.M. Chalk and W.M. Strong (1984) Effect of nitrapyrin on nitrous oxide emission from fallow soils fertilized with anhydrous ammonia. Fertilizer Research 5:411-421.

Smith, C.J. and P.M. Chalk (1980) Gaseous nitrogen evolution during nitrification of ammonia fertilizer and nitrite transformations in soils. Soil Science Society of America Journal 44:277-282.

U.S. EPA (Environmental Protection Agency) (1988) Stabilization report, workshop on agriculture and climate change, February 29- March 1, 1988. Washington, D.C., U.S.A.

Nitrous Oxide and Methane Emission from Gulf Coast Wetlands

R.D. DELAUNE, W.H. PATRICK, C.W. LINDAU and C.J. SMITH

Laboratory for Wetland Soils and Sediments, Center for Wetland Resources
Louisiana State University, Baton Rouge, Louisiana 70803, U.S.A.

ABSTRACT

Recent studies have provided estimates of methane and nitrous oxide emission from U.S. Gulf Coast Mississippi River deltaic plain marshes and rice soils. Seasonal emissions were measured using chambers placed over the soil surface. Interstitial waters of fresh and brackish surface peats or marsh soils contained considerable concentrations of methane. Methane was evolved at the annual rates of 4, 73 and 100 g C m^{-2} y^{-1} from salt, brackish and fresh water marshes, respectively. Emission was inversely related to salinity and sulfate concentrations. Annual methane emission from the northern Gulf of Mexico coastal wetlands is estimated to be 1.5×10^{12} g CH_4-C. Nitrous oxide was evolved at the rates of 31, 48 and 55 mg N m^{-2} y^{-1}, respectively, from salt, brackish, and fresh water marshes. Nitrous oxide emission increased significantly as the result of partial drainage or the addition of inorganic nitrogen. It is estimated that Gulf Coast wetlands contribute 3.3×10^9 g N_2O annually to the atmosphere.

Rice culture along the Louisiana Gulf Coast can also contribute both nitrous oxide and methane to the atmosphere. Root rhizosphere and oxidized surface of flooded rice soils are zones of active nitrification-denitrification and subsequent N_2O production. Fertilization with urea-N led to emission of 74 to 171 g N ha^{-1} of N_2O over a period of 105 days following nitrogen addition. The above estimates are probably minimum values since recent research shows that appreciable amounts of biologically produced gases remain entrapped in small gas pockets.

BACKGROUND

Natural wetlands bordering the Gulf of Mexico contains 14200 km^2 of marsh (Alexander et al., 1988) representing 58.1 percent of total coastal marsh area in the continental United States. The majority (9800 km^2) of the Gulf Coast marsh are found in Louisiana, largely in the Mississippi River deltaic plain. Louisiana coastal marshes extend inland from the Gulf of Mexico for distances ranging from 24 to 80 km and reach their greatest width in southeastern Louisiana (Chabreck, 1972). Due to the tremendous width of these marshes, distinct vegetation or marsh types can be identified along a salinity gradient extending inland from the coast.

In addition to natural wetlands, rice paddy soils are found along adjacent coastal plain soil bordering the natural wetlands in Louisiana

Soils and the Greenhouse Effect. Edited by A.F. Bouwman

and Texas. Over 300,000 ha of rice is grown in areas bordering the Gulf of Mexico (Rice Journal, 1988).

In this paper, we summarize research findings on methane and nitrous oxide production and emission from Louisiana Gulf Coast marsh and paddy soils.

GASEOUS FLUX DATA

Methane flux (DeLaune et al., 1983) and nitrous oxide flux (Smith et al., 1983) from marsh soil to atmosphere have been measured in fresh, brackish and salt marsh using chambers placed over the marsh surface. In these studies vegetation was clipped prior to determining gaseous fluxes. Interstitial water of fresh and brackish peat contains considerable methane. Methane concentrations of 8.8 mg CH_4 per liter in interstitial water of peats from brackish marshes have been measured (DeLaune et al., 1986). Seasonal emissions of methane from fresh, brackish and salt marshes were 4.3, 73, and 160 g C m^{-2} y^{-1} respectively (Table 19.18.1). Methane emissions from the marsh surface to the atmosphere was greater during the summer months as compared to the winter months. As shown, methane emission was inversely related to salinity and sulfate concentrations. Extrapolation of methane flux data from these sites in Louisiana to the entire area of coastal wetlands of the Gulf Coast indicates that the equivalent of 1.5×10^{12} g CH_4- y^{-1} is being evolved to the atmosphere.

Table 19.18.1 Methane and nitrous oxide fluxes from Louisiana Gulf Coast marshes (modified from DeLaune et al., 1983 and Smith et al., 1983)

Site	Measurement Period (days) *CH_4	**N_2O	Salinity PPT	CH_4 emission g C m^{-2} y^{-1}	N_2O emmission mg N m^{-2} y-1
Salt	507	750	10-15	4.3	31
Brackish	516	750	4-5	73	48
Fresh	493	750	<0.5	160	55

*   Methane flux determined 13 times during the sampling period
**   Nitrous oxide flux determined 17 times during the sampling period

Reported nitrous oxide flux measurements (Table 19.18.1), also using chambers placed over the marsh surface, have shown that the N_2O emission from the Gulf Coast fresh, brackish, and salt marsh to be equivalent to an annual loss of 31, 48 and 55 mg N_2O-N m^{-2} (Smith et al., 1983). Measured N_2O emission increased slightly with distance from the coast. Louisiana Gulf Coast marsh contain little nitrate which could account for the relatively low N_2O emissions reported. However, a large potential for denitrification or N_2O emission exists if extraneous sources (e.g., run-off from agricultural areas, secondary sewage treatment) of nitrogen are introduced into these marshes. Up to 12 fold increase in N_2O emission was reported following the addition of the equivalent of 0.6 g NO_3^--N m^{-2} to cores taken from these marshes (Smith et al., 1983). Flooding and draining cycles increase N_2O evolution in these marsh soils. Extrapolation of these reported N_2O flux data as representative of the marsh areas of the Gulf Coast region indicates that the equivalent of 3.3×10^9 g N y^{-1} would be evolved to the atmosphere.

The presence of marsh vegetation has been shown to enhance rates of denitrification in Louisiana Gulf Coast marshes (Smith and DeLaune, 1984). Such increases in denitrification would also increase nitrous oxide production. The increased denitrification was attributed to the oxidized rhizosphere. *Spartina alterniflora* and other wetland plants found along the Gulf Coast possess the ability to transport oxygen to the roots. Such translocation of oxygen to the root zone in flooded soil would tend to increase the area of zones suitable for nitrate formation. Once the nitrate has formed it can then move into reduced zone where it can be denitrified.

Denitrification and resultant N_2O and N_2 evolution to the atmosphere was observed to be a major removal mechanism of nitrate and ammonium nitrogen entering swamp forest bordering fresh water marshes along the Louisiana Gulf Coast (Lindau et al., 1988). Nitrous oxide evolution as a result of added nitrogen was dependent on the source of nitrogen (Table 19.18.2). Nitrous oxide evolution to the atmosphere was observed the second day following application of NO_3^--N at the rate of 10 g N m^{-2}. Emission lasted for approximately 2 weeks, apparently the period of time needed for all the NO_3^--N to be denitrified. By contrast it took 15 days before N_2O was evolved from adjacent plots which received 10 g N m^{-2} of NH_4^+-N. This lag occurred because of the time required for nitrification of NH_4^+-N.

Table 19.18.2 Nitrous oxide emission from flooded swamp forest receiving the addition 10g N m^{-2} in two forms (($NH_4)_2SO_4$ or KNO_3)

Sampling time (days)	N_2O Evolved mg m^{-2} min^{-1}	
	NO_3^--N Source	NH_4^+-N Source
0	ND	ND
2	49	ND
5	61	ND
8	14	ND
15	ND	12
21	ND	26
27	ND	26

*ND: Non detectable

Waterlogged soils on which lowland rice is grown in the Louisiana and Texas Gulf Coast also create conditions where nitrification-denitrification can occur. Nitrous oxide emission using chambers placed over flooded rice soil fertilized with urea-N have determined N_2O fluxes as affected by rate and methods of application (Table 19.18.3). The application of urea-N led to an increase in nitrous oxide evolution; however, the emissions over a 105 day sampling period were low ranging from 74 to 171 g N ha^{-1}. The nitrous oxide emissions were found to be correlated with the exchangeable NH_4^+-N content of the soil and NO_3 content of the flood water (Smith et al., 1982).

Table 19.18.3 Nitrous oxide emission from Louisiana Gulf Coast rice soils as influenced by rate and method of application of urea-N (modified from Smith et al., 1982)

Method of application	Urea Application Rate kg N-ha^{-1}	Evolved Nitrous Oxide g N ha^{-1}
Drilled	0	74
	90	108
	180	171
Top Dressed	90	109
	180	90

These reported nitrous oxide emissions from rice soils are likely underestimated. Recent studies (Lindau et al., 1988) have shown that under laboratory conditions up to 20 percent of applied urea-N and 40% of KNO_3 was trapped as N_2 in rice soil 1 month after application of ^{15}N labelled nitrogen source. Since significant quantities of applied $^{15}N_2$ can be denitrified and trapped with the soil layer are N_2, it is most likely that appreciable N_2O oxide is also trapped (Reddy et al., 1989). Detailed field and laboratory experiments are needed to quantify the significance of N_2O entrapment and relationship to atmospheric emission from rice paddies. Preliminary field results showed 25, 28, 37, and 25 μg N_2O m^{-2} were entrapped in the saturated soil 5, 10, 15 and 20 days, respectively after KNO_3 addition (12 g N m^{-2}) to the floodwater of a rice field. This compares to 0, 5, 36 and 0 μg N_2O m^{-2} entrapped 5, 10 15 and 20 days, respectively, after urea addition (Lindau and DeLaune, unpublished data). Such entrapment would be of greater significance in fertilized rice cultures rather than in coastal marshes. Appreciable N_2O is likely lost upon drainage of rice soil which would release any entrapped N_2O. Coastal marshes are not subject to fertilization and entrapment of gaseous nitrogen may not be as important.

Nitrous oxide emission can be enhanced from rice soils by subjecting these soils to alternate anaerobic and aerobic conditions (Smith and Patrick, 1983). Continuous aerobic soil amended with $(NH_4)_2SO_4$ produced 0.8 mg N_2O-N g soil. Alternate anaerobic-aerobic cycles increased N_2O emission to 7.2 mg N_2O-N. Evolution increased further when duration of anaerobic-aerobic cycle was increased from 7-7 day to 14-14 days. No N_2O was evolved under continuous anaerobic conditions.

CONCLUSIONS

U.S. Gulf Coast marsh and rice soils contribute both methane and nitrous oxide to the atmosphere. Considerable methane is evolved from fresh and brackish marshes. Measured nitrous oxide emissions are low. Such emissions may be underestimated in fertilized rice soils due to entrapment in gas pockets. Additional studies are needed for quantifying methane fluxes from flooded rice soils along the Louisiana Gulf Coast. These sites potentially could evolve considerable methane to the atmosphere.

ACKNOWLEDGEMENT

Supported in part by NSF Grant (BSR-8414000).

REFERENCES

Alexander, C.E., M.A. Boutman and D.W. Field (1986) An inventory of coastal wetlands of the USA. U.S. Dept. of Commerce - Washington, D.C. 25 pp.

Chabreck, R.H. (1972) Vegetation, water and soil characteristics of the Louisiana coastal region. Louisiana Agriculture Experiment Station Bull. 664, Baton Rouge, LA USA.

DeLaune, R.D., C.S. Smith and W.H. Patrick, Jr. (1983) Methane release from Gulf Coast Wetlands. Tellus 35B:8-15.

DeLaune, R.D., C.J. Smith and W.H. Patrick, Jr. (1986) Methane production in Mississippi River deltaic plain peat. Organic Geochemistry 5:193-197.

Lindau, C.W., W.H. Patrick, Jr., R.D. DeLaune, K.R. Reddy and P.K. Bollich (1988) Entrapment of nitrogen-15 dinitrogen during soil denitrification. Soil Science Society of America Journal 52:538-540.

Lindau, C.W., R.D. DeLaune and G.L. Jones (1988) Fate of added nitrate and ammonium nitrogen entering a Louisiana Gulf Coast swamp forest. Journal Water Pollution Control Federation 60:386-390.

Reddy, K.R., W.H. Patrick, Jr. and C.W. Lindau (1989) Nitrification-denitrification at the plant root-sediment interface in wetlands. Liminology Oceanography. (In press).

Rice Journal (1988) 1988 Southern United States Rice Acreage Survey. Rice Journal 91:17-23.

Smith, C J., R.D. DeLaune, and W.H. Patrick, Jr. (1983) Nitrous oxide emission from Gulf Coast wetlands. Geochimica Et Cosmochimica Acta 47:1805-1814.

Smith, C.J. and R.D. DeLaune (1984) Influences of the rhizosphere of Spartina alterniflora on nitrogen loss from a Louisiana Gulf Coast salt marsh. Environmental and Experimental Botany 26:91-93.

Smith, C.J. and W.H. Patrick, Jr. (1983) Nitrous oxide emissions as affected by alternate anaerobic and aerobic conditions from soil suspensions enriched with ammonium sulfate. Soil Biology and Biochemistry 15:693-697.

Smith, C.S., M. Brandon and W.H. Patrick, Jr. (1982) Nitrous oxide emission following urea-N fertilizer of wetland rice. Soil Science Plant Nutrition 28:161-171.

Emission of Nitrous Oxide Dissolved in Drainage Water from Agricultural Land

K. MINAMI[1] and A. OHSAWA[2]

[1]*National Institute of Agro-Environmental Sciences*
Tsukuba 305, Japan
[2]*The University of Tokyo, Tokyo 113, Japan*

ABSTRACT

There is wide-spread concern that increased crop production to meet the food requirements of an expanding world population promotes emissions of nitrous oxide from nitrogen-fertilized soils to the atmosphere and thereby contributes to the greenhouse effect and the destruction of ozone layer. This concern has stimulated research to assess nitrous oxide emission from drainage water from agricultural land and to identify mechanisms that influence the emission of nitrous oxide form drainage water.

This paper will be organized as follows: first the field site studies and the simple methods for estimating nitrous oxide emission from drainage water will be described; then the results of the field measurements during one year will be presented. Finally, we discuss three mechanisms involved in the nitrous oxide emission process from drainage water; 1) emission of nitrous oxide from dissolved nitrous oxide in drainage water, 2) emission of nitrous oxide through nitrification in drainage water and 3) emission of nitrous oxide through denitrification in sediments.

There is great concern that increased crop production to meet the food requirements of an expanding world population promotes emissions of nitrous oxide from nitrogen-fertilized soils to the atmosphere and thereby contributes to global warming and the destruction of the ozone layer (Ramanathan et al., 1985; Cicerone, 1987; and Dickinson and Cicerone, 1986).

Although many studies have been conducted at several locations in the world to assess the contribution of agriculture to global nitrous oxide emission from soils, very little attention has been paid to the N_2O emission from drainage water from agricultural land and N_2O dissolved in ground water due to the application of fertilizer and sewage (Dowdell et al., 1979: Minami and Fukushi, 1984; and Ronen et al., 1988). As N_2O is fairly soluble in water (1.0 ml gas per ml water at 5°C), we examined the possibility that significant quantities are really leached from the soil towards drainage- and groundwater.

The objectives of the work reported here are to gather reliable information regarding the concentration of N_2O in drainage water and the effect of nitrogen fertilization on N_2O emission from drainage water, and to clarify the mechanism of N_2O production.

Soils and the Greenhouse Effect. Edited by A.F. Bouwman

MATERIAL AND METHODS

Field studies

Field study sites were located at Yachio-cho (36°09'N, 139°54'E), near Tsukuba, Japan, where vegetable crops were grown on a soil formed in volcanic ash (Andosol) to which a large amount of chemical fertilizer (500 Kg N ha^{-1} y^{-1}) had been applied. Figure 19.19.1 shows the location of the sites for measuring N_2O flux and sampling water or sediment in the drainage ditches. Paddy fields are located in the southern part along the drainage ditch. The samples were obtained from the points numbered 1 through 6. The drainage ditch at point number 1 through 4, and 5 through 6 were approximately 3m wide and about 3m deep, and 1m wide by about 0.5m deep, respectively. There is an impervious layer at 2m depth from the soil surface, so that all the percolating water in this area flows from the vegetable fields to the drainage ditch. The average annual precipitation is about 1400mm.

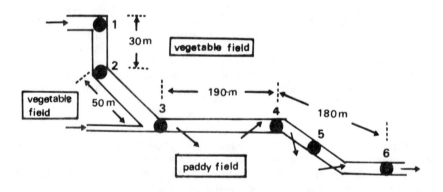

Figure 19.19.1 Location of measuring points

Analytical Methods

The atmosphere, water and sediments in the study sites were sampled at one month intervals for determination of N_2O and the activity of denitrifying bacteria. A chamber system was used for the measurement of N_2O flux from the water surface. The system is similar to that previously used for the measurements of N_2O emission in paddy fields from water surfaces (Minami and Fukushi, 1984).

Nitrous oxide level in the air samples and dissolved N_2O in drainage water were determined by a gas-chromatographic technique using xenon in the air as an internal standard (Minami and Fukushi, 1984). Ammonium and nitrate were determined by the steam-distillation method and nitrite nitrogen by the modified Griess-Llosvay method described in Bremner (1965). Denitrification potential was determined by the method of Fukushi and Minami (1984).

RESULTS AND DISCUSSIONS

Figure 19.19.2 shows the seasonal variation of N_2O dissolved in drainage water in measuring points 1 through 6. The distance to point 1 increases to 450m at point 6 (see Figure 19.19.1). The amounts of N_2O dissolved in drainage water decreased with increasing distance to point N^o1. The amount of N_2O dissolved in the water samples at point N^o1 ranged from 40 to 120 ng/l and at point N^o6 from 0.2 to 8 ng/l during the season. This indicates that dissolved N_2O in drainage water in the stream evolved from the water surface to the atmosphere. Part of the N_2O in the water infiltrated into the paddy soil where it is reduced to N_2 and thus the N_2O concentration is decreased (Minami and Fukushi, 1984). There were no significant differences in the content of dissolved N_2O through the season, but the amount of N_2O dissolved was 10 to 500 times higher in the drainage water than in fresh water in equilibrium with the atmosphere.

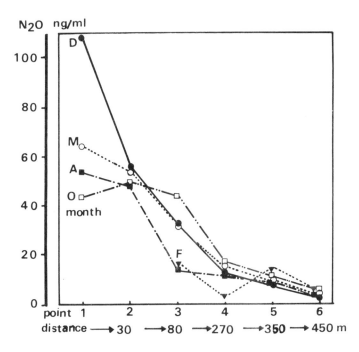

Figure 19.19.2 Seasonal variation of dissolved N_2O in drainage water at measuring points 1 - 6.

Figure 19.19.3 shows the seasonal variation of N_2O flux at points 2 and 6, and the vegetable field. The observed N_2O flux from the drainage water was comparable to or even higher than from the vegetable field, showing that N_2O evolved from drainage ditches should be recognized to be one of the major sources of atmospheric N_2O .

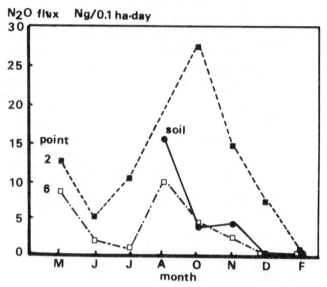

Figure 19.19.3 Seasonal variation of N_2O flux from drainage water and vegetable soil

It is generally recognized that the solubility of N_2O is higher in an acidic solution than in alkaline solutions. The content of N_2O dissolved in the drainage water showed the same tendency. It was higher in the more acidic solution, as shown in Figure 19.19.4.

It is generally believed that much of the N_2O produced is the result of denitrification of NO_3^- under anaerobic conditions in soils or in aerobic soils having anaerobic microsites. However, recent studies have shown that N_2O is produced in soils even under aerobic conditions in the presence of nitrifiable forms of N (Bremner and Blackmer, 1978). To investigate the occurrence of biological nitrification in drainage water, 20 ml of water sampled at point 2 was incubated under aerobic conditions at 25°C for 30 days after treatment with ammonium N. The data presented in Figure 19.19.5 show that N_2O was produced during oxidation of ammonium to nitrite or nitrate in drainage water, indicating that N_2O evolved from drainage water is derived partly from nitrification.

We also studied the relationship between nitrate and dissolved N_2O concentration in drainage water. Nitrate and N_2O flowing out with the drainage water to the drainage ditch are produced during nitrification in soils under aerobic conditions following ammonium N fertilization.

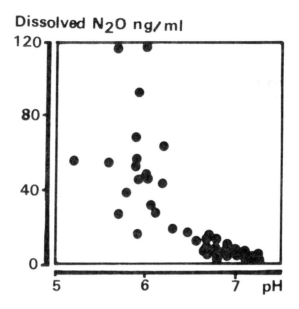

Figure 19.19.4 Relationship between pH and dissolved N_2O in drainage water.

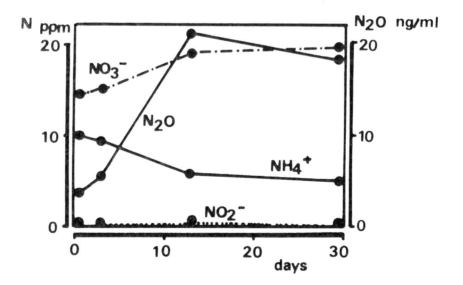

Figure 19.19.5 Nitrification and produced N_2O during incubation in drainage water at 25°C (NH_4-N was added in 20 ml sample).

The amount of dissolved N_2O was well correlated to nitrate concentration in drainage water as shown in Figure 19.19.6. From the above results, it seems likely that dissolved N_2O in drainage water stems from N_2O in the effluent water from vegetable fields.

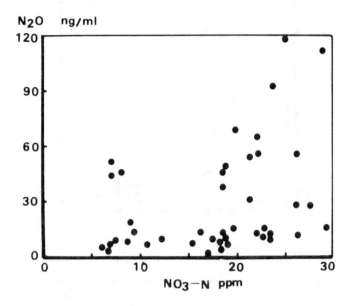

Figure 19.19.6 Relationship between NO_3-N concentration and dissolved N_2O in drainage water.

It is possible that nitrate is denitrified by denitrifying bacteria in drainage ditch sediments. To investigate whether biological denitrification occurs in the sediments, we studied the number of denitrifying bacteria and denitrification potential in sediment material, as shown in Table 19.19.1. The number of denitrifying bacteria and denitrification potential in the sediments were higher than in the soil of the vegetable field, and these values were approximately the same as those of reduced paddy soils. The results suggest that N_2O may be produced during denitrification in the sediment.

We conclude from the results of these investigations that N_2O evolved from drainage water is derived from the dissolved N_2O in the drainage water from agricultural land, the N_2O produced during nitrification in drainage water and the N_2O produced during denitrification in the sediments.

Table 19.19.1 The most-probable-number of denitrifying bacteria and denitrification potential for the sediments and vegetable soil. MPN = most probable number

Point	Month	MPN	Denitrification potential (mgN/100g day)	C(%)	N(%)	C/N
2	Sept.	7.6×10^6	17.03	6.7	0.51	13.0
	Nov.	6.2×10^5	19.08	6.0	0.48	12.6
	Dec.	4.9×10^5	9.71	6.1	0.47	13.1
4	Nov.	2.1×10^6	11.84	6.0	0.55	12.0
6	Nov.	6.5×10^5	9.86	6.2	0.50	12.3
Soil	Dec.	2.8×10^5	1.07	6.8	0.41	16.6

REFERENCES

Bremner, J.M. (1965) Inorganic forms of nitrogen. In: C.A. Black et al. (Eds.), Methods of Soil Analysis, Part 2, p. 1191-1206. American Society of Agronomy, Inc. Publisher, Madison.

Bremner, J.M. and A.M. Blackmer (1978) Nitrous oxide: emission from soils during nitrification of fertilizer nitrogen. Science 199:295-296.

Cicerone, R.J. (1987) Changes in stratospheric ozone. Science 237:35-42.

Dickinson, R.E. and R.J. Cicerone (1986) Future global warming from atmospheric trace gases. Nature 319:109-115.

Dowdell, R.J., J.R. Burford and R. Crees (1979) Losses of nitrous oxide dissolved in drainage water from agricultural land. Nature 278:342-343.

Fukushi, S. and K. Minami (1984) Methods for measuring denitrification potential of soils and its application to problems in the field. Bulletin of the National Institute of Agricultural Sciences B36:1-18.

Minami, K. and S. Fukushi (1984) Methods for measuring N_2O flux from water surface and N_2O dissolved in water from agricultural land. Soil Science and Plant Nutrition 30:495-502.

Ramanathan, V., R.J. Cicerone, H.B. Singh and J.K. Kiehl (1985) Trace gas trends and their potential role in climatic change. Journal of Geophysical Research 90:5547-5566.

Ronen, D., M. Magaritz and E. Almon (1988) Contaminated aquifers are a forgotten component of the global N_2O budget. Nature 335:57-59.

EXTENDED ABSTRACT 19.20

Climate Change and Patterns of Denitrification in the Willamette Basin of Western Oregon, USA

D.P. TURNER[1], D.D. MYROLD[2] and J.D. BAILEY[1]

[1] NSI Technology Services, Inc.
US EPA Environmental Research Laboratory
200 SW 35th, Corvallis OR 97333, USA
[2] Department of Soil Science, Oregon State University
Corvallis OR 97331, USA

ABSTRACT

The strong seasonality of precipitation in the Pacific Northwest Region of the United States has a large influence on temporal patterns of microbial activity. In this study we investigated the possible influence of a doubled CO_2 climate on rates of denitrification and associated N_2O production in soils of the Willamette Basin (2.9×10^6 ha) in western Oregon. N_2O is an efficient greenhouse gas and increased biogenic production may represent a positive feedback to global warming. A multiple regression equation based on in situ incubations of Willamette Valley soils was used in combination with historical and predicted precipitation and temperature data to estimate monthly denitrification for a range of soils. The main influence of the climate change was on potential winter emissions which approximately doubled due to warmer soil temperatures. Earlier summer drying under the $2\times CO_2$ climate caused slightly decreased potential emissions in June and July. The predicted annual increases in N_2O flux were small relative to the pools of total N in these soils and thus could probably be sustained for decades to centuries even in the absence of increased N fixation.

INTRODUCTION

The general circulation models (GCM) currently used to predict effects of increasing concentrations of radiatively important trace gases (RITG) on climate do not generally incorporate potential feedbacks mediated by soil microorganisms. Faster rates of microbial respiration and denitrification associated with warmer temperatures have the potential to significantly amplify the climate forcing associated with anthropogenic emissions of the trace gases. This is particularly true in temperate and boreal latitudes, where the greatest increases in air temperature are predicted to occur (Manabe and Wetherald, 1980) and where large reservoirs of soil carbon and nitrogen are potentially available as substrates.

Soils and the Greenhouse Effect. Edited by A.F. Bouwman
© 1990 John Wiley & Sons Ltd.

Nitrous oxide (N_2O) is a relatively efficient RITG (Ramanthan et al., 1985) produced by microbially mediated denitrification and nitrification as well as by burning of biomass and fossil fuel. The increasing flux from industrial combustion along with land use changes and increased biomass burning probably accounts for most of the observed increase in the atmospheric concentration. As global warming begins to be manifest, however, there is a potential for additional increases from biogenic sources due to effects of changes in soil temperature and moisture status on microbial metabolism.

Climate may be a particularly important variable relative to N_2O emissions in regions such as the Pacific Northwest of the United States because of the strong seasonality of precipitation. Under current climatic conditions, a period of drought reduces soil microbial activity in the summer whereas in the cool wet winters, activity is limited by low temperatures. In this study, we develop an approach to modelling potential rates of denitrification in soils of the Willamette Basin in western Oregon. The approach was used to compare denitrification under the historical climate and that under the climate predicted for a doubled CO_2 atmosphere. Given the uncertainty in predicted climate, the study is intended to indicate the sensitivity of the system to climate change rather than accurately predict emissions.

METHODS

Soils in the Willamette Basin (2.9×10^6 ha) were grouped into four aggregates based on soil characteristics and location (Tables 19.20.1 and 19.20.2). The percent carbon and water holding capacity for each grouping is based on the range and mean of values for the major soil series within each group, compiled from county-level USDA Soil Conservation Survey data and Huddleston (1982). The areal basis for each grouping was determined by planimetry from the Oregon General Soil Map (SCS, 1986).

Each soil group essentially corresponded to an altitudinal range and thus with a climatic regime. Historical climate (30 year means of observed monthly temperature and precipitation) was acquired from established weather stations located within the altitudinal ranges (Quinlan et al., 1987). Climate for a doubled CO_2 atmosphere was predicted based on output from the Geophysical Fluid Dynamics Laboratory (GFDL) model (Wetherald and Manabe, 1988) and using the technique described in Smith and Tirpak (1989). In this approach, the climate change factors determined by the GCM are applied to the historical climate data.

Rates of denitrification were estimated from a multiple regression equation (19.20.1) derived from a two year study of denitrification rates in Willamette Valley soils (Myrold, 1988). In that study, monthly in situ incubations (24 hrs) of 0-20 cm mineral soil (with acetylene) were monitored for N_2O and CO_2 production as well as water content,

Table 19.20.1 Soil groups within the Willamette Basin

Location (weather station)	Elevation	% of Basin	Vegetation
Valley Floor (Corvallis)	< 150 m	26	grasses, shrubs, mostly cultivated
Foothills (Cottage Grove)	150-600 m	21	Douglas-fir, red alder white oak
Mid-elevations (McKenzie Bridge)	600-1200 m	27	Douglas-fir Western hemlock Noble fir
High elevations (Government Camp)	>1200 m	24	Noble fir Mountain hemlock

Table 19.20.2 Soil characteristics. WHC refers to soil column water holding capacity

Location	% Carbon (0-20cm)	WHC (mm)	Nitrate-N ($\mu g\ g^{-1}$)
Valley Floor	2.2	275	2.5
Foothills	3.5	150	7.0
Mid-elevations	4.5	200	0.05
High elevations	10.0	150	0:05

temperature and nitrate concentration. When these variables were included, the multiple regression accounted for 43% of the observed variation in denitrification rate.

$$\begin{aligned}\log(\text{denitrification}) =\ &6.98(\text{soil water content}) \\ &+ 0.776\ [\log\ (\text{respiration rate})] \\ &+ 0.324\ [\log\ (\text{nitrate concentration})] \\ &+ 0.0278\ (\text{soil temperature}) \\ &- 2.71\end{aligned} \tag{19.20.1}$$

Denitrification (g $N_2O\ ha^{-1}\ d^{-1}$) was calculated on a monthly basis for each soil type and climate scenario. In situ incubations with acetylene may overestimate N_2O production since acetylene blocks transformation of N_2O to N_2, but the method does indicate relative N_2O source strength. Incubations without acetylene were not carried out and the tremendous variability in the N_2O fluxes, measured with and without acetylene (e.g. Robertson and Tiedje, 1984), precludes any generalizations that might cover these soils.

Water content (WC) was calculated from field observations of maximum water content (MWC) for each soil grouping (D. Myrold, personal communication) and the outputs from a water budget model (Mather, 1985). The water budget model takes inputs of soil column water holding capacity (WHC) along with monthly precipitation and temperature and gives estimates of water storage at monthly intervals based on potential and actual evapotranspiration. WHC of the soil profiles was taken from Soil Survey data.

Respiration was calculated by determining a base respiration rate and modifying it with a temperature factor and a soil moisture factor. The base rate was calculated using a linear regression (equation 19.20.2) of %C against mean annual respiration (R) over a range of Oregon soils with a similar mean annual temperature (11°C) (D. Myrold, personal communication).

$$R \ (kg \ C \ ha^{-1} \ d^{-1}) = 2.37 \times \%C + 1.11; \ (r^2 = .995) \quad (19.20.2)$$

For incorporating temperature effects, soil temperature was assumed equal to monthly air temperature from the GCM outputs and a Q_{10} (11-21°C) of 2.0 (Flanagan and Veum, 1974) was used to modify the base respiration rate. The range of the temperature modified rates was close to the observed range of respiration rates for these soils (Myrold, 1988). The temperature modified respiration rate was then multiplied by a soil moisture factor between 0 and 1.0. That factor was based on a calculated percent water filled pores (%WFP) and an empirical relationship between %WFP and microbial respiration from the literature (Linn and Doran, 1984).

Monthly soil temperature was also assumed to be equal to monthly air temperature for the multiple regression equation. This assumption does not reflect lags occurring between the two over short time frames but it provided a first approximation of potential soil temperatures

Estimated nitrate values were based on observations in Willamette Basin soils (D. Myrold, personal communication). Concentrations were held constant within soil groupings because the seasonality of N mineralization, nitrification and N uptake are not well quantified for these soils.

RESULTS

Denitrification rates under both climatic regimes were highest in the foothills (Table 19.20.3), mainly because the N input from symbiotic nitrogen fixation associated with red alder promotes high nitrate levels. The predicted $2 \times CO_2$ climate was warmer (4.9 °C) and wetter (+3.5%) than the historical climate. Over the entire Basin, denitrification was 1.7 times more active under the $2 \times CO_2$ climate (Table 19.20.3), with consistently higher rates during the winter (Figure 19.20.1).

Table 19.20.3. Estimated annual denitrification rates in Willamette Basin soils (0-20 cm)

Location	Areal Basis 10^3 ha	Annual Sum kg N ha^{-1} y^{-1}		Basin Sum $\times 10^3$ kg N y^{-1}	
		Historical	2×CO$_2$	Historical	2×CO$_2$
Valley	775.2	1.04	1.78	806	1379
Foothills	635.6	1.82	3.54	1156	2248
Mid-elevations	795.4	0.28	0.44	223	350
High elevations	724.0	0.78	1.07	565	774
			Total:	2750	4751

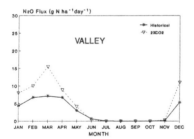
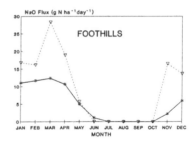
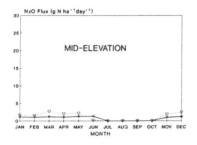
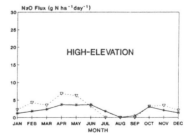

Figure 19.20.1 Estimated daily denitrification rates by soil group.

The differences in denitrification rate between the climatic regimes were a function of both higher temperatures and altered soil moisture status. The effect of a doubled CO_2 climate on soil moisture conditions was to keep the WC consistently lower during the summer, to bring the water storage down to 0 approximately one month earlier, and to maintain a moisture deficit in the soil slightly later in the year (Figure 19.20.2). These differences meant an earlier cessation of denitrification in the summer but a stronger increase in emissions in the fall.

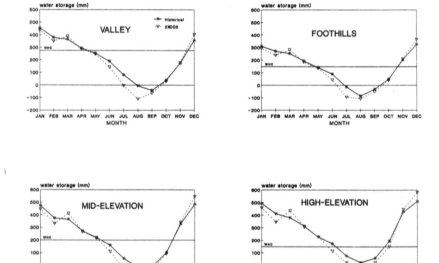

Figure 19.20.2 Annual water budgets by soil group. WHC refers to water holding capacity. Amounts above WHC (runoff) and below 0 (deficits) are not carried over from month to month.

DISCUSSION

For denitrification in soils of the Pacific Northwest, it is apparent that the stimulation of microbial metabolism caused by the warmer winter temperatures is likely to be the most significant effect of the doubled CO_2 climate. Earlier summer drying will reduce potentiel N_2O emissions (denitrification activity) to a small extent, but the period during which this is a significant effect is short relative to the periods during which denitrification is enhanced by temperature. The small differences in annual precipitation suggest that soils would not be significantly more anaerobic.

One critical question is whether higher rates of N_2O emission could be sustained over decadal and longer time frames. Agricultural and especially forest soils in the Pacific Northwest typically have large pools of carbon (e.g. 90-130 Mg ha^{-1}) and nitrogen (2-20 × 10^3 kg ha^{-1}) and it appears likely that increased fluxes on the order of a few kg N and C per ha per year could be readily sustained. Thus, a warmer climate and associated increases in N_2O emissions from these soils would produce a positive feedback to global warming.

This study suggests that the combination of Soil Survey data on soil characteristics, experimental data on responses of microbial metabolism to environmental variables, and climate data from historical records and GCMs can yield important insights into current and potential rates of microbially mediated trace gas emissions.

REFERENCES

Flanagan, P.W. and A.K. Veum (1974) Relationships between respiration, weight loss, temperature and moisture in organic residues on tundra. In: A.J. Holding, O.W. Heal, S.F. MacLean, Jr., and P.W. Flanagan (Eds.), Soil Organisms and Decomposition in Tundra, p 249-277. Stockholm: Tundra Biome Steering Committee.

Huddleston, J.H. (1982) Soils of Oregon: Summaries of Physical and Chemical Data. Oregon State University Extension Service. Special Report 662.

Linn, D.M and J.W. Doran (1984) Effect of water-filled pore space on carbon dioxide and nitrous oxide production in tilled and non-tilled soils. Soil Science Society American Journal 48:1267-1272.

Manabe, S. and R.T. Wetherald (1980) On the distribution of climate change resulting from an increase in CO_2-content of the atmosphere. Journal of Atmospheric Science 176:914-915.

Mather, J.R. (1985) The water budget and the distribution of climates, vegetation, and soils. Publications in Climatology 38(2). C.W. Thornthwaite Associates, Centerton, New Jersy and the University of Delaware Center for Climatic Research, Newark, Delaware. 36 pp.

Myrold, D.D. (1988) Denitrification in ryegrass and winter wheat cropping systems of Western Oregon. Soil Science Society American Journal 52:412-416.

Quinlan, F.T., T.R. Karl and C.N. Williams, Jr. (1987) United States historical climatology network (HCN) serial temperature and precipitation data NDP-019, Carbon Dioxide Information Analysis Center, Oak Ridge National Laboratory, Oak Ridge, Tennessee.

Ramanathan, V., R.J. Cicerone, H.B. Singh and J.T. Kiehl (1985) Trace gas trends and their potential role in climate change. Journal of Geophysical Research 90:5547-5566.

Robertson, G.P. and J.M. Tiedje (1984) Denitrification and nitrous oxide production in successional and old-growth Michigan forests. Soil Science Society of America Journal 48:383-389.

SCS Map. (1986) General Soil Map, State of Oregon (scale 1:500,000). 4-R-39694. USDA Soil Conservation Service.

Smith, J.B. and D.A. Tirpak (1989) Potential effects of global climate change on the U.S. (Draft June 1989). U.S. Environmental Protection Agency. Washington D.C.

Wetherald, R.T. and S. Manabe (1988) Cloud feedback processes in a general circulation model. Journal of the Atmospheric Sciences 45:1397-1415.

EXTENDED ABSTRACT 19.21

A Global Soils and Terrain Database: a Tool to Quantify Global Change

M.F. BAUMGARDNER

Agronomy Department, Purdue University
Agricultural Experiment Station, West Lafayette, IN 47907, USA

ABSTRACT

During the past three decades several areas of science and technology have evolved which now make it possible and feasible to design and implement global resource information management systems. One of these areas is our conceptualization and understanding of the Earth system. Another is the rapid development of Earth observation capabilities. A third area is the computer revolution which provides the capability to store, retrieve, analyze and manipulate masses of data about the Earth system. This paper addresses the problem of the design and implementation of a world soils and terrain digital database at a scale of 1:1 million. It is proposed that a database of natural resources (soils, geology, hydrology, climatology, vegetation, land use, other) containing map and attribute data at this scale can be an essential tool for understanding and quantifying processes of global change. Such a database can also for the first time provide decision-makers and policy-makers uniform, standardized, organized, easily accessible information for more rational management of Earth resources.

INTRODUCTION

In a very real sense most of us soil scientists are also information scientists. We are involved in generating and/or disseminating information. One of our responsibilities is to deliver useful, timely and accurate information to decision-makers and policy-makers.

Today we soil scientists are caught up, as are all other environmental scientists, in the challenge of defining and understanding global change. Our intimate knowledge about soils and soil processes tells us that soils are an important component of the Earth system, that soils are involved in many complex ways in the greenhouse effect. We have become proud inheritors of a discipline or division of natural science which began one hundred years ago. During the past century thousands of soil scientists around the world have contributed to our vast body of knowledge about the properties and processes of soils.

Most of our research has been conducted at the micro scale, and many of us have difficulty relating what we have done on small field plots, in the greenhouse or in the laboratory to the meso and macro scales. How do we relate this valuable accumulated body of knowledge

Soils and the Greenhouse Effect. Edited by A.F. Bouwman
© 1990 John Wiley & Sons Ltd.

to global change, to the greenhouse effect? How do we relate what we know about variability and measurements made within a 10 cm pixel to the variability within a 10 m pixel, a 100 meter pixel or a 1000 meter pixel? To quantify the relationships among the variations over a wide range of spatial scales is a very complex problem, but it is one which must be addressed seriously by participants in this Conference and our colleagues in research around the world.

There are many reasons for our inability to date to measure with any precision and predict with any degree of certainty the rates of global change and the consequences which can be expected from such change. Many global models have been constructed and many interesting scenarios related to global change have been proposed. Perhaps the most serious limitation in our ability to model global change is the dearth, or in many cases absence, of credible spatial and temporal data about Earth resources at the global scale.

This paper attempts to address the challenge of building a database at the global scale which can contribute to our quantizing the contributions of soils and terrain to the greenhouse effect and other processes of global change.

BACKGROUND

In 1985 a provisional working group was established by the International Society of Soil Science (ISSS) to consider the feasibility and desirability of developing a world soils and terrain digital database at a map scale of 1:1M. A background paper in support of this concept was written by Sombroek (1985) and distributed to more than 60 soils and terrain scientists around the world for their consideration and comments. This background paper served to focus the discussions of the 40 participants in an International Workshop on the Structure of a World Soil Resources Map Annex Digital Database (Baumgardner and Oldeman, 1986). As an outcome of this Workshop held in Wageningen, The Netherlands, in January 1986, a proposal to develop a World SOils and TERrain (SOTER) Digital Database at a scale of 1:1M was written (ISSS, 1986).

THE SOTER PROJECT

At the 13th International Soils Congress in Hamburg, West Germany in August 1986, the SOTER Proposal was endorsed and the provisional Working Group was given formal status and charged with implementing the SOTER Project. During the months which followed the Congress, contacts were made with many potential national and international funding agencies to solicit support for the Project.

Because of their strong support of activities in global databases for environmental sciences, officials of the United Nations Environment Programme (UNEP) expressed an interest in SOTER, especially if the

Project could make a significant contribution to the assessment of degradation of global soils and terrain resources.

Fifteen soil scientists representing the SOTER Working Group were invited by UNEP to an Expert Group Meeting on the Feasibility and Methodology of Global Soil Degradation Assessment. This meeting was held at UNEP Headquarters in Nairobi, Kenya, in May 1987. As a result of this meeting a UNEP Project Document entitled "Global Assessment of Soil Degradation" was prepared, and in September 1987 a contract was awarded by UNEP for Phase 1 of the SOTER Project (UNEP, 1987).

There are two primary tasks under the UNEP contract. The first is to produce a general soil degradation map of the world at a scale of 1:15M. The second is to develop a soils and terrain digital database at a scale of 1:1M for an area of approximately 250,000 km^2 which includes portions of Argentina, Brazil and Uruguay. The main thrust of this paper is to discuss the objectives and approach to the development of a global soils and terrain digital database at a scale of 1:1M.

Objectives

The long range objective of the SOTER Project is to produce a world soils and terrain digital database containing digitized map unit boundaries and their attribute (descriptive) database. The database has the following characteristics:

1. General average map scale, or accuracy, of 1:1M;
2. compatible with global databases of other environmental resources and features;
3. amenable to updating and purging of obsolete and/or irrelevant data;
4. accessible to a broad array of international, regional, and national decision-makers and policy-makers;
5. transferable to and useable by developing countries for national database development at larger scales (greater detail).

Specific short range objectives are required in the initial phases of the Project to provide a logical and orderly sequence of activities to produce an operational world soils and terrain digital database. Emphasis will be on research, development and testing of methodologies in the field and in the laboratory and demonstration of the uses of the database. Specific short term objectives are as follows:

1. Development of an implementation plan;
2. adoption of a universal legend for the SOTER Database;
3. development of guidelines for correlation of soils and terrain mapping units;
4. definition of soils and terrain parameters and specifications to be included in the Database;
5. development of a detailed set of specifications and logic which define the minimum set of capabilities/functions required for the Database;

6. selection of three specific areas of 250,000 sq km each in developing countries for initial database construction;
7. acquisition and correlation of all relevant maps and data about the selected areas essential for the Database;
8. input of data, including digitized maps, into the Database;
9. test and demonstration of the reliability, accuracy and utility of the Database;
10. conduct of an assessment of current geographic information systems and development of recommendations on the optimal system for the SOTER Project; and
11. documentation of results, conclusions and recommendations from the initial phase of the SOTER Project.

A Universal Legend for the SOTER Database

An international committee of soil scientists was appointed in January 1986 to develop a universal legend for soils and terrain data to be entered into the SOTER Database. A draft version was distributed in March 1988 and revised in January 1989. The manual entitled "SOTER Procedures Manual for Small Scale Map and Database Compilation" (Shields and Coote, 1989) describes procedures for compiling and coding the following kinds of data for entry into the SOTER Database (Figure 19.21.1):

Polygon file   (15 attributes)
Terrain component file   (31 attributes)
Soil layer file   (73 attributes)
Soil degradation file

The Manual also presents coding forms on which to enter all the attribute file data, which have been translated into the universal legend from whatever soil classification system used in the mapping of soils of a particular area or country.

Selection of Base Map for SOTER Database

In 1984 a joint Working Group of the International Geographical Union (IGU) and the International Cartographic Association (ICA) was established to explore the feasibility of developing a standard global data set. This project, entitled World Digital Database for Environmental Science (WDDES), concluded that the Operational Navigation Chart (ONC) series produced by the US Defense Mapping Agency provides consistent global map coverage of high cartographic quality at a scale of 1:1M. The IGU/ICA Working Group recommended the use of digitized ONCs as the best available 1:1M base map for input into world databases for registration and overlay of other natural resource data, including soils and terrain (Bickmore, 1987).

Representatives of the SOTER Project, having participated in many of the deliberations of the IGU/ICA Working Group, accepted the recommendations of that Working Group and have made the decision to use digitized ONCs as the base map for SOTER.

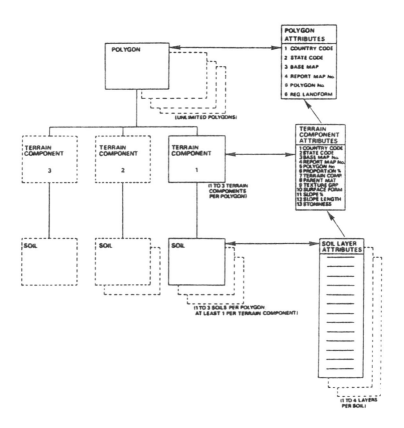

Figure 19.21.1. Schema of relationships among components of SOTER Database (Baumgardner and van de Weg, 1989).

PROGRESS REPORT ON THE SOTER PROJECT

LASOTER (Latin American SOTER) Pilot Area. Scheduled for completion in 1990, database development for the Latin American Pilot Area (Figure 19.21.2) is on schedule. In March 1988 a regional workshop was held in Montevideo to develop an implementation plan and to train soil scientists from Argentina, Brazil and Uruguay to use the universal SOTER legend and procedures manual for correlating soil maps of different classification systems to a uniform system of description and attribute entry into the database (Peters, 1988).

This workshop was followed by two separate correlation field trips by participants of the three countries and an external soil correlator into

Figure 19.21.2. Location of Latin American Pilot Area of SOTER Project (Baumgardner and van de Weg, 1989)

the pilot areas of each country. By mid–December 1988 acquisition of all map and attribute data from each of the three countries for the Pilot Area was complete. All attribute data have been coded and entered into the SOTER attribute files. Entry of polygon or map data into the SOTER Database will be done in the latter half of 1989 after selection of a geographic information system (GIS) for SOTER. Once the polygon and attribute files are in place, testing and demonstration of utility of the database will be conducted.

NASOTER (North American SOTER) Pilot Area. Work was begun on a US-Canadian Pilot Area following a workshop and implementation planning meeting in Ottawa in March 1989. This pilot area includes the state of Montana USA and the southern portion of the Canadian provinces of Alberta and Saskatchewan. A cooperative effort by the US

Soil Conservation Service and the Land Resource Research Centre of Agriculture Canada, the SOTER Database for this Pilot Area is scheduled for completion in 1990.

WASOTER (West African SOTER) Pilot Area. Negotiations are proceeding now to define an area which will involve six countries in West Africa in the development, testing and demonstration of the utility of a SOTER Database for that area.

Within the next two or three years it is anticipated that the SOTER Project may expand into other areas of the world. Particular interest has been expressed for cooperative implementation of the SOTER concept in an area of the Middle East, India, Southeast Asia, northern South America, Central Europe, and Western Europe.

EXPECTED RESULTS OF THE SOTER PROJECT

In general, the overriding objective of the SOTER Project is i) to improve the capability to deliver accurate, timely and useful information about soils and terrain resources to decision-makers and policy-makers, and ii) to overlay these data onto and combine with other resources data (topography, vegetation, geology, hydrology, land use, climate) in a global geographic information management system.

In the attempt at the global scale to quantify the role of soils, both as a source and sink, in fluxes of greenhouse gases and the resultant effects on global change, a World Soils and Terrain Digital Database will provide an important tool for accomplishing this task. Other expected results are as follows:

1. Orderly arrangement of global soils and terrain resource information;
2. Incorporation of high quality, standardized soils and terrain database into a global geographic information system;
3. Improvement in standardization and compatibility of reporting soils and terrain data/information;
4. Improvement in accessibility of soils and terrain and related resource information;
5. Dynamic resource information system with updating and purging capabilities;
6. Information service for national resource planning in developing countries; and
7. System model for technology transfer.

As the world is being caught up in the "information revolution," there is an increasing need to find innovative and more effective methods for using and transferring this technology, which can be applied to any spatial scale for decision-making in resource management problems (Table 19.21.1).

Table 19.21.1. Relationship among map scales, area of mapping units, and user requirements

Scale	Hectares in Mapping Unit of 1 cm^2	0,25 cm^2	Scale/Detail required by different users
1:5,000,000*	250,000	62,500	I,R ***
1:2,500,000	62,000	15,625	I,R,N
1:1,000,000**	10,000	2,500	I,R,N,P
1:500,000	2,500	625	R,N,P
1:250,000	625	156.25	N,P
1:100,000	100	25	N,P
1:50,000	25	6.25	N,P
1:25,000	6.25	1.56	P,L
1:10,000	1	0.25	L

*   Scale of FAO/Unesco Soil Map of the World
**   Average scale of SOTER Database
*** I = international; P = provincial or state; R = regional (multi-national); L = local; N = national.

The SOTER Project can provide an excellent vehicle for training a cadre of specialists, especially in developing countries, for using the database, providing new data and developing new uses of the database. The operational World Database can also serve as a model for the design and construction of in-country databases with sufficient detail and scale (accuracy) for local and provincial use.

ACKNOWLEDGEMENTS

Contribution from the Agronomy Department, Purdue University, Agricultural Experiment Station, West Lafayette, IN 47907. Journal article number 12294.

REFERENCES

Baumgardner, M.F. and L.R. Oldeman (eds.) (1986) Proceedings of an International Workshop on the Structure of a Digital International Soil Resources Map Annex Database. SOTER Report 2. International Soil Reference and Information Centre, Wageningen, The Netherlands.
Baumgardner, M.F. and R.F. van de Weg (1989) Space and time dimensions of a world soils and terrain digital database. In J. Bouma and A.K. Bregt (eds.). Land Qualities in Space and Time. Pudoc. Wageningen, The Netherlands.
Bickmore, D.P. (1987) Report on World Digital Database for Environmental Science--An IGU/ICA Project. International Geographical Union/International Cartographic Association Working Group. Oxford, UK.

ISSS (1986) Project Proposal. World Soils and Terrain Digital Database at a Scale of 1:1M. International Society of Soil Science, Wageningen, The Netherlands.

Peters, W.L. (ed.) (1988) Proceedings of the First Regional Workshop on a Global Soils and Terrain Digital Database and Global Assessment of Soil Degradation. SOTER Report 3. International Soil Reference and Information Centre. Wageningen, The Netherlands.

Shields, J.A. and D.R. Coote (1989) SOTER Procedures Manual for Small Scale Map and Database Compilation (Revised). International Soil Reference and Information Centre. Wageningen, The Netherlands.

Sombroek, W.G. (1985) Toward a Global Soil Resources Inventory at Scale 1:1M. Working Paper and Preprint Series 84/4. International Soil Reference and Information Centre. Wageningen, The Netherlands.

UNEP (1987) Global Assessment of Soil Degradation. United Nations Environment Programme Project Document. Nairobi, Kenya.

A Hierarchy of Soil Databases for Calibrating Models of Global Climate Change

N.B. BLISS

TGS Technology, Inc., EROS Data Center
Sioux Falls, South Dakota, U.S.A.

ABSTRACT

Digital soil maps at three scales are linked to a detailed database of soil attributes in the United States. A wide variety of interpretive soil maps can be produced from complex queries of multiple factors. Many of the soil properties are of interest in studies of global change, including parameters for calibrating general circulation models of the atmosphere. The techniques can be extended to global soil databases, can be combined with data on climate and vegetation to produce interpretations of albedo, soil carbon, and evapotranspiration, and can aid in evaluations of the impact of climate change.

A hierarchy of databases for digital soil maps is used to provide detailed attribute information aggregated to small scale generalized maps. This approach will lead to an improvement in the use of soils information in general circulation models of the atmosphere that are being used to study the potential effects of CO_2 increase on global warming. It will also be useful in evaluating potential impacts of climate change on human activity.

A hierarchy of soil databases is being developed in the United States by the Soil Conservation Service (Reybold and TeSelle, 1989). The U.S. Geological Survey has assisted in developing techniques to analyze and display the data in several nationwide databases in order to produce small-scale interpretive maps of detailed soil properties.

Detailed soil mapping for county-sized soil survey areas typically has been done at scales of about 1:20,000. Many of these maps were not compiled on an orthophoto base, and progress will be slow for converting these maps into digital form. The attributes for these detailed maps were entered into a nationwide database called the Soil Interpretations Record.

This Soil Interpretations Record Data Base contains information on about 25 primary soil properties for over 15,000 soil series used in the United States. The properties and derived interpretations are represented in over 150 data fields in the database for each soil phase. Examples of properties include slope, flooding, water table, depth to rock, depth to

Soils and the Greenhouse Effect. Edited by A.F. Bouwman
© 1990 John Wiley & Sons Ltd.

pan, and percent organic matter. For each soil layer, properties include texture, particle size distribution, liquid limit, bulk density, permeability, available water capacity, pH, and salinity.

Interpretations include USDA soil taxonomy class, USDA capability class, average crop yields, range site names, erosion factors, potential native plant communities, and suitability for construction and recreation.

A State Soil Geographic database (STATSGO) is being developed at a scale of 1:250,000 with nationwide coverage (except Alaska) and should be complete at the end of 1989. The maps are compiled by using transects across the more detailed soil surveys (where available) and then retaining the percentage composition of the phases of soil series that make up the general map unit (Bliss, 1987). Interpretive maps are based on an analysis of the detailed properties and interpretations (Bliss and Reybold, 1989). For example, one possible map-theme would be soils having a depth to bedrock of less than 50 cm. Each component (soil phase) is tested according to the criterion that defines the theme of the map. The percentage of the area that meets the criterion is then classified into a limited number of map categories to construct the map legend. For the thematic map representing depth to bedrock of less than 50 cm, the map classes may be 0 to 25 percent, 25 to 50 percent, 50 to 75 percent, and 75 to 100 percent. The map units that have a large proportion of the area covered with soil phases having shallow bedrocks will be depicted in the colors of the high percentage classes.

The National Soil Geographic database (NATSGO) was developed using a 1:7,500,000 scale map of Major Land Resource Areas (MLRA) to provide the spatial units for aggregating information from the National Resources Inventory (NRI). The 841,000 sample points for the 1982 NRI are used to characterize the polygons on the map according to land use, vegetation cover, soils, cropping history, and erosion. Each of the sample points is labelled with the phase of the soil series, and this is linked to the Soil Interpretations Record database described above. The approach allows the aggregation of field level observations at a national scale. Hundreds of soil phases are used as components of each MLRA polygon on the interpretive maps. In addition to producing interpretive maps of single factors, such as the available water capacity, it is possible to produce maps that make use of the co-occurrence of several attributes. For example, Figure 19.22.1 is a NATSGO map showing the yield of corn on highly erodible lands in comparison with the yield of corn on all lands (Bliss, 1989).

At a global scale, the Food and Agriculture Organization (FAO) Soil Map of the World is also available as a digital database. The original map was compiled at a scale of 1:5,000,000. The legend uses a classification of 106 primary taxonomic units, qualified with information on texture, slope, and phase that result in approximately 5,000 map units to cover the whole world (Food and Agriculture Organization (FAO)/United Nations Educational, Scientific and Cultural

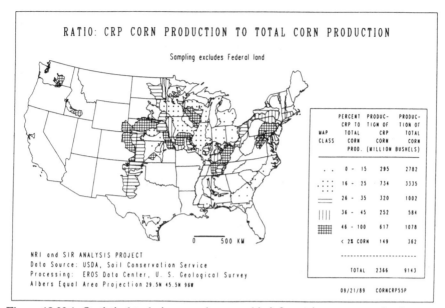

Figure 19.22.1 Statistical techniques and geographic information system technology are combined to produce interpretive maps based on the map of Major Land Resource Areas, the 1982 National Resources Inventory, and the Soil Interpretations Record database.

Organization (UNESCO), 1974-81). Unpublished and non-digital sources provide soil information for many parts of the world at a scale of 1:1,000,000. The Soil and Terrain (SOTER) database of the world is being developed under the auspices of the International Society of Soil Science to provide a revised legend and global coverage at a nominal scale of 1:1,000,000 (Shields and Coote, 1989).

Some of the general circulation models (GCM) of the atmosphere use soil information to model the movement of water between the soil and the atmosphere and to parameterize the albedo of the soil surface (Dickinson et al., 1986). Several of the databases that are currently used to characterize soils for the GCM's were developed by manual interpretation of the FAO Soil Map of the World (Zobler, 1989; Wilson and Henderson-Sellers, 1985). The aggregation techniques resulted in some loss of information in moving to the scale of the gridboxes used for modelling climate change. The techniques developed for handling attribute information in STATSGO and NATSGO are currently being adapted for use with a vector digitized version of the FAO Soil Map of the World (Environmental Systems Research Institute, 1984) to extract the attributes of interest to the climate modellers.

Models of global climate change require accurate descriptions of the properties of the land surface. Digital soil geographic databases contain information that can be used to help calibrate components of the models dealing with energy balance, water balance, and carbon balance at the land surface.

Energy balance calculations include terms for surface albedo, which is a function of both soil and vegetation properties. Data on soil color when wet and dry may contribute to understanding albedo. Modelling energy transfers due to evapotranspiration requires modelling changes in soil moisture, which in turn requires data on the soil water holding capacity and hydraulic conductivity.

Water balance calculations account for runoff, infiltration, evapotranspiration, and deep percolation. Soil databases can contribute information on particle size distribution and the depth to impermeable layers, leading to more accurate modelling of the soil water balance. The quantity and type of vegetation is also important for these estimates, and knowledge of soil productivity can aid in quantifying the vegetation response that is observed with remotely sensed vegetation indices.

Carbon balance calculations account for historical, current, and potential transfers of carbon between vegetation, soil, atmosphere, and oceans. The soil represents a significant repository for the active carbon in the global cycle. Deforestation provides a net contribution of carbon dioxide to the atmosphere and afforestation provides a net removal. Although the contributions of soil and vegetation to increases in atmospheric carbon dioxide are considerably less than those from burning fossil fuels, accurate land surface inventories will improve the calibration of models of these factors. Soil maps are important for assessing the potential for afforestation projects, should policies call for forest plantation programmes for sequestering carbon.

Other greenhouse gases may be more accurately modelled if data on soils and land use are used in the stratification of the land surface. The use of nitrogen fertilizers can contribute nitrous oxide to the atmosphere, and the locations where fertilizer is applied (on agricultural land) are correlated with information in the soil databases. Soil maps often record the locations of wet soils where organic matter accumulation and methane release can take place.

When the calibrated global climate change models are run in a simulation mode, they produce scenarios of altered climatic conditions. To interpret the impacts of such altered climates on human activities, such as agricultural production, knowledge of soil properties and productivity will be needed. For example, a global warming may cause shifts in the optimum climatic zone for grain production (Rosenzweig, 1985). Soil databases are important for assessing the potential productivity of the soils in the location of the new climatic optimum.

ACKNOWLEDGMENTS

Work performed under U.S. Geological Survey contract 14-08-0001-22521.

REFERENCES

Bliss, N.B. (1987) Structuring the soils-5 data into a relational database. Proceedings, Seventh Annual Environmental Systems Research Institute User Conference, April 27 - May 1, 1987, Palm Springs, California, Environmental Systems Research Institute, Redlands, California, U.S.A.

Bliss, N.B. and W.U. Reybold (1989) Small-scale digital soil maps for interpreting natural resources. Journal of Soil and Water Conservation, January-February 1989: 30-34.

Bliss, N.B. (1989) A National Natural Resource Data Base: techniques for linking the major Land Resource Area Map, the 1982 National Resources Inventory and the Soil Interpretations Record Data Bases in a geographic information system. EROS Data Center, U.S. Geological Survey, Sioux Falls, South Dakota, U.S.A. (in press).

Dickinson, R.E., A. Henderson-Sellers, P.J. Kennedy and M.F. Wilson (1986) Biosphere-Atmosphere Transfer Scheme (BATS) for the NCAR Community Climate Model. National Center for Atmospheric Research (NCAR) Technical Note NCAR/TN-275+STR, Boulder, Colorado.

Environmental Systems Research Institute (1984) Final Report: United Nations Environment Programme (UNEP)/Food and Agriculture Organization (FAO) World and Africa GIS Data Base. Environmental Systems Research Institute, Redlands, California, U.S.A.

Food and Agriculture Organization (FAO)/United Nations Educational, Scientific, and Cultural Organization (UNESCO) (1974-81) Soil Map of the World, volumes 1-10, Paris.

Reybold, W.U. and G.W. TeSelle (1989) Soil geographic databases. Journal of Soil and Water Conservation, January-February 1989:28-29.

Rosenzweig, C. (1985) Potential CO_2-induced climate effects on North American wheat-producing regions. Climatic Change 7:367-389.

Shields, J.A. and D.R. Coote (1989) SOTER procedures manual for small scale map and database compilation including proposed procedures (for discussion) for interpretation of soil degradation status and risk. Land Resource Research Centre, Agriculture Canada, Ottawa.

Wilson, M.F. and A. Henderson-Sellers (1985) A global archive of land cover and soils data for use in general circulation climate models. Journal of Climatology 5:119-143.

Zobler, L. (1989) A world soil hydrology file for global climate modelling, in: Proceedings, International Geographic Information Systems (IGIS) Symposium, The Research Agenda, November 15-18, 1987, Arlington, Virginia, U.S.A. Association of American Geographers.

Compilation of a Soil Dataset for Europe

H. GROENENDIJK

International Soil Reference and Information Centre
P.O.Box 353, 6700 AJ Wageningen, The Netherlands

ABSTRACT

A soil data set for Europe was compiled from various soil maps on scale 1:1,000,000 and smaller. Soil data were stored on the basis of ½ × ½ degree grid. For each grid cell three dominant soil types were recorded, with information on topsoil texture, stoniness and slope. The soil types were grouped according to the FAO revised legend for the Soil Map of the World. The data set covers the European territory up to the 44th degree of longitude.

As a contribution to an agro-climate study, a simple method was set up to estimate the available water-holding capacity of soils in Europe, based on information from the data set under consideration.

INTRODUCTION

The terminology used on soil maps makes soil information little accessible to workers from disciplines other than soil science. Especially relating different map legends and classification systems is difficult and time consuming. Moreover, users of soil data in other disciplines are seldom interested in the soil itself, but need information on e.g. available water for crop growth, resistance to soil erosion, trafficability.

The soil data set presented in this paper unites information from various maps. The flexible organization of the data allows easy handling and manipulation. The soil data set can contribute to the assessment of land qualities for the European territory as a whole.

The compilation of the data set was actuated as part of an agro-climate research project. The data set will be used to estimate the available water-holding capacity of soils in Europe, as input in a waterbalance.

METHOD AND MATERIALS

Starting material

Soil information was taken from maps present in the map collection of the International Soil Reference and Information Centre. Where available, maps on scale 1:1,000,000 were chosen. The detail of

Soils and the Greenhouse Effect. Edited by A.F. Bouwman

information on this scale is convenient for representing major differences in soil type and soil properties. However, not for all European countries maps on the desired scale are available. For Norway and Poland maps on scale 1:2,000,000 were used, while for Albania, Bulgaria, Csechoslovakia, GDR and Yugoslavia the only available map is the 1:5,000,000 FAO-Unesco Soil Map of the World. Therefore not all the stored data refer to the same detail of information.

Data handling

Storing information on the basis of a grid. The data were assembled on the basis of a $\frac{1}{2} \times \frac{1}{2}$ degree grid. The soil maps were overlain by a lattice and from each grid cell the principal information was taken. The extent of the territory represented by a grid cell varies with latitude. In lat.50 a grid cell refers to approximately $(55 \times 35)km^2$. The chosen grid size is appropriate when maps on scale 1:1,000,000 are used. The information can be stored without losing too much detail. On the other hand the number of grid cells in the data set is not too large to handle with microcomputers. The European territory up to the 44th degree of latitude comprises approximately 4,500 grid cells.

Selection of soil units and characteristics presented. The mapping units on maps of scale 1:1,000,000 usually not only contain the soil type, but also information on the particle size distribution (texture) and the dominant slope. On a number of soil maps the surface stoniness, depth of a shallow hard rock contact and saline/sodic properties are designated as phases. Part of this supplementary information was stored in the data base. The data are arranged in classes adopted from the EC soil map (CEC, 1985). The following information for each of the three distinguished soil types per grid cell was included: topsoil texture (5 classes), slope (4 classes), surface stoniness (2 classes) and presence of a shallow lithic contact.

The different systems of soil classification used on the various national soil maps were transcribed into the system of the revised legend of the FAO-Unesco Soil Map of the World (FAO, 1988). Although not a classification system, the FAO legend serves well to describe soil types and place the different national classification systems under a common denominator.

Structure of the data base. The soil data were stored with the help of a spreadsheet program. Each record of the spreadsheet contains the information of an entire grid cell. All records are made up by 18 fields, representing the geographical situation of the grid cell (latitude, longitude, country) and information on the three distinguished soil types (for each scil type: soil name, relative proportion, topsoil texture, slope, presence of a stony and/or lithic phase). Except for latitude and longitude, all information is stored using codes. The arrangement of the data is illustrated in Table 19.23.1.

Table 19.23.1 Example of data arrangement in the soil data set (data are coded)

lat.	long.	nation	soil1	%1	text.1	slope1	stone1	soil2	%2	text.2	slope2	stone2	soil3	%3	text.3	slope3	stone3
49.5	5.5	9	74	50	13	2	0	71	30	7	3	3	77	20	13	3	0
49.5	6	9	79	50	13	3	0	71	40	7	3	3	51	10	4	1	0
49	-4.5	0	171	60	9	2	0	178	20	9	2	0	72	20	9	3	0
49	-4	0	171	40	9	2	0	72	40	9	2	0	178	20	9	2	0
49	-3.5	9	171	50	9	2	0	72	30	9	3	0	178	20	9	2	0
49	-3	0	171	80	9	2	0	178	10	9	2	0	0	10	0	0	0
49	-2.5	0	170	60	9	2	0	70	40	10	3	2	0	0	0	0	0
49	-2	0	71	40	9	1	2	170	30	9	1	2	51	30	4	1	0
49	-1.5	9	72	50	9	2	0	171	30	9	2	0	178	20	9	2	0
49	-1	9	170	40	13	3	2	71	40	13	3	2	72	20	9	2	0
49	-0.5	9	71	50	13	3	2	170	30	3	2	2	74	20	13	3	0
49	0	9	70	60	13	3	0	90	20	9	1	2	178	20	13	3	0
49	0.5	9	90	60	9	1	2	178	30	9	1	2	171	10	9	1	2
49	1	9	171	60	9	2	1	71	30	13	3	2	90	10	9	1	2
49	1.5	9	171	34	7	2	0	178	33	9	2	0	90	33	9	1	2
49	2	9	178	50	9	2	0	171	50	9	2	0	0	0	0	0	0
49	2.5	9	171	60	9	2	0	90	30	9	1	0	51	10	4	1	0
49	3	9	90	40	9	1	2	178	30	9	1	0	23	30	7	2	3
49	3.5	9	23	40	7	2	3	90	40	9	1	2	52	20	4	1	0
49	4	9	23	70	7	2	3	74	20	7	2	3	52	10	4	1	0

APPLICATION: ESTIMATION OF THE AVAILABLE WATER-HOLDING CAPACITY (AWC)

The concept of AWC refers to the maximum amount of plant-available water that can be stored in the soil profile. Usually an upper limit, the field capacity, and a lower limit, the permanent wilting point, are defined as soil constants. It is generally accepted that the water content at permanent wilting point equals the water content at pF 4.2. The amount of water at field capacity, however, is determined at different pF-values, of which pF 1.7, pF 2.0 and pF 2.3 are most commonly used.

AWC is regarded as a static variable: the amount of available water stored in the soil profile at the beginning of the growing season. Principal determinants of AWC are physical soil properties, such as soil structure, texture, bulk density and pore size distribution. Only soil texture is indicated on maps of scale 1:1,000,000 and smaller. Although relatively important, other determinants can not be recorded, because their spatial variability is too high. Therefore, AWC is estimated using direct relations between texture and available reserve and considering the root depth of crops. For stony soils a correction factor is introduced. Figure 19.23.1 presents the procedure used to estimate AWC for each soil type, starting from the information stored.

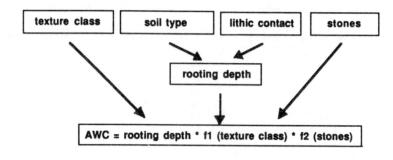

f1 = function supplying the available water depending on the texture class

f2 = reduction factor for stony soils

Figure 19.23.1 Calculation of AWC.

Direct relations between texture classes and availabe water have been worked out for a variety of national systems of texture classification (e.g. Hall et al, 1977; Jamagne et al, 1977; Wösten et al, 1987). King and Daroussin (1988) determined mean values of available water (cm/cm soil) for each of the five texture classes of the EC soil map (CEC, 1985).

Besides the presence of hard rock at a shallow depth, information included in the soil type can indicate unfavourable root conditions. In a coarse textures Podzol for example, crop root depth will rarely exceed 50 cm (Madsen and Platou, 1983). AWC is determined by multiplying the available reserve for coarse textured soil materials (0.110 cm/cm soil, from figures of King and Daroussin, 1988) with the maximum root depth. In this way, AWC for coarse textured Podzols is estimated at 55 mm.

Thus, the AWC can be estimated for each soil type and each grid cell from the information in the data base using simple algorithms.

DISCUSSION

In the soil data set under consideration information from different sources is combined to the same denominator, i.e. the FAO (1988) legend. Although set up to estimate AWC, it also enables evaluation of other soil characteristics for the complete European territory. Data storage on the basis of a grid allows easy manipulation of the information with a personal computer and enables accessibility to potential users.

The data are stored on a small scale, resulting in a considerable reduction of the original soil information. For each grid cell the three dominant soil types were selected. As most mapping units consist of soil associations, first all major soil types with associated types were dissected and compared. They were regrouped to form three soil types, considering the principal differentiating soil properties and soil forming processes. Moreover, part of the geographical information is lost. The %-distribution of three major soil units in a grid cell is stored, but their distribution pattern can not be reproduced using a $\frac{1}{2} \times \frac{1}{2}$ degree lattice.

The data set can be used in environmental studies on a supranational scale. The considerable reduction of the original soil information makes the data set unsuited to more detailed surveys.

REFERENCES

CEC (1985) Soil Map of the European Communities 1:1.000.000. ECSC, EEC, EAEC, Brussels-Luxembourg.

FAO (1988) Soil Map of the World. Revised legend. World Soil Resources Report 60, FAO, Rome.

FAO-Unesco (1981) Carte Mondiale des sols, Volume V, Europe. Unesco, Paris.

Gupta, S.C. and W.E. Larson (1979) Estimating soil water retention characteristics from particle size distribution, organic matter percent, and bulk density. Water Resources Research 15:1633-1635.

Hall, D.G.M., M.J. Reeve, A.J. Thomasson and V.F. Wright (1977) Water retention, porosity and density of field soils. Soil survey, Technical Monograph no. 9. Harpenden: Rothamsted Experimental Station.

Jamagne, M., R. Betremieux, J.C. Begon, A. Mori (1977) Quelques données sur la variabilité dans le milieu naturel de la réserve en eau des sols. Bulletin Technique d'Information, 324/325:627-641.

King, D. and J. Daroussin (1988) Test for estimating the available soil moisture reserve using the European Community soil map on the scale of 1:1.000.000. Paper presented at the EC-workshop "Application of Computerized EC-Soil Maps and Climate Data", 15-16th Nov.1988, Wageningen.

Letey, J. (1985) Relationship between soil fysical properties and crop production. In: Advances in soil science, Volume 1:277-294, Springer-Verlag, New York.

Madsen, H.B. and S.W. Platou (1983) Land use planning in Denmark: The use of soil physical data in irrigation planning. Nordic Hydrology 14:267-276.

Rawls, W.J., D.L. Brakensiek and K.E. Saxton (1982) Estimation of soil water properties. Transactions of the American Society of Agricultural Engineers, 25:1316-1320.

Wösten, J.H.M., M.H. Bannink and J. Beuving (1987) Waterretentie- en doorlatendheidskarakteristieken van boven- en ondergronden in Nederland: De Staringreeks. Stiboka rapp.no. 1932, Wageningen.

Surface Reflectance and Surface Temperature in Relation with Soil Type and Regional Energy Fluxes

W.G.M. BASTIAANSSEN and M. MENENTI

The Winand Staring Centre for Integrated Land, Soil and Water Research
P.O. Box 125, 6700 AC Wageningen, the Netherlands

ABSTRACT

Understanding the coupling between surface reflectance and surface temperature is essential for the classification of soil types and the calculation of related energy fluxes. The use of satellite observation of surface reflectance and surface temperature is investigated for an area in the North-African desert. Surface reflectance in arid regions varies considerably since both wet and extreme saline top soils exist. An empirical equation relating surface reflectance with sun zenith angle, optical depth of the atmosphere and soil characteristics such as moisture content and roughness is presented. Particular attention has been dedicated to the definition of a reference surface reflectance. A table with reference reflectances of different soil types is presented. Classification of soil types can be done by the combined use of normalized surface reflectance and surface temperature. Such classification can be applied to determine the spatial distribution of actual evaporation when the relationship between soil hydraulic and surface properties is described. The specific evaporation rate for a hydraulic class can be simulated by numerical models of water transport in the unsaturated zone. Another approach is by mapping the different terms of the surface energy balance. It is shown that under restrictive conditions, the effective resistance for transport of heat in air can be determined with measurements of surface reflectance and surface temperature and applied to calculate the sensible heat flux. Observed relationships between soil heat and surface properties give a parameterization of the soil heat flux as a function of net radiation. The method presented here does provide an operational procedure to classify soil types, depth of the ground water table, specification evaporation rates and mapping of instantaneous energy fluxes with satellite observations.

INTRODUCTION

Vast aquifer systems are present in the North-African deserts. These fossil ground water systems are feeding unoccupied natural depressions and oases. It was shown by Menenti (1984) that the natural evaporation of ground water exceeds the man-made extraction. The determination of ground water losses through bare soil evaporation is necessary for proper planning of ground-water development.

Soils and the Greenhouse Effect. Edited by A.F. Bouwman

Hypersaline, structured playa soils consist of a mixture of sand and salts. The soil dependent daily behaviour of surface reflectance and of the terms of the energy balance results in variations of surface temperature. Field and satellite observations of surface reflectance and temperature did demonstrate the existence of correlations between these variables. The physical meaning of these relationships has been explained in detail by Menenti et al. (1989a).

This paper has the following objectives: (i) defining a normalized surface reflectance, (ii) classification of surface types, and (iii) relating surface with hydraulic soil properties.

SIMPLE PARAMETERIZATIONS OF THE SURFACE ENERGY BALANCE

The energy balance at the earth surface consists of the following components:

$$M_R^* + M_{SH} + M_{LH} + M_{G,0} = 0 \tag{1}$$

where: M_R^* = net radiation (W m^{-2})
$\quad M_{SH}$ = sensible heat flux (W m^{-2})
$\quad M_{LH}$ = latent heat flux (W m^{-2})
$\quad M_{G,0}$ = soil heat flux (W m^{-2}).

Under hyperarid conditions the liquid–vapour phase transition takes place below the soil surface. In this case the energy balance equation reads:

$$M_R^* + M_{SH} + M_{G,0} = 0 \tag{2}$$

at the surface ($M_{LH} = 0$ here) and:

$$M_{G,0} + M_G + M_{LH} = 0$$

at the evaporation front.

For a zero depth of the evaporation front, M_G is identical with the soil heat flux at the surface ($M_{G,0}$). Fluxes towards the reference surface are counted positive. Net radiation depends on surface reflectance:

$$M_R^* = (1-\rho_o) M_{sw}^- + \epsilon'\sigma T_a^4 - \epsilon\sigma T_o^4 \tag{3}$$

where: ρ_o = surface reflectance (-)
$\quad M_{sw}^-$ = incoming shortwave solar radiation (W m^{-2})
$\quad T_a$ = air temperature (K)
$\quad T_o$ = surface temperature (K)
$\quad \epsilon'$ = apparent emissivity of the atmosphere (-)
$\quad \epsilon$ = emissivity of the surface (-)
$\quad \sigma$ = Stefan-Boltzmann constant (W m^{-2} K^{-4}).

With $M_{sw}^- = 1000$ W m^{-2}, an increment of $\delta\rho_o = 0.30$ e.g. from wet ($\rho_o = 0.10$) to dry ($\rho_o = 0.40$) soil gives a difference of $\delta M_R^* = 300$ W m^{-2}. This affects surface temperature. Surface reflectance depends linearly on surficial soil water content (Idso et al., 1975). Deposition of dew and the subsequent evaporation gives large daily changes in surface reflectance.

Because of surface roughness, reflectance depends on sun zenith angle, with the daily variation being affected by the ratio of direct to diffuse solar irradiance (Menenti et al., 1989b). The following empirical surface reflectance model represents the effects on surface reflectance ρ_o mentioned above:

$$\rho_o = \rho_o{'} \; m_i f_i(\Phi_{su}) \; \{g_i(\tau)\}^{\sin \Phi_{su}} \tag{4}$$

where: Φ_{su} = solar zenith angle (-)
g_i = a surface dependent function describing the effect of surface roughness (-)
τ = optical depth of the atmosphere (-)
m_i = factor representing the relation between ρ_o and soil water content (-)
$f_i(\Phi_{su})$ = a function accounting for dew deposition
$\rho_o{'}$ = surface reflectance at $\Phi_{su} = 0$ and $m_i f_i(\Phi_{su}) = 1$ (-).

The soil dependent m-factor expresses the decrease of surface reflectance ($\rho_o{'}$) with increasing surficial soil water content and is defined as:

$$m = \rho_o{'}(\text{sunrise})/\rho_o{'}(\text{afternoon}) \tag{5}$$

The function $f_i(\Phi_{su})$ describes the transition from wet to dry surface conditions. If one defines reference values like $m_i f_i(\Phi_{su}) = 1.0$, $g_i(\tau) = 1.6$ and $\Phi_{su} = 0$, it is possible to calculate normalized reflectance ($\rho_o{''}$) values.

An analytical relationship between surface reflectance and temperature can be derived by substituting eq.(3) and a transfer equation $M_{SH} = \rho_a \, C_p/r_{ah} \, (T_o - T_a)$ for sensible heat flux (M_{SH}) into the surface energy balance equation (1):

$$T_o = C^{-1} \, \{M_{sw}^{-}(1 - \rho_o) + \epsilon{'}\sigma T_o^{4} + \rho_a C_p T_a/r_{ah} + 3\epsilon\sigma T_o^{'4} - M_G - M_{LH}\} \tag{6}$$

with $C = 4 \, \epsilon\tau T_o^{'3} + \rho_a \, C_p/r_{ah}$

where: ρ_a = air density (Kg m^{-3})
C = a constant (W m^{-2}K^{-1})
$T_o{'}$ = reference surface temperature (K).
C_p = air specific heat (J Kg^{-1} K^{-1})
r_{ah} = resistance for heat transport in air (s m^{-1})

Equation (6) has been linearized by means of a Taylor expansion with T_o as independent variable.

Both field and satellite observations indicate that there may be a relationship between ρ_o and T_o, which is illustrated in Figure 19.24.1. At low reflectance values the increase in excess energy as a result of decreasing M_{LH} is not offset by the decrease in M_R^*, resulting from increasing reflectance (Menenti et al., 1989b). So T_o will increase along with ρ_o.

At high reflectance values the increase in excess energy, due to the decrease in M_{LH}, is less than the decrease in M_R^*. Thus T_o will decrease with increasing ρ_o.

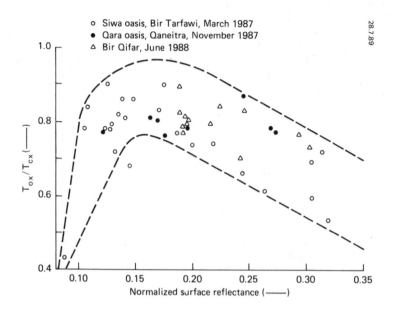

Figure 19.24.1 Field measurements of maximum surface radiation temperature, T_{ox}, divided by the maximum temperature of a reference panel, T_{cx}, plotted versus normalized surface reflectance; broken lines give general trend of data points (extended after Menenti et al., 1989a).

Explicit formulae can be derived from equations (1) and (6) to relate the slope of the increasing, respectively decreasing, branch of $T_o = T_o(\rho_o)$ to the fluxes and surface properties which appear in these equations. Assuming constant T_a, M_{sw}^{-}, M_{LH} and M_G above non-homogeneous soil surfaces, the slope a of the decreasing branch of $T_o(\rho_o)$ is:

$$a = -4\ \epsilon\sigma\ T_o^{*3}/M_{sw}^{-} - \rho_a\ C_p/(r_{ah}\ M_{sw}^{-}) \tag{7}$$

It was shown that the observed trend can be applied to derive an effective transport resistance for sensible heat in air ($r_{ah}\text{eff}$) over non-homogeneous surfaces (Menenti et al., 1989a). Values of $r_{ah}\text{eff}$ have been verified experimentally. Hence, M_{SH} can be calculated on the basis of $r_{ah}\text{eff}$, without measurements of wind speed.

Another simple and useful parameterization of the surface energy balance can be obtained by relating the ratio $M_{G,0}/M_R^*$ to surface and soil properties. $M_{G,0}/M_R^*$ has been related to ρ_o'' and soil thermal properties. (Figure 19.24.2).

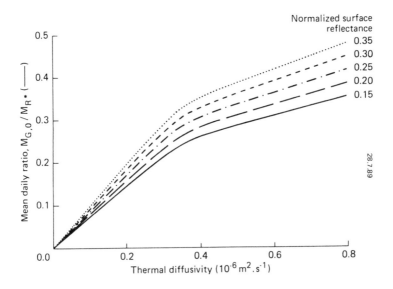

Figure 19.24.2 Mean daily ratio $M_{G,0}/M_R{}^*$ plotted versus the apparent soil thermal diffusivity for different normalized surface reflectance as observed during field experiments in the Western Desert of Egypt.

Experiments have shown that the mean daily ratio $(M_{G,0}/M_R{}^*)$ can be expressed (Bastiaanssen, 1988) as:

$$M_{G,0}/M_R{}^* = (0.84\rho_o{}'' + 0.35) (p - \sqrt{p^2 - 4\,pp}) \qquad (8)$$

where: p $= 0.995\ a' + 0.185$
       pp $= 0.179\ a'^2 + 0.14\ a'$
       $a'\ (m^{-2}\ s^{-1}) =$ soil thermal diffusivity in the range $0.2 \leq a' \leq 1.10^{-6}\ m^{-2}\ s^{-1}$.

CORRELATION OF SURFACE AND SUB-SURFACE SOIL PROPERTIES

The extremely slow process of soil formation in the depressions of the Sahara, where ground water is present at shallow depths, has brought about a relationship between depth of the ground water table, soil hydraulic properties and surface reflectance properties. This interrelation will be illustrated by means of the results of field experiments in the Western Desert of Egypt and in the Libyan Desert.

The parameters appearing in equation (4) were determined for all available measurements of surface reflectance. The value of the normalized surface reflectance $\rho_o{}''$ was calculated and the results averaged over different locations with similar soil or vegetation type (Table 19.24.1).

Table 19.24.1 Normalized surface reflectance of different soil and vegetation types; measurements have been collected at different locations in Libya and Egypt; standard deviation of data points from eq. (4) and number of daily data sets available for each surface type are indicated

Code	Surface type	Location	Mean ρ_o''	Stand. dev.	Repli- cations
A	White salt crust	Idri	0.414	0.012	2
B	Brine	Idri,Siwa,Qara,Qifar	0.310	0.009	7
C	Bare coarse sand	Tarfawi,Sharib,Siwa,Qifar	0.309	0.012	10
D	Sandcrust	Sharib,Qaneitra	0.240	0.002	5
E	Sandcrust on limestone	Qara	0.269	0.004	2
F	Limestone	Qaneitra,Qifar	0.253	0.005	3
G	Sandy puffy	Qifar	0.302	0.008	4
H	Light brown hard puffy	Qaneitra,Siwa,Qifar,Qara	0.244	0.014	6
I	Hard puffy salt crust	Idri,Siwa,Qifar	0.211	0.012	9
J	Hard puffy salt crystals	Siwa,Qifar	0.192	0.005	9
K	Clayey hard puffy	Qifar	0.191	0.012	6
L	Grey brown soft puffy	Qaneitra,Siwa	0.171	0.005	8
M	Brown soft puffy	Siwa,Qaneitra	0.134	0.007	6
N	Dark grey soft puffy	Siwa	0.113	0.011	5
O	Very wet soft puffy	Siwa	0.088	-	1
P	Hummocky with polygons	Siwa,Qifar	0.179	0.005	4
Q	Brown hummocky	Siwa,Qara,Qifar	0.154	0.021	8
R	Sandy soft clay	Qaneitra	0.129	-	1
S	Palms	Idri	0.220	0.002	4
T	Acacia	Tarfawi	0.172	0.010	3

The normalized surface reflectance of puffy soil surfaces gradually changes from $\rho_o'' = 0.09$ for saturated puffy soils (O) to $\rho_o'' = 0.30$ for dry sandy soils with a puffy structure (G). This is related with the hydrological situation.

To understand the role of ground water availability, the ρ_o''-values have been correlated with observed ground water depths (Figure 19.24.3). The following regression equation describes the linear trend between the depth of the ground water table and ρ_o'' (r = 0.88), with the exception of group A and B:

$$Z_{gw} = -18.9 + 285.8 \, \rho_o'' \tag{9}$$

where: ρ_o'' = ≤ 1
      Z_{gw} = depth of ground water table (cm).

The average error of the estimates of the depth of the ground water table is 9 cm. This is an acceptable error, especially when no direct observations are available.

Vapour diffusion and thermal convection of moist soil air above the evaporation front are the dominant mass transport processes in desert soils. The depth of the evaporation front depends on the functions

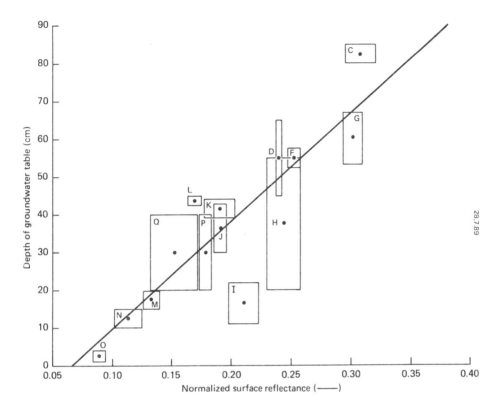

Figure 19.24.3 Observed relationships (correlation coefficient r = 0.9) between depth of the ground water table and the normalized surface reflectance; letters indicate soil type; for explanation of codes see Table 19.24.1.

$h_m(\theta)$ and $K(h_m)$, which are used to describe the relations between soil water content (θ), pressure head (h_m), and the unsaturated capillary conductivity (K) (Menenti, 1984; Bastiaanssen et al., 1989). The $h_m(\theta)$ and $K(h_m)$ relationships have been fitted by means of the S-shape model presented by Van Genuchten et al., (1989). A cluster analyses technique has been applied to the parameters of this model.

The classes having similar soil hydraulic properties have been compared with the classes of surface types with similar reflectance properties and the mean depth and range of the ground water table. The results are presented with Table 19.24.2.

Table 19.24.2 Comparison of soil surface reflectance with soil hydraulic properties of desert soils

Hydraulic class	Ground water table (cm)		Corresponding classes* of normalized surface reflectance
	average	range	
1	79	30-150	C,G,J,P
2	53	30- 65	D,P,Q
3	74	20-200	C,H,I,N,O
4	40	15- 45	K,M
5	16	15- 20	M
6	60	25- 95	H,I,L,R

* see Table 19.24.1

White salt crust (A), brines (B), limestone (F) and a mixture of sandcrust and limestone (E) were not assigned to any hydraulic class since they could not be studied with tensiometers. Some classes (e.g. C,I,M) of normalized surface reflectance correspond with more than one hydraulic class.

CONCLUSIONS

The new surface reflectance model has been shown to properly describe the impact of soil and atmospheric conditions on the daily variation of surface reflectance. The equilibrium surface temperature can be explicitly related to surface reflectance. The observed relationships between T_o and ρ_o, e.g. with satellite measurements give estimations of an effective aerodynamic resistance of non-homogeneous land surfaces. Surface reflectance in arid regions appears to have a good correlation with the hydrological situation such as the depth of the ground water table. The capillary transport properties such as hm(θ) and k(hm) relationships, affect soil surface properties in a quasi-steady state situation. Consequently, hydraulic properties are related to surface reflectance. On the basis of these relationships maps showing the depth of the ground water table can be made and applied as bottom boundary condition for models simulating water flow through the unsaturated zone, such as EVADES.

REFERENCES

Bastiaanssen, W.G.M. (1988) New empirical aspects of the Bowen-ratio energy balance method. Nota ICW 1914. The Winand Staring Centre, Wageningen, The Netherlands. 41 pp.
Bastiaanssen, W.G.M., P. Kabat and M. Menenti (1989) A new simulation model of bare soil evaporation in arid regions (EVADES). Nota ICW 1938. The Winand Staring Centre, The Netherlands. 74 pp.

Genuchten, M.Th. Van, F. Koveh, W.B. Russel and S.R. Yates (1989) Direct and indirect methods for estimating the hydraulic properties of unsaturated soils. In: J. Bouma and A.K. Bregt (eds.): Land qualities in space and time; proceedings of a symposium organizedby the International Society of Soil Science (ISSS), Wageningen, The Netherlands, 22-26 August 1988. Pudoc, Wageningen, pp 61-73.

Idso, S.B., R.D. Jackson, B.A. Kimball and F.S. Nakayama (1975) The dependence of bare soil albedo on soil water content. Journal of Applied Meteorology 14:109-113.

Menenti, M. (1984) Physical aspects and determination of evaporation in deserts applying remote sensing techniques. Report ICW 10 (special issue), The Winand Staring Centre, Wageningen. 202 pp.

Menenti, M., W.G.M. Bastiaanssen, D. Van Eick and M.H. Abd El Karim (1989a) Linear relationships between surface reflectance of temperature and their application to map actual evaporation of groundwater. Advances in Space Research 9:165-176.

Menenti, M., W.G.M. Bastiaanssen and D. Van Eick (1989b) Determination of hemispherical reflectance with Thematic Mapper measurements. Remote Sensing of Environment 27 (in press).

Relationships between Vegetation Index, Crop Photosynthesis and Crop "Stomatal" Conductance Simulated with a Radiative Transfer Model

A. OLIOSO[1] and F. BARET[2]

[1]*Laboratoire Commun de Télédétection CEMAGREF-ENGREF*
B.P. 5093, F-34033 Montpellier Cedex 1, France
[2]*Station de Bioclimatologie, INRA*
B.P. 91, F-84140 Montfavet, France

ABSTRACT

The SAIL radiative transfer model is used to calculate normalized difference vegetation index ND which combines red and infrared reflectances. This model is also used to compute canopy photosynthesis and canopy "stomatal" conductance by integrating light response of leaves over the whole canopy. The sensitivity of the relationships between ND and the parameters of canopy light response function were analyzed. These relations appear to be very sensitive to soil reflectance at low LAI. They are also affected by canopy structure and by sun elevation. They become unpredictable for LAI greater than 3 when ND reaches asymptotic value.

INTRODUCTION

Crop productivity studies using remote sensing often relate the vegetation index (a combination of red (r) and near infrared (nir) reflectances) to biomass, leaf area index or radiation absorption by the canopy. Recently, theoretical works (Sellers, 1985; Choudhury, 1987; Hope, 1988) show that canopy photosynthesis or canopy "stomatal" conductance can also be assessed by using the vegetation index. But these relations depend on canopy structure, optical properties of soil, and particularly on the amount of incident radiation. In this paper we present a crop photosynthesis and conductance model derived from the ·SAIL radiative transfer model. This model is used to analyze the sensitivity of the relationships between vegetation index and the parameters (independent of irradiance) of the light response function of crop photosynthesis or conductance.

Soils and the Greenhouse Effect. Edited by A.F. Bouwman

Models presentation

The SAIL model was developed by Verhoef (1984, 1985) to compute bidirectional reflectance of an homogeneous canopy. It was successfully tested several times (e.g. Goel and Thompson, 1984). We used it to calculate the normalized difference vegetation index (ND) :

$$ND = (nir - r)/(nir + r) \qquad (19.25.1)$$

The SAIL radiative transfer model can also simulate hemispherical reflectance and transmittance. Here, we used it to compute the distribution of irradiance in the photosynthetic active radiation range (PAR = 400 to 700 nm) inside the canopy. Thus, photosynthesis and "stomatal" conductance of a canopy can be calculated by summing light responses of leaves over the whole canopy. The light response function F_1 of leaf photosynthesis is given by equation 19.25.2. The same formulation applies to leaf stomatal conductance (Table 19.25.1) :

$$F_1 = F_1min - (F_1min - F_1max).f_1(PAR) \qquad (19.25.2)$$

where subscript 1 refers to leaves. F_1min and F_1max are minimum and maximum values of F_1. $f_1(PAR)$ is a function of the normal (relative to leaf) flux density of incident PAR and of one parameter F_1o that determines initial response (when PAR = 0) (see Table 19.25.1). Since the $f_1(PAR)$ function is non-linear the canopy integration must take into account radiation attenuation inside the canopy, orientation of leaves to the incident PAR flux and proportions of sunlit and shaded leaves. Calculations are made by dividing the canopy into thin layers (LAI about 0.15). In each layer the leaf angle distribution is discretized into 13 zenith and 45 azimuth classes. The proportion of sunlit area is obtained from the proportion of direct radiation extinction.

Table 19.25.1 Characteristics of the light response of leave photosynthesis or "stomatal" conductance

	Photosynthesis	Conductance
Units	gCO_2 $m^{-2}h^{-1}$ or $mgCO_2$ $m^{-2}s^{-1}$	m s^{-1}
F_1min	dark respiration rate $F_1min < 0$	cuticular conductance $F_1min > 0$
F_1max	maximum rate of net photosynthesis	Maximal stomatal conductance

$f_1(PAR) = \quad PAR/(F_1o + PAR)$
Hesketh, 1963; Jarvis, 1975

$(F_1max - F_1min)/F_1o$ determines the gradient of the initial response at low irradiance

Comparison with field data

There have been few experimental studies where leaf and canopy photosynthesis or conductance were measured simultaneously. Using values of $F_l min$, $F_l max$ and $F_l o$ from Biscoe et al. (1975b), net photosynthesis was calculated for a barley canopy with a leaf area index of 2.9, for a clear day and a cloudy day. The simulated and observed net photosynthesis (Biscoe et al., 1975a) are compared in Figure 19.25.1. In this example the model gives a good description of diurnal behaviour of crop photosynthesis. However we observe that simulated values often overestimate experimental data. This could be explained by :

1. the variation of leaf parameters. They were measured two days after the whole canopy measurements and it is known that they can vary in a wide range over short time periods;
2. the use of a single light response function for all the leaves whereas this function depends on leaf position inside the canopy (e.g. Biscoe et al., 1975b; Angus and Wilson, 1976);

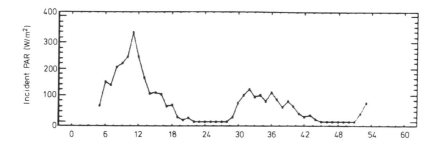

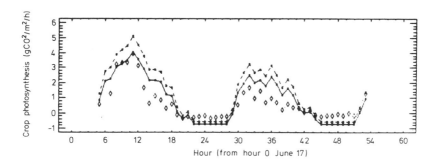

Figure 19.25.1 Comparison of simulations with field data
a- incident PAR evolution for the two days (June 17 and 18, Biscoe et al., 1975a).
b- observed barley crop photosynthesis; simulations for Flmax = 2.90 $gCO_2 m^{-2} h^{-1}$; for Flmin = -0.20 $gCO_2 m^{-2} h^{-1}$; for Flo = 50 $m^2 W^{-1}$. Simulations for Flmax = 2.35 $gCO_2 m^{-2} h^{-1}$; for Flmin = -0.25 $gCO_2 m^{-2} h^{-1}$; for Flo = 50 $m^2 W^{-1}$.

3. the assumptions of the SAIL model. Particularly we assume a homogeneous canopy that may be not enough realistic in the case of this barley crop.

Sensitivity analysis of ND, canopy photosynthesis and conductance

The normalized difference and the canopy response were computed for a wide range of input parameters and variables (see Table 19.25.2). Our simulations agree with other works which show that light response of the canopy can be formulated by the same equation as the leaf response equation (eq. 19.25.2):

$$F_c = F_c min - (F_c min - F_c max).f_c(PAR) \qquad (19.25.3)$$

where: subscript c refers to canopy
$f_c(PAR) = PAR/(F_c o + PAR)$
$F_c min = LAI \times F_l min.$

$F_c min$ is independent of radiative processes and we take $F_l min = 0$ to reduce the number of parameters in equation 19.25.3. In that case and for photosynthesis equation 19.25.3 describes gross photosynthesis.

Table 19.25.2 Parameters and input variables used for simulations

Leaf optical properties:	Reflectance	Transmittance
red	0.075	0.007
near infrared	0.520	0.440
PAR	0.100	0.050

Leaf Area Index LAI: 0.0   0.25   0.5   1.0   2.0   4.0   8.0   16.0

Mean leaf angle inclination ttl: 20°   30°   40°   50°   60°   70°

Soil red reflectance rsol: 0.05   0.10   0.15   0.20   0.25   0.30

NB: near infrared reflectance value is deduced from the red value and the soil line equation nir = 1.16.r + 0.067 and PAR value from rpar = 0.66.r

Solar zenith angle tts: 10°   20°   30°   40°   50°   60°   70°   80°

Diffuse proportion: 0.20

View orientation: vertical viewing

Incident PAR (Wm^{-2}): 0 to 700 by 50 step

Leaf response function parameters $F_l min$, $F_l max$ and $F_l o$:
0.0   1.0   180.0
(consistent with mgCO$_2$/m^2/s and cm/s photosynthesis and conductance units)

We estimated $F_c$max and F_co by using a non linear regression method (least square minimization) for each set of simulation parameters. These adjustments yield a mean coefficient R^2 greater than 0.999. A variance analysis shows that LAI variations explain more than 80% of $F_c$max, F_co and ND variations. For this reason we have made a variance analysis for each LAI (Table 19.25.3). Soil reflectance (rsol) appears to have only little effect on $F_c$max and no effect on F_co. These parameters are mainly influenced by sun position (tts) at low LAI and by leaf inclination (ttl) at high LAI.

ND is very sensitive to soil reflectance when LAI is low, but this sensitivity decreases as LAI increases (Table 19.25.3). At high values of LAI, ND only depends on canopy structure and is close to 0.92.

Relationships between ND and light canopy response parameters

The relationship between ND and the crop photosynthesis or the crop "stomatal" conductance is strongly affected by the level of incident radiation (Sellers, 1985; Hope, 1988). Therefore we relate ND to the parameters of light crop response ($F_c$max and F_co) that are independent of irradiance.

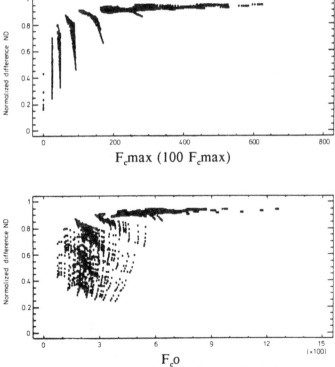

Figure 19.25.2   ND vs. $F_c$max and ND vs. F_co for our simulation parameters (Table 19.25.2)

Figure 19.25.2 shows that :
- for values of ND greater than 0.8, variations of $F_c o$ and $F_c max$ are independent of ND. It is explained by the rapid saturation of ND when LAI increases (Table 19.25.3).
- for values of ND lower than 0.8, the input factors tts, ttl and rsol (Table 19.25.2) strongly influence the relationships between ND and $F_c max$ or $F_c o$.

We study the influence of these variation factors by introducing the equations (19.25.4) and (19.25.5) :

$$ND = ND\infty + (NDs - ND\infty).exp(-k1.F_c max) \qquad (19.25.4)$$

$$ND = ND\infty + (NDs - ND\infty).exp(-k2.(F_c o - F_c os)) \quad (19.25.5)$$

where: $ND\infty$ = the asymptotic value of ND when LAI tends to ∞. According to the previous analysis $ND\infty = 0.92$.
 NDs = the value of ND for bare soils (LAI = 0).
 k1,k2 = statistically adjusted coefficients.
 $F_c os$ = the extrapolated value of $F_c o$ for bare soils.

Table 19.25.3 variance analysis of simulated ND and parameters $F_c max$ and $F_c o$ for each LAI

LAI	0.25	0.5	1	2	4	8	16
Fcmax×100							
Mean	24.1	45.4	83.5	140.9	204.0	261.9	362.2
Standard deviation	0.4	2.0	5.9	13.4	24.3	42.8	102.9
tts	10.6	74.9	81.2	78.6	49.6	6.1	16.9
ttl	55.9	13.6	3.1	0.1	26.2	81.9	80.7
rsol	9.0	4.7	3.7	3.9	4.1	2.2	0.1
tts*ttl	21.5	5.2	10.7	15.4	18.8	9.0	1.7
others	3.0	1.6	1.3	2.0	1.3	0.8	0.6
Fco							
Mean	210.6	236.7	275.5	344.2	429.9	523.3	677.0
Standard deviation	68.1	66.5	62.3	54.7	57.5	98.6	198.6
tts	58.9	56.2	50.0	33.1	14.2	24.0	35.7
ttl	12.5	14.4	19.0	33.8	62.9	70.1	61.1
rsol	0.2	0.1	0.0	0.3	1.6	0.8	0.1
tts*ttl	28.1	29.1	30.7	32.4	20.9	4.9	3.0
others	0.2	0.3	0.4	0.4	0.2	0.1	

Table 19.25.3   (cont'd)

ND

Mean	0.44	0.57	0.73	0.85	0.90	0.9	0.92
Standard deviation	0.11	0.10	0.08	0.04	0.01	0.0	0.02

tts	12.3	22.9	30.7	35.7	33.8	4.1	2.5
ttl	9.4	18.6	28.0	21.4	31.1	94.4	96.9
rsol	75.3	50.5	23.0	4.8	0.2	0.1	0.0
tts*ttl	2.3	6.8	13.7	28.2	29.9	1.4	0.6
others	0.7	1.2	4.6	9.9	5.0	0.0	0.0

tts: solar zenith angle; ttl: mean leaf inclination angle; rsol: soil reflectance; Values are in per cent of the variance.

Simulated data are adjusted to the equations (19.25.4) and (19.25.5) for each set of simulation parameters (tts,ttl,rsol). The coefficients R^2 are higher than 0.995. Variance analysis of parameters k1, k2 and F_cos are presented in Table 19.25.4. These parameters are not very sensitive to soil optical properties. These properties mainly affect ND (Table 19.25.3) and this effect is taken into account by introducing NDs. k1 and F_cos depend on sun position (tts), but also on leaf inclination angle especially for low solar zenith angles (Table 19.25.4 and Figure 19.25.3).

Table 19.25.4   Variance analysis of parameters k1, k2 and F_cos

parameter	k1	k2	F_cos
mean	0.0151	0.0146	188.2
standard deviation	0.0047	0.0002	75.4
tts	68.0	10.2	59.3
ttl	9.6	37.3	14.3
rsol	7.7	6.3	0.2
tts*ttl	13.1	16.5	25.6
ttl*rsol	0.2	28.4	0.4
others	1.4	1.3	0.2

tts: solar zenith angle; ttl: mean leaf inclination angle; rsol:soil reflectance; Values are in per cent of the variance.

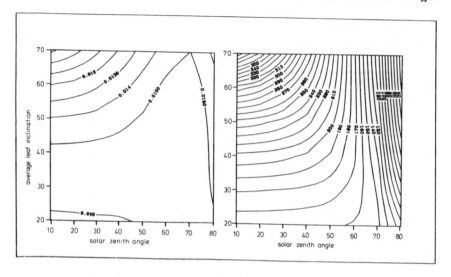

Figure 19.25.3 Dependency of k1, k2 and $F_c$os to average leaf inclination and solar zenith angle: k2 (left), $F_c$os (right)
NB : k1 shows the same behaviour than $F_c$os

Discussion and conclusion

Relations between ND and crop light response parameters are influenced by canopy characteristics, such as leaf inclination, and over all by external factors such as soil optical properties and sun elevation. Sun position is usually known at the time of ND measurement and its effect can be quantified (Figure 19.25.3). Soil optical properties mainly affect ND. Its influence can be taken into account by introducing the ND value of bare soil NDs (equations 19.25.4 and 19.25.5). The influence of canopy structure on the relationships between ND and F_co or $F_c$max only occurs at low solar zenith angles (Figure 19.25.3). But the main problem appears when LAI is greater than 3 (when ND is almost constant and > 0.85), whereas F_co and $F_c$max are strongly affected by leaves inclination and by increase of LAI. Consequently, when ND reaches saturation values (0.9), the relations become unpredictable.

We establish the relationships between ND and F_co or $F_c$max for a standard leaf light response function (Table 19.25.2). We must notice that the leaf parameters F_lo and $F_l$max can widely vary with factors such as aging, previous and current conditions of development (such as light, temperature, water status, CO_2 concentration and humidity (e.g. Biscoe et al., 1975b; Jarvis, 1975; Angus and Wilson, 1976; Avissar et al., 1985)).

A mechanistic approach of productivity by means of photosynthesis and "stomatal" conductance is more realistic than the simple Monteith's description of biomass production which takes only into account the canopy PAR absorption and the efficiency of its conversion into biomass (Monteith, 1977). But the full representation of the involved processes requires a lot of detailed information, which is not always available. It is necessary to know :

- the determinism of leaf response parameters ($F_l$max, F_lo) and respiration ($F_l$min or $F_c$min) (from physiological and climatological data, thermal infrared remote sensing);
- how parameters of canopy light response function (F_co and $F_c$max) vary between ND measurements. Variations will occur as a result of incident radiation geometry (solar angle, diffuse proportion) and leaf parameters.

Improvement of vegetation index in order to reduce soil effect at low LAI and to obtain a better information on canopy structure at high LAI is also required. It will be then interesting to compare this complex description of canopy processes with the simple and more empiric ones (Asrar et al., 1985; Baret and Olioso, 1989) at different scales (local to global).

REFERENCES

Angus, J.F. and J.H. Wilson (1976) Photosynthesis of barley and wheat leaves in relation to canopy models, Photosynthetica 10:367-377

Asrar, G., E.T. Kanemasu, R.D. Jackson and P.J. Pinter (1985) Estimation of total above ground phytomass production using remotely sensed data. Remote Sensing of Environment 17:211-220

Avissar, R., P. Avissar, Y. Mahrer and B.A. Bravdo (1985) A model to simulate response of plant stomata to environmental conditions. Agricultural and Forest Meteorology 34:21-29.

Baret, F. and A. Olioso (1989) Estimation de l'énergie photosynthétiquement active absorbée par une culture de blé à partir de données radiométriques, to be published in Agronomie

Biscoe, P.V.,R.K. Scott and J.L. Monteith (1975a) Barley and its environment. III.Carbon balance of the crop. Journal of Applied Ecology 12:269-293

Biscoe, P.V., J.N. Gallagher, E.J. Littleton, J.L. Monteith and R.K. SCOTT (1975b) Barley and its environment. IV. Sources of assimilate for the grain. Journal of applied Ecology 12:295-318

Choudhury, B.J. (1987) Relationships between vegetation indices, radiation absorption, and net photosynthesis evaluated by a sensitivity analysis. Remote Sensing of Environment 22:209-233

Goel, N.S.and R.L. Thompson (1984) Inversion of vegetation canopy reflectance models for estimating agronomic variables. V.Estimation of leaf area index and average leaf angle using measured canopy reflectances. Remote Sensing of Environment 16:69-85

Hesketh, J.D. (1963) Limitation to photosynthesis responsible for difference among species. Crop Science 3:393-496

Hope, A.S. (1988) Estimation of wheat canopy resistance using combined remotely sensed spectral reflectance and thermal observations. Remote Sensing of Environment 24:369-383

Jarvis, P.G. (1976) The interpretation of the variations in leaf water potential and stomatal conductance found in canopies in the field, Phil. Transactions Royal Society, London B273:593-610

Monteith, J.L. (1977) Climate and the efficiency of crop production in Britain, Phil.Transactions Royal Society, London B281:277-294

Sellers, P.J. (1985) Canopy reflectance, photosynthesis and transpiration. International Journal of Remote Sensing 6:1335-1372

Verhoef, W. (1984) Light scattering by leaf layers with application to canopy reflectance modelling : the SAIL model. Remote Sensing of Environment 16:125-141

Verhoef, W. (1985) Earth observation modelling based on layer scattering matrices. Remote Sensing of Environment 17:165-178

Soil Moisture and the Greenhouse Effect: the Changing Conditions of Agriculture and Water Management in Hungary

K. SZESZTAY

Ernö u. 15, Budapest 1096, Hungary

ABSTRACT

It is through soil moisture that water accomplishes her major planetary role: the support of plant and animal life. Within the evolution of terrestrial life forms nature has elaborated a sophisticated and efficient mechanism for the harmonization of the regulatory processes of the soil-water-vegetation (SWV) system (Eagleson, 1982) within the broad framework of the lithosphere-atmosphere-biosphere interaction (Budyko et al., 1985). Man is now intervening into this regulatory mechanism at a rapidly increasing scale and without a clear and reliable knowledge of the long term consequences of his interventions. The expectable significant and rapid change of climate due to the increasing greenhouse effect is a good, and perhaps close to the last, opportunity for learning how to restrain our technological and economic capabilities according to the limits set by the carrying capacity of the biosphere. As frank assessments usually reveal an increasing urgency of appropriate social responses (Rind et al., 1988) the conventional step-by-step analytical research must frequently be supplemented by an combined with management-oriented empirical or conditional short-cut solutions (Orlóci and Pintér, 1981).

INTERACTIONS BETWEEN SOIL MOISTURE AND THE GREENHOUSE EFFECT

As depicted schematically in Figure 19.26.1 the interrelation between soil moisture and the greenhouse effect is a cyclic one. Evapotranspiration (ET) as a greenhouse gas emission (alongside with CO_2 and other greenhouse gases) contributes to the building up of the greenhouse effect and becomes a climate formation factor. Climate shapes land surface meteorology which in turn regulates the soil moisture regime. Soil moisture feeds ET and closes the cycle. At the land surface this cycle can be closed in two ways: directly from local precipitation ($4\rightarrow5\rightarrow1$ in Figure 19.26.1) and indirectly, i.e. from precipitation which reaches the given locality from outside through the processes of runoff and groundwater ($4\rightarrow6\rightarrow5\rightarrow1$ and $4\rightarrow7\rightarrow5\rightarrow1$).

Soils and the Greenhouse Effect. Edited by A.F. Bouwman

CLIMATE AND LAND SURFACE HYDROLOGY

In humid regions where actual evapotranspiration (ETA) is always equal to potential evapotranspiration (ETP) local climate rather precisely controls land surface hydrology. Under such conditions the aridity factor a (defined as the ration of long term ETP to long term precipitation P) and the runoff factor r = R:P (where R is long term runoff + groundwater recharge) are in a simple quasi-functional relationship: $r = 1-a$ (see the strait line and the data around section C-D in Figure 19.26.2). This means that in humid regions soil moisture interacts with the greenhouse effect through the 4→5→1 direct atmospheric pathway of Figure 19.26.1.

In arid regions where soil moisture availability limits ET during certain periods of time and ETA<ETP the local soil moisture regime (and through it local ETA and R) is influenced by both, the local climate as well as the heat and water advection, i.e. by the excess waters of the broader environment recharging riverflow and groundwater flow systems. Under such conditions long term ETA and R values can considerably differ from those corresponding to local atmospheric conditions (see the significant scattering of the data along the statistically defined equalizing curve in section B-C of Figure 19.26.2).

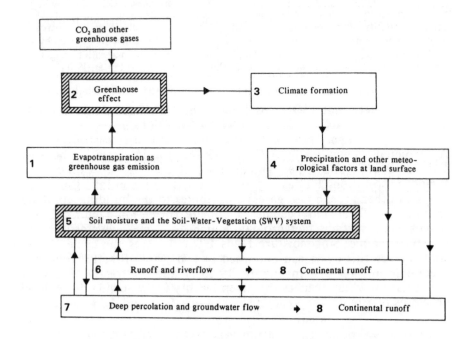

Figure 19.26.1 Major pathways of the interaction between soil moisture and the greenhouse effect.

Such advection effects can be particularly significant in large and long river systems crossing several climatic zones, such as the Nile, AmuDarya, Indus, Tarim, Kerulem, Colorado, Parana and others, as well as in temperate and semi-arid flatlands surrounded by high and humid mountain ranges, such as the Hungarian Great Plains framed by the Carpathians and affected by a large-scale hydrogeological teleconnection system as depicted schematically in Figure 19.26.3.

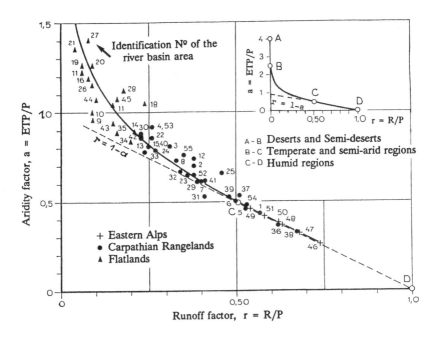

Figure 19.26.2 Interrelation between aridity factor a and runoff factor r after Budyko with adaptation to river basin data of the Carpathians (Szesztay, 1967).

The inclusion of such "inter-gridbox" effects that characterize the soil moisture formation and evapotranspiration processes (and, in fact, the global hydrological cycle as a whole) seems to be one of the crucial issues in the further development of climate models in general, and of General Circulation Models in particular. The envisaged doubling of the present day (mostly 8°×10° or 4°×5°) grid point densities within the next few years makes such initiatives both desirable and possible (Abramopoulos et al., 1988). Until these improvements bring widely applicable results useful information can be obtained on expectable future redistribution of regional soil moisture regimes through paleogeographical reconstructions as repeatedly advocated by Budyko (see e.g. Budyko et al., 1985) and demonstrated recently by Vinnikov et al. (1988).

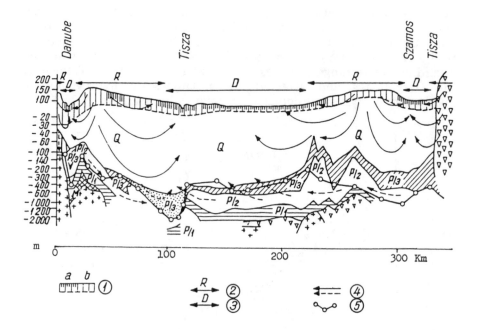

Figure 19.26.3   Schematical flow diagram of groundwaters in the Hungarian Basin (Erdélyi, 1979). 1 = Aquitard (a) and aquifer (b) layers; 2 = Recharge areas; 3 = Discharge areas; 4 = Dominant flow directions; 5 = Freshwater/saltwater interface.

THE SOIL-WATER-VEGETATION (SWV) SYSTEM AND CLIMATIC CHANGE

Natural vegetation communities have learned during millions of years of dynamic adaptation to respond efficiently to changing climate and soil conditions. This adaptation mechanism seems to be able to maintain smooth functioning and long term stability of the SWV systems and the constituting sub-systems, and furthermore to give place and rise to a continuing evolution of the vegetation communities in terms of their capacity to preserve and foster terrestrial life. This development of plant communities seems to be based on two long term strategies (Eagleson, 1982). In case of unlimited soil moisture recharge, when ETA = ETP and the cycle of Figure 19.26.1 follows the 4→5→1 direct atmospheric pathway, vegetative activity tends to maximize biomass productivity under the given radiation and heat energy constraints. However, under water-stressed conditions, when ETA<ETP and the soil moisture cycle of Figure 19.26.1 includes also the indirect riverflow and groundwater flow pathways, plant communities tend to develop a canopy density which would minimize moisture stress at the root zone under the given climate and soil conditions.

Up to now, natural and man-established SWV systems have both evolved on a trial and error basis, with one significant difference: natural evolution has usually experimented in small steps and with essentially no time constraint; human interventions in the form of deforestation and farming practices were, however, abrupt and with little or no regard to the long term equilibrium and development of the disturbed SWV systems. The results are written on the face of our planet by degradation of millions of hectares of previously fertile and flourishing lands through the processes of erosion, salinization, alkalinization, water logging and other forms of collapsing equilibrium within the disturbed SWV system.

With the expected significant and extremely rapid changes in the planetary heat and water regime during the coming decades and centuries man must now be able to understand and adapt the long term land use strategies of nature. He also must find ways and tools for applying and supplementing these strategies under the extreme conditions of the predicted rapid climatic changes which have probably never yet occurred during the previous history of the biosphere (Budyko et al., 1985).

SOIL MOISTURE MANAGEMENT IS RISK MANAGEMENT

Recent investigations seem to indicate that the natural strategy of maintaining the stability and flexibility of the SWV systems can also be conceptualized in terms of risk management (Parry, 1985). For given climate and soil conditions plant communities tend to evolve according to well defined margins of deficiencies and excesses of heat and water whereby the extent and frequency of these limiting extremes vary from species to species and also according to the phenological phases of plant development. In order to identify these margins and apply them in predictions or planning one has to quantify the annual and seasonal variability of heat and water availabilities in probabilistic terms. Such quantifications can be made relatively easily for radiation and heat factors on the basis of regular meteorological observations on sunshine and air temperature. For quantifying critical deficiencies and excesses of soil moisture the required continuing and detailed observational data are, however, not available on a regular basis. Modelling and simulation of the soil moisture balance seems to be, therefore, the only feasible solution to this effect.

HEAT-CONTROLLED AND WATER-CONTROLLED ET REGIMES

With regard to possibilities and ways of quantifying evapotranspiration as the key element of the soil moisture balance equation there seem to be important differences between the above mentioned two regimes of the SWV systems. When soil moisture is unlimited and the system maximizes biomass production under given radiation and heat constraints the use of water in the form of ET can be well approximated for a given plant species and a given phase of phenological evolution through atmospheric data alone. The possibility of such approximation is probably a reflection of the coincidence of factors and mechanism guiding stomatal regulation of ET with those promoting maximal biomass productivity. The acceptability of such approximation seems to be supported by the quasi-functional relationship between runoff and climate in humid regions (section C-D of Figure 19.26.2), as well as by the great number of empirical irrigation water requirement formulae proposed and applied all around the world.

When, however, soil moisture supply becomes restricted and the life processes of the plant communities become water-controlled, stomatal regulation must act in compensation of and contrary to atmospheric influences. In such cases ET can not be quantified by atmospheric data alone; it becomes the residual effect of a rather sophisticated stomatal regulation trying to compromise between the requirements of water use efficiency and biomass production. Under such conditions heat and water advection through riverflow and groundwater flow become important or dominant factors of local ET.

POLICY-ORIENTED MODELLING OF SOIL MOISTURE REGIMES

Considering the relative simplicity of quantifying heat-controlled ET and noting the presently unsurmountable theoretical and practical difficulties of doing the same under water-controlled conditions Órlóci and Pintér (1981) have designed a soil moisture simulation methodology which extends the advantages of the heat-controlled conditions to a partial and conditional quantification of the water-controlled ET regime. The proposed procedure starts with a well watered beginning of the growing season and it is based on a step by step compilation of the soil moisture balance equation of the root zone for continuous ten day periods. Wherever the soil moisture content drops below the level of unrestricted water availability the root zone is hypothetically filled up to field capacity. The amount of irrigation water needed to this effect is registered in the simulation procedure as a measure of natural (climatic) water deficiency event. Similarly, the validity of the simple heat-controlled ET equations (empirically determined for the major agricultural crops of Hungary by S. Szalóky and others) can be extended to the dry seasons whereby the accumulated amounts of the hypothetical irrigations provide a quantitative measure of the dry weather events corresponding to the *actual* local conditions.

For the purpose of an agricultural baseline for the State Water Plan for Hungary (NWA, 1984) some 2000 computerized soil moisture simulations have been made for the country's farmlands according to 13 crop varieties, 7 soil types, 4 categories of groundwater depth and 3 levels of farming technologies. The simulations are based on the 1928-

1978 data of 23 meteorological stations characterizing climatic differences. The major results were summarized in the form of maps quantifying deficiencies and excesses of the soil moisture balance according to major crop types for selected characteristic periods. One strongly simplified example of such maps is given in Figure 19.26.4.

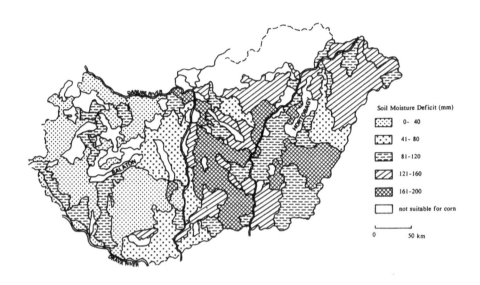

Figure 19.26.4 Aggregate water deficit within the root zone of corn during the growing season with a 80% probability of occurence as derived from the 1928-1978 data of 23 climate stations (NWA 1984, p.176).

The above simulation is not a full description of the natural SWV systems. It does not clarify how stomatal regulation minimizes moisture stress within the root zone and it does not quantify the water-controlled ET regime. Yet it can provide indirect solution for practical questions as it offers a comprehensive and quantitative description of soil moisture management measures which can keep the SWV system within the desirable range of heat-controlled ET regime. In this way the simulation procedure identifies long term equilibrium criteria for the given SWV system and it can be envisaged for wide scale use in the allocation of available farmland resources among the required major crops, and in the selection of optimal combination of soil moisture regulation measures and cultivation technologies for given land conditions and crop type. Although originally not envisaged, the proposed methodology could also be applied in climate impact assessments by repeating the simulations for various assumed or predicted greenhouse effects.

REFERENCES

Abramopoulos, F., C. Rosenzweig and B. Choudhury (1988) Improved ground hydrology calculations for global climate models (GCMs): Soil water movement and evapotranspiration. Journal of Climate Vol.1, N°9, Sept.1988, American Meteorological Society.

Budyko, M.I. (1956) Heat balance of the Earth surface. Gidrometeoizdat, Leningrad (in Russian).

Budyko, M.I., A.B. Ronov and A.L. Yanshin (1985) History of the atmosphere. Gidrometeoizdat, Leningrad (in Russian).

Dyck, S. (1988) Biospheric aspects of the hydrological cycle. Report of the meeting of the SC/IGBP Co-ordinating Panel, Postdam/Babelsberg 7-9 June 1988.

Eagleson, P.S. (1982) Ecological optimality in water-limited natural sol-vegetation systems. Water Resources Research, Vol.18, N°2, April 1982.

Erdélyi, M. (1979) Hydrodynamics of the Hungarian Great Plains. Research Institute for Water Resources "VITUKI", Publication N°18, Budapest. (in Hungarian and English).

NWA (1984) Master Plan for the water management of Hungary. Third elaboration. National Water Authority, Budapest. (in Hungarian).

Orlóci, I. and Á. Pintér (1981) Analysis of the interrelation between crop production and water management. Institute for Water Management, Budapest, August 1981. (in Hungarian with English summary).

Parry, M.L. (1985) The impact of climatic variations on agricultural margins. In: Kates et al. (ed): Climate Impact Assessment, SCOPE 27, Wiley and Sons, Chichester.

Rind D., A. Roseinzweig and C. Roseinzweig (1988) Modelling the future: A joint venture. Nature, Vol.334, N°6182, August 1988.

Rosenberg, N.J., J.A. Kimball, Ph. Martin and Ch.F. Cooper (1988) Climatic change, CO_2 enrichment and evapotranspiration. Draft paper distributed at the IIASA Task Force meeting on Regional Climate Scenarios, 20-22 January 1989, Laxenburg, Austria.

Szesztay, K. (1967) The water balance. In: T. Puskás (ed.): The surface waters of Hungary, p. 103-120. Research Institute for Water Resources "VITUKI", Budapest. (in Hungarian).

Vinnikov K.Ya., N.A. Lemeshko and N.A. Speranskaya (1988) Changes of soil wetness induced by global warming. Paper for the IIASA Task Force meeting on Regional Climate Scenarios, 20-22 January 1989, Laxenburg, Austria.

Index